Metabolic Engineering for Bioactive Compounds

Vipin Chandra Kalia • Adesh Kumar Saini
Editors

Metabolic Engineering for Bioactive Compounds

Strategies and Processes

Springer

Editors
Vipin Chandra Kalia
Microbial Biotechnology and Genomics
CSIR - Institute of Genomics and Integrative Biology (IGIB)
New Delhi
India

Adesh Kumar Saini
Faculty of Applied Sciences and Biotechnology
Shoolini University of Biotechnology and Management Sciences
Solan, Himachal Pradesh
India

ISBN 978-981-10-5510-2 ISBN 978-981-10-5511-9 (eBook)
DOI 10.1007/978-981-10-5511-9

Library of Congress Control Number: 2017951913

Printed on acid-free paper

This Springer imprint is published by Springer Nature
The registered company is Springer Nature Singapore Pte Ltd.
The registered company address is: 152 Beach Road, #21-01/04 Gateway East, Singapore 189721, Singapore

Dedicated to our parents

Preface

The nineteenth and twentieth centuries have progressed well on chemical and physical sciences, providing a huge reservoir of scientific knowledge and technologies. Even today, we see tremendous and rapid progress in the fields based on ion electronics and information technologies. Biological sciences have been slow to develop and had taken a back seat for long. Our society has exploited chemical and physical science-based products to such an extent that the ill effects are surfacing: high energy inputs, toxic nature, high pollution, and slow or non-degradability. This has opened up the opportunities for scientists especially biologists to look for alternative processes, which circumvent these limitations. Biological processes have all the potential good qualities but are rather inefficient and uneconomical to start with. In order to improve their efficiency and make them economical, there has been a rapidly growing interest in bioactive compounds. Bioactive compounds are defined primarily with reference to their association and usage. In a short span of time, these compounds have found potential applications in a wide array of fields: medicine, agriculture, food, cosmetics, etc. The bioactivity of these compounds is limited by their small quantities and poor bioavailability. The growth of this field has been facilitated by the development of nanobiotechnology. It has emerged as a highly efficient tool for designing of diverse compounds and efficient biosynthesis. In humans, these compounds help in improving the biological activity of the organism. As food components, these don't act as nutrients but can positively affect human health. The sources of bioactive substances range from bacteria, mushroom, plants, and animals. Metabolic engineering for bioactive compounds is being worked out by heterologous gene expression in a wide range of organisms. The model systems being employed for metabolic manipulation for bioactive molecules such as single-cell protein, antibody generation, metabolites, proteases, chaperones, therapeutic proteins, nanomaterials, polymeric conjugates, dendrimers, and nano-assemblies include *Escherichia coli*, *Agrobacterium*, and *Saccharomyces cerevisiae* and cell lines. The scope of these works is in important areas of prevention, diagnosis, and treatment of diseases, e.g., immunotherapy for curing various diseases like cancer, allergy, autoimmune diseases, etc.

Society supports scientific adventures with the hope that these will find their applications for human welfare. Young students are our future stars, who need to be nourished with novel scientific findings. These are expected to open up new vistas for the curious minds—to learn the basic principles of science and how to translate them into products. This book has been brought out to satisfy the young minds and provide economic benefits. In this book, the learned scientific community has shared their immense knowledge and expertise, which they have gained through hard and dedicated efforts targeted to understand the biological world. This book is a true reflection of the sincerity with which the scientific community is curious to create write-ups which can be adapted by the young minds. This effort is likely to benefit the world as a single entity. I am sincerely thankful to all the authors, who did all that is necessary to bring this compilation to this stage. I am heavily indebted to all of them. I am not sure if this acknowledgment is sufficient to justify the worthiness of their efforts. My inspiration to serve the humanity burgeons from the faith shown in me and the constant support of my family and friends, the pillar of my life: the late Mrs. Kanta Kalia and Mr. R.B. Kalia (parents); Amita (wife); Sunita, Veena, Priti, and Shruti (sisters); Ravi, Kamal, Virender, Raman (Nippy), and Satyendra (brothers); Daksh and Bhrigu (sons); my teachers (Dr. P. Usha Sarma); and my young friends—Anil, Balvinder, Bela, Bhartendu, Devanshi, Meera, Mukesh, Neelam, Rajan, Rajiv, Sadhana, Sanjeev, Satyendra, Shunty of DeshBandhu76 clan, and Chinoo. I must also acknowledge the support of my student friends—Vikas, Neena, Asha, Sadhana, Sanjay, Mamtesh, Subhasree, Shikha, Jyotsana, and Rahul.

New Delhi, India Vipin Chandra Kalia

Contents

Editors' Biography

Vipin Chandra Kalia is an emeritus scientist of the CSIR Institute of Genomics and Integrative Biology, Delhi. He holds an M.Sc. and Ph.D. in Genetics, both from the Indian Agricultural Research Institute, New Delhi. Further, he is a fellow of the National Academy of Sciences, National Academy of Agricultural Sciences, and Association of Microbiologists of India.

His main areas of research are microbial biodiversity, bioenergy, biopolymers, quorum sensing and quenching, and drug discovery. He has published more than 100 papers in scientific journals including *Nature Biotechnology*, *Biotechnology Advances*, and *Bioresource Technology*.

He has authored 14 book chapters. He has edited seven books and is currently the editor in chief of the *Indian Journal of Microbiology* (since 2013) and editor of several journals, including the *Journal of Microbiology and Biotechnology* (Korea) and *Dataset Papers in Microbiology*. He is a member of numerous professional associations, including the Society of Biological Chemists of India and the Biotech Research Society of India (BRSI). He can be contacted at vckalia@igib.res.in and vc_kalia@yahoo.co.in.

Adesh Kumar Saini is a professor and the director of the Center of Research on Himalayan Sustainability and Development, Shoolini University, India. He received his Ph.D. in Microbiology from the CSIR Institute of Genomics and Integrative Biology and completed his postdoctoral research at the National Institutes of Health. His main areas of research are microbial biodiversity, sustainable development of the Himalayan ecosystem, and yeast cell biology. He has published more than 25 papers in prominent scientific journals, including *The EMBO Journal* and the *Journal of Biological Chemistry*. His research projects aim at plant-growth-promoting rhizobacteria (PGPR), postfire management in the pine forests, and revival of forest productivity. He can be contacted at sainiade@gmail.com and adeshsaini@shooliniuniversity.com.

Contributors

Anshu Alok Department of Biotechnology (DBT), National Agri-Food Biotechnology Institute, Mohali, Punjab, India

Meenakshi Balhara Centre for Biotechnology Maharshi Dayanand University, Rohtak 124001, Haryana, India

Saurabh Bansal Department of Biotechnology and Bioinformatics, Jaypee University of Information Technology, Solan, Himachal Pradesh, India

Gopal Singh Bisht Department of Pharmacy, Jaypee University of Information Technology, Solan, Himachal Pradesh, India

Renu Chaudhary Centre for Biotechnology, Maharshi Dayanand University, Rohtak, Haryana, India

Anil Kumar Chhillar Centre for Biotechnology, MD University, Rohtak, Haryana, India

Anil A. Chuturgoon Discipline of Medical Biochemistry and Chemical Pathology, School of Laboratory Medicine and Medical Sciences, College of Health Sciences, University of KwaZulu-Natal, Durban, South Africa

Andrés Gárriz Laboratorio de Interacciones Planta-microorganismo, Instituto de Investigaciones Biotecnológicas-Instituto Tecnológico de Chascomús/Universidad Nacional de General San Martín-Consejo Nacional de Investigaciones Científicas y Técnicas (IIB-INTECH/UNSAM-CONICET), Chascomús, Buenos Aires, Argentina

Naseem A. Gaur Yeast Biofuel Group, International Centre for Genetic Engineering and Biotechnology (ICGEB), New Delhi, Delhi, India

Ritu Ghildiyal Department of Biotechnology and Bioinformatics, Jaypee University of Information Technology, Solan, Himachal Pradesh, India

María Elisa Gonzalez Laboratorio de Interacciones Planta-microorganismo, Instituto de Investigaciones Biotecnológicas-Instituto Tecnológico de Chascomús/ Universidad Nacional de General San Martín-Consejo Nacional de Investigaciones Científicas y Técnicas (IIB-INTECH/UNSAM-CONICET), Chascomús, Buenos Aires, Argentina

Kinam Gupta Department of Biotechnology and Bioinformatics, Jaypee University of Information Technology, Solan, Himachal Pradesh, India

Vipin Chandra Kalia Microbial Biotechnology and Genomics, CSIR—Institute of Genomics and Integrative Biology (IGIB), New Delhi, India

Savita Khatri Centre for Biotechnology, Maharshi Dayanand University, Rohtak, Haryana, India

Jitesh Kumar Department of Biotechnology (DBT), National Agri-Food Biotechnology Institute, Mohali, Punjab, India

Manish Kumar Centre for Biotechnology, MD University, Rohtak, Haryana, India

Manoj Kumar Amity Institute of Microbial Technology, Amity University, Noida, Uttar Pradesh, India

Reena Kumari Animal Biotechnology Laboratory, Faculty of Applied Sciences and Biotechnology, Shoolini University, Solan, Himachal Pradesh, India

Bisma Malik Department of Bioresources, University of Kashmir, Srinagar, Jammu and Kashmir, India

María Marina Laboratorio de Interacciones Planta-microorganismo, Instituto de Investigaciones Biotecnológicas-Instituto Tecnológico de Chascomús/Universidad Nacional de General San Martín-Consejo Nacional de Investigaciones Científicas y Técnicas (IIB-INTECH/UNSAM-CONICET), Chascomús, Buenos Aires, Argentina

Deepak Pathania Department of Environmental Sciences, Central University of Jammu, Jammu, Jammu and Kashmir, India

Neetu Phougat Centre for Biotechnology, MD University, Rohtak, Haryana, India

Fernando Luis Pieckenstain Laboratorio de Interacciones Planta-microorganismo, Instituto de Investigaciones Biotecnológicas-Instituto Tecnológico de Chascomús/ Universidad Nacional de General San Martín-Consejo Nacional de Investigaciones Científicas y Técnicas (IIB-INTECH/UNSAM-CONICET), Chascomús, Buenos Aires, Argentina

Tanveer Bilal Pirzadah Department of Bioresources, University of Kashmir, Srinagar, Jammu and Kashmir, India

Poonam Department of Biotechnology and Bioinformatics, Jaypee University of Information Technology, Solan, Himachal Pradesh, India

Ashu M. Poswal Faculty of Applied Sciences and Biotechnology, Shoolini University, Solan, Himachal Pradesh, India

Reiaz Ul Rehman Department of Bioresources, University of Kashmir, Srinagar, Jammu and Kashmir, India

Fernando Matias Romero Laboratorio de Interacciones Planta-microorganismo, Instituto de Investigaciones Biotecnológicas-Instituto Tecnológico de Chascomús/ Universidad Nacional de General San Martín-Consejo Nacional de Investigaciones Científicas y Técnicas (IIB-INTECH/UNSAM-CONICET), Chascomús, Buenos Aires, Argentina

Franco Ruben Rossi Laboratorio de Interacciones Planta-microorganismo, Instituto de Investigaciones Biotecnológicas-Instituto Tecnológico de Chascomús/ Universidad Nacional de General San Martín-Consejo Nacional de Investigaciones Científicas y Técnicas (IIB-INTECH/UNSAM-CONICET), Chascomús, Buenos Aires, Argentina

Adesh K. Saini Faculty of Applied Sciences and Biotechnology, Shoolini University of Biotechnology and Management Science, Solan, Himachal Pradesh, India

Reena V. Saini Animal Biotechnology Laboratory, Faculty of Applied Sciences and Biotechnology, Shoolini University, Solan, Himachal Pradesh, India

Jata Shankar Department of Biotechnology and Bioinformatics, Jaypee University of Information Technology, Solan, Himachal Pradesh, India

Arush Sharma School of Chemistry, Shoolini University, Solan, Himachal Pradesh, India

Deepika Sharma Department of Pharmacy, Jaypee University of Information Technology, Solan, Himachal Pradesh, India

Shivam Sharma Jaypee University of Information Technology, Solan, Himachal Pradesh, India

Rahul Shrivastava Department of Biotechnology and Bioinformatics, Jaypee University of Information Technology, Solan, Himachal Pradesh, India

Hemant Sood Jaypee University of Information Technology, Solan, Himachal Pradesh, India

Department of Biotechnology and Bioinformatics, Jaypee University of Information Technology, Solan, Himachal Pradesh, India

Swadeep Sood School of Chemistry, Shoolini University, Solan, Himachal Pradesh, India

Ankit Srivastava Kusuma School of Biological Sciences, Indian Institute of Technology, New Delhi, Delhi, India

Raman Thakur Department of Biotechnology and Bioinformatics, Jaypee University of Information Technology, Solan, Himachal Pradesh, India

Charlette Tiloke Discipline of Medical Biochemistry and Chemical Pathology, School of Laboratory Medicine and Medical Sciences, College of Health Sciences, University of KwaZulu-Natal, Durban, South Africa

Department of Chemistry, Faculty of Applied Sciences, Durban University of Technology, Durban, South Africa

Subodh Verma National Institute of Plant Genome Research, New Delhi, Delhi, India

Paulina Vignatti Laboratorio de Interacciones Planta-microorganismo, Instituto de Investigaciones Biotecnológicas-Instituto Tecnológico de Chascomús/ Universidad Nacional de General San Martín-Consejo Nacional de Investigaciones Científicas y Técnicas (IIB-INTECH/UNSAM-CONICET), Chascomús, Buenos Aires, Argentina

Farnaz Yusuf Yeast Biofuel Group, International Centre for Genetic Engineering and Biotechnology (ICGEB), New Delhi, Delhi, India

Part I

Introduction to Bioactive Compounds

The Dawn of the Era of Bioactive Compounds

1

Vipin Chandra Kalia

Abstract

Microbes and plants are a major source of naturally occurring compounds. However, a few of them especially the secondary metabolites are among those termed as bioactive compounds. These bioactive compounds have unique biological activities, which can be assigned to unique chemical structures, uncommon chemical groups and structural elements. Due to these specific properties, bioactive compounds have been exploited for diverse biotechnological applications. These compounds have strong flavours and odours with usage as antibacterials and for antineoplastic, anticancer and antiviral activities. A few strategies to enhance their production and efficacy involve metabolic engineering of the process through the use of tools such as molecular biology, nanotechnology, bioinformatics, etc.

Keywords

Bioactive compounds • Antibacterials • Antineoplastic • Anticancer • Antiviral • Nanotechnology

1.1 Introduction

The immense progress of chemical and physical sciences during the nineteenth and twentieth centuries has provided a vast reservoir of knowledge and technologies. A few associated limitations were the high-energy inputs to run these technologies and pollution caused by their toxic nature and slow or non-degradability. This has forced

---

V.C. Kalia
Microbial Biotechnology and Genomics, CSIR - Institute of Genomics and Integrative Biology (IGIB), Delhi University Campus, Mall Road, New Delhi 110007, India
e-mail: vckalia@igib.res.in; vc_kalia@yahoo.co.in

V.C. Kalia, A.K. Saini (eds.), *Metabolic Engineering for Bioactive Compounds*,
DOI 10.1007/978-981-10-5511-9_1

scientists to look for alternative and bio-processes. There has been a rapidly growing interest in bioactive compounds (Pessione et al. 2017). In a short span, these types of compounds found potential applications in wide array of fields: medicine, agriculture, food, cosmetics, etc. This growth was facilitated by the development of nanobiotechnology, which allowed designing of diverse compounds and efficient synthesis. The bioactivity of these compounds is affected by their quantity and bioavailability. Bioactive compounds are defined primarily in the context of their association and usage. In humans, these compounds help in improving the biological activity of the organism. As food components, these don't act as nutrients but can affect human health (Guaadaoui et al. 2014). The sources of bioactive substances range from bacteria, mushroom, plants to animals.

1.2 Bioactive Compounds

Bioactive compounds are distinguished from other "inactive" naturally occurring compounds primarily because of their biological activities such as antimicrobial, antitumor, pharmacological, antiviral, etc. These specific properties can be assigned to unique chemical structures, uncommon chemical groups and structural elements—macrolactones, cyclopeptide skeletons, etc. (Shukla 2015).

1.2.1 From Plants

1.2.1.1 Glycosides

These secondary metabolites (aglycone) are bound to saccharides or uronic acid (the glycone component). The glycosides are grouped as cardiac, cyanogenic and anthraquinone glycosides and saponins. Flavonoids (tannins) also exist as glycosides. These compounds can be isolated from *Digitalis purpurea*, *Convallaria majalis*, *Prunus* spp., *Narthecium ossifragum* and members of Brassicaceae, Fabaceae and Fagaceae (Bernhoft 2010).

1.2.1.2 Terpenoids and Phenylpropanoids

These compounds have strong flavours and odours with usage in herbal treatments especially as antibacterials and for antineoplastic and antiviral activities. Most members of Lamiaceae are well-known producers of these secondary metabolites (Bernhoft 2010).

Other important compounds are diterpenoids, resins, lignans, alkaloids, furocoumarins and naphthodianthrones, proteins and peptides, carotenoids and phytosterols (Bernhoft 2010).

1.2.2 From Medicinal Plants

Traditional medicinal plants are used by around 80% of the population around the world for primary health care. The plants contain compounds, which are effective in treating chronic and infectious diseases, such as cancers. They also act as antimicrobial, antioxidant, antidiarrheal and analgesic and heal wounds. Worldwide around 20,000 medicinal plants are known for their health benefits (Sasidharan et al. 2011).

1.2.3 From Microbes

Bioactive compounds being sought from microbes are largely for their antibacterial and antimicrobial properties, insecticidal activity, antagonistic to pathogens, etc. These activities help in prevention of infectious diseases and food spoilage (Shukla 2015). Although carotenoids are secondary metabolites which are synthesized by algae, plants, fungi and bacteria, however, microbes as a source of carotenoids—lutein, β-carotene, lycopene, astaxanthin and zeaxanthin—have recently been recognized (Saini and Keum 2017).

1.2.3.1 Bacteria (*Bacillus* sp. and *Pseudomonas* sp.)

Bacteria produce insecticidal compounds. Protobiotic bacteria produce lactic acid, hydrogen peroxide and acetic acids to inhibit pathogens. Lactic acid bacteria produce antimicrobials such as bacteriocin, organic acids, gamma-aminobutyric acid (GABA), hydrogen peroxide, lactoperoxidase, diacetyl, and ethanol, which are anticarcinogenic and anti-cholesterol, improve tolerance to lactose, help to reduce intestinal disorders, inhibit pathogen, etc. Soil bacteria such as *Serratia* spp. are antagonistic towards worms, yeasts and others. *Microbacterium mangrove*, *Sinomonas humi* and *Monashia flava* have antibacterial, neuroprotective and anticancer properties (Shukla 2015; Azman et al. 2017; Pessione et al. 2017; Sanchart et al. 2017). *Lactobacillus* spp. with applications in the field of action against food spoilage by fungi were isolated from traditional pickles. *Lactobacillus brevis* strain also showed antioxidant and probiotic characteristics (Arasu et al. 2015). *Lactobacillus* spp., *Aquabacterium* sp., *Methylibium* sp., *Piscinobacter* sp. and *Staphylococcus xylosus* isolated from goat milk and cheese produced bacteriocins, which could inhibit the growth of a wide range of pathogens—*Escherichia coli*, *Staphylococcus aureus*, *Serratia marcescens*, *Listeria monocytogenes*, *L. innocua*, *Bacillus cereus*, *Shigella flexneri*, *Pseudomonas aeruginosa*, *Enterobacter cloacae* and *Klebsiella pneumoniae* (Hernández-Saldaña et al. 2016). *Lactococcus*, *Paracoccus*, *Pseudomonas* and *Streptomyces* strains produce heterocyclic compounds including fatty acid methyl ester, fatty alcohols, aromatic hydrocarbon, long-chain alkanes and phenazines having antibacterial, antifungal, anticancer and aminopeptidase inhibitory potential (Begum et al. 2016; Varsha et al. 2016). Bioactive compounds which promote plant growth are produced by bacteria such as *Bacillus*, *Viridibacillus*, *Pseudomonas*, etc. (Peña-Yam et al. 2016; Thakur et al. 2017). *Bacillus subtilis* strain could degrade feathers and produce antifungal compounds (Go et al. 2015).

Marine bacteria such as *Brevibacillus laterosporus* produced lipopeptide—tauramamide and its esters. These could selectively act against *Enterococcus* sp. (a gram-positive human pathogen) and to some extent also against MRSA (multidrug-resistant *Staphylococcus aureus*) (Debbab et al. 2010). Marine bacterium *Vibrio ruber* is known to produce the pigment prodigiosin, with broad antimicrobial activity, and can induce autolytic activity in *Bacillus subtilis* (Pessione et al. 2017). *Bacillus* sp. associated with marine sponge *Tedania anhelans* as a promising source of antioxidant has been explored recently (Balakrishnan et al. 2015).

1.2.3.2 Actinomycetes (*Streptomyces* sp.)

Actinomycetes produce (1) istamycin, aminoglycoside with antibiotic activity; (2) altemicidin, alkaloid with anticancer activity; (3) marinone, naphthoquinone; and (4) asplasmomycin with antibiotic activities. *Marinispora* isolate was reported to produce a few novel bioactive compounds—lipoxazolidinone—which had broad-spectrum antibacterial activity against *Haemophilus influenza*. Other isolates of the genus yielded chlorinated bisindole pyrroles and lynamicins, which were effective against *S. aureus* isolates—MSSA and MRSA, *Staphylococcus epidermidis* and *Enterococcus faecalis* (Macherla et al. 2007; McArthur et al. 2008; Debbab et al. 2010).

1.2.3.3 Fungi

Penicillium, yeasts, slime moulds, *Nigrospora* sp. and *Aspergillus* sp. produced metabolites like nigrospoxydons and dehydroxychlorofusarielin, which showed antibacterial activity against *S. aureus*, MRSA, *Microsporum gypseum*, etc. (Debbab et al. 2010). *Saccharomyces cerevisiae* could boost the growth and immunity of *Labeo rohita* juveniles (Bandyopadhyay et al. 2015).

1.2.3.4 Microscopic Algae (Seaweeds, Dinoflagellates, Diatoms, Etc.)

Microscopic algae produce bioactive molecules, which have major applications in the area of nutraceutical, food industry, cosmetic and pharmaceutical sectors (Shukla 2015; Lauritano et al. 2016).

1.3 The Future

Nature has a vast reservoir of products, most of which have nutritional values and medicinal properties. In spite of the fact that a large number of highly diverse plants, microbes and marine organisms have found to be bioactive, there is still scope for new secondary metabolites (Karumuri et al. 2015; Park et al. 2015; Shiva Krishna et al. 2015; Jeyanthi and Velusamy 2016; Pessione et al. 2017). Their role in chemotherapy of cancers has provoked the investigators to look into the chemistry of the naturally produced substances. Metabolic engineering for bioactive compounds is targeted for increasing biosynthetic diversity. Compounds in this category range from acrylamidine to macrocyclic colubricidin. Among the strategies which will promote the role of bioactive molecules, nanotechnology has been proving quite effective. Recent works have demonstrated nanoparticles as potential therapeutic agents to benefit human beings against pathogens (Bose and Chatterjee 2015; Dobrucka and Długaszewska 2015; Szweda et al. 2015; Wadhwani et al. 2016; Ahiwale et al. 2017).

Bacterial polymers such as polyhydroxyalkanoates (PHAs) are produced under nutritional stress conditions (Reddy et al. 2003; Kalia et al. 2003, 2007, 2016; Kumar et al. 2009, 2013, 2014, 2015a, b, c, 2016; Patel et al. 2015a, b, 2016; Singh et al. 2009, 2013, 2015). PHAs can be metabolized to produce (*R*)-3-hydroxyoctanoic acid (3HA) (Ray and Kalia 2017a, b). 3HA can be derivatized into halo and unsaturated methyl and benzyl esters. These derivatives with carboxylic group were reported to have antimicrobial activity against bacteria and fungi. In addition, they were reported to have anti-proliferative effect on mammalian cell lines. The range of the activity was found to extend even in acting as quorum sensing inhibitors against opportunistic pathogen *P. aeruginosa* PAO1 (Radivojevic et al. 2016).

1.4 Opinion

In recent times, molecular biology tools have become handy in drug discovery by finding new and potent metabolites. Strategies to counter microbial pathogenic attack by using secondary metabolites to regulate microbial interactions need to be developed. Novel imaging techniques—molecular eyes—are going to prove to be among the best technological advances for searching biologically active products. Integration of genomics, proteomics and metabolomics is going to supplement other strategies to develop safe, efficient and acceptable leads in natural products leading from laboratory to clinics and to bedside. Metabolic engineering for bioactive compounds is being worked out by heterologous gene expression in a wide range of organisms. The model systems being employed for metabolic manipulation for bioactive molecules such as single-cell protein, antibody generation, metabolites, proteases, chaperones, therapeutic proteins, nanomaterials, polymeric conjugates, dendrimers and nanoassemblies include *Escherichia coli*, *Agrobacterium*, *Saccharomyces cerevisiae* and cell lines, etc. The scope of these works is in important areas of prevention, diagnosis and treatment of diseases, e.g. immunotherapy for curing various diseases like cancer, allergy, autoimmune diseases, diabetes, etc.

Acknowledgements The author wish to thank the Director of CSIR-Institute of Genomics and Integrative Biology, Delhi, India, OLP1126 (CSIR-HRD ES21(1022)/16/EMR-II) for providing the necessary funds, facilities and moral support.

References

Ahiwale SS, Bankar AV, Tagunde S, Kapadnis BP (2017) A bacteriophage mediated gold nanoparticle synthesis and their anti-biofilm activity. Indian J Microbiol 57(2):188–194. doi:10.1007/s12088-017-0640-x

Arasu MV, Al-Dhabi NA, Rejiniemon TS, Lee KD, Huxley VAJ, Kim DH, Duraipandiyan V, Karuppiah P, Choi KC (2015) Identification and characterization of *Lactobacillus brevis* P68 with antifungal, antioxidant and probiotic functional properties. Indian J Microbiol 55:19–28. doi:10.1007/s12088-014-0495-3

Azman CA-S, Othman I, Fang C-M, Chan K-G, Goh B-H, Lee L-H (2017) Antibacterial, anticancer and neuroprotective activities of rare actinobacteria from mangrove forest soils. Indian J Microbiol 57:177–187. doi:10.1007/s12088-016-0627-z

Balakrishnan D, Bibiana AS, Vijayakumar A, Santhosh RS, Dhevendaran K, Nithyanand P (2015) Antioxidant activity of bacteria associated with the marine sponge *Tedania anhelans*. Indian J Microbiol 55:13–18. doi:10.1007/s12088-014-0490-8

Bandyopadhyay P, Mishra S, Sarkar B, Swain SK, Pal A, Tripathy PP, Ojha SK (2015) Dietary *Saccharomyces cerevisiae* boosts growth and immunity of IMC *Labeo rohita* (Ham.) juveniles. Indian J Microbiol 55:81–87. doi:10.1007/s12088-014-0500-x

Begum IF, Mohankumar R, Jeevan M, Ramani K (2016) GC–MS analysis of bioactive molecules derived from *Paracoccus pantotrophus* FMR19 and the antimicrobial activity against bacterial pathogens and MDROs. Indian J Microbiol 56:426–432. doi:10.1007/s12088-016-0609-1

Bernhoft A (2010) A brief review on bioactive compounds in plants. In: Bernhoft A (ed) Bioactive compounds in plants—benefits and risks for man and animals. Proceedings of the symposium held at the Norwegian Academy of Science and Letters, Oslo, 13–14 Nov 2008. pp 11–7

Bose D, Chatterjee S (2015) Antibacterial activity of green synthesized silver nanoparticles using Vasaka (*Justicia adhatoda* L.) leaf extract. Indian J Microbiol 55:163–167. doi:10.1007/s12088-015-0512-1

Debbab A, Aly AH, Lin WH, Proksch P (2010) Bioactive compounds from marine bacteria and fungi. Microb Biotechnol 3:544–563. doi:10.1111/j.1751-7915.2010.00179.x

Dobrucka R, Długaszewska J (2015) Antimicrobial activities of silver nanoparticles synthesized by using water extract of *Arnicae anthodium*. Indian J Microbiol 55:168–174. doi:10.1007/s12088-015-0516-x

Go T-H, Cho K-S, Lee S-M, Lee O-M, Son H-J (2015) Simultaneous production of antifungal and keratinolytic activities by feather-degrading *Bacillus subtilis* S8. Indian J Microbiol 55:66–73. doi:10.1007/s12088-014-0502-8

Guaadaoui A, Benaicha S, Elmajdoub N, Bellaoumi M, Hamal A (2014) What is a bioactive compound? A combined definition for a preliminary consensus. Int J Nutrition Food Sci 3:174–179. doi:10.11648/j.ijnfs.20140303.16

Hernández-Saldaña OF, Valencia-Posadas M, de la Fuente-Salcido NM, Bideshi DK, Barboza-Corona JE (2016) Bacteriocinogenic bacteria isolated from raw goat milk and goat cheese produced in the Center of México. Indian J Microbiol 56:301–308. doi:10.1007/s12088-016-0587-3

Jeyanthi V, Velusamy P (2016) Anti-methicillin resistant *Staphylococcus aureus* compound isolation from halophilic *Bacillus amyloliquefaciens* MHB1 and determination of its mode of action using electron microscope and flow cytometry analysis. Indian J Microbiol 56:148–157. doi:10.1007/s12088-016-0566-8

Kalia VC, Chauhan A, Bhattacharyya G, Rashmi (2003) Genomic databases yield novel bioplastic producers. Nat Biotechnol 21:845–846. doi:10.1038/nbt0803-845

Kalia VC, Lal S, Cheema S (2007) Insight into the phylogeny of polyhydroxyalkanoate biosynthesis: horizontal gene transfer. Gene 389:19–26. doi:10.1016/j.gene.2006.09.010

Kalia VC, Prakash J, Koul S (2016) Biorefinery for glycerol rich biodiesel industry waste. Indian J Microbiol 56:113–125. doi:10.1007/s12088-016-0583-7

Karumuri S, Singh PK, Shukla P (2015) *In silico* analog design for terbinafine against *Trichophyton rubrum*: a preliminary study. Indian J Microbiol 55:333–340. doi:10.1007/s12088-015-0524-x

Kumar P, Patel SKS, Lee JK, Kalia VC (2013) Extending the limits of *Bacillus* for novel biotechnological applications. Biotechnol Adv 31:1543–1561. doi:10.1016/j.biotechadv.2013.08.007

Kumar P, Mehariya S, Ray S, Mishra A, Kalia VC (2015a) Biodiesel industry waste: a potential source of bioenergy and biopolymers. Indian J Microbiol 55:1–7. doi:10.1007/s12088-014-0509-1

Kumar P, Mehariya S, Ray S, Mishra A, Kalia VC (2015b) Biotechnology in aid of biodiesel industry effluent (glycerol): biofuels and bioplastics. In: Kalia VC (ed) Microbial factories. Springer, New Delhi, pp 105–119. doi: 10.1007/978-81-322-2598-0

Kumar P, Ray S, Kalia VC (2016) Production of co-polymers of polyhydroxyalkanoates by regulating the hydrolysis of biowastes. Bioresour Technol 200:413–419. doi:10.1016/j.biortech.2015.10.045

Kumar P, Ray S, Patel SKS, Lee JK, Kalia VC (2015c) Bioconversion of crude glycerol to polyhydroxyalkanoate by *Bacillus thuringiensis* under non-limiting nitrogen conditions. Int J Biol Macromol 78:9–16. doi:10.1016/j.ijbiomac.2015.03.046

Kumar P, Singh M, Mehariya S, Patel SKS, Lee JK, Kalia VC (2014) Ecobiotechnological approach for exploiting the abilities of *Bacillus* to produce co-polymer of polyhydroxyalkanoate. Indian J Microbiol 54:151–157. doi:10.1007/s12088-014-0457-9

Kumar T, Singh M, Purohit HJ, Kalia VC (2009) Potential of *Bacillus* sp. to produce polyhydroxybutyrate from biowaste. J Appl Microbiol 106:2017–2023. doi:10.1111/j.1365-2672.2009.04160.x

Lauritano C, Andersen JH, Hansen E, Albrigtsen M, Escalera L, Esposito F, Helland K, Hanssen KØ, Romano G, Ianora A (2016) Bioactivity screening of microalgae for antioxidant, anti-inflammatory, anticancer, anti-diabetes, and antibacterial activities. Front Mar Sci. doi:10.3389/fmars.2016.00068

Macherla VR, Liu J, Sunga M, White DJ, Grodberg J, Teisan S (2007) Lipoxazolidinones A, B, and C: antibacterial 4-oxazolidinones from a marine actinomycete isolated from a Guam marine sediment. J Nat Prod 70:1454–1457. doi:10.1021/np0702032

McArthur KA, Mitchell SS, Tsueng G, Rheingold A, White DJ, Grodberg J (2008) Lynamicins A-E, chlorinated bisindole pyrrole antibiotics from a novel marine actinomycete. J Nat Prod 71:1732–1737. doi:10.1021/np800286d

Park J-M, Radhakrishnan R, Kang S-M, Lee I-J (2015) IAA producing *Enterobacter* sp. I-3 as a potent bio-herbicide candidate for weed control: a special reference with lettuce growth inhibition. Indian J Microbiol 55:207–211. doi:10.1007/s12088-015-0515-y

Patel SKS, Kumar P, Singh S, Lee JK, Kalia VC (2015a) Integrative approach for hydrogen and polyhydroxybutyrate production. In: Kalia VC (ed) Microbial factories waste treatment. Springer, New Delhi, pp 73–85. doi:10.1007/978-81-322-2598-0_5

Patel SKS, Kumar P, Singh S, Lee JK, Kalia VC (2015b) Integrative approach to produce hydrogen and polyhydroxybutyrate from biowaste using defined bacterial cultures. Bioresour Technol 176:136–141. doi:10.1016/j.biortech.2014.11.029

Patel SKS, Lee JK, Kalia VC (2016) Integrative approach for producing hydrogen and polyhydroxyalkanoate from mixed wastes of biological origin. Indian J Microbiol 56:293–300. doi:10.1007/s12088-016-0595-3

Peña-Yam LP, Esaú Ruíz-Sánchez E, Barboza-Corona JE, Reyes-Ramírez A (2016) Isolation of Mexican *Bacillus* species and their effects in promoting growth of chili pepper (*Capsicum annuum* L. cv Jalapeño). Indian J Microbiol 56:375–378. doi:10.1007/s12088-016-0582-8

Pessione E, Maxxoli R, Riedel K (2017) Bioactive compounds from microbes. Front Microbiol. doi:10.3389/fmicb.2017.00392

Radivojevic J, Skaro S, Senerovic L, Vasiljevic B, Guzik M, Kenny ST, Maslak V, Nikodinovic-Runic J, O'Connor KE (2016) Polyhydroxyalkanoate-based 3-hydroxyoctanoic acid and its derivatives as a platform of bioactive compounds. Appl Microbiol Biotechnol 100:161–172. doi:10.1007/s00253-015-6984-4

Ray S, Kalia VC (2017a) Biological significance of degradation of polyhydroxyalkanoates. In: Kalia VC, Kumar P (eds) Microbial applications, Bioremediation and bioenergy, vol Vol 1. SpringerNature, New York. 978-3-319-52665-2

Ray S, Kalia VC (2017b) Microbial cometabolism and polyhydroxyalkanaote co-polymers. Indian J Microbiol 57:39–47. doi:10.1007/s12088-016-0622-4

Reddy CSK, Ghai R, Kalia VC (2003) Polyhydroxyalkanoates: an overview. Bioresour Technol 87:137–146. doi:10.1016/S0960-8524(02)00212-2

Saini RK, Keum Y-S (2017) Progress in microbial carotenoids production. Indian J Microbiol 57:129–130. doi:10.1007/s12088-016-0637-x

Sanchart C, Rattanaporn O, Haltrich D, Phukpattaranont P, Maneerat S (2017) *Lactobacillus futsaii* CS3, a new GABA-producing strain isolated from Thai fermented shrimp (*Kung–Som*). Indian J Microbiol 57:211–217. doi:10.1007/s12088-016-0632-2

Sasidharan S, Chen Y, Saravanan D, Sundram KM, Latha LY (2011) Extraction, isolation and characterization of bioactive compounds from plants' extracts. Afr J Tadit Complement Altern Med 8:1–10

Shiva Krishna P, Sudheer Kumar B, Raju P, Murty MSR, Prabhakar Rao T, Singara Charya MA, Prakasham RS (2015) Fermentative production of pyranone derivate I from marine *Vibrio* sp. SKMARSP9: isolation, characterization and bioactivity evaluation. Indian J Microbiol 55:292–301. doi:10.1007/s12088-015-0521-0

Shukla R (2015) Studies on bioactive compounds from different microorganisms. Int J Sci Eng Res 6:1225–1233. ISSN229-5518

Singh M, Kumar P, Patel SKS, Kalia VC (2013) Production of polyhydroxyalkanoate co-polymer by *Bacillus thuringiensis*. Indian J Microbiol 53:77–83. doi:10.1007/s12088-012-0294-7

Singh M, Kumar P, Ray S, Kalia VC (2015) Challenges and opportunities for customizing polyhydroxyalkanoates. Indian J Microbiol 55:235–249. doi:10.1007/s12088-015-0528-6

Singh M, Patel SKS, Kalia VC (2009) *Bacillus subtilis* as potential producer for polyhydroxyalkanoates. Microb Cell Factories 8:38. doi:10.1186/1475-2859-8-38

Szweda P, Gucwa K, Kurzyk E, Romanowska E, Dzierzanowska-Fangrat K, Jurek AZ, Kuś PM, Milewski M (2015) Essential oils, silver nanoparticles and Propolis as alternative agents against fluconazole resistant *Candida albicans, Candida glabrata* and *Candida krusei* clinical isolates. Indian J Microbiol 55:175–183. doi:10.1007/s12088-014-0508-2

Thakur R, Sharma KC, Gulati A, Sud RK, Gulati A (2017) Stress-tolerant *Viridibacillus arenosi* strain IHB B 7171 from tea rhizosphere as a potential broad-spectrum microbial inoculant. Indian J Microbiol. doi:10.1007/s12088-017-0642-8

Varsha KK, Nishant G, Sneha SM, Shilpa G, Devendra L, Priya S, Nampoothiri KM (2016) Antifungal, anticancer and aminopeptidase inhibitory potential of a phenazine compound produced by *Lactococcus* BSN307. Indian J Microbiol 56:411–416. doi:10.1007/s12088-016-097-1

Wadhwani SA, Shedbalkar UU, Singh R, Vashisth P, Pruthi V, Chopade BA (2016) Kinetics of synthesis of gold nanoparticles by *Acinetobacter* sp. SW30 isolated from environment. Indian J Microbiol 56:439–444. doi:10.1007/s12088-016-0598-0

Developments and Diversity of Proteins and Enzymes

2

Ankit Srivastava, Saurabh Bansal, and Jata Shankar

Abstract

Proteins form an integral part of a large consortium of industrially and therapeutically important products. Thus, efficient strategies for producing high-quality proteins and enzymes at industrial scale have gained importance in the past few decades. However, the heterogeneous production of recombinant proteins at industrial setups requires remotion and optimization of certain drawbacks such as protein aggregation, contamination, unstable cultures and less productive fermentation process. Therefore, the ultimate aim is to develop bioprocess that could be utilized for efficient production of important proteins. This has led to the development of structural biology, protein engineering and bioinformatics tools for identifying and rectifying these issues at molecular level. In the postgenomic era, a huge variety of protein families are yet to be functionally characterized. In this context, the discovery of new proteins with varied functionalities can only be achieved using modern-day methodologies that incorporate sequences, structure and activity-based approaches. Later, the positive outcomes could be easily directed towards the development of stable recombinant culture and productive fermentation process alongside a cost-effective downstream process. This chapter discusses the strides in the protein/enzyme production and their biotechnological perspectives along with the limitations in scale-up. Besides, several methodologies for selecting efficient protein molecules such as directed evolution approaches as well as for studying their physicochemical characteris-

A. Srivastava
Kusuma School of Biological Sciences, Indian Institute of Technology,
Hauz Khas, New Delhi 110016, India

S. Bansal (✉) • J. Shankar
Department of Biotechnology and Bioinformatics, Jaypee University of Information Technology, Solan 173234, India
e-mail: saurab.bansal02@gmail.com

V.C. Kalia, A.K. Saini (eds.), *Metabolic Engineering for Bioactive Compounds*,
DOI 10.1007/978-981-10-5511-9_2

tics such as single-molecule techniques are also described. In addition, this chapter also delineates production of therapeutic proteins and proteases and applications of chaperones in industrial production process.

Keywords
Therapeutics • Bioprocess • Directed evolution • Protein engineering • Interactions Single-molecule biophysics • Fluorescence • Atomic force microscopy • Biochemical diversity • Protease • Down-stream processing • Chaperones • Protein folding

2.1 Protein Science: Journey from Cell to Industry

The past two decades had been crucial for the development of industrial production process of recombinant proteins. Several new strategies have been developed and implemented for producing large amounts of high-quality proteins of varied commercial applications. A huge number of new proteins from a range of origins such as virus, bacteria, fungus, plants and higher animals have been produced using genetically modified organisms (GMO) (Ahmad et al. 2012; Buiatti et al. 2013). The post-genomic era has opened new vistas in identifying alternate sources of scarce proteins which were earlier found difficult to produce at small or industrial scales. These proteins/enzymes are now readily produced in forms resembling closely to the actual native proteins. This has also cut down the amount of protein and protein-based products that were earlier obtained directly from the human or animal tissues. During the past decade, the US Food and Drug Administration (FDA) has been approving 15 new recombinant proteins on an average each year (Mullard 2014). The protein production at industrial scale differs from traditional small molecules that are chemically synthesized. In this case, recombinant DNA technology plays the pivoting role along with protein expression systems directed in various living systems. Basically, successful protein/enzyme production process depends heavily upon the following five factors, viz. (a) production cost, (b) control on actual product processing such as native post-translational modifications, (c) processing time for producing protein from the gene of interest, (d) host-based nuisances in protein production and (e) control of protein aggregation (Rai and Padh 2001; Gupta and Shukla 2015).

The postgenomic boom in genetic information has led to an enormous increase in a number of innovative biological products and biosimilars that mostly comprise proteins. This has fuelled the demand of customized production cell lines with high level of productivity. Nevertheless, heterologous protein production at industrial scale still requires much optimization and differs significantly case by case. Based on widely characterized genetic architecture, *E. coli* has been a widely employed host in protein products that do not require post-translational modifications (Chen 2012). Besides safe genetic manipulations, the simplified purification strategies for producing recombinant proteins using *E. coli* at industrial scales also make them a preferred host. On the other hand, several protein products have been industrially produced using yeast as an alternative host (Celik and Calık 2012). This host is typically free from pyrogens and viral and oncogenic DNA and thus is a preferred for production of pharmaceutical proteins (Mattanovich et al. 2011). Apart from this, baculoviruses have evolved as popular systems for recombinant protein overproduction in eukaryotic cells (Fornwald et al. 2016). Owing to the similar cellular milieu, these insect cell lines promote protein modifications and processing similar to higher eukaryotic cells and thus act as preferred hosts for producing qualitatively pure proteins (Radner et al. 2012). These are generally employed when the product specificity is essential and aids in its clinical efficacy. Collectively, the design of heterologous protein production process depends heavily on the level of purification, stabilization and formulation required for a particular protein product.

The other major drawback in industrial protein production is attributed to protein aggregation and precipitation (Vázquez-Rey and Lang 2011). Protein folding is a complex phenomenon that incorporates two major components, i.e. foldases, that accelerate folding and chaperones that prevent formation of insoluble protein aggregates. In an industrial bioprocess, due to the presence of numerous cell stress factors such as heat shock, nutrient depletion or other environmental contaminants, the protein products tend to aggregate (den Engelsman et al. 2011). These aggregates due to their toxic nature further lead to loss in biomass and hence significantly affect the efficacy of the production process. Thus, recovery and renaturation steps are added to obtain optimized amounts of proteins/enzymes in an economically feasible fashion. Thus, several molecular techniques such as utilization of protease-deficient strains as well as bioprocess strategies such as high-temperature culturing and engineering of fusion peptides/proteins that promote protein solubilization have been widely used (Clark 2001; Rajan et al. 2011; Levy-Sakin et al. 2014). Incorporation of all these schemes into a unified bioprocess scale-up will lead to achieve the goal of obtaining high-quality proteins/enzymes in excellent quantities.

2.2 Protein Families: From Biotechnology Perspective

The biotechnological advances in recent times have been found to heavily depend upon new sequencing technologies and advances in proteomics that have enabled large-scale identification and functional characterization of new proteins and enzymes (Ideker et al. 2001; van Beilen and Li 2002; Hartmann et al. 2014). Nevertheless, the process of identification is often complicated owing to the fact that sequence similarities do not warrant similar functional characteristics. Since, a small modification or mutation in a protein could severely affect its structure and native functionality (Wang and Moult 2001). Besides, several microbes with identified functions may not essentially harbour an expected gene in their genomes. Also, there are many protein families that are yet to be discovered based on sequence-based or activity-based approaches. The newly sequenced genomes have created ample sequence information about umpteen of proteins/enzymes spread across the taxa (Loman et al. 2012). Thus, it is essential to annotate these genes and understand how these proteins function in the living cells. This would essentially result in novel protein candidates that could be used in the biotechnological perspective. Since the experimental verification requires time as well as large amount of technical infrastructure, recently several computational algorithms have been developed to partially predict the functions of unknown proteins. Although highly error prone, the bioinformatics approaches essentially cut down the large list and help in creating a precise catalogue of gene products (Weckwerth 2011). Thus, the discovery of new enzymes and proteins helps in cataloguing functional homologues that could serve as alternatives to the proteins that are industrially and therapeutically important. This essentially is an important prerequisite for fuelling research in protein science for identifying new therapeutics as well as commercially important proteins/enzymes.

A large number of environmental factors influence the survival of microorganisms with fitting genetic machinery that helps them sustain. Extreme environmental stress directs the evolution of thermostable, psychrophilic and halophilic proteins as well as proteins with specific functions such as degradation of xenobiotics and other chemical substances (Elleke et al. 2013). In the recent past, several databases and automatic annotation tools have been developed that include IMG-M, MG-RAST, COG/KOG, TIGRFAM, CDD and PROSITE (Kunin et al. 2008). The new protein families' characterization and the novel functionalities identification of the already existing protein families are the major outcome of an integrated pipeline consisting of biochemical, structural, genomic and meta-omics data (Thomas et al. 2012). This pipeline essentially helps in characterizing huge number of newly identified scaffolds that await functional assignment. In the coming years, these strategies would help in identifying novel proteins and enzymes with varied functions that could be developed into a wide variety of biotechnological products such as therapeutic molecules, drug delivery scaffolds, enzymes and various other commercial products.

2.3 Protein Production: Scaling Up and Limitations

Since proteins are important as therapeutics, as catalysts and as diagnostics tools, their demand has continuously increased (Tripathi et al. 2009). Therefore, the ultimate goals of the development of bioprocess and its scale-up are:

1. The production of desired proteins at minimal cost
2. Maximize the volumetric productivity, i.e. the maximum amount of protein obtained in given volume in the least period of time

Thus, nowadays, most of the proteins are the recombinant proteins obtained through heterogeneous production (Xia et al. 2015). The heterogeneous production of recombinant proteins requires a genetically and biochemically stable recombinant culture with higher yield, a high productive fermentation process and cost-effective downstream process. Various strategic choices may be used to the process for enhancing the protein production while designing the purification protocols and strategies, and these can be listed as follows:

1. Selection of right expression host for the protein(s): bacteria, yeast, insect cells or human cells.
2. Selection of suitable expression vector.
3. Selection of suitable strain(s) if bacterial expression is used.
4. Whether the expression of full-length protein is required or a fragment thereof.

5. Whether the tagging of protein is required. If yes, which one is the best?
6. Identification of a good purification strategy and its common pitfalls.

Since every protein is in itself different, unfortunately, the answer to the each of these questions a priori cannot be fitted with every protein (Structural Genomics Consortium et al. 2008). Therefore, the strategies and the purification protocols need to be designed individually for each protein while keeping in mind of its intended use. The above given questions have already been discussed earlier in the book. After fixing all these queries, the next step is the scaling up of the process. Thus, the development of successful scaling-up process is the key factor for the commercial success of any proteins.

The scale-up and optimization of the protein production via biochemical activities of microorganisms, including bacteria, yeasts, algae and moulds, and of animal and plant cell systems require sophisticated engineering skills. To achieve the protein production at commercial level needs the increase in size of fermenter without affecting the overall production efficiency of the process. For this purpose, various fermentation parameters need to be considered which are directly linked with the biochemical activities of the culture and so the protein production as the protein is directly linked with the biochemical activities of the cells. Majorly mixing, mass transfer and heat transfer play the critical role during scale-up of the process (Schmidt 2005). Theoretically the following criteria are assumed suitable bases for the scaling up of the process (Ju and Chase 1992):

1. Constant fermenter geometry (constant height to diameter ratio; *H*/*D*)

   While selecting constant H/D as a basis for the scale-up, the surface to volume ratio declines dramatically during scale-up. It will decrease the relative contribution of surface aeration to oxygen supply and dissolved carbon dioxide removal as compared to sparging contribution. This factor can be critical when the shear-sensitive cultures like animal cells are being used for the protein production which leads to significant decrease in the volumetric productivity. In contrast, this factor will not affect the protein production from the microbial culture as in this case the surface aeration is unimportant (Shuler and Kargi 2002).
2. Constant power per liquid volume

   To minimize the cost of final protein product and to earn the maximum profit, the protein production at commercial level must be cost-effective which is directly linked with the power consumption during aeration and mixing of the system. Therefore, while scaling up, constant power per liquid volume can be an important factor. For geometrically similar, fully baffled vessels with turbulent conditions, it may be noted that if scale-up should be based on maintaining a constant power input per unit volume considering no gassing in the system, one may have the following:

$$N_1 = N_2 \left(D_2 / D_1\right)^{2/3}$$

where N = impeller speed, D = tank diameter and subscripts 1 and 2 represent lab scale and scaled-up fermenter.

For the gassing system,

$$N_1 = N_2 \left(D_2 / D_1\right)\left(Q_1 / Q_2\right)^{1/14}$$

where Q = volumetric gas flow rate.

3. Constant volumetric mass transfer coefficient (K_La)

   Volumetric mass transfer coefficient is an important criterion for an aerobic fermentation. Most of the protein production is carried out under the aerobic condition. During the scale-up of a process for the aerobic fermentation, K_La needs to be increased in the way which it can fulfil the oxygen demand of the biomass during protein production. In taking constant K_La as basis of scale-up criteria, the following equation can be used during the scaling up of fermenter for the protein production:

$$N_1 = N_2 \left(D_2 / D_1\right)^{13/21} \left(Q_2 / Q_1\right)^{5/42}$$

4. Constant impeller tip speed

   The impeller tip speed is important not only for the well mixing of the reactor content but also for the culture protection from shear damage. If the animal cell culture or the bacterial cells are the source for the protein production and as these cells are sensitive to the shear damage, this factor needs to be optimized carefully while scaling up the process; otherwise, it may lead to the less protein production due to the loss of active biomass. Therefore, the equal impeller tip speed in both the laboratory and full-scale plant reactor is one of the key scale-up criteria. The typical impeller tip speed ranges from 5 to 7 m/s:

$$N_1 = N_2 \left(D_{i2} / D_{i1}\right)$$

where D_i = impeller diameter.

5. Constant mixing quality (constant Reynolds number)

   Constant Reynolds number in other words indicates the constant mixing quality. The constant mixing quality for the optimum protein production is important as it affects both the mass and heat transfer in the system which may lead to be a limiting factor for the biomass production and so protein production. Therefore, constant Reynolds number can also be used as the basis criteria for the scale-up:

$$t_{m2} = t_{m1} \left(D_2 / D_1\right)^{11/18}$$

where t_m = mixing time.

6. Constant mixing rate number

Minimum mixing time is desirable for creating the homogenous environment in the reactor. Mixing time is often used as scale-up criteria. Mixing time (t_m) can be defined as the minimum time required for the homogeneous distribution of a small volume of pulsating masses in the bulk of the liquid. It is important to adjust the proper mixing conditions in the large vessels during the scale-up. However, the use of mixing time as a scale-up criteria basis has its own limitations as to maintain the t_m constant in both the vessels increases the power expenditure significantly which is not suggestible as it increases the production cost significantly. Thus, as a whole, it can be understood now that the major limitations of scale-up are the mass, heat and momentum transfer deficiencies with the increase in size of the bioreactors which affect the overall production of the protein (Palomares and Ramírez 2003). Mass transfer becomes more important when the nutrient or chemical inducers like IPTG are to be added while the fermentation is on run. With the increasing size of reactor, mixing time increases significantly (Junker 2004). For example, mixing time in large bioreactors containing animal or plant cell culture (10,000 l) can be in the order of 10^3 s. Thus, approximately 17 min would be required for the homogenous distribution of nutrient or inducer in the reactor under the specified conditions. However, it can be solved by adding these chemical inducers or nutrients though multiple ports to the reactors (Ozturk 1996).

Other important mass transfer limitation is related to the oxygen supply to the cells in aerobic culture system due to low solubility of oxygen in water (Shuler and Kargi 2002). The problem gets aggravated with the increase of cell densities due to higher oxygen demand for supporting the cell growth as compared to the oxygen supply. The effective mixing in reactor to generate homogenous environment is necessary to enhance the volumetric productivity of the protein; otherwise, it may lead to the alcoholic or acid fermentation in bacteria, yeast and animal cell cultures (Palomares and Ramírez 1996; O'Beirne and Hamer 2000). Therefore, during scaling up of the process, the bioreactor should be designed to have a higher K_La to increase the oxygen transfer rate (OTR) as much as possible. In general, the dissolved oxygen tension (DOT) should be higher than 20% (with respect to air saturation) to support the growth of the culture. The limitation of oxygen transfer in bulk liquid to the cell may also be due to diffusion through additional resistances, such as cell aggregates or pellets and immobilized cells. In such cases, a 50% DOT in the bulk liquid may be required to sustain the growth of aggregated or immobilized cells in the reactor (Yegneswaran et al. 1991).

As a whole, the primary limitations of the scale up are gas supply and heat removal. For geometrically similar tanks, shear, mixing time and K_La are impossible to maintain identically simultaneously in both the large and small tanks. Thus, scale-up is difficult as various gradients such as DO gradient, substrate gradient, temperature gradient, pH gradient and CO_2 gradient do occur in the large reactor which led to the induction of multiple physiological responses by the cells. Thus, such oscillating environment in large reactors leads to a heterogeneous cell growth. Consequently, biomass, productivity and yield get lowered significantly, and the by-product formation gets enhanced as compared to small-scale reactors (Bylund et al. 1998; Enfors et al. 2001; Käß et al. 2014). Therefore, a successful conversion of lab scale process to commercial scale requires a lot of time and efforts which is the major limitation for the rapid development of a successful scaling-up process (Neubauer et al. 2013).

2.4 Directed Evolution Approaches

Native proteins many a times are required to be modified and altered to an extent that they meet the industrial demands. Directed evolution has been long proven as an effective approach for enhancing the properties of proteins for commercial and therapeutic applications (Cobb et al. 2012; Goldsmith and Tawfik 2012; Porter et al. 2016). This technique relies on evolution of proteins in laboratories through the manifestation of diverse environmental strains. This is followed by screening and identification of protein variants with desired properties (Kumar and Singh 2013). During the course of evolution, iterative mutations and natural selection have helped organisms to survive in the changing environments. To direct useful phenotypes by providing artificial environmental constraints forms the base of directed evolution approaches. This artificial selection process has been utilized since long for selective breeding of crops as well as domestication of animals. Thus, directed evolution has been established as a highly effective and broadly applicable methodology for obtaining modified gene products. This technique basically mimics the natural evolution process and has been hence termed as 'directed evolution' or 'in vitro evolution' (Zhao and Zha 2004) (Table 2.1).

Earlier, the traditional genetic screens utilized physicochemical agents that randomly modified genomic DNA. This methodology was called random mutagenesis that relied upon agents that mainly included several chemical compounds such as ethyl methanosulphonate (EMS), nitrous acids, 2-aminopurines and even ultraviolet (UV) radiations (Kaur and Sharma 2006). These artificial mutagenesis agents actually augment the rate of mutations during the DNA replication process that result in modified gene products. Later, due to relatively lower mutation rates and lack of controls, in vivo random mutagenesis strategies were employed which are still in infancy (Ravikumar et al. 2014). Further, several workers reported the use of error-prone PCR (epPCR) that utilizes DNA polymerases with mutational bias resulting in a large mutational scope (Leung et al. 1989; Camps et al. 2003). Another important methodology that was developed later included DNA shuffling (Stemmer 1994). This was the first homologous recombination method that involves digestion of particular gene by DNases followed by the reassembly of fragments by PCR. Although the fragments reassemble based on sequence homology, the entire gene also incorporates recombinations during the crossover events (Coco et al. 2001). Owing to high rate of point and deletion mutation, a library of mutants could be generated for screening new mutants. Later, a modification to this technique was developed that utilizes recombination of a family of related genes from various species (Crameri et al. 1998). This was termed as family shuffling and involved

Table 2.1 Comparison of different direct evolution approaches

Approach	Methods	Advantages	Disadvantages
Chemical mutagenesis	Ethyl methanesulphonate (EMS), nitrous acid, UV radiation, etc.	Mutations are dose dependent	Chemical hazard, low and uneven mutations
Mutator strains	Phage-assisted continuous evolution (PACE), XL1-red *E. coli*, etc.	Comparatively simple	Low and uneven mutations
Error-prone PCR (ep-PCR)	Taq polymerase with varying dNTPs and divalent ion (Mg^{2+}, Mn^{2+}) concentrations, etc.	High and even mutation rate	Non-even sampling of codons
Homologous recombinations	DNA shuffling, family shuffling, StEP, RACHITT, NExT, etc.	Ease of identification of useful or futile mutations Sequence reshuffling for attaining functional diversity in protein families	Overdependence on sequence homology
Nonhomologous recombinations	ITCHY, SHIPREC, NRR, overlap extension PCR, etc.	Ease of shuffling poorly related sequences.	Poor general applicability

StEP Staggered extension process, *RACHITT* random chimeragenesis on transient templates, *NExT* nucleotide exchange and excision technology, *ITCHY* incremental truncation for the creation of HYbrid enzymes, *SHIPREC* sequence homology-independent protein recombination, *NRR* non-homologous random recombination

naturally occurring homologous genes. The technique was found effective in producing improved enzymes as well as structurally stable proteins using a single recombination-selection cycle. More recently, focussed mutagenesis approaches have been developed that utilize synthetic DNA oligonucleotides which contain one or more degenerate codons for targeted sites (Acevedo-Rocha and Reetz 2014). These mutation-inducing oligonucleotides are incorporated into a gene library using restriction digestion or gene assembly protocol. This results in simultaneous saturation mutagenesis that has higher probabilities of recombination and an increased ability to produce a large library of mutants that could be screened of desired property.

With further advent of technology, random chimeragenesis on transient templates (RACHITT) was developed which was an improved version of the DNA shuffling technique (Coco et al. 2001). In this case, fragmented DNA anneals directly to a transient polynucleotide scaffold which is further digested. This helps in increasing number of fragment recombinations as well as crossover frequencies. Further modification to this technique in the form of nucleotide exchange and excision technology (NExT) has been proposed that incorporates treatment with uracil-DNA-glycosylase (UDG) and piperidine or apurinic/apyrimidinic lyases leading to higher frequency of recombinations and larger library sizes (Müller et al. 2005). Further, several nonhomologous recombination methods have also been devised that help in shuffling genes with significantly lower sequence identity. The major technique is the incremental truncation for the creation of hybrid enzymes (ITCHY). Here, the domains from two parent enzymes are randomly fused to generate chimaera library that is screened later for novel activities. The technique involves exonuclease III for the controlled digestion of DNA to generate truncated fragments of parent genes in maximum numbers which further undergoes blunt-end ligation to form chimaeras (Ostermeier et al. 1999).

Several screening procedures for directed evolution approaches have been also developed alongside. Spatial separation screening involves expression in model unicellular organisms such as *E. coli* or yeast followed by screening as colonies against environmental constraints. Besides this, spatial separation technique has also been recently utilized for screening individual mutants. Several workers have utilized fluorescent readouts, high-performance liquid chromatography (HPLC), nuclear magnetic resonance (NMR), gas chromatography as well as mass spectroscopy for directly monitoring substrate utilization or product formation in colonies (Cobb et al. 2012). Apart from these, earlier fluorescence-activated cell sorting (FACS) had also been used for screening mutant colonies (Fulwyler 1965). The technique relies on a non-diffusing fluorescent reporter (fluorescent-labelled antibodies or tags) for the identification and isolation of cells containing desired gene variants from heterogeneous cell mixture. The current directed evolution protocols have been effectively utilized for enhancing solubility, substrate affinity, catalytic efficiency, thermostability and activity for multiple substrates in a large variety of proteins.

2.5 Single-Molecule Studies of Proteins

Force represents a crucial functional and structural parameter that defines the role of biological macromolecules in their native environment. This has made nanomechanical single-molecule experiments as important tools for probing variety of biomolecular events such as protein folding, protein-ligand interactions, DNA-protein interactions, molecular motors and enzyme catalysis (Mehta et al. 1999; Deniz et al. 2008). These experimental tools amalgamate material sciences with molecular sciences for precise measurement of miniscule regulatory forces that determine function in the life forms. The outcome gives important insights into the intra- and intermolecular interactions that influence biological macromolecules. Looking at the historical perspective, earliest reported single-molecule experiments were the patch clamp-based conductance measurement of ion channels (Sakmann and Neher 1984). However, this technique has its limitations in the choice of systems that could be studied. In due course, spectroscopy provided a more versatile and convenient means to study single molecules (Joo et al. 2008). The commercial availability of sensitive optical detectors made it easier to monitor molecules that emit photons upon excitation. Besides, this also helped in precise observation of photon emission events with high sensitivity and time resolution. Thus, single-molecule fluorescence-based studies became more relevant with the advent of better imaging and biomolecule labelling techniques (Tian et al. 2017; Chen and Ting 2005; Peterman et al. 2004). Later, owing to its functional operation at biomolecular length

Table 2.2 Comparison of single-molecule techniques used for studying biomolecules

Method	Principle technique	Force (nN)	Resolution (nm)	Applications
Single-molecule FRET	Fluorescence	None	2–10	Protein folding and interactions
Magnetic tweezers	Electromagnet and magnetic bead	0.001–100	1–10	Stretching and deformations in DNA and proteins
Optical tweezers	Infrared laser and microbead	0.1–150	0.1–5	Molecular motors, actin assembly
Microneedles	Optical fibre	1–1000	10	DNA and protein deformations
Atomic force microscopy (AFM)	Cantilever and piezoelectric surface	5–1000	0.1–1	Protein folding, interactions, DNA super structures, DNA unwinding
Biomembrane force probes (BFP)	Lipid vesicle and suction pipette	0.01–1000	10	Protein-ligand and protein-lipid interactions

scales, atomic force microscopy was also widely employed in single-molecule-based studies. Currently, several techniques exist that can take into account a wide range of biomolecular forces. These include optical tweezers, glass microneedles, magnetic beads and biomembrane force probes (BFP) (Table 2.2). Thus, single-molecule study has gradually emerged as an important tool for investigating structure-function correlation of biomolecules.

2.5.1 Single-Molecule FRET

Current understanding of protein folding phenomenon has taken an interesting turn with the advent of single-molecule fluorescence techniques. Single-molecule FRET (Förster resonance energy transfer) represents a very precise and controlled methodology to understand the folding of individual protein molecules (Haran 2003). The relaxation rates observed during the folding kinetics of proteins give only the theoretical steps defining the folding states of a protein (Michalet et al. 2006). Considering a protein exhibiting simple two-state folding behaviour, it can be assumed that the measured spectroscopic property (during unfolding/refolding) is a linear combination of averaged values that demarcate conformational states populated by folded and unfolded states at different time points. On the other hand, theoretical studies provide direct information about the behaviour of single protein molecule under the same experimental condition (Snow et al. 2005). Thus, the single-molecule FRET experiments are aimed at direct visualization of folding events that directly correspond to an individual protein moiety. The methodology involves the attainment of trajectories of FRET efficiency or the distance versus time plot during the time scale of transition between unfolded and folded states or vice versa (Lee et al. 2005).

Unlike other spectroscopic techniques, FRET provides quantitative details of the variation in intermolecular distances. Thus, FRET is sometimes called the 'spectroscopic ruler' in the areas of biophysical studies on biomolecules (Piston and Kremers 2007). Nevertheless, doubts on preciseness of this distance calculation as well as efficiency in multiple experiments remained. This incertitude was dealt in detail by the works utilizing polyproline as spacer between donor and acceptor chromophores. It was observed that the average FRET efficiency at shorter distances was slightly lower than theoretically calculated by Forster theory and way much higher for the longer distances (Watkins et al. 2006). The plausible explanation to this divergence could be the negligible information about orientational averaging during donor lifetime as well as the requirement of interchromophore distance longer than the sizes of chromophores. This early evidences paved way for the development of more accurate single-molecule FRET methodologies. Recently, more precise calculations indicated that molecular simulations could be used in alignment with the FRET-based measurements (Reif and Oostenbrink 2014; Best et al. 2015).

A classical single-molecule protein folding experiment monitored by FRET identifies photon bursts from freely diffusing protein molecules under increasing denaturant concentration. The major advantage of this simple experiment is that it is free from any artefacts that may arise from surface interactions. Each subpopulation of unfolded, partially folded and folded protein moieties slowly interconvert, and the FRET efficiency distribution displays peaks corresponding to the mean efficiency of each subpopulation (Haran 2003; Forman and Clarke 2007). Basically, the peak numbers in the FRET efficiency distribution precisely show the number of thermodynamic states during this simple protein refolding/unfolding experiment. In most efficiency distribution experiments consisting of slowly refolding and unfolding rates, only two major peaks are obtained at an intermediate denaturant concentration. These two peaks represent the unfolded and folded subpopulations of the protein. Similarly, several protein systems have been studied that include cold shock protein (CspTm) (Nettels et al. 2007), Acyl-CoA-binding protein (ACBP) (Laurence et al. 2005), Chymotrypsin inhibitor 2 (CI2) (Deniz et al. 2000), RNase HI (Kuzmenkina et al. 2006) and immunity protein Im9 (Tezuka-Kawakami et al. 2006). Further, workers have also utilized other applications such as microfluidic devices in association with single-molecule FRET measurements to elucidate the protein folding phenomenon more precisely (Lipman et al. 2003).

2.5.2 AFM-Based Force Spectroscopy

Biomolecular interactions are majorly regulated by multiple weak but specific forces that define the affinity of interacting partners. Single-molecule studies have identified topological features and mechanical characteristics of these forces and have thus provided mechanistic insights into numerous biological interactions. With the advent of scanning probe microscopy (SPM) and later the atomic force microscopy (AFM), force spectroscopy emerged as a new tool for studying biomolecular systems (Liu and Wang 2011). The AFM utilizes a sharp tip of a flexible cantilever that probes the mechanical characteristics as well as numerous other parameters of a sample (Carvalho and Santos 2012). Most current day AFM machines rely upon computer-controlled piezoceramic stage that holds the sample and the cantilever that moves over the sample. The deflection force acting between the sample surface and the cantilever is utilized to precisely calculate topological and mechanical features of the sample. This deflection is generated due to lateral (XY) and vertical (Z) displacements of the sample with respect to probe (Dufrene 2002). The sensitivity and resolution of the calculated force depend on the spring constant of cantilever and the laser power that monitors the movement of cantilever. There are two general modes of cantilever motion, 'tapping' mode and the 'contact' mode. In 'tapping' mode, the cantilever is oscillated above the sample surface near its resonance frequency, whereas the 'contact' mode majorly involves constant approach-retraction cycle of cantilever close to sample surface (Garcia and Perezb 2002).

The cantilever motion is utilized to calculate the force, and thus a typical force versus displacement (FD) curve is generated. The variation of force in an FD curve describes the movement of probe towards the sample surface. The movement of probe close or away from the surface is equated with the elastic force of the surface that defines the interaction of the probe and the sample (Anczykowski et al. 1999; Butt et al. 2005).

This interaction is defined on the basis of Hooke's law:

$$F = -k\mathrm{d}$$

where F is the force acting on the cantilever, k is the spring constant of the cantilever and d is deflection.

The cantilever force constants are calibration dependent and vary with different materials used to prepare the probes. Basically, the probe/sample separation affects the different forces that act on the cantilever. Several complex phenomena such as receptor-ligand, protein-protein, DNA-ligand and DNA-protein interactions have been analysed using single-molecule force spectroscopy measurements (Churnside et al. 2012). The biotin/streptavidin interaction is a classical ligand/receptor interaction that has been studied in detail using AFM-based force spectroscopy. The unusually high affinity of the pair as well as the availability of structural information of the interaction has aided in precise calculation of forces acting between them (Grubmuller et al. 1996; Yuan et al. 2000). Several groups have studied the variation in these forces by using mutant protein as well as different analogues of biotin. Besides, this intermolecular force measurements involving protein/protein interactions have been greatly studied using numerous antigen/antibody interactions (Carvalho and Santos 2012). In this case, owing to the diversity in protein pairs, there were tremendous variations in the force measurements. The resulting information provided ample information about the acting forces in different protein scaffolds and molecular pairs and their correlation with their physiological functions.

Another complex phenomenon studied using single-molecule force spectroscopy is the protein folding/unfolding. Several research groups have utilized AFM to study the stretching of single- or multi-domain proteins (Fisher et al. 1999; Besta et al. 2003). In these cases, the FD curves were correlated with the enthalpic and entropic parameters associated with protein folding. In most cases, the proteins generate initial entropic stretching that further leads to domain-wise unfolding generating continuous cycles of deflections. This cycle of deflections is followed by entropic stretching everytime a new domain is unfolded. The compilation of entire entropic and enthalpic cycles is used to define folding/unfolding pathway of the protein.

2.5.3 Optical Tweezers

Optical tweezers form another important tool for single-molecule studies of proteins. Basically, optical tweezers utilize the force of laser radiation pressure to trap small particles. Following this, other associated techniques are used to manipulate the trapped particles and study the variation of forces (Novotny et al. 1997; Curtis et al. 2002). Initially, this methodology was used for manipulating and studying micron-sized dielectric particles. Nevertheless, now optical tweezers have been implemented in studying biological systems to a good extent. Mainly, the ability to manipulate single molecules at nanometre scale and to measure forces on these molecules has made this technique an important biophysical tool. In general, the radiation pressure is defined as a force per unit area on a molecule due to change in the momentum of light. All light is composed of photons having a consistent momentum (p). And if there is a change in momentum of light due to refraction with a molecule, the difference between the momentum flux entering and leaving the object can be used to define the total force. Thus, force due to refraction of light can be used to trap and manipulate molecules and helps in making quantitative measurements (Smith et al. 2003). Conformational changes in molecules can generate change in these measurements and thus can provide opportunity to demonstrating their conformational space under particular physicochemical condition. In other words, these measurements can be used to study structural transitions in molecules, define presence or absence of intermediates or even predict the presence of an alternate pathway. Moreover, the wealth of kinetic and thermodynamic information can be obtained that include rates of transitions, free-energy calculations, identification and calculation of energy barriers and even the full energy landscape showing the conformational dynamics of a protein (Wang et al. 1998).

Recently, many workers have used optical tweezers to obtain insights into the conformational paradigm of intrinsically disordered proteins (IDPs). The technique has helped in probing the low-energy fluctuations and marginally stable structures of IDPs. A very important example is the study of α-synuclein, an IDP that is associated with the pathophysiology of Parkinson's disease (Neupane et al. 2014). Several transient metastable structures of this protein have been identified and studied using optical tweezers. This has helped in showing quasi-equilibrium fluctuations in α-synuclein structure that are short-lived and very unstable that may be attributed to the previously predicted molten globule-like conformation. Besides, several groups have also studied transient states of proteins in their heterogeneous mixtures such as during misfolding and aggregation (Stangner et al. 2013; Heidarsson et al. 2014;

Mashaghi et al. 2014). Recent work on dimeric prion protein has shown the presence of multiple intermediates with comparatively slower diffusion rates that indicates a rugged or rough energy landscape of misfolding (Yu et al. 2015). Similarly, other studies involving optical tweezers have shown the presence of misfolded intermediates in case of slow refolding luciferase enzyme and have helped in correlating calcium concentration and misfolding of neuronal calcium sensor-1 (NCS-1) (Heidarsson et al. 2014). Overall, optical tweezers form an interesting tool to study the dynamics and function of proteins under both physiological and non-physiological conditions.

2.6 Proteases in Biotechnology

Proteases form one of the largest family of enzymes that constitute about 2% of the human genome and account for more than 60% of the total enzyme market. In general, proteases are enzymes that perform proteolysis in a specific or non-specific manner. These proteins mainly hydrolyse peptide bonds and are commonly found associated with protein catabolic pathways. These catabolic pathways in turn form a part of variety of important physiological processes such as cell division, apoptosis, signal transduction, food digestion, blood clotting, etc. Besides, several disease aetiologies are found associated with variety of proteases playing a part in growth and progression of pathogens including the replication of viruses (Neurath and Walsh 1976). Thus, these proteins harbour a diversity of structural and functional features that help them carry out critical purposes inside and outside the cells. Despite several evolutionarily linked diversifications, all proteases essentially function to cleave the amide bond by a nucleophilic attack on the carbonyl group that results in hydrolysis. These proteases naturally exist as inactive precursors called zymogens. These zymogens get converted to their active forms by particular environmental triggers. These triggers could be a variety of factors such as pH, salt, ions, small molecules, peptides or specific ligands that induce specific conformational changes in the protein (Anwar and Saleemuddin 1998; Gupta et al. 2002).

Owing to the huge demand of industrial proteases, sources other than plants and animals have garnered much interest in the scientific and industrial community. This increased interest has led to the discovery of numerous microbial proteases as an alternative source. Basically, the broad biochemical diversity of microbial enzymes as well as their high yield has led to their high acceptability and success at the commercial scale. Additionally, protein engineering approaches to modify the cleavage specificities, solubility and stability of proteases have further broadened their commercial landscape (Eijsink et al. 2004; Bansal et al. 2012). These engineered next-generation proteases are tailored to increase substrate selectivity as well as lend multiple substrate hydrolyzing capabilities. Besides other industrial applications, proteases are also advantageous therapeutic agents that have been employed in the treatment of several diseases such as cancer, AIDS, cardiovascular disease and other infections (Rao et al. 1998; Chanalia et al. 2011).

2.6.1 Classification of Proteases

Proteases are broadly classified based on their site of action and by the nature of active site residues. The Enzyme Commission (EC) classifies proteases into group 3 (hydrolases) and subgroup 4 (peptide bond hydrolases). On the basis of site of action, proteases are further categorized as endopeptidases and exopeptidases. The exopeptidases specifically attack on the peptide bonds proximal to the start or end of protein substrate, i.e. either at the amino- or carboxy-terminal. Thus, based on their N- or C-terminus specificity, these enzymes are further divided into aminopeptidases or carboxypeptidases. Besides, the endopeptidases cleave peptide bonds inside the polypeptide chain mainly away from the C- or N-termini.

Further, these endopeptidases are classified into six major groups based on the nature of active site residues involved in their catalytic mechanism (Garcia-Carreon and Del Toro 1997). These are called aspartate, glutamate, cysteine, serine, threonine and metalloproteases. The serine, cysteine and threonine proteases involve nucleophilic attack on the substrate and produce covalently linked acyl-enzyme intermediates. On the other hand, aspartate, glutamate and other metalloproteases activate water molecules that act as nucleophile for breaking the peptide bond. Additionally, based on pH optima, proteases are also categorized as acidic, neutral or alkaline proteases.

2.6.2 Protease Sources

Proteases are ubiquitously found in all forms of life and play a variety of physiological functions. Owing to their large presence in the living milieu, initially, these proteases were commercially obtained from several plant and animal sources directly. However, due to low agricultural facilities specifically for industrial production of proteases, plant sources are limited. Similarly, animal origin proteases such as trypsin, pepsin and rennins are extracted directly in bulk quantities from animal sources (Gupta et al. 2002). Nevertheless, their production also depends on livestock availability and ethical and operational limitations. Thus, low production from plant and animal sources and increasing industrial demands led to identification of other alternative sources of proteases (Table 2.3).

Thus, owing to their huge biochemical diversity, microbial sources are slowly becoming commercial generators of various proteases. With the advent of new protein engineering approaches, it has become easier to optimize and regulate microbial growth and hence enzyme production. Besides the ease of bulk production, the other major advantage of using microbial enzymes is their similar characteristics compared to original animal and plant enzymes.

2.6.3 Engineered Proteases

The proteases have been commercially utilized in several applications and are often required to work under extreme and non-physiological conditions such as high temperature, pH, presence of chaotropes, chelating agents and detergents (Gupta et al.

Table 2.3 Few important commercial proteases and their source of production

Product	Company	Protease type	Source
Protex 30L	Genencor/DuPont	Serine protease	*Bacillus subtilis*
Neutrase 0.8L	Novozymes	Protease	*B. amyloliquefaciens*
Protex 6L	Genencor/DuPont	Serine protease	*B. amyloliquefaciens*
Ronozyme ProAct	DSM/Novozymes	Alkaline serine protease	*B. licheniformis*
Esperase 8.0L	Novozymes	Protease	*Bacillus* spp.
FNA	Genencor/DuPont	Serine protease	*B. amyloliquefaciens*
Versazyme	Novus	Keratinase	*B. licheniformis*
Arazyme-One-Q	Insect Biotech Co.	Metalloprotease	*Serratia proteamaculans* HY-3
Alcalase 2.4	Novozymes	Subtilisin	*Bacillus* spp.
Savinase	Novozymes	Subtilisin	*Bacillus* spp.
Porperase 1600L	Genencor/DuPont	Serine protease	*B. alcalophilus*
Allzyme FD	Alltech	Serine protease	*Aspergillus niger*

2002). This leads to lowered efficacies and degradation of proteases in industrial setups. Thus, proteases with improved stability and functions have been produced using modern-day genetic engineering, protein engineering and directed evolution approaches. Beside this, the proteases have also been used as therapeutics as they are capable of cleaving various physiological substrates. However, very few proteases have been used as therapeutics because of narrow substrate specificity, stability and activity. The development of high-throughput, bacterial and yeast-based methods has facilitated the redesigning of proteases to the modified one with novel specificities, lesser toxicities and higher stability (Guerrero et al. 2016).

In leather and detergent industries, proteases with high thermostability are required for optimal activity and yield. In general, high temperatures affect the native fold of proteins leading to aggregation and loss of activity. Protease engineering approaches have been successful in enhancing thermostability of proteases to a good extent (Mitrea et al. 2010; Zhao et al. 2016). These approaches mainly rely on availability of thermostable homologue of particular enzyme/protease that is used to identify crucial residues and interactions. These critical residues are then engineered into the mesophilic protein scaffolds to mimic their thermostable counterparts. Another widely used approach is the introduction of non-natural disulphides in the protein structures to confer more stability. The addition of extra-stabilizing disulphides helps in lowering the entropy of the protein and thus keeping the native and active scaffold of the protein intact. An important example is the production of engineered thermostable subtilisin E that incorporates an unnatural disulphide based on structural comparisons with a thermophilic subtilisin-type protease, aqualysin I (Bryan 2000).

Similarly, metal binding sites in many enzymatic proteins have been found to stabilize and help increase the activity. Introduction of calcium-binding pockets by substituting Gly131 and Pro172 by Asp residues in subtilisin increased electrostatic interactions and stabilized the protein (Pantoliano et al. 1988). On the contrary, industries such as detergent and laundries require proteins to function under highly chelating conditions. Thus, proteins with high stability aided with better catalytic function in absence of metal ligands have been produced by a combination of protein engineering and directed evolution. A subtilisin variant Sbt88 with 1000 times greater stability was characterized based on deletion of calcium-binding loop followed by improvement of activity by directed evolution (Strausberg et al. 1995). Also there are several industrial applications that require proteases to function in organic solvents. Several methodologies that mainly include error-prone PCR and directed evolution have been utilized to screen proteases with improved activity in organic solvents.

Besides other industrial applications, proteases have also been extensively used in wide range of therapeutics and have been engineered to suit particular application. Engineered factor VIIa (FVIIa) with improved pro-coagulant and anti-fibrinolytic activity by residue mutations and increasing its enzymatic efficacy by increased localization affinity are such few examples (Persson et al. 2001). Another important aspect of engineering therapeutic proteases has been to reduce their non-specificity and thus the side effects. On these lines, bifunctional thrombin represents an excellent example where the protein not just acts as an activator of multiple pro-coagulant substrates but also acts as an anticoagulant agent (Hanson et al. 1993). Further, thrombin has also been engineered to produce a more efficient protein having higher anticoagulant function. An engineered double mutant thrombin (W215A/E217A) with compromised pro-coagulant activity and very slow (>2000-fold) fibrinogen cleaving activity has been developed that specifically activates protein C in the presence of thrombomodulin (Gandhi et al. 2009). Another important factor associated with therapeutic proteases or proteins in general is their half-lives under physiological milieu. Several factors such as protease digestion and endogenous inhibition have been attributed for their short half-lives in organismic environment. Among several examples, designing inhibitors of tissue plasminogen activator (t-PA) has been shown to prevent systemic activation of plasminogen and helped in promoting thrombolytic property of t-PA (Madison et al. 1989). Furthermore, reducing immunogenicity of therapeutic proteases has also been a subject of interest to increase their overall clinical efficacy. Thus, novel protein engineering approaches along with modern-day computational biochemistry and synthetic biology have helped in channelizing the development of better and high-quality proteases with improved commercial and industrial applications (Table 2.3).

2.6.4 Production Process

The animal- and plant-based enzymes are directly produced from the animal tissues and extracts. Initially, the tissue and organs are processed in an enzyme-extracting solution to obtain the crude proteases. Later, several steps of extraction and concentration are involved to get proteases of typical potency (Jaswal et al. 2008). Nowadays, most major proteases and enzymes are obtained at commercial scales using microbial and fungal strains. These protease-producing strains are screened using plate assay method, and their production capability is estimated quantitatively. At the next stage, media design and optimization are done to help in inducing the production of particular enzyme/protease. The optimization process involves continuous evaluation of process parameters, such as media composition, volume, salt, pH, inoculum age, fermentation time, temperature and carbon, nitrogen and other organic supplements (Vishalakshi et al. 2009).

Depending on the type of microbial growth, the nature of fermentation process, i.e. solid or submerged, is chosen to obtain maximal yields of protease. Solid-state fermentation (SSF) has been extensively used for several *Aspergillus*-based enzymes previously (Olama and el-Sabaeny 1993). In general this fermentation methodology involves growth of microbial cultures in moist solid and undissolved substrates containing media with negligible water. On the other hand, submerged fermentation (SmF) has been also utilized widely where the substrate is kept suspended or solubilized in the media (Sandhya et al. 2005). In addition, nitrogen-rich medium supplemented with glucose also helps in increasing protease production. Utilization of batch, fed-batch, continuous and semi-continuous cultures has been described for protease production at industrial scale. The optimization process for all these production processes involves traditional statistical optimization methods as well as newer radial basis function (RBF), response surface methodology (RSM) and artificial neural network (ANN) (Nagata and Chu 2003; Desai et al. 2008).

SSF has been shown to be more cost-effective than SmF and hence is more widely utilized at commercial scale. Moreover, the yield efficiency and product recovery in SSF are higher as compared to SmF. *Aspergillus* sp. and *Mucor pusillus* showed higher productivity for proteases through SSF using wheat bran as substrate (Sandhya et al. 2005). Besides, a combination of more than two or more different substrates has been found to give higher productivity of proteases. At molecular level, the differences in production process of SSF and SmF have been elucidated. In SSF, the nutrients are firstly released from substrate and then utilized by the fungus, whereas in SmF, the nutrients are easily accessible as they are released in excess from media and result in nutrient repression affecting the enzyme production directly. Besides both these major fermentation techniques, several newer modifications have been studied and reported in the recent past. For example, membrane-surface liquid culture (MSLC) method for the production of neutral protease by *A. oryzae* has been developed that utilizes microporous membrane surface for growth of culture (Yasuhara et al. 1994).

2.6.5 Applications

Proteases play a variety of physiological functions that spread through cellular-, tissue- and organ-level organization and encompass important responses such as haemostasis, inflammation and digestion. Besides, proteases have been also found associated with several disease aetiologies such as AIDS, cancer and prion diseases and thus serve as potential targets for developing new therapeutics (Craik et al. 2011). In the recent past, the microbial proteases have been extensively used in the treatment of several pathological conditions such as cardiovascular disorders, inflammation and cancers. Thus, proteases with their high level of specificity as well as diversity have been used as therapeutic targets as well as therapeutic agents.

Their therapeutic applications range from digestive aids, to antileukemic agents, to thrombolytic agents used in cardiovascular therapy. Also, proteases have found huge applications in preparation of burn and wound treatment creams, as well as contact lens and denture cleaners (Ogunbiyi et al. 1986). Besides wide therapeutic usage, proteases are also used in detergents, food industry, leather treatment and other bioremediation processes.

In detergent industry, mostly, the alkaline proteases are used as supplement to laundry detergents in households as well as industrial setups. The addition of proteases helps in the release of tough and coagulated proteinaceous depositions such as from body secretions, meat, eggs, fish, blood, etc. The utility of detergents in wider systems requires higher washing temperatures, and hence thermostable proteases have been largely applied. Similarly, proteases find applications in leather industry where they have been mainly applied in hair removal and clearing sticky and adherent subcutaneous tissue layers. However, due to certain disadvantages such a low or non-functionality at different pH and temperature ranges and non-specific degradation of collagenous matter, the application of proteases is limited. Interestingly, microbial proteases have also found a relatively newer, cost-effective and pollution-free application in recovering silver from photographic and x-ray films (Shankar et al. 2010).

In food industry, proteases form one of the earliest groups of enzymes that were identified and applied in several processes. Proteases have been routinely used in preparing soya hydrolysates, baking, cheese making as well as making meat tenderizers. Cheese manufacture is one of the major applications where proteases have been used extensively (Cheong et al. 1993). Cheese making involves milk coagulation that is carried out using either of the three enzyme categories, viz. animal rennet, microbial enzymes and recombinant chymosin. Both animal and microbial milk-coagulating proteases belong to the same family of aspartate proteases. The animal-based milk-coagulating proteases or Rennet is a mix of chymosin and pepsin and is very effective. However, owing to the shortage of calf rennet due to high market demands, alternate sources of milk-coagulating proteases were identified. Initially characterized microbial proteases came with two major drawbacks, i.e. presence of nonspecific proteases producing bitterness of cheese on long-term storage and relatively lower yields. Thus, novel milk-coagulating proteases generated recombinantly were preferred that could specifically hydrolyse peptide bonds to generate para-k-casein and macropeptides from milk (Feijoo-Siota et al. 2014). Evidently, chymosin formed one of the most widely used recombinant proteases with high yields and negligible nospecificity during cheese production (Kim et al. 2004). Similarly, in baking industry, several endo- and exoproteases have been used for modifying wheat gluten and to increase the bakery dough. The limited proteolysis by these proteases helps in effective treatment of wheat dough leading to increased loaf volumes for producing large variety of bakery products. In this case, the bacterial proteases as well as proteases from *Aspergillus oryzae* have been used widely for improving and facilitating the baking process (Theron and Divol 2014).

There are several other applications of proteases which include protein sequencing, peptide synthesis as well as structural analysis of proteins. Moreover, proteases find application in other basic science protocols such as general protein digestion, removal of affinity tags, tissue extraction and separation and several cell culture methodologies. The rise in engineered proteases in recent times has brought down the cost of production in major industries such as pharmaceuticals, bakery, detergent, leather and food. Protein engineering helps in incorporating desirable characteristics such as heat stability, pH range, substrate specificity, increased activity and storage time. Proteases from various sources such as bacterial, fungal and viral have been engineered to produce suitable application-based enzymes.

2.7 Chaperones in Biotechnology

Molecular chaperones have been an area of growing interest in the field of medical, pharmaceutical and industrial biotechnology (Henderson et al. 1996; Arakawa et al. 2006). Chaperone function has been found critical in both normal and stress conditions in all living cells. The term chaperone dates back to 1978 where it was used to describe the function of a nuclear protein called nucleoplasmin that helps in chromatin assembly by preventing nonspecific interaction of histone proteins with DNA (Schlieker et al. 2002). Later, these were categorized as a class of proteins that assist in post-translational modification of other proteins. The current understanding of chaperones describes them as a highly conserved protein family that primarily functions to assist folding of nascent polypeptides and prevent their misfolding and aggregation (Hartl and Hayer-Hartl 2002). This stabilizing function of chaperones has been responsible for protecting proteins under variety of stress conditions including heat, chemicals, toxins, free radicals, radiations, etc. Chaperones binding to exposed hydrophobic sites in polypeptides help in preserving their correct scaffold thereby preventing their aggregation. Their role in aggregation prevention has been probed and studied in several neurodegenerative and systemic disorders such as Alzheimer's, Parkinson's, diabetes, etc. (Henderson et al. 1996; Barral et al. 2004). Besides, the role and application of chaperones in folding proteins of therapeutic and industrial importance have been areas of intense research.

The chaperoning function is not just associated with protein-based chaperones but has also been successfully achieved using certain chemical scaffolds and co-solvents too. These co-solvents include several osmolytes and small molecules such as arginine that help in raising the osmotic pressure against environmental water stresses and result in protein stabilization (Arya et al. 2014; Dandage et al. 2015; Srivastava et al. 2015). The well-characterized genetic architecture as well as the ease of fermentation has made *E. coli* as an important industrial organism for the production of proteins of various origins. It has been shown that chaperones can help in improving the refolding process obtained from inclusion bodies and may also be expressed in *E. coli* to reduce the formation of inclusion bodies (Thomas and Baneyx 1997). Thus, in vivo overexpression of chaperones along with protein of interest has been utilized as a successful strategy to obtain proteins of industrial and commercial interest with high yields (Villaverde and Carrio 2003; Ventura and Villaverde 2006). Besides, there had been several other strategies where the yields of industrial proteins have been raised by addition of chemical chaperones at different steps of purification process (Mannen et al. 2001; Rajan et al. 2011).

Moreover, several studies have also demonstrated that chaperones coexist and cooperate with each other during the refolding process in vivo (Hartl and Hayer-Hartl 2002; Genevaux et al. 2007). Hence, there has been a rise in newer strategies that involve expression of a combination of chaperones rather than a single chaperone. However, there are several limitations associated with chaperone overexpression, and hence including more steps in the in vitro refolding and purification process has been favoured in the recent years (Hoffmann and Rinas 2004; de Marco et al. 2007). This has led to utilization of protein chaperones as well as chemical and pharmacological chaperones in the industrial setups. Besides, it has been also found that immobilization of chaperones has been also identified as an important tool for retaining and reutilization of chaperones for longer usage. As a whole, chaperones have been identified as crucial tools for improving recombinant production of therapeutic and industrial important proteins that are aggregation prone.

2.7.1 Chaperone-Assisted Protein Folding

Molecular chaperones bind selectively to nascent polypeptides and partially folded protein intermediates and help them fold into their native and active structure. Thus, chaperones effectively demarcate folded and unfolded polypeptides based on their hydrophobic features (McHaourab et al. 2009). Although the primary sequence contains inherent information on protein folding, the crowded cellular milieu makes a nascent polypeptide very prone to misfolding and aggregation. At this stage, chaperones play a crucial role in stabilizing and assisting the initial protein scaffold to correctly fold into its active form. Besides, molecular chaperones are often involved in compartmental translocations, proteolytic degradation and removal of misfolded and aggregation-prone proteins (Young et al. 2003; Young et al. 2004).

In general, chaperones are found in both prokaryotic and eukaryotic systems and are similar in their basic structural features. The major differences occur in quaternary organization involving the number of subunit, monomer associations and their active oligomeric states. Nevertheless, all chaperones function in a similar fashion and bind to aggregation prone and partially folded protein intermediates to prevent their aggregation. Out of several identified systems, the best characterized chaperone system considers a chamber-based folding of a nascent polypeptide. The GroEL-GroES chaperonin system represents the best example of this chamber-based protein folding (Mayhew et al. 1996). Here, the GroES molecule acts as the lid of the folding chamber in the form of GroEL and helps in creating an isolated environment for a nascent polypeptide. The folding is assisted in the presence of ATP, and the phenomenon is explained via *cis*- or *trans*-mechanism (Hayer-Hartl et al. 2015). The folding mechanism in most chaperone systems is ATP dependent and is considered highly efficient and error-free. The protein folding funnel describes large number of protein conformations with very high entropy at the top. During the folding process, the entropy slowly decreases owing to the number of unstable states receding into native or near-native conformations (Baldwin and Rose 2013).

Here, it is worth mentioning that smaller proteins generally undergo spontaneous folding and large proteins mostly require chaperone assistance for correct folding.

Chaperone assistance actually helps in reducing the population of unfolded conformers that represent different energy levels. Thus, a chaperone-assisted channelization of the energy funnel helps the protein population to attain native structure. In most cases, along with the main chaperone, several co-chaperones smaller in sizes are also present. A major example is the DnaK chaperone system that involves formation of ATP-bound DnaJ complex with DnaJ co-chaperonin (Rosenzweig et al. 2013). The process follows ATP hydrolysis by water and addition of another subunit GprE that releases both ADP and DnaJ. This propagates the release of folded protein from the chaperone complex. Besides these in vivo chaperone systems that have been utilized for recombinant protein production, several chemicals and solvents have been also used to refold proteins in vitro (Bruzdziak et al. 2013). This group of chemical chaperone consists mainly of osmolytes and non-osmolytic small molecules. Chemical chaperones also function by sticking and stabilizing the exposed hydrophobic patches of partially folded proteins and hence facilitate their folding (Rajan et al. 2011). Thus, they form a key step in obtaining correctly refolded proteins in industrial setups. Osmolytes augment the thermodynamic stability and thus aid in protein folding. Besides, solvent additives like arginine also help in recovering active protein during large-scale production of bacterially expressed proteins.

2.7.2 Classes of Chaperones

Molecular chaperones are categorized into two major classes, viz. group I type chaperones found in bacteria, mitochondria and plastids and the group II type chaperones found only in archaeal and eukaryotic cytoplasm (Johnson and Craig 1997). Besides, there is another group of co-chaperones that assist chaperones in selecting protein targets and regulate their association and dissociation. Based on function and mechanism of action, molecular chaperones have been further classified into three subclasses. The 'folding' chaperones (e.g. DnaK and GroEL) depend on ATP-driven conformation change for polypeptide substrate refolding. The 'holding' chaperones (e.g. Hsp 31 and IbpA) bind to the partially unfolded protein intermediates until the 'folding' chaperones are available during stress conditions. Besides, the third class of chaperones (e.g. ClpB) acts on already aggregated forms of proteins and helps their resolubilization. On the other hand, there are two separate classes of nonbiological chaperones that include 'chemical chaperones' consisting of mainly osmolytes (e.g. trehalose) and 'pharmacological chaperones' that include mainly protein-stabilizing ligands and small molecules (e.g. arginine).

2.7.3 Aspects in Industrial Protein Production

The well-characterized genetics as well as fermentation technology of *E. coli* has made it the most commonly used organism for the heterologous production of proteins from various sources. However, the major bottleneck with *E. coli*-based

Table 2.4 Some important chaperones expressed in specific host cells for assisting protein folding

Chaperone	Host
GroEL/ES	*E. coli*
BiP	*Insect cells*
DegP	*E. coli*
Peptidylprolyl *cis-trans* isomerise	*Insect cells*
DnaK	*E. coli*
ClpB	*E. coli*
Hsp 70	*Insect cells*
HtbG	*E. coli*
Calnexin	*Insect cells*
Human protein disulphide-isomerase (PDI)	*CHO cells*
Polyubiquitin	*Kluyveromyces lactis*
DegP	*E. coli*

protein production is the formation of insoluble inclusion bodies (IB). Although IBs help in greatly increasing the overall yield of the recombinant protein, the process of obtaining the protein in its function form is often problematic (Ventura and Villaverde 2006). The exact mechanism behind the formation of these depositions is not known, but it is assumed that eukaryotic nascent polypeptides generally have higher exposure of hydrophobic sites and thus have higher affinity for aggregation (Fahnert et al. 2004). The other important factor is the inefficient protein chaperoning and stabilization system in the bacteria. This led to large number of reports supporting the usage of chaperone overexpression as an immediate solution to this problem (Georgiou and Valax 1996; Thomas and Baneyx 1996; Carrio and Villaverde 2002) (Table 2.4).

Molecular chaperones like GroEL/ES or DnaK have been co-expressed with desired proteins to obtain considerable yields (Chen et al. 2003; Gupta et al. 2006). Nevertheless, it was found that expression of single chaperones does not have significant increase in the amount of soluble protein. Further, it was found that large chaperones actually require small co-chaperones that function in consort to 'hold' and 'refold' the protein. It was shown that application of bi-chaperone systems such as ClpB and DnaK could be a better resort in obtaining soluble protein in significantly higher yield as compared to their individual co-expression (Mogk et al. 1999; Mogk et al. 2003). The eukaryotic counterparts of both these proteins (Hsp70 and Hsp40) have been also shown to improve refolding of recombinant proteins as compared to co-expression of large chaperones such as Hsp90 alone (Glover and Lindquist 1998; Genest et al. 2013). Molecular chaperones are currently utilized in biotechnology industry for their ability to prevent as well as reverse the protein aggregation. Several complex systems such as the *E. coli* DnaK-DnaJ-GrpE system have been utilized to suppress protein aggregation and effectively perform disaggregation (Zolkiewski 1999).

On the other hand, owing to their complex mechanism of action, the precise calculation of aggregation kinetics as well as affinities of chaperone systems (bi-, tri- or more) to different aggregate types are not yet characterized. This fall back has led to several model propositions for defining the actual phenomenon. Two major models are widely accepted that explain the mechanism of action of multiple chaperone systems. The DnaK/ClpB system (chaperone/co-chaperone system) has been studied in detail and has been utilized to develop these models (Mogk et al. 1999). According to the first model, ClpB makes initial contacts with the protein aggregates and later gets attached to DnaK. This model is based on the finding that DnaK has relatively lower binding affinity of binding to smaller protein aggregates. On the other hand, the second model suggests that the ClpB/DnaK complex directly attaches with the protein which is solely determined by specificity of individual chaperone components. Nevertheless, at present, there is still a dearth of significantly active chaperone systems that could be applied in the industrial setups effectively.

2.8 Therapeutic Proteins

There has been a surge in protein-based therapeutics during the past few decades. The major reason behind this surge is the critical role of proteins in cellular machinery and elucidation of mechanisms of their function in great detail. This led to the development of an entirely new class of biotherapeutics that is produced from living systems or their products (Carter 2011). Protein-based biotherapeutics extensively accounts for monoclonal antibodies, recombinant proteins, fusion proteins and peptides. Owing to their broad-spectrum specifications, therapeutic proteins have been successfully utilized in treating cancers, cardiovascular diseases, diabetes, anaemia, haemophilia, etc. (Leader et al. 2008). Nevertheless, these proteins have their own demerits and limitations such as low oral stability due to enzymatic degradation, low physiological half-life and poor shelf life. Thus, the area of protein therapeutics is continuously being researched upon to make them better and more efficient. Next-generation protein therapeutics relies heavily upon protein engineering that helps in modifying these proteins of different origins to suit human immunological response. Currently, these emerging new pharmaceutical products account for more than 200 commercial products that mainly include therapeutics, diagnostics and vaccines (Pavlou and Reichert 2004) (Table 2.5).

2.8.1 Classes and Source

Protein therapeutics have been broadly classified based on their pharmacological activity. They can be used as replacement to abnormal or deficient protein and as delivery systems for certain drugs, effectors or toxins, for enhancing a particular metabolic pathway, adding a new function or activity, and interfere or inhibit certain pathological molecule or an organism (Caravella and Lugovskoy 2010). Besides this function-based categorization, they have been also grouped based on molecular type such as blood factor, antibody, anticoagulant, enzyme, scaffold, fusion protein, growth factors, etc. Overall, protein-based therapeutics is an ever-growing field that relies on producing newer, engineered, efficient and immunologically less-reactive protein-based pharmaceuticals.

Table 2.5 Important therapeutic proteins along with their trade name and usage

Therapeutic protein	Commercial name	Treatment
Insulin	Humulin, Novolin	Diabetes mellitus
Growth hormone (GH)	Genotropin, Norditropin, Nutropin, Omnitrope, etc.	Growth hormone deficiency, Turner's syndrome
Factor VIII	Bioclate, Recombinate, Helixate, ReFacto, etc.	Haemophilia A
Protein C	Ceprotin	Venous thrombosis, protein C deficiency
Lactase	Lactaid	Gas, bloating, diarrhoea due to lactose indigestion
Human albumin	Albumarc, Flexbumin, AlbuRx, Albutein, etc.	Low albumin or loss of albumin
Erythropoietin	Epogen, Procrit	Anaemia
Sargramostim	Leukine	Leukopaenia, AIDS
Human follicle-stimulating hormone (FSH)	Gonal-F, Follistim	Assisted reproduction
Human chorionic gonadotropin (HCG)	Ovidrel	Assisted reproduction
Interferon-β1a (rIFN-β)	Avonex, Rebif	Multiple sclerosis
Alteplase (tissue plasminogen activator, tPA)	Activase	Myocardial infarction, pulmonary embolism, etc.
Urokinase	Abbokinase	Pulmonary embolism
Factor VII a	NovoSeven	Haemophilia A/B
Papain	Accuzyme, Panafil	Various ulcers, cyst and wounds
L-Asparaginase	ELSPAR	Acute lymphocytic leukaemia (ALL)
Streptokinase	Streptase	Myocardial infarction, pulmonary embolism, deep vein thrombosis
Bevacizumab	Avastin	Colorectal cancer, non-small-cell lung cancer
Natalizumab	Tysabri	Multiple sclerosis

2.8.2 Production of Therapeutic Proteins

Therapeutic proteins are produced commercially through various methodologies that include chemical synthesis, animal or insect cell culture, microbial fermentation and transgenic plants and animals (Carter 2011). These methodologies individually have complicated downstream steps and are equally time consuming. In fact, the initial identification and characterization of a probable protein-based therapeutic candidate may take several years that also include their clinical testing. Animal or insect cell cultures have been extensively utilized to produce a number of therapeutic proteins such as erythropoietin, factor VIII, tissue plasminogen activator (tPA), etc. (Ikonomou et al. 2003). This production process involves optimal control of several parameters such as oxygen, serum, glucose and nutrients and also prevention of any contamination that may lead to cell death. On the other hand, few therapeutic proteins have been also successfully produced using microbial hosts like bacteria and yeast. The ease of scalability and well-characterized genetics of these tiny host cells help in producing proteins of interest at a significantly lower cost. Several microbial host cells have been successfully implemented in protein therapeutics production such as *E. coli*, *Saccharomyces cerevisiae*, *Pseudomonas* spp. and *Aspergillus niger* (Demain and Vaishnav 2009). However, in this case, the downstream recovery of proteins poses a major problem in obtaining optimal yields. Hence, several bioprocess improvizations along with protein engineering approaches such as addition of signal peptides, addition of stabilizing charged amino acids and mutations have been utilized for improving production yields.

Transgenics are another very widely used production methodology for obtaining therapeutic proteins of good quality with higher yields (Larrick and Thomas 2001). Several animals such as rabbit, pig, cow and goat are used as transgenic animals for producing therapeutic proteins. These transgenic animals have many important advantages that include lower cost, high expression and reproduction facility. These advantages are utilized maximally by genetically engineering these animals to initiate the synthesis of desired proteins. The process involves microinjection of recombinant DNA molecules into proembryonic stages which are transferred into recipient animal, and positive transgenic animals are screened later. Based on the rate of transgenesis, the animals are used as bioreactors for producing desired therapeutic proteins. Similarly, transgenic plants have also been utilized for producing specific proteins in different tissue loci and plant products. For example, leaf and stem tissues of tobacco and *Arabidopsis* spp., seeds of rice, maize and tobacco, carrot roots and fruits of tomatoes and strawberries (Doran 2000). Besides, many workers have also utilized single-cell cultures of algae such as *Chlorella* and *Chlamydomonas*, hairy root cultures and transformed chloroplasts of various plants for producing desired therapeutic proteins. The major advantage of using plant-based expression systems is their ability of producing very high yields of proteins at significantly low production cost (Cramer et al. 1999).

After obtaining proteins from different sources, the next step is to purify and remove proteinaceous and non-proteinaceous contaminants that may alter the biological activity of the desired protein. Several chromatographic methods such as affinity, ion exchange and ion exchange chromatography have been utilized to achieve high purity grade therapeutic proteins. Before formulating the purified proteins, a safer and consistent delivery system is also incorporated for obtaining maximal efficacy. These delivery systems are chosen to prevent the proteins from getting altered or degraded by temperature and moisture and are also required to reduce immunogenicity and increase shelf life and physiological half-life (Pisal et al. 2010). Finally, other parameters of the delivery vehicle such as pH, solvent and preservatives are chosen, and the therapeutic formulation is finally packaged.

2.8.3 Engineered Proteins as Therapeutics

Protein engineering has been extensively utilized to rationally design as well as modify already available therapeutic proteins to increase their efficacy and target specificity. In fact protein engineering is a tool that helped in the inception of protein-based therapeutics and has greatly revolutionized their clinical usage. Engineered natural proteins, fusion proteins and PEG conjugated and antibody-based therapeutics are only few examples that have been widely utilized for producing effective therapeutics (Carter 2011). Insulin is the first therapeutic protein produced through recombinant DNA technology that was commercialized. This synthetic counterpart (Humulin, Eli Lilly Inc.) was approved by the US Food and Drug Administration (FDA) in 1982 for the treatment of diabetes mellitus (Johnson 1983). The two major advantages were its rapid action after meals and also maintenance of low levels of insulin between meals. The engineered insulin incorporating stabilizing amino acid mutations was similar in action to the endogenous insulin in the body. Later, several companies modified and produced more efficient insulin analogues that had better physiological half-lives, reduced risk of hyperglycaemia and ease of administration. These rationalizations were employed in commercialization of several other natural proteins such as erythropoietin and tissue plasminogen activator (tPA) and different cytokines such as interferon-α (Andersen and Krummen 2002).

Several other protein-based therapeutics have been produced in mammalian cells with additional glycosylation sites that help in augmenting the in vivo half-life. For example, engineered darbepoetin (Aranesp) is administered to anaemic patients for enhancing the production of RBCs (Egrie and Browne 2002). Similarly, epoetin-α has been also produced to contain extra N-linked glycosylation sites for better physiological half-life. Besides this, the advent of fusion proteins has opened a new era of protein-based therapeutics. Fusion proteins are created to combine properties of two or more different proteins and have been commercially successful in countering many diseases such as cancer and multiple sclerosis (Dingermann 2008). An excellent example is the fusion of Fc region of immunoglobulins to target proteins that help in increasing their serum retention and half-life. Apart from increasing half-lives, conjugation with Fc region enables high-throughput purification and increased solubility and stability of therapeutic proteins. Important examples of Fc-fusion proteins that have been approved after clinical trials include etanercept (Enbrel) incorporating a TNF receptor 2 and romiplostim, a thrombopoietin (TPO) receptor agonist that is composed of peptide-Fc fusion (Bibila and Robinson 1995; Chapman et al. 1999).

Other than this, polyethylene glycol (PEG) conjugation has also been shown to improve clinical efficacy of therapeutic proteins. This has given rise to an entirely new engineering technology called 'PEGylation' that has been employed in improving efficacy of several proteins (Ginn et al. 2014). PEGylated proteins have higher molecular volume that prevents their early renal clearance leading to an increased half-life. Several PEGylated proteins have been commercialized successfully as efficient therapeutic proteins that include analogues of interferon-α, interferon-β1a, erythropoietin, human GCSF (granulocyte colony-stimulating factor) and L-asparaginase, etc.

Moreover, there has been a rise in production of 'engineered protein scaffolds' that act as alternative to antibody-based therapeutics since they recognize specific protein targets (Gebauer and Skerra 2009). These scaffolds are produced as a library of small and highly soluble moieties and later screened using phage or yeast systems against specific protein targets. A number of these scaffolds are now increasingly adding up into the next-generation therapeutics (Skerra 2000; Lofblom et al. 2011). Several examples include avimer (based on LDLR-A modules), Adnectins (based on type III fibronectin domains), affibodies (based on synthetic Z domain of staphylococcal protein) and DARPins (based on ankyrin repeat proteins). These scaffolds owing to their small sizes are now utilized for creating multi-targeting and multivalent next-generation protein-based therapeutics. Thus, the clinical success of protein-based therapeutics has revolutionized the clinical development of several new, engineered and efficient protein drugs against several pernicious diseases.

Conclusion

The postgenomic era has paved new pathways into the discovery and development of novel proteins and protein/enzyme-based products. This huge information base has aided protein engineering and later bioprocess to emerge as more efficient and product oriented. The molecular details of protein functions have been elucidated using modern-day single-molecule studies. Utilizing this functional information along with structural biology, several new modifications in industrial protein production have been implemented. In this context, traditional methodologies including directed evolution approaches are now aided with highly curated sequence information as well as more precise molecular biology tools. Besides, the industrial bioprocess setup for commercial proteins/enzymes now includes chaperone-based methodologies. Both at the in vitro as well as in vivo levels, protein recovery is greatly enhanced using biological and chemical chaperones. Overall, the present scenario in protein and enzyme production at industrial scale looks highly competitive and exacting due to increasingly high stimulus of information on sequence, structural and functional aspects of proteins.

References

Acevedo-Rocha CG, Reetz MT (2014) Assembly of designed oligonucleotides: a useful tool in synthetic biology for creating high-quality combinatorial DNA libraries. Methods Mol Biol 1179:189–206. doi:10.1007/978-1-4939-1053-3_13

Ahmad P, Ashraf M, Younis M, Hu X, Kumar A, Akram NA, Al-Qurainy F (2012) Role of transgenic plants in agriculture and biopharming. Biotechnol Adv 30:524–540. doi:10.1016/j.biotechadv.2011.09.006

Anczykowski B, Gotsmann B, Fuchs H, Cleveland JP, Elings VP (1999) How to measure energy dissipation in dynamic mode atomic force microscopy. Appl Surf Sci 140:376–382. doi:10.1016/S0169-4332(98)00558-3

Andersen DC, Krummen L (2002) Recombinant protein expression for therapeutic applications. Curr Opin Biotechnol 13:117–123. doi:10.1016/S0958-1669(02)00300-2

Anwar A, Saleemuddin M (1998) Alkaline proteases: a review. Bioresour Technol 64:175–183. doi:10.1016/S0960-8524(97)00182-X

Arakawa T, Ejima D, Kita Y, Tsumoto K (2006) Small molecule pharmacological chaperones: from thermodynamic stabilization to pharmaceutical drugs. Biochim Biophys Acta 1764:1677–1687. doi:10.1016/j.bbapap.2006.08.012

Arya P, Srivastava A, Vasaikar SV, Mukherjee G, Mishra P, Kundu B (2014) Selective interception of gelsolin amyloidogenic stretch results in conformationally distinct aggregates with reduced toxicity. ACS Chem Neurosci 5:982–992. doi:10.1021/cn500002v

Baldwin RL, Rose GD (2013) Molten globules, entropy-driven conformational change and protein folding. Curr Opin Struct Biol 23:4–10. doi:10.1016/j.sbi.2012.11.004

Bansal S, Srivastava A, Mukherjee G, Pandey R, Verma AK, Mishra P, Kundu B (2012) Hyperthermophilic asparaginase mutants with enhanced substrate affinity and antineoplastic activity: structural insights on their mechanism of action. FASEB J 26:1161–1171. doi:10.1096/fj.11-191254

Barral JM, Broadley SA, Schaffar G, Hartl FU (2004) Roles of molecular chaperones in protein misfolding diseases. Semin Cell Dev Biol 15:17–29. doi:10.1016/j.semcdb.2003.12.010

Best RB, Hofmann H, Nettels D, Schuler B (2015) Quantitative interpretation of FRET experiments via molecular simulation: force field and validation. Biophys J 108:2721–2731. doi:10.1016/j.bpj.2015.04.038

Besta RB, Brockwellb DJ, Toca-Herreraa JL, Blakeb AW, Smithc AD, Radford SE, Clarke J (2003) Force mode atomic force microscopy as a tool for protein folding studies. Anal Chim Acta 479:87–105. doi:10.1016/S0003-2670(02)01572-6

Bibila TA, Robinson DK (1995) In pursuit of the optimal fed-batch process for monoclonal antibody production. Biotechnol Prog 11:1–13. doi:10.1021/bp00031a001

Bruzdziak P, Panuszko A, Stangret J (2013) Influence of osmolytes on protein and water structure: a step to understanding the mechanism of protein stabilization. J Phys Chem B 117:11502–11508. doi:10.1021/jp404780c

Bryan PN (2000) Protein engineering of subtilisin. Biochim Biophys Acta 1543:203–222. doi:10.1016/S0167-4838(00)00235-1

Buiatti M, Christou P, Pastore G (2013) The application of GMOs in agriculture and in food production for a better nutrition: two different scientific points of view. Genes Nutr 8:255–270. doi:10.1007/s12263-012-0316-4

Butt HJ, Cappella B, Kappl M (2005) Force measurements with the atomic force microscope: technique, interpretation and applications. Surf Sci Rep 59:1–152. doi:10.1016/j.surfrep.2005.08.003

Bylund F, Collet E, Enfors SO, Larsson G (1998) Substrate gradient formation in the large-scale bioreactor lowers cell yield and increases by-product formation. Bioprocess Eng 18:171–180. doi:10.1007/s004490050427

Camps M, Naukkarinen J, Johnson BP, Loeb LA (2003) Targeted gene evolution in *Escherichia coli* using a highly error-prone DNA polymerase I. Proc Natl Acad Sci U S A 100:9727–9732. doi:10.1073/pnas.1333928100

Caravella J, Lugovskoy A (2010) Design of next-generation protein therapeutics. Curr Opin Chem Biol 14:520–528. doi:10.1016/j.cbpa.2010.06.175

Carrio MM, Villaverde A (2002) Construction and deconstruction of bacterial inclusion bodies. J Biotechnol 96:3–12. doi:10.1016/S0168-1656(02)00032-9

Carter PJ (2011) Introduction to current and future protein therapeutics: a protein engineering perspective. Exp Cell Res 317:1261–1269. doi:10.1016/j.yexcr.2011.02.013

Carvalho FA, Santos NC (2012) Atomic force microscopy-based force spectroscopy—biological and biomedical applications. IUBMB Life 64:465–472. doi:10.1002/iub.1037

Celik E, Calık P (2012) Production of recombinant proteins by yeast cells. Biotechnol Adv 30:1108–1118. doi:10.1016/j.biotechadv.2011.09.011

Chanalia P, Gandhi D, Singh JD (2011) Applications of microbial proteases in pharmaceutical industry: an overview. Rev Med Microbiol 22:96–101. doi:10.1097/MRM.0b013e3283494749

Chapman AP, Antoniw P, Spitali M, West S, Stephens S, King DJ (1999) Therapeutic antibody fragments with prolonged in vivo half-lives. Nat Biotechnol 17:780–783. doi:10.1038/11717

Chen R (2012) Bacterial expression systems for recombinant protein production: *E. coli* and beyond. Biotechnol Adv 30:1102–1107. doi:10.1016/j.biotechadv.2011.09.013

Chen I, Ting AY (2005) Site-specific labelling of proteins with small molecules in live cells. Curr Opin Biotechnol 16(1):35–40

Chen Y, Song J, Sui SF, Wang DN (2003) DnaK and DnaJ facilitated the folding process and reduced inclusion body formation of magnesium transporter CorA over expressed in *Escherichia coli*. Protein Expr Purif 32:221–231. doi:10.1016/S1046-5928(03)00233-X

Cheong C, Chun SS, Kim YH (1993) Production and properties of an alkaline protease from *Pseudomonas sp.* SJ-320. Korean Biochem J 26:479–484

Churnside AB, Sullan RM, Nguyen DM, Case SO, Bull MS, King GM, Perkins TT (2012) Routine and timely sub-picoNewton force stability and precision for biological applications of atomic force microscopy. Nano Lett 12:3557–3561. doi:10.1021/nl301166w

Clark ED (2001) Protein refolding for industrial processes. Curr Opin Biotechnol 12:202–207. doi:10.1016/S0958-1669(00)00200-7

Cobb RE, Si T, Zhao H (2012) Directed evolution: an evolving and enabling synthetic biology tool. Curr Opin Chem Biol 16:285–291. doi:10.1016/j.cbpa.2012.05.186

Coco WM, Levinson WE, Crist MJ, Hektor HJ, Darzins A, Pienkos PT, Squires CH, Monticello DJ (2001) DNA shuffling method for generating highly recombined genes and evolved enzymes. Nat Biotechnol 19:354–359. doi:10.1038/86744

Craik CS, Page MJ, Madison EL (2011) Proteases as therapeutics. Biochem J 435:1–16. doi:10.1042/BJ20100965

Cramer CL, Boothe JG, Oishi KK (1999) Transgenic plants for therapeutic proteins: linking upstream and downstream strategies. Curr Top Microbiol Immunol 240:95–118. doi:10.1007/978-3-642-60234-4_5

Crameri A, Raillard SA, Bermudez E, Stemmer WP (1998) DNA shuffling of a family of genes from diverse species accelerates directed evolution. Nature 391:288–291. doi:10.1038/34663

Curtis JE, Koss BA, Grier DG (2002) Dynamic holographic optical tweezers. Opt Commun 207:169–175. doi:10.1016/S0030-4018(02)01524-9

Dandage R, Bandyopadhyay A, Jayaraj GG, Saxena K, Dalal V, Das A, Chakraborty K (2015) Classification of chemical chaperones based on their effect on protein folding landscapes. ACS Chem Biol 10:813–820. doi:10.1021/cb500798y

de Marco A, Deuerling E, Mogk A, Tomoyasu T, Bukau B (2007) Chaperone-based procedure to increase yields of soluble recombinant proteins produced in *E. coli*. BMC Biotechnol 12:7–32. doi:10.1186/1472-6750-7-32

Demain AL, Vaishnav P (2009) Production of recombinant proteins by microbes and higher organisms. Biotechnol Adv 27:297–306. doi:10.1016/j.biotechadv.2009.01.008

den Engelsman J, Garidel P, Smulders R, Koll H, Smith B, Bassarab S, Seidl A, Hainzl O, Jiskoot W (2011) Strategies for the assessment of protein aggregates in pharmaceutical biotech product development. Pharm Res 28:920–933. doi:10.1007/s11095-010-0297-1

Deniz AA, Laurence TA, Beligere GS, Dahan M, Martin AB, Chemla DS, Dawson PE, Schultz PG, Weiss S (2000) Single-molecule protein folding: diffusion fluorescence resonance energy transfer studies of the denaturation of chymotrypsin inhibitor 2. Proc Natl Acad Sci U S A 97:5179–5184. doi:10.1073/pnas.090104997

Deniz AA, Mukhopadhyay S, Lemke EA (2008) Single-molecule biophysics: at the interface of biology, physics and chemistry. J R Soc Interface 5:15–45. doi:10.1098/rsif.2007.1021

Desai KM, Survase SA, Saudagar PS, Lele SS, Singhal RS (2008) Comparison of artificial neural network (ANN) and response surface methodology (RSM) in fermentation media optimization: case study of fermentative production of scleroglucan. Biochem Eng J 41:266–273. doi:10.1016/j.bej.2008.05.009

Dingermann T (2008) Recombinant therapeutic proteins: production platforms and challenges. Biotechnol J 3:90–97. doi:10.1002/biot.200700214

Doran PM (2000) Foreign protein production in plant tissue cultures. Curr Opin Biotechnol 11:199–204. doi:10.1016/S0958-1669(00)00086-0

Dufrene YF (2002) Atomic force microscopy, a powerful tool in microbiology. J Bacteriol 184:5205–5213. doi:10.1128/JB.184.19.5205-5213.2002

Egrie JC, Browne JK (2002) Development and characterization of darbepoetin alfa. Oncology (Williston Park) 16:13–22

Eijsink VGH, Bjorkb A, Gaseidnesa S, Sirevagb R, Synstada B, van den Burgc B, Vriendd G (2004) Rational engineering of enzyme stability. J Biotechnol 113:105–120. doi:10.1016/j.jbiotec.2004.03.026

Elleke FB, van der Oost J, Vos MW, van Kranenburg WR (2013) Sustainable production of bio-based chemicals by extremophiles. Curr Biotechnol 2:360–379. doi:10.2174/18722083113076660028

Enfors SO, Jahic M, Rozkov A, Xu B, Hecker M, Jürgen B, Krüger E, Schweder T, Hamer G, O'Beirne D, Noisommit-Rizzi N, Reuss M, Boone L, Hewitt C, McFarlane C, Nienow A, Kovacs T, Trägårdh C, Fuchs L, Revstedt J, Friberg PC, Hjertager B, Blomsten G, Skogman H, Hjort S, Hoeks F, Lin HY, Neubauer P, van der Lans R, Luyben K, Vrabel P, Manelius AL (2001) Physiological responses to mixing in large scale bioreactors. J Biotechnol 85:175–185. doi:10.1016/S0168-1656(00)00365-5

Fahnert B, Lilie H, Neubauer P (2004) Inclusion bodies: formation and utilisation. Adv Biochem Eng Biotechnol 89:93–142. doi:10.1007/b93995

Feijoo-Siota L, Blasco L, Rodríguez-Rama JL, Barros-Velázquez J, Miguel T, Sánchez-Pérez A, Villa TG (2014) Recent patents on microbial proteases for the dairy industry. Recent Adv DNA Gene Seq 8:44–55. doi:10.2174/2352092208666141013231720

Fisher TE, Oberhauser AF, Carrion-Vazquez M, Marszalek PE, Fernandez JM (1999) The study of protein mechanics with the atomic force microscope. Trends Biochem Sci 24:379–384. doi:10.1016/S0968-0004(99)01453-X

Forman JR, Clarke J (2007) Mechanical unfolding of proteins: insights into biology, structure and folding. Curr Opin Struct Biol 17:58–66. doi:10.1016/j.sbi.2007.01.006

Fornwald JA, Lu Q, Boyce FM, Ames RS (2016) Gene expression in mammalian cells using BacMam, a modified Baculovirus system, Baculovirus and insect cell expression protocols. In: Murhammer DW (ed). Methods Mol Biol 1350:95–116. doi:10.1007/978-1-4939-3043-2_5

Fulwyler MJ (1965) Electronic separation of biological cells by volume. Science 150:910–911. doi:10.1126/science.150.3698.910

Gandhi P, Page M, Chen Z, Bush-Pelc L, Di Cera E (2009) Mechanism of the anticoagulant activity of thrombin mutant W215A/E217A. J Biol Chem 284:24098–24105. doi:10.1074/jbc.M109.025403

Garcia R, Perezb R (2002) Dynamic atomic force microscopy methods. Surf Sci Rep 47:197–301. doi:10.1016/S0167-5729(02)00077-8

Garcia-Carreon FL, Del Toro MAN (1997) Classification of proteases without tears. Biochem Educ 25:161–167. doi:10.1016/S0307-4412(97)00005-8

Gebauer M, Skerra A (2009) Engineered protein scaffolds as next-generation antibody therapeutics. Curr Opin Chem Biol 13:245–255. doi:10.1016/j.cbpa.2009.04.627

Genest O, Reidy M, Street TO, Hoskins JR, Camberg JL, Agard DA, Masison DC, Wickner S (2013) Uncovering a region of heat shock protein 90 important for client binding in *E. coli* and chaperone function in yeast. Mol Cell 49:464–473. doi:10.1016/j.molcel.2012.11.017

Genevaux P, Georgopoulos C, Kelley WL (2007) The Hsp70 chaperone machines of *Escherichia coli*: a paradigm for the repartition of chaperone functions. Mol Microbiol 66:840–857. doi:10.1111/j.1365-2958.2007.05961.x

Georgiou G, Valax P (1996) Expression of correctly folded proteins in *Escherichia coli*. Curr Opin Biotechnol 7:190–197. doi:10.1016/S0958-1669(96)80012-7

Ginn C, Khalili H, Lever R, Brocchini S (2014) PEGylation and its impact on the design of new protein-based medicines. Future Med Chem 6:1829–1846. doi:10.4155/fmc.14.125

Glover JR, Lindquist S (1998) Hsp104, Hsp70, and Hsp40: a novel chaperone system that rescues previously aggregated proteins. Cell 94:73–82. doi:10.1016/S0092-8674(00)81223-4

Goldsmith M, Tawfik DS (2012) Directed enzyme evolution: beyond the low-hanging fruit. Curr Opin Struct Biol 22:406–412. doi:10.1016/j.sbi.2012.03.010

Grubmuller H, Heymann B, Tavan P (1996) Ligand binding: molecular mechanics calculation of the streptavidin-biotin rupture force. Science 271:997–999. doi:10.1126/science.271.5251.997

Guerrero JL, Daugherty PS, O'Malley MA (2016) Emerging technologies for protease engineering: new tools to clear out disease. Biotechnol Bioeng. doi:10.1002/bit.26066

Gupta P, Aggarwal N, Batra P, Mishra S, Chaudhuri TK (2006) Co-expression of chaperonin GroEL/GroES enhances in vivo folding of yeast mitochondrial aconitase and alters the growth characteristics of *Escherichia coli*. Int J Biochem Cell Biol 38:1975–1985. doi:10.1016/j.biocel.2006.05.013

Gupta R, Beg QK, Lorenz P (2002) Bacterial alkaline proteases: molecular approaches and industrial applications. Appl Microbiol Biotechnol 59:15–32. doi:10.1007/s00253-002-0975-y

Gupta SK, Shukla P (2015) Advanced technologies for improved expression of recombinant proteins in bacteria: perspectives and applications. Crit Rev Biotechnol 18:1–10. doi:10.3109/07388551.2015.1084264

Hanson SR, Griffin JH, Harker LA, Kelly AB, Esmon CT, Gruber A (1993) Antithrombotic effects of thrombin-induced activation of endogenous protein C in primates. J Clin Invest 92:2003–2012. doi:10.1172/JCI116795

Haran G (2003) Single-molecule fluorescence spectroscopy of biomolecular folding. J Phys Condens Matter 15:R1291–R1317. doi:10.1088/0953-8984/15/32/201

Hartl FU, Hayer-Hartl M (2002) Molecular chaperones in the cytosol: from nascent chain to folded protein. Science 295:1852–1858. doi:10.1126/science.1068408

Hartmann EM, Durighello E, Pible O, Nogales B, Beltrametti F, Bosch R, Christie-Oleza JA, Armengaud J (2014) Proteomics meets blue biotechnology: a wealth of novelties and opportunities. Mar Genomics 17:35–42. doi:10.1016/j.margen.2014.04.003

Hayer-Hartl M, Bracher A, Hartl FU (2015) The GroEL-GroES Chaperonin machine: a Nano-cage for protein folding. Trends Biochem Sci S0968-0004:00140–00141. doi:10.1016/j.tibs.2015.07.009

Heidarsson PO, Naqvi MM, Otazo MR, Mossa A, Kragelund BB, Cecconi C (2014) Direct single-molecule observation of calcium-dependent misfolding in human neuronal calcium sensor-1. Proc Natl Acad Sci U S A 111:13069–13074. doi:10.1073/pnas.1401065111

Henderson B, Nair SP, Coates ARM (1996) Molecular chaperones and disease. Inflamm Res 45:155–158. doi:10.1007/BF02285154

Hoffmann F, Rinas U (2004) Stress induced by recombinant protein production in *Escherichia coli*. Adv Biochem Eng Biotechnol 89:73–92. doi:10.1007/b93994

Ideker T, Galitski T, Hood L (2001) A new approach to decoding life: systems biology. Annu Rev Genomics Hum Genet 2:343–372. doi:10.1146/annurev.genom.2.1.343

Ikonomou L, Schneider YJ, Agathos SN (2003) Insect cell culture for industrial production of recombinant proteins. Appl Microbiol Biotechnol 62:1–20. doi:10.1007/s00253-003-1223-9

Jaswal RK, Kocher GS, Virk MS (2008) Production of alkaline protease by *Bacillus circulans* using agricultural residues: a statistical approach. Indian J Biotechnol 7:356–360

Johnson IS (1983) Human insulin from recombinant DNA technology. Science 219:632–637. doi:10.1126/science.6337396

Johnson JL, Craig EA (1997) Protein folding in vivo: unraveling complex pathways. Cell 90:201–204. doi:10.1016/S0092-8674(00)80327-X

Joo C, Balci H, Ishitsuka Y, Buranachai C, Ha T (2008) Advances in single-molecule fluorescence methods for molecular biology. Annu Rev Biochem 77:51–76. doi:10.1146/annurev.biochem.77.070606.101543

Ju LK, Chase GG (1992) Improved scale-up strategies of bioreactors. Bioprocess Eng 8:49–53. doi:10.1007/BF00369263

Junker BH (2004) Scale-up methodologies for *Escherichia coli* and yeast fermentation processes. J Biosci Bioeng 97:347–364. doi:10.1016/S1389-1723(04)70218-2

Käß F, Junne S, Neubauer P, Wiechert W, Oldiges M (2014) Process in homogeneity leads to rapid side product turnover in cultivation of *Corynebacterium glutamicum*. Microb Cell Factories 13:1–11. doi:10.1186/1475-2859-13-6

Kaur J, Sharma R (2006) Directed evolution: an approach to engineer enzymes. Crit Rev Biotechnol 26:165–199. doi:10.1080/07388550600851423

Kim SY, Gunasekaran S, Olson NF (2004) Combined use of chymosin and protease from *Cryphonectria parasitica* for control of meltability and firmness of cheddar cheese. J Dairy Sci 87:274–283. doi:10.3168/jds.S0022-0302(04)73166-5

Kumar A, Singh S (2013) Directed evolution: tailoring biocatalysts for industrial applications. Crit Rev Biotechnol 33:365–378. doi:10.3109/07388551.2012.716810

Kunin V, Copeland A, Lapidus A, Mavromatis K, Hugenholtz P (2008) A bioinformatician's guide to metagenomics. Microbiol Mol Biol Rev 72:557–578. doi:10.1128/MMBR.00009-08

Kuzmenkina EV, Heyes CD, Nienhaus GU (2006) Single-molecule FRET study of denaturant induced unfolding of RNase H. J Mol Biol 357:313–324. doi:10.1016/j.jmb.2005.12.061

Larrick JW, Thomas DW (2001) Producing proteins in transgenic plants and animals. Curr Opin Biotechnol 12:411–418. doi:10.1016/S0958-1669(00)00236-6

Laurence TA, Kong XX, Jager M, Weiss S (2005) Probing structural heterogeneities and fluctuations of nucleic acids and denatured proteins. Proc Natl Acad Sci U S A 102:17348–17353. doi:10.1073/pnas.0508584102

Leader B, Baca QJ, Golan DE (2008) Protein therapeutics: a summary and pharmacological classification. Nat Rev Drug Discov 7:21–39. doi:10.1038/nrd2399

Lee NK, Kapanidis AN, Wang Y, Michalet X, Mukhopadhyay J, Ebright RH, Weiss S (2005) Accurate FRET measurements within single diffusing biomolecules using alternating-laser excitation. Biophys J 88(4):2939–2953

Leung DW, Chen E, Goeddel DV (1989) A method for random mutagenesis of a defined DNA segment using a modified polymerase chain reaction. Technique 1:11–15

Levy-Sakin M, Berger O, Feibish N, Sharon N, Schnaider L, Shmul G, Amir Y, Buzhansky L, Gazit E (2014) The influence of chemical chaperones on enzymatic activity under thermal and chemical stresses: common features and variation among diverse chemical families. PLoS One 9:e88541. doi:10.1371/journal.pone.0088541

Lipman EA, Schuler B, Bakajin O, Eaton WA (2003) Single-molecule measurement of protein folding kinetics. Science 301:1233–1235. doi:10.1126/science.1085399

Liu S, Wang Y (2011) A review of the application of atomic force microscopy (AFM) in food science and technology. Adv Food Nutr Res 62:201–240. doi:10.1016/B978-0-12-385989-1.00006-5

Lofblom J, Frejd FY, Stahl S (2011) Non-immunoglobulin based protein scaffolds. Curr Opin Biotechnol 22:843–848. doi:10.1016/j.copbio.2011.06.002

Loman NJ, Constantinidou C, Chan JZ, Halachev M, Sergeant M, Penn CW, Robinson ER, Pallen MJ (2012) High-throughput bacterial genome sequencing: an embarrassment of choice, a world of opportunity. Nat Rev Microbiol 10:599–606. doi:10.1038/nrmicro2850

Madison EL, Goldsmith EJ, Gerard RD, Gething MJ, Sambrook JF (1989) Serpin-resistant mutants of human tissue-type plasminogen activator. Nature 339:721–724. doi:10.1038/339721a0

Mannen T, Yamaguchi S, Honda J, Sugimoto S, Nagamune T (2001) Expanded-bed protein refolding using a solid-phase artificial chaperone. J Biosci Bioeng 91:403–408. doi:10.1016/S1389-1723(01)80159-6

Mashaghi A, Mashaghi S, Tans SJ (2014) Misfolding of luciferase at the single-molecule level. Angew Chem Int Ed Engl 53:10390–10393. doi:10.1002/anie.201405566

Mattanovich D, Branduardi P, Dato L, Gasser B, Sauer M, Porro D (2011) Recombinant protein production in yeasts. Methods Mol Biol 824:329–358. doi:10.1007/978-1-61779-433-9_17

Mayhew M, da Silva AC, Martin J, Erdjument-Bromage H, Tempst P, Hartl FU (1996) Protein folding in the central cavity of the GroEL-GroES chaperonin complex. Nature 379:420–426. doi:10.1038/379420a0

McHaourab HS, Godar JA, Stewart PL (2009) Structure and mechanism of protein stability sensors: chaperone activity of small heat shock proteins. Biochemist 48:3828–3837. doi:10.1021/bi900212j

Mehta AD, Rief M, Spudich JA, Smith DA, Simmons RM (1999) Single-molecule biomechanics with optical methods. Science 283:1689–1695. doi:10.1126/science.283.5408.1689

Michalet X, Weiss S, Jager M (2006) Single-molecule fluorescence studies of protein folding and conformational dynamics. Chem Rev 106:1785–1813. doi:10.1021/cr0404343

Mitrea DM, Parsons LS, Loh SN (2010) Engineering an artificial zymogen by alternate frame protein folding. Proc Natl Acad Sci U S A 107:2824–2829. doi:10.1073/pnas.0907668107

Mogk A, Schlieker C, Friedrich KL, Schönfeld HJ, Vierling E, Bukau B (2003) Refolding of substrates bound to small Hsps relies on a disaggregation reaction mediated most efficiently by ClpB/DnaK. J Biol Chem 278:31033–31042. doi:10.1074/jbc.M303587200

Mogk A, Tomoyasu T, Goloubinoff P, Rüdiger S, Roder D, Langen H, Bukau B (1999) Identification of thermolabile *Escherichia coli* proteins: prevention and reversion of aggregation by DnaK and ClpB. EMBO J 18:6934–6949. doi:10.1093/emboj/18.24.6934

Mullard A (2014) 2013 FDA drug approvals. Nat Rev Drug Discov 13:85–89. doi:10.1038/nrd4239

Müller KM, Stebel SC, Knall S, Zipf G, Bernauer HS, Arndt KM (2005) Nucleotide exchange and excision technology (NExT) DNA shuffling: a robust method for DNA fragmentation and directed evolution. Nucleic Acids Res 33:e117. doi:10.1093/nar/gni116

Nagata Y, Chu K H (2003) Optimization of a fermentation medium using neural networks and genetic algorithms. Biotechnol Lett 25: 1837–1842. doi: 10.1023/A:1026225526558.

Nettels D, Gopich IV, Hoffmann A, Schuler B (2007) Ultrafast dynamics of protein collapse from single-molecule photon statistics. Proc Natl Acad Sci U S A 104:2655–2660. doi:10.1073/pnas.0611093104

Neubauer P et al (2013) Consistent development of bioprocesses from microliter cultures to the industrial scale. Eng Life Sci 13:224–238. doi:10.1002/elsc.201200021

Neupane K, Solanki A, Sosova I, Belov M, Woodside MT (2014) Diverse metastable structures formed by small oligomers of a-synuclein probed by force spectroscopy. PLoS One 9:e86495. doi:10.1371/journal.pone.0086495

Neurath H, Walsh KA (1976) Role of the proteolytic enzymes in biological regulation (a review). Proc Natl Acad Sci U S A 73:3825–3832

Novotny L, Bian RX, Xie XS (1997) Theory of nanometric optical tweezers. Phys Rev Lett 79:645–654. doi:10.1103/PhysRevLett.79.645

O'Beirne D, Hamer G (2000) Oxygen availability and the growth of *Escherichia coli* W-3110: a problem exacerbated by scale up. Biprocess Eng 23:487–494. doi:10.1007/s004499900185

Ogunbiyi L, Riedhammer TM, Smith FX (1986) Method for enzymatic cleaning and disinfecting contact lenses. US Patent 4614549

Olama ZA, el-Sabaeny AH (1993) Lipase production by *Aspergillus niger* under various growth conditions using solid state fermentation. Microbiologia 9:134–141

Ostermeier M, Shim JH, Benkovic SJ (1999) A combinatorial approach to hybrid enzymes independent of DNA homology. Nat Biotechnol 17:1205–1209. doi:10.1038/70754

Ozturk SS (1996) Engineering challenges in high density cell culture systems. Cytotechnology 22:3–16. doi:10.1007/BF00353919

Palomares LA, Ramírez OT (1996) The effect of dissolved oxygen tension and the utility of oxygen uptake rate in insect cell culture. Cytotechnology 22:225–237. doi:10.1007/BF00353943

Palomares LA, Ramírez OT (2003) Bioreactor scaleup. In: Spier RE (ed) The encyclopedia of cell technology. Wiley, New York, pp 174–183. doi:10.1002/0471250570.spi023

Pantoliano MW, Whitlow M, Wood JF, Rollence ML, Finzel BC, Gilliland GL, Poulos TL, Bryan PN (1988) The engineering of binding affinity at metal ion binding sites for the stabilization of proteins: subtilisin as a test case. Biochemist 27:8311–8317. doi:10.1021/bi00422a004

Pavlou AK, Reichert JM (2004) Recombinant protein therapeutics-success rates, market trends and values to 2010. Nat Biotechnol 22:1513–1519. doi:10.1038/nbt1204-1513

Persson E, Kjalke M, Olsen O (2001) Rational design of coagulation factor VIIa variants with substantially increased intrinsic activity. Proc Natl Acad Sci U S A 98:13583–13588. doi:10.1073/pnas.241339498

Peterman EJ, Sosa H, Moerner WE (2004) Single-molecule fluorescence spectroscopy and microscopy of biomolecular motors. Annu Rev Phys Chem 55:79–96. doi:10.1146/annurev.physchem.55.091602.094340

Pisal DS, Kosloski MP, Balu-Iyer SV (2010) Delivery of therapeutic proteins. J Pharm Sci 99:2557–2575. doi:10.1002/jps.22054

Piston DW, Kremers GJ (2007) Fluorescent protein FRET: the good, the bad and the ugly. Trends Biochem Sci 32:407–414. doi:10.1016/j.tibs.2007.08.003

Porter JL, Rusli RA, Ollis DL (2016) Directed evolution of enzymes for industrial biocatalysis. Chembiochem 17:197–203. doi:10.1002/cbic.201500280

Radner S, Celie PH, Fuchs K, Sieghart W, Sixma TK, Stornaiuolo M (2012) Transient transfection coupled to baculovirus infection for rapid protein expression screening in insect cells. J Struct Biol 179:46–55. doi:10.1016/j.jsb.2012.04.013

Rai M, Padh H (2001) Expression systems for production of heterologous proteins. Curr Sci 80:1121–1128

Rajan RS, Tsumoto K, Tokunaga M, Tokunaga H, Kita Y, Arakawa T (2011) Chemical and pharmacological chaperones: application for recombinant protein production and protein folding diseases. Curr Med Chem 18:1–15. doi:10.2174/092986711793979698

Rao MB, Tanskale AM, Ghatger MS, Deshpande VV (1998) Molecular and biotechnological aspects of microbial proteases. Microbiol Mol Biol Rev 62:597–635

Ravikumar A, Arrieta A, Liu CC (2014) An orthogonal DNA replication system in yeast. Nat Chem Biol 10:175–177. doi:10.1038/nchembio.1439

Reif MM, Oostenbrink C (2014) Molecular dynamics simulation of configurational ensembles compatible with experimental FRET efficiency data through a restraint on instantaneous FRET efficiencies. J Comput Chem 35:2319–2332. doi:10.1002/jcc.23756

Rosenzweig R, Moradi S, Zarrine-Afsar A, Glover JR, Kay LE (2013) Unraveling the mechanism of protein disaggregation through a ClpB-DnaK interaction. Science 339:1080–1083. doi:10.1126/science.1233066

Sakmann B, Neher E (1984) Patch clamp techniques for studying ionic channels in excitable membranes. Annu Rev Physiol 46:455–472. doi:10.1146/annurev.ph.46.030184.002323

Sandhya C, Sumantha A, Szakacs G, Pandey A (2005) Comparative evaluation of neutral protease production by *Aspergillus oryzae* in submerged and solid-state fermentation. Process Biochem 40:2689–2694. doi:10.1016/j.procbio.2004.12.001

Schlieker C, Bukau B, Mogk A (2002) Prevention and reversion of protein aggregation by molecular chaperones in the *E. coli* cytosol: implications for their applicability in biotechnology. J Biotechnol 96:13–21. doi:10.1016/S0168-1656(02)00033-0

Schmidt FR (2005) Optimization and scale up of industrial fermentation processes. Appl Microbiol Biotechnol 68:425–435. doi:10.1007/s00253-005-0003-0

Shankar S, More SV, Laxman RS (2010) Recovery of silver from waste x-ray film by alkaline protease from *Conidiobolus Coronatus*. Kathmandu Univ J Sci Eng Technol 6:60–69. doi:10.3126/kuset.v6i1.3311

Shuler ML, Kargi F (2002) Bioprocess engineering. Basic concepts, 2nd edn. Prentice Hall, Upper Saddle River

Skerra A (2000) Engineered protein scaffolds for molecular recognition. J Mol Recognit 13:167–187. doi:10.1002/1099-1352(200007/08)13:4<167::AID-JMR502>3.0.CO;2-9

Smith SB, Cui Y, Bustamante C (2003) Optical-trap force transducer that operates by direct measurement of light momentum. Methods Enzymol 361:134–162

Snow CD, Sorin EJ, Rhee YM, Pande VS (2005) How well can simulation predict protein folding kinetics and thermodynamics? Annu Rev Biophys Biomol Struct 34:43–69. doi:10.1146/annurev.biophys.34.040204.144447

Srivastava A, Arya P, Goel S, Kundu B, Mishra P, Fnu A (2015) Gelsolin amyloidogenesis is effectively modulated by curcumin and emetine conjugated PLGA nanoparticles. PLoS One 10:e0127011. doi:10.1371/journal.pone.0127011

Stangner T, Wagner C, Singer D, Angioletti-Uberti S, Gutsche C, Dzubiella J, Hoffmann R, Kremer F (2013) Determining the specificity of monoclonal antibody HPT-101 to tau-peptides with optical tweezers. ACS Nano 7:11388–11396. doi:10.1021/nn405303u

Stemmer WP (1994) Rapid evolution of a protein in vitro by DNA shuffling. Nature 370:389–391. doi:10.1038/370389a0

Strausberg SL, Alexander PA, Gallagher DT, Gilliland GL, Barnett BL, Bryan PN (1995) Directed evolution of a subtilisin with calcium-independent stability. Nat Biotechnol 13:669–673. doi:10.1038/nbt0795-669

Structural Genomics Consortium, China Structural Genomics Consortium, Northeast Structural Genomics Consortium, Gräslund S, Nordlund P, Weigelt J, Hallberg BM, Bray J, Gileadi O, Knapp S, Oppermann U, Arrowsmith C, Hui R, Ming J, dhe-Paganon S, Park HW, Savchenko A, Yee A, Edwards A, Vincentelli R, Cambillau C, Kim R, Kim SH, Rao Z, Shi Y, Terwilliger TC, Kim CY, Hung LW, Waldo GS, Peleg Y, Albeck S, Unger T, Dym O, Prilusky J, Sussman JL, Stevens RC, Lesley SA, Wilson IA, Joachimiak A, Collart F, Dementieva I, Donnelly MI, Eschenfeldt WH, Kim Y, Stols L, Wu R, Zhou M, Burley SK, Emtage JS, Sauder JM, Thompson D, Bain K, Luz J, Gheyi T, Zhang F, Atwell S, Almo SC, Bonanno JB, Fiser A, Swaminathan

S, Studier FW, Chance MR, Sali A, Acton TB, Xiao R, Zhao L, Ma LC, Hunt JF, Tong L, Cunningham K, Inouye M, Anderson S, Janjua H, Shastry R, Ho CK, Wang D, Wang H, Jiang M, Montelione GT, Stuart DI, Owens RJ, Daenke S, Schütz A, Heinemann U, Yokoyama S, Büssow K, Gunsalus KC (2008) Protein production and purification. Nat Methods 5:135–146. doi:10.1038/nmeth.f.202

Tezuka-Kawakami T, Gell C, Brockwell DJ, Radford SE, Smith DA (2006) Urea-induced unfolding of the immunity protein Im9 monitored by spFRET. Biophys J 91:L42–L44. doi:10.1529/biophysj.106.088344

Theron LW, Divol B (2014) Microbial aspartic proteases: current and potential applications in industry. Appl Microbiol Biotechnol 98:8853–8868. doi:10.1007/s00253-014-6035-6

Thomas JG, Baneyx F (1996) Protein misfolding and inclusion body formation in recombinant *Escherichia coli* cells over expressing heat-shock proteins. J Biol Chem 271:11141–11147. doi:10.1074/jbc.271.19.11141

Thomas JG, Baneyx F (1997) Divergent effects of chaperone overexpression and ethanol supplementation on inclusion body formation in recombinant *Escherichia coli*. Protein Expr Purif 11:289–296. doi:10.1006/prep.1997.0796

Thomas T, Gilbert J, Meyer F (2012) Metagenomics—a guide from sampling to data analysis. Microb Inform Exp 2:3. doi:10.1186/2042-5783-2-3

Tian H, Fürstenberg A, Huber T (2017) Labeling and single-molecule methods to monitor G protein-coupled receptor dynamics. Chem Rev 117(1):186–245

Tripathi NK, Sathyaseelan K, Jana AM, Rao PVL (2009) High yield production of heterologous proteins with *Escherichia coli*. Def Sci J 59:137–146

van Beilen JB, Li Z (2002) Enzyme technology: an overview. Curr Opin Biotechnol 13:338–344. doi:10.1016/S0958-1669(02)00334-8

Vázquez-Rey M, Lang DA (2011) Aggregates in monoclonal antibody manufacturing processes. Biotechnol Bioeng 108:1494–1508. doi:10.1002/bit.23155

Ventura S, Villaverde A (2006) Protein quality in bacterial inclusion bodies. Trends Biotechnol 24:179–185. doi:10.1016/j.tibtech.2006.02.007

Villaverde A, Carrio MM (2003) Protein aggregation in recombinant bacteria: biological role of inclusion bodies. Biotechnol Lett 25:1385–1395. doi:10.1023/A:1025024104862

Vishalakshi N, Lingappa K, Amena S, Prabhakar M, Dayanand A (2009) Production of alkaline protease from *Streptomyces gulbargensis* and its application in removal of blood stain. Ind. J Biotechnol 8:280–285

Wang MD, Schnitzer MJ, Yin H, Landick R, Gelles J, Block SM (1998) Force and velocity measured for single molecules of RNA polymerase. Science 282:902–907. doi:10.1126/science.282.5390.902

Wang Z, Moult J (2001) SNPs, protein structure, and disease. Hum Mutat 17:263–270. doi:10.1002/humu.22

Watkins LP, Chang H, Yang H (2006) Quantitative single-molecule conformational distributions: a case study with poly-(L-proline). J Phys Chem A 110:5191–5203. doi:10.1021/jp055886d

Weckwerth W (2011) Green systems biology—from single genomes, proteomes and metabolomes to ecosystems research and biotechnology. J Proteome 75:284–305. doi:10.1016/j.jprot.2011.07.010

Xia J, Wang G, Lin J, Wang Y, Chu J, Zhuang Y, Zhang S (2015) Advances and practices of bioprocess scale-up. Adv Biochem Eng Biotechnol 152:137–151. doi:10.1007/10_2014_293.

Yasuhara A, Ogawa A, Tanaka T, Sakiyama T, Nakanishi K (1994) Production of neutral protease from *Aspergillus oryzae* by a novel cultivation method on a microporous membrane. Biotechnol Tech 8:249–254. doi:10.1007/BF00155416

Yegneswaran PK, Thompson BG, Gray MR (1991) Effect of dissolved oxygen control on growth and antibiotic production by *Strepmomyces clavuligerus* fermentations. Biotechnol Prog 7:246–250. doi:10.1021/bp00009a008

Young JC, Agashe R V, Siegers K, Hartl FU (2004) Pathways of chaperone-mediated protein folding in the cytosol. Nat Rev Mol Cell Biol 5:781–791. doi:10.1038/nrm1492

Young JC, Barral JM, Ulrich Hartl F (2003) More than folding: localized functions of cytosolic chaperones. Trends Biochem Sci 28:541–547. doi:10.1016/j.tibs.2003.08.009

Yu H, Dee DR, Liu X, Brigley AM, Sosova I, Woodside MT (2015) Protein misfolding occurs by slow diffusion across multiple barriers in a rough energy landscape. Proc Natl Acad Sci U S A 112:8308–8313. doi:10.1073/pnas.1419197112

Yuan C, Chen A, Kolb P, Moy T V (2000) Energy landscape of streptavidin-biotin complexes measured by atomic force microscopy. Biochemistry 39:10219–10223. doi:10.1021/bi992715o

Zhao H, Zha W (2004) Evolutionary methods for protein engineering. Marcel Dekker, Inc., New York, pp 353–373

Zhao HY, Wu LY, Liu G, Feng H (2016) Single-site substitutions improve cold activity and increase thermostability of the dehairing alkaline protease (DHAP). Biosci Biotechnol Biochem 80:2480–2485. doi:10.1080/09168451.2016.1230005

Zolkiewski M (1999) ClpB cooperates with DnaK, DnaJ, and GrpE in suppressing protein aggregation. A novel multi-chaperone system from *Escherichia coli*. J Biol Chem 274:28083–28086. doi:10.1074/jbc.274.40.28083

Molecular Farming Approach Towards Bioactive Compounds

3

Savita Khatri, Reena V. Saini, and Anil Kumar Chhillar

Abstract

Medicinal plants are the chief significant resource of therapeutic drugs for health-care of the world's population. Because of their enormous structural diversity and vast array of pharmacological functions, secondary metabolites have become an attractive and imperative basis for research. Bioprocessing has been successful for enhanced secondary metabolite production in numerous cases. Various plant cell culture techniques characterize a probable renewable source of therapeutic compounds, which cannot be produced by chemical synthesis. Therefore, it has been reported as an important step in the direction of industrial production via plant biotechnology. Various biotechnological tools are significant to select, multiply, improve and analyse medicinal plants. In this regard, genetic transformation is a prevailing tool for improved production of novel secondary metabolites. Functional genomics provides high-throughput analysis of multiple genes and their expression that becomes necessary for understanding the regulatory mechanism and biochemical pathways associated with secondary metabolites. Combinatorial biosynthesis is another approach for the production of novel natural products. Here, we present an update on biotechnological approaches that are used for enhanced production of novel bioactive compounds in medicinal plants. The review also gives perspectives on upcoming research in this area.

Keywords

Medicinal plants • Secondary metabolites • Phytochemicals

S. Khatri • A.K. Chhillar (✉)
Centre for Biotechnology, Maharshi Dayanand University, Rohtak, Haryana, India
e-mail: anil.chhillar@gmail.com

R.V. Saini
Animal Biotechnology Laboratory, Faculty of Applied Sciences and Biotechnology, Shoolini University, Solan, Himachal Pradesh, India

V.C. Kalia, A.K. Saini (eds.), *Metabolic Engineering for Bioactive Compounds*,
DOI 10.1007/978-981-10-5511-9_3

3.1 Introduction

Bioactive compounds represent various pharmacological or toxicological activities in human and animals. Primary and secondary metabolites are the major compounds synthesized by medicinal plants. Primary metabolites possess fundamental processes like respiration, photosynthesis and development. Secondary metabolites are produced besides the primary biosynthetic compounds with different array of functions. Medicinal plants produce a variety of secondary metabolites including glycosides, alkaloids, tannins, flavonoids, volatile oils, terpenoids, saponins, steroids and resins (Verpoorte and Memelink 2002). Due to the remarkable structural diversity in secondary metabolites, they appear to be an endless resource of novel pharmacological compounds. It is estimated that around one lakh therapeutic compounds are discovered which have plant origin (Verpoorte et al. 1999; Afendi et al. 2012). In a detailed analysis regarding new approved medicines by the Food and Drug Administration (FDA) between 1981 and 2010, 34% of medicines including anticancer drugs and immunosuppressants were based on natural products or their derivatives (Harvey et al. 2015). Nowadays, advancements in various methodologies and techniques are adopted to improve the chemical diversity of phytochemicals (Wong et al. 2016). To improve the natural treasure of these drugs, further study is required to understand the biochemical pathways and associated enzymes, mode of action at target sites, and large-scale production of secondary metabolites under controlled conditions (Julsing et al. 2006). Various plants explants, including leaves, stems, meristems, roots, etc. can be used to generate secondary metabolites under the aseptic culture conditions (DiCosmo and Misawa 1995). Metabolite production can be enhanced via various parameters such as improvement of strains, selection of vigour cell lines and optimization of culture media. However, plant cultures fail to manufacture the desired compounds. Deficient knowledge about the mechanism of metabolite production pathways is the main limitation of cell culture (Gandhi et al. 2015). Molecular breeding, by applying genetic engineering, is a promising approach. Some research has been done during the past 10 years in this area; a major restraint has been the lack of characterization of secondary metabolic pathways and intermediate products (Verpoorte and Memelink 2002). One of the major objectives of molecular farming is to build up eco-friendly means for scaling up active pharmaceutical ingredient. In this regard, effective molecular techniques have been discovered to unravel bacterial pathways to manufacture new metabolites (Prather and Martin 2008). This could be achieved due to simplicity, lower redundancy and easy accessibility of whole genome of huge number of microorganisms. Moreover, biotechnological innovations have also displayed a main function in crop development to generate priceless proteins used as enzymes and antibodies (Desai et al. 2010). However, only restricted function of biotechnology has been employed in medicinal plants. In medicinal plants, this is usually due to less knowledge about molecular tool and protocols for transgenic approaches and marker-assisted

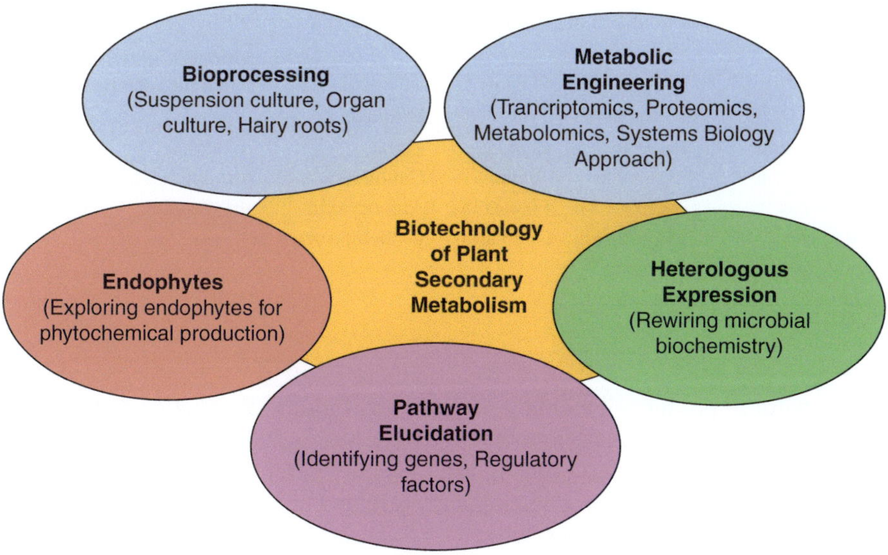

Fig. 3.1 Biotechnology of plant secondary metabolism (Gandhi et al. 2015)

selection. In contrast, the presence of large EST libraries and standardized protocols for most crop species for transgenesis has displayed a key function in yield improvement. However, for secondary metabolite production from medicinal plants, uses of hairy root cultures and bioreactors have turned out to be accepted (Srivastava and Srivastava 2007). Furthermore, lesser time and expenditure of EST sequencing have made it achievable to disclose the molecular pathways of bioactive metabolites. The development of various "omics" technologies, along with advanced spectrometry methods, has extended the secondary metabolite biosynthesis (Chung et al. 2016). An outline of biotechnological applications in metabolite production is represented in Fig. 3.1 (Gandhi et al. 2015).

3.2 Bioactive Compounds and Their Sources

A bioactive compound is "a compound with the capability to interact with one or more component(s) of the living tissue by giving a broad range of feasible effects". The origin of these bioactive compounds can be natural (terrestrial/aquatic, plant, animal/microorganisms) or synthetic (partially or totally). Bioactive compounds have a wide range of applications in different areas including geology, botany, modern pharmacology, agro-industry, cosmetics, food technology and nano-biotechnology (Guaadaoui et al. 2014). Various plants and microorganisms (fungi, bacteria) have confirmed to be a tremendous source of

Table 3.1 Classification of secondary metabolites (Garcia et al. 2013)

Terpenes		Phenols		Nitrogen- and sulfur-containing compounds	
Type	Example	Type	Example	Type	Example
Monoterpenes	Farnesol	Lignan	Lignan	Alkaloids	Nicotine
Sesquiterpenes	Limonene	Tannins	Gallotannin	Atropine	
Diterpenes	Taxol	Flavonoids	Anthocyanin	glucosinolates	Sinigrin
Triterpenes	Digitogenin	Coumarins	Umbelliferone		
Tetraterpenoids	Carotene				
Sterols	Spinasterol				

novel natural products including peptide antibiotics, polyketides and other classes of different bioactive compounds (O'Keefe 2001). Some metabolites from microbes are used as antineoplastic agents, antimicrobial agents and bio-insecticides. The marine environment has presented to be a rich source of natural bioactive compounds (Haefner 2003).

3.2.1 Bioactive Compounds from Medicinal Plants

Plants produce various bioactive secondary metabolites to survive in adverse conditions since the time immemorial. These herbals played an imperative function to survive against insects, herbivores and phytopathogens and to facilitate pollination and reproduction. Secondary metabolites can be classified into three broad categories (Table 3.1). These compounds are synthesized in specialized cells only during a specific growth phase in a particular season, making their extraction and purification relatively complicated (Verpoorte et al. 2002). Examples of commercially valuable plant secondary metabolites are presented in Table 3.2. These compounds are frequently utilized for the production of pharmaceuticals (Shahidi 2009). Therefore, due to the broad array of pharmacological activities, researchers focus on these compounds.

3.2.2 Biosynthetic Pathway of Secondary Metabolites

Secondary metabolites are produced along with primary metabolites (carbohydrates, amino acids, proteins, and lipids) within the plants. They can be named as side products of biochemical pathways in the plant cells and not needed for daily plant functions (Ncube and Van Staden 2015). An outline of the pathways for secondary metabolites is represented in Fig. 3.2. Secondary metabolites are biosynthesized in enzyme-catalysed reactions using building block elements of plants in different ways. In plants, shikimic acid pathway, mevalonic acid pathway, alkaloid pathway, phenolic acid pathway and flavonoid pathways are crucial for the biosynthesis of metabolites (Hussain et al. 2012).

Table 3.2 Plant-derived pharmaceutical importance of secondary metabolites (Hussain et al. 2012)

Secondary metabolites	Uses	Plant species	Chemical structure
Ajmalicine	Antihypertensive	*Catharanthus roseus*	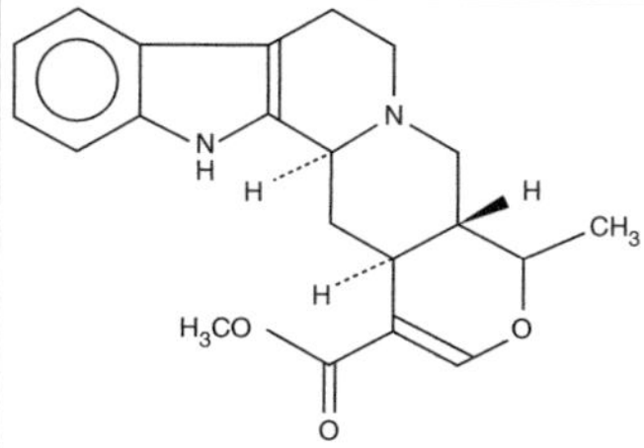
Artemisinin	Antimalarial	*Artemisia annua*	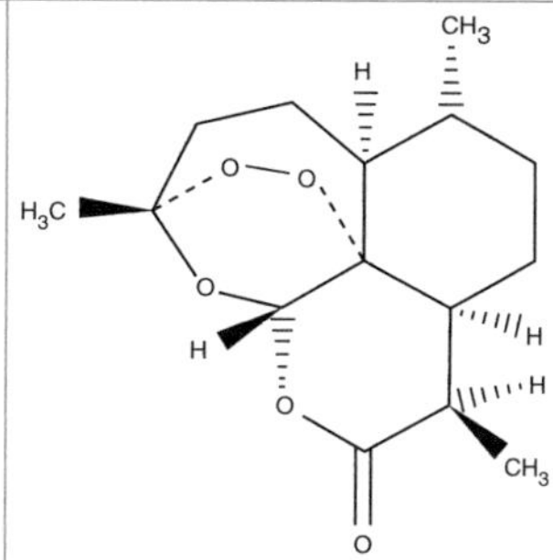
Berberine	Intestinal ailment	*Camellia japonica*	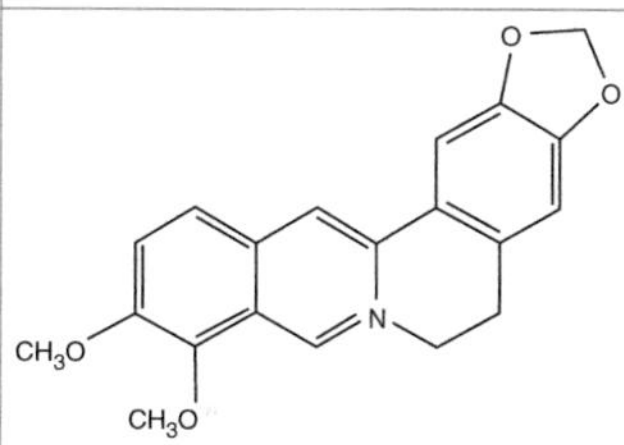
Camptothecin	Antitumour	*Camptotheca acuminate*	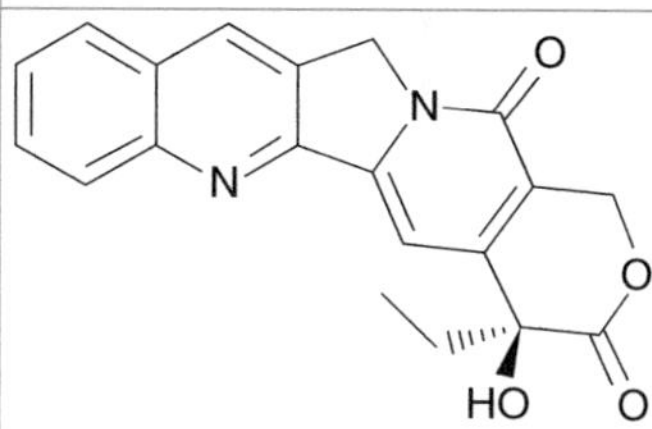
Morphine	Sedative	*Papaver somniferum*	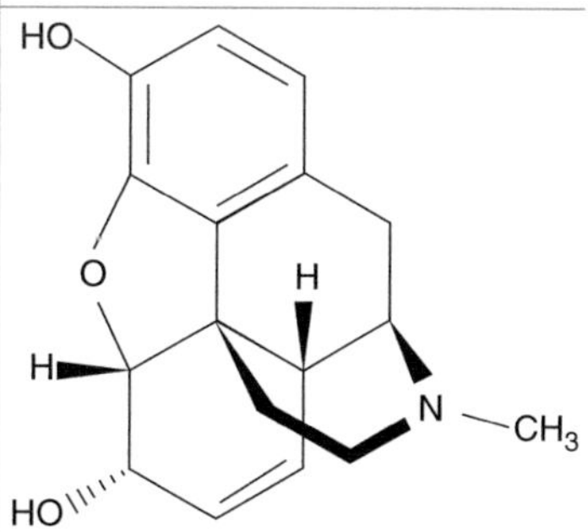

(continued)

Table 3.2 (continued)

Secondary metabolites	Uses	Plant species	Chemical structure
Quinine	Antimalarial	*Cinchona ledgeriana*	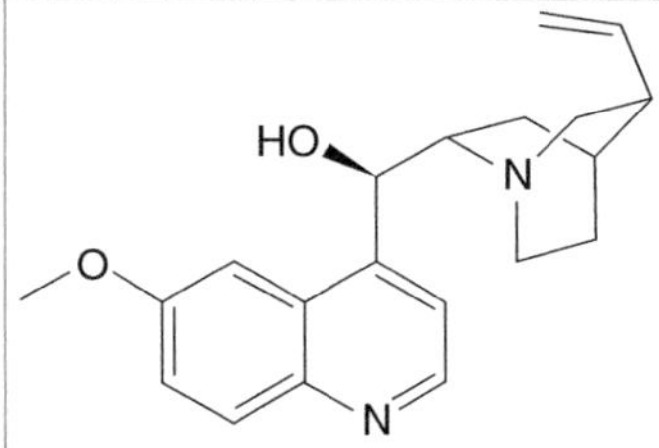
Sanguinarine	Antiplaque	*Sanguinaria canadensis*	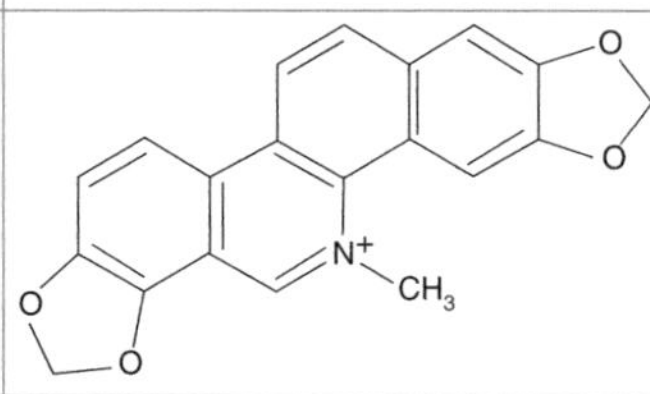
Diosgenin	Steroidal precursor	*Dioscorea deltoidea*	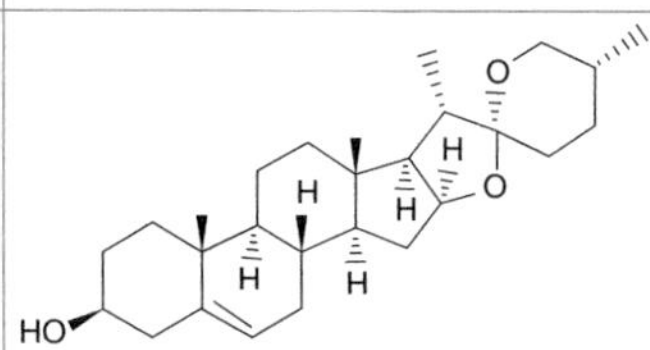
Digoxin	Heart stimulant	*Digitalis lanata*	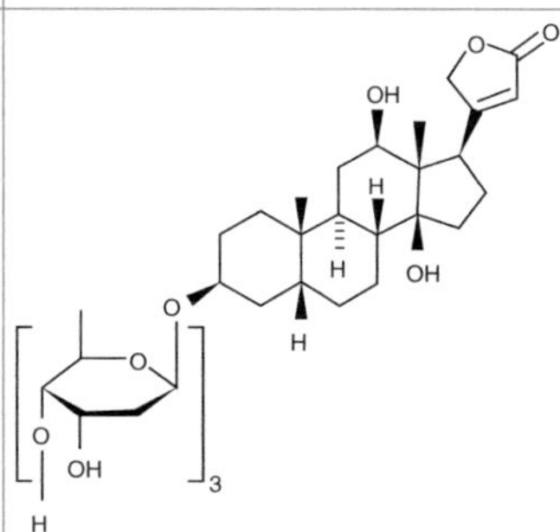
Colchicine	Antitumour	*Colchium autumnale*	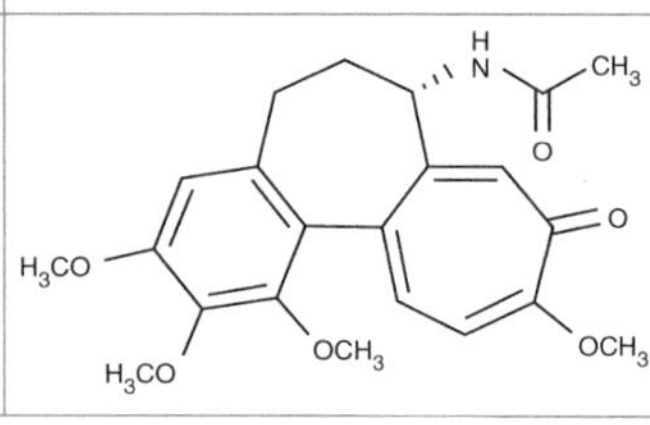

Table 3.2 (continued)

Secondary metabolites	Uses	Plant species	Chemical structure
Codeine	Sedative	*Papaver somniferum*	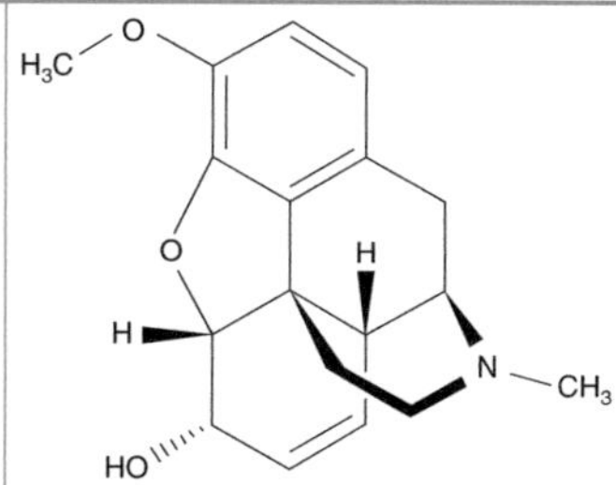
Shikonin	Antimicrobial	*Lithospermum erythrorhizon*	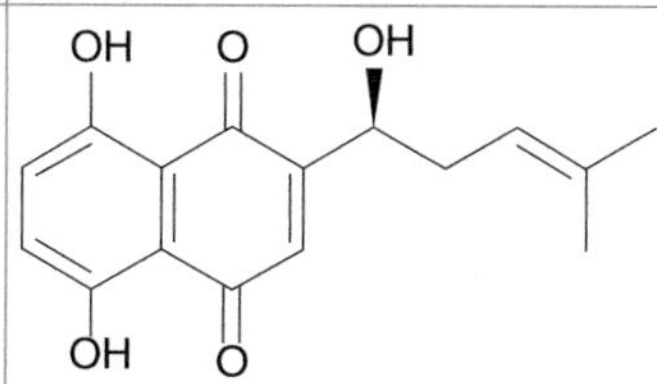
Taxol	Against cancer	*Taxus brevifolia*	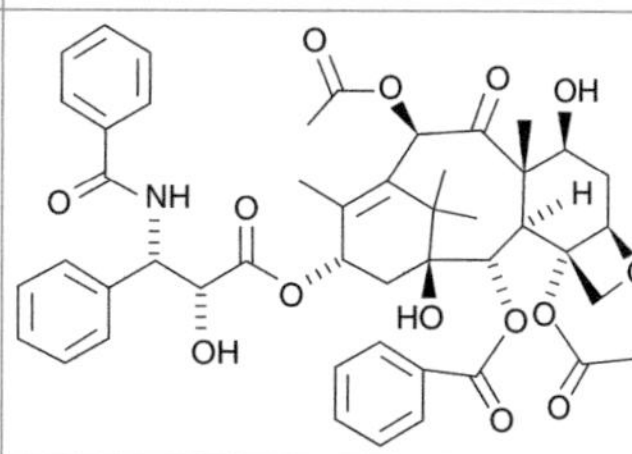
Vincristine	Against leukemia	*C. roseus*	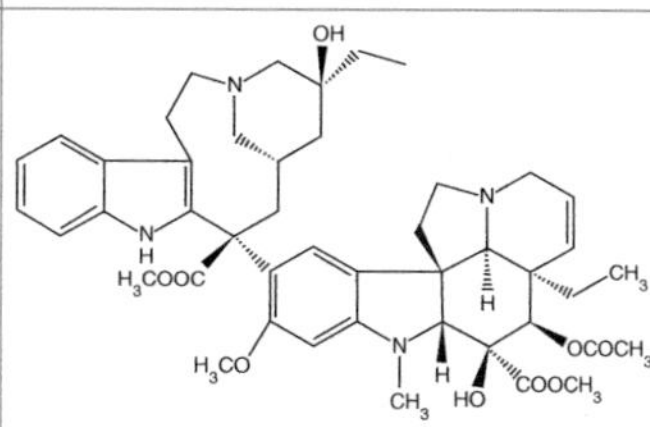
Vinblastine	Against leukemia	*C. roseus*	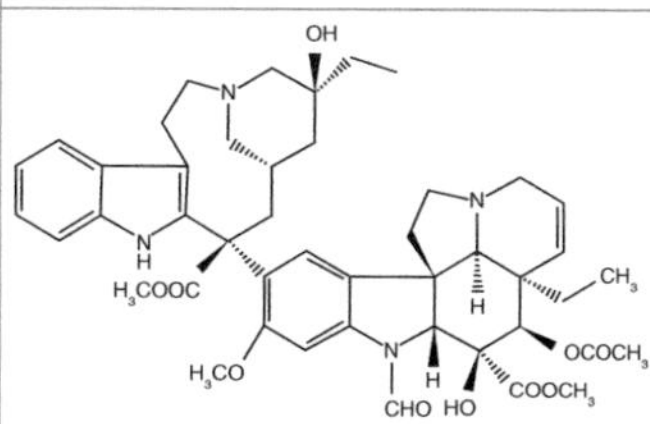

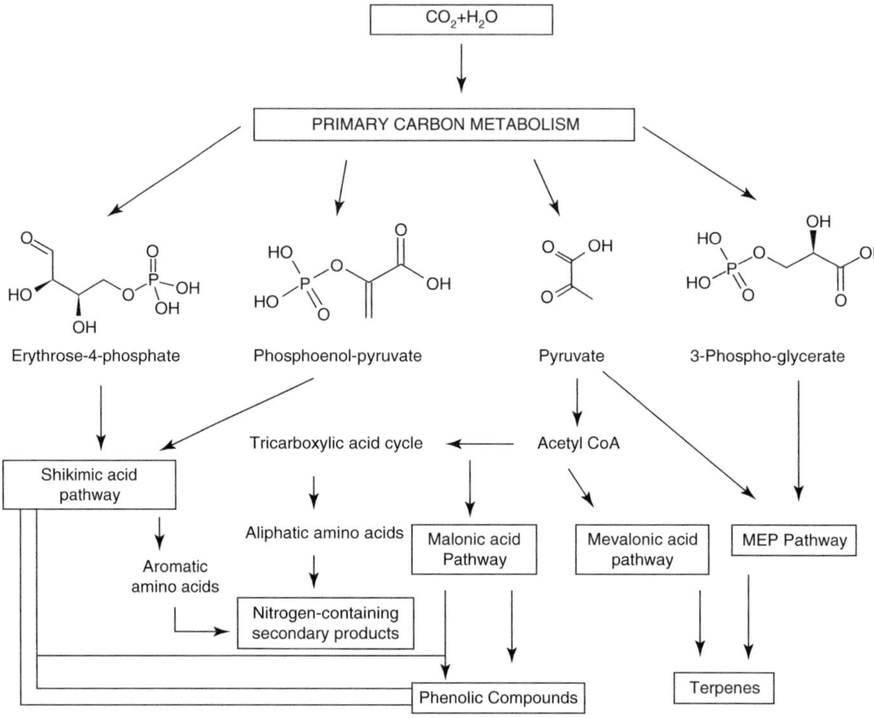

Fig. 3.2 Pathways for the biosynthesis of secondary metabolites (Ncube and Van Staden 2015)

3.3 Bioprocessing

Particular plant organs at different developmental stages produce secondary metabolites with low yields (Chemler and Koffas 2008). Therefore, the commercial production of various metabolites such as artemisinin, Taxol, etc. is extremely complicated (Hashimoto et al. 1988; Corey et al. 1988; Heinstein and Chang 1994). Industrial large-scale tissue culture represents as a substitute for the commercial production of several photochemicals. Increased demand of metabolites, long processing time and reduced area availability can be overcome by bioprocessing in large-scale production (Nalawade and Tsay 2004) (Fig. 3.3). However, there are some limitations including cost, scalability, safety and compound authenticity. These limitations have prompted research into alternatives such as plant-based "molecular farming" approach with lesser cost and toxicity (Xu et al. 2011).

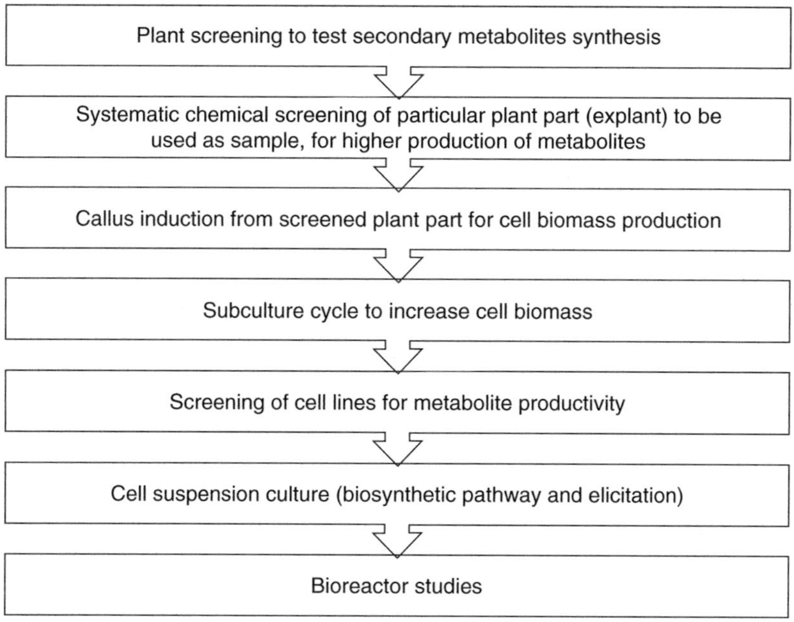

Fig. 3.3 Steps required for the production of secondary metabolites from plant cell culture (Bourgaud et al. 2001)

3.4 Metabolic Engineering

Metabolic engineering (ME) is defined as the directed enhancement of product formation or alteration of particular biochemical pathway with the help of recombinant DNA (rDNA) technology (Fig. 3.4) (Demain and Adrio 2008). Therefore, metabolic engineering has become an interesting research area in biotechnology to obtain natural bioactive compounds, driven by the increasing understanding biosynthetic pathways of secondary metabolites and recent developments in molecular genetics (Wenzel and Muller 2005; Wilkinson and Micklefield 2007). The major goals of molecular farming approach include (a) increased production of particular compounds in plants or (b) transfer of a pathway to other plants and microbes (Verpoorte et al. 2002; Crozier et al. 2009). In this context, the information about the entire biosynthetic pathway and their associated enzymes still needs to be elucidated (Oksman-Caldentey and Saito 2005).

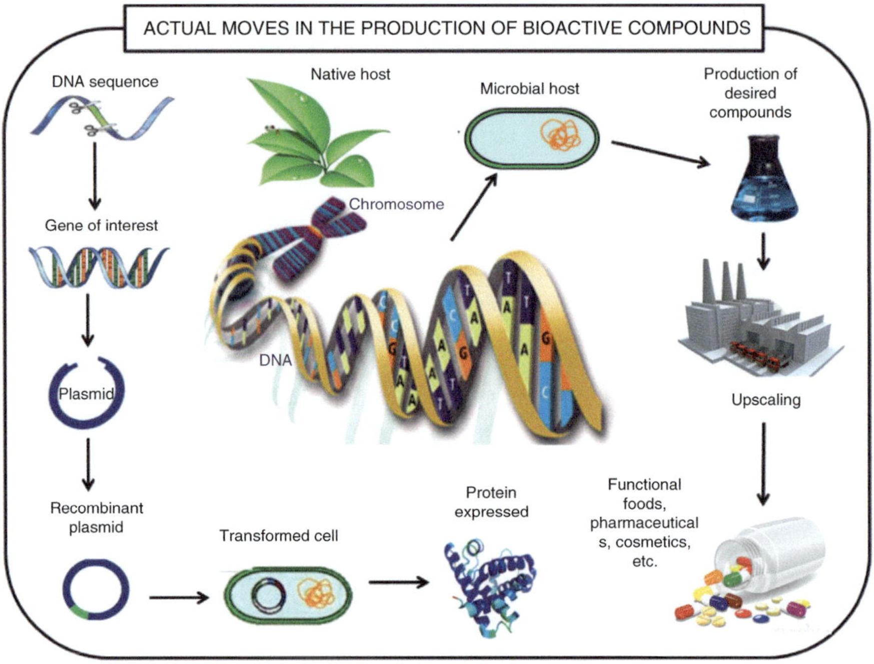

Fig. 3.4 Systematic representation of production of natural bioactive compounds through metabolic engineering (Gil-Chavez et al. 2013)

3.5 Elucidation of Molecular Mechanism of Secondary Metabolite Pathways

Various methods were used to find out the specific biochemical pathways of metabolites, their related enzymes, specific genes that encode the regulatory enzymes and their cofactors. Detection of genes concerned in biosynthetic pathways is very imperative element of biotechnology of medicinal plants. Various molecular techniques used to explore secondary metabolite routes have transformed over time, through better advancements. These molecular strategies can be classified into pre- and post-genomic eras.

3.5.1 Precursor Labelling

The characterization of enzymes and genes involved in biosynthesis could usually be done by means of labelling. For example, condensation of isopentenyl pyrophosphate and dimethylallyl pyrophosphate led to the production of all terpenoids (Poulter et al. 1981). Previously, the belief was that the mevalonate biosynthetic pathway generates isopentenyl pyrophosphate. However, labelling with ^{13}C led to revelation of another pathway for production of particular terpenoids (Rohmer

1999). Now it is well recognized that sesquiterpenes (C15) and triterpenes (C30) are synthesized from cytosolic mevalonate pathway while mono- (C10), di- (C20), and tetra-terpenes (C40) are produced by plastidial methylerythritol phosphate (MEP) pathway (Dudareva et al. 2005).

3.5.2 Elucidation of Microbial Biochemistry

Cloning and expression of genes in a microorganism are essential factors in revelation of metabolite routes to verify its biochemical activity and to consign its function in metabolite production. Co-expression of additional gene along with others in a microorganism leads to the formation of metabolic cluster. For example, co-expression of three enzymes, including 4-coumaroyl:CoA ligase from *L. erythrorhizon*, phenylalanine ammonia-lyase from *Rhodotorula rubra* and curcuminoid synthase from *Oryza sativa* in *E. coli*, showed the making of curcuminoids in recombinant forms of *Escherichia coli* (Hwang et al. 2003). Heterologous expression of biosynthetic pathway in a microbial host is significantly complicated like CYP71AV1, CPR1, CYB5, ADH1 and ALDH1 enzymes that are involved in a pathway which is responsible for the production of artemisinin from *A. annua*, which were reconstituted heterologously for the production of artemisinin in *S. cerevisiae* (Ro et al. 2006; Paddon et al. 2013). Therefore, heterologous microorganisms play a crucial role for elucidation and production of phytochemicals.

3.6 Strategies in Pre-genomic Time

3.6.1 Biochemical Means

Firstly, a hypothetical model is formed on the basis of known biosynthetic pathways. Then, the enzyme is purified by chromatography and its activity is examined by systems devoid of cells. The sequencing of purified enzyme is done and polymerase chain reaction is used for amplification of its cDNA (partial). The cloning of cDNA is done by the designing of RACE (rapid amplification of cDNA ends) primers. Proteins that are expressed at various sites are examined for their enzymatic activity with the help of purified substrates. For example, phenylalanine aminomutase (Taxol biosynthesis) and phenylalanine ammonia (phenylpropanoid biosynthesis) were purified from *Taxus chinensis* Roxb. and *Pinus taeda* L., respectively, by this method (Steele et al. 2005; Whetten and Sederoff 1992).

3.6.2 Expression Strategy Through Library Screening

This method is initiated by formation of mutants of defective pathways of secondary metabolites. These mutants are grouped by complementation and map-based gene cloning. After cloning, the proteins are expressed in heterologous systems and then screened for their relevant enzyme-based assay. Different pathway for the production of b-carotene in chloroplasts was discovered with these methods (Ronen et al. 2000). On the other hand, purified substrates are used to examine the activity of essential enzyme from library of cDNA which are expressed functionally. After this, clones are sequenced. For example, cytochrome P_{450} enzymes have been cloned by this technique (Schoendorf et al. 2001).

3.6.3 Homology-Based Cloning

Known sequences from above approaches and their associated enzymes distribute substantial sequence identity both at gene and protein levels, which might be involved for primer designing and gene cloning in novel plants. Cloning time is reduced by this technique that is required in synthesis of secondary metabolites. For example, conversion of farnesyl diphosphate to b-caryophyllene is done by b-caryophyllene synthase which was cloned from *A. annua* by this method (Cai et al. 2002).

3.7 Strategies in Post-genomic Time

3.7.1 Differential Expression, Expressed Sequence Tag Libraries and Next-Generation Sequencing

Various advances in elucidation approaches include gene sequencing at lower costs and advancement in proteomic tools with various bioinformatics methods, which have transformed the viewpoint to know biosynthetic pathways at a molecular level. The production of secondary metabolites in specific plant organs (Lommen et al. 2006) is studied by performing transcriptomics analysis which is based on differential expression. Uses of various elicitors also represent a prospect for this approach (Zulak et al. 2005). One such method is suppression subtractive hybridization. Where the genomic criteria pre-exist, analysis of differential expression may be performed by microarray methods. Various genes which are required to manufacture various anticancer compounds such as vincristine and vinblastine from *Catharanthus roseus* have been characterized by these approaches (Rischer et al. 2006; Miettinen et al. 2014).

3.7.2 Functional Genomics

In the area of functional genomics, reverse genetics has become an accepted technique. Various functional genomics strategies based upon RNA profiling (transcriptomics), protein profiling (proteomics) and metabolome study (metabolomics) are dominant for prediction of pathways involved in secondary metabolite production

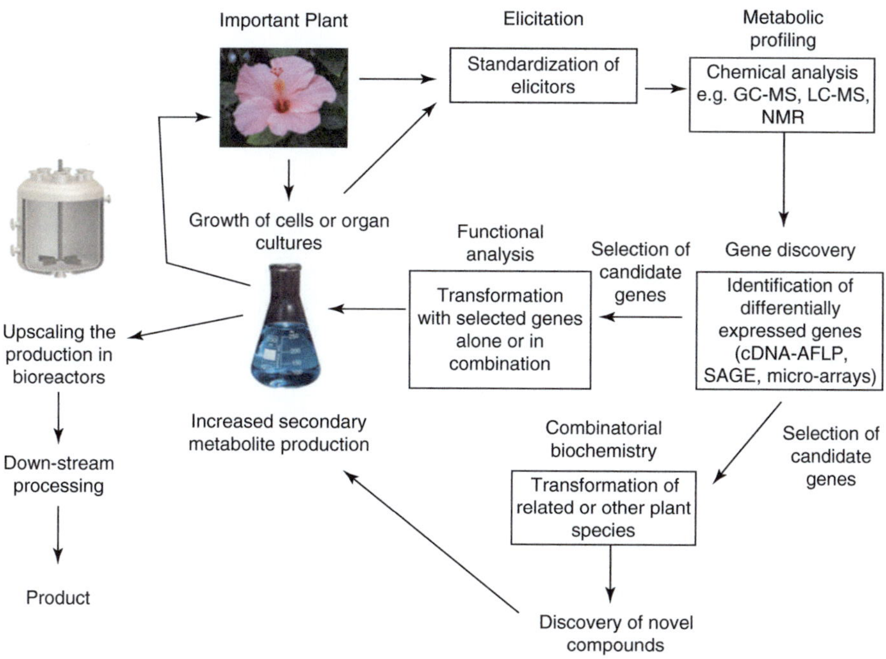

Fig. 3.5 An outline of functional genomics approach (Oksman-Caldentey and Inze 2004)

(Decker et al. 2000; Yamazaki and Saito 2002; Suzuki et al. 2002; Guterman et al. 2002). An outline of functional strategy approach for secondary metabolite production is depicted in Fig. 3.5. This technique relies on the high-throughput screening of genes responsible for the synthesis of important secondary metabolite.

3.8 Strategies Involved in Secondary Metabolite Production

There are various transformation approaches including biolistics method, electroporation and *Agrobacterium tumefaciens*-mediated transformation, which significantly make it effortless to attain transient expression at high levels (Khan and Rahman 2016).

3.8.1 *Agrobacterium tumefaciens*-Mediated Transformation

In today's era, it is the most accepted method for the enhanced synthesis of secondary metabolites *in planta* (Khan and Rahman 2016). Various examples of secondary metabolite production through *Agrobacterium tumefaciens*-mediated transformation are illustrated in Table 3.3.

Table 3.3 *Agrobacterium tumefaciens*-mediated genetic transformation resulting in significant alteration in secondary metabolites in plant

Host plant	Gene	Isolated from	Inferences	References
Withania somnifera	Sterol glucosyltransferase (SGTL1) gene	*Withania somnifera*	Increase in glycosylated withanolide and sterols	Saema et al. (2015)
Artemisia annua	HDR and ADS genes. Artemisinic aldehyde Δ11 (13) reductase (DBR2) gene	*A. annua*	Facilitates higher production of artemisinin content than non-transgenic lines	Yuan et al. (2014)
Artemisia annua L.	HMG-Co A reductase gene (HMGR); amorpha-4,11-diene synthase (ads) gene	*Catharanthus roseus* (L) G. Don; *A. annua*	Transgenic line was found to contain 7.65-fold higher (1.73 mg/gDW) artemisinin than the non-transgenic plant	Alam and Abdin (2011)
Artemisia annua	Cytochrome P450 monooxygenase (cyp71av1) and cytochrome P450 reductase (cpr) genes	*A. annua*	Thirty-eight percent higher accumulation of artemisinin content	Shen et al. (2012)
Eucommia ulmoides Oliver	EuIPI gene	*Eucommia ulmoides*	Three to fourfold increase in the total content of transpolyisoprenes	Chen et al. (2012)
Artemisia annua	3-Hydroxy-3-methylglutaryl CoA reductase (HMGR) gene	*Catharanthus roseus* (L.) G. Don	Increase of 22.5% artemisinin content	Aquil et al. (2009)
Eleutherococcus senticosus Rupr. and Maxim. plants	PgSS1 gene (a squalene synthase gene)	*Panax ginseng*	Phytosterols (beta-sitosterol and stigmasterol) as well as triterpene levels were increased by 2- to 2.5-folds	Seo et al. (2005)
Catharanthus roseus	Deacetylvindoline-4 O-acetyltransferase (DAT) gene	*Catharanthus roseus*	Increased yield of vindoline in transgenic plants	Wang et al. (2012)
Bacopa monnieri	Cryptogein gene	–	Accumulation of saponin in transgenic plants maximally up to 1.4–1.69%	Majumdar et al. (2012)
Panax ginseng	PgDDS Dammarenediol-II synthase	*Panax ginseng*	Production of a tetracyclic triterpenoid; dammarenediol-II	Han et al. (2014)
Panax notoginseng (Burk) F. H.	PnFPS, PnSS, PnSE1, PnSE2, and PnDS	*Panax notoginseng*	Biosynthesis of triterpene saponin	Niu et al. (2014)
Kale (*Brassica oleracea* var. *acephala*)	AtMYB12 transcription factor	*Arabidopsis thaliana*	Severalfold increase in both total phenolics content and flavonol accumulation	Lannenpaa (2014)

3.8.2 Combinatorial Biosynthesis of Secondary Metabolites

Combinatory biosynthesis is an innovative mode to unite metabolic routes of diverse species at molecular level. Currently, *E. coli* and *S. cerevisiae* are used as model species for combinatory synthesis because of their well-developed transformation and expression systems. Besides, the whole genome sequence is also available. Biosynthetic pathways are partially involved in the microbes (Julsing et al. 2006). The excellent example of combinatorial biosynthesis is the grouping of the product of one species and the enzymes from another species (Hranueli et al. 2005; Moore et al. 2005).

3.9 Examples of Metabolic Engineering

It is necessary for metabolic engineering to understand the metabolite production pathways and their associated enzymes in plants. A functional genomics approach, integrating various approaches (metabolomics, proteomics and transcriptomics), facilitates the biotechnologist to enhance the production of secondary metabolites (Yang et al. 2014). Various examples of metabolic engineering for the production of desired bioactive compounds are shown in Table 3.4.

3.10 Future Perspectives

Bioactive compounds are produced either directly or indirectly by a huge number of pharmaceutical industries at larger scale. Commercialization of various culture techniques provides immense opportunity for manufacture of phytocompounds. Elicitation of plant cell culture may be hopeful as it showed favourable outcomes in antibiotics production. Innovative advancements in culture techniques could present new areas for the cost-effective production of bioactive compounds from rare plant species. Different effective strategies in plant tissue culture including media optimization, precursor feeding, cell immobilization, hairy root cultures and elicitation have been tried but failed to produce the desired compounds in noticeable amount. This reflects the poor understanding of biosynthetic regulation in plant cell cultures. Molecular farming represents a substitute technique for large-scale production of herbal drugs with lesser cost and in an eco-friendly way. A combination of bioprocessing and genetic engineering could assist to make the plant culture more fruitful. On the other hand, it necessitates an improved knowledge about metabolite pathways and their control. It is recognized that some pathways communicate well in suspension culture than others. This might possibly be synchronized by particular transcription factors that do not communicate in undifferentiated cells. DNA methylation may control the transcription of a number of enzymes involved in biosynthetic pathways. DNA microarrays serve as high-throughput screening tools for the instantaneous investigation of multiple genes and their expression that becomes indispensable for understanding about regulatory mechanism and biochemical

Table 3.4 Strategies employed in metabolic engineering for improving secondary metabolite production

Metabolite	Plant	Strategy	Biosynthetic gene(s)	Outcome	References
Alkaloids	*Catharanthus roseus* (L.) G. Don	Overexpression of pathway gene	*Strictosidine synthase*	Increased content	Whitmer et al. (1998)
Scopolamine	*Hyoscyamus niger* L.	Overexpression of pathway gene	*Hyoscyamine 6b-hydroxylase and putrescine N-methyltransferase*	Increased content	Zhang et al. (2004)
Flavonoids	*Solanum lycopersicum* L.	Overexpression of pathway gene	*Chalcone isomerase*	Increased content	Muir et al. (2001)
Artemisinin	*Artemisia annua* L.	Overexpression of pathway gene	*Farnesyl diphosphate synthase*	Increased content	Chen et al. (2000)
Menthol	*Mentha* × *piperita* L.	Inhibiting competitive pathway (through gene silencing/RNAi)	*Menthofuran synthase*	Increased content	Mahmoud and Croteau (2001)
Pinoresinol	*Forsythia koreana* (Rehder) Nakai	Inhibiting competitive pathway (through gene silencing/RNAi)	*Pinoresinol/lariciresinol reductase* (PLR)	Increased content	Kim et al. (2009)
Reticuline	*Papaver somniferum* L.	Inhibiting competitive pathway (through gene silencing/RNAi)	*Codeinone reductase*	Increased content	Allen et al. (2004)
Apocarotenoids and flavonoids	*Solanum lycopersicum* L.	Engineering regulatory mechanism (through gene silencing/RNAi)	De-etiolated1 (DET1)	Increased content	Davuluri et al. (2005)
Flavonoids	*Zea mays* L.	Engineering regulatory mechanism (through overexpression)	Leaf color (Lc)	Increased content	Li et al. (2007)
Anthocyanins	*Zea mays* L.	Engineering regulatory mechanism (through overexpression)	C1 and R (transcription factors)	Increased content	Grotewold et al. (1998)

Anthocyanins	*Arabidopsis thaliana* (L.) Heynh.; *Nicotiana tabacum* L.	Engineering regulatory mechanism (through overexpression)	C1 and R (transcription factors)	Increased content	Lloyd et al. (1992)
Anthocyanins	*Solanum tuberosum* L.	Engineering regulatory mechanism (through overexpression)	WD40-repeat gene (StAN11)	Increased content	Li et al. (2014)
Anthocyanins	*Fragaria × ananassa*, cv. Sachinoka	Engineering regulatory mechanism (through overexpression)	FaPHOT2 (phototropin)	Increased content	Kadomura-Ishikawa et al. (2013)
Alkaloids	*Catharanthus roseus* (L.) G. Don	Engineering regulatory mechanism (through overexpression)	Octadecanoid-derivative responsive catharanthus AP2-domain protein (ORCA3) and geraniol 10-hydroxylase (G10H)	Increased content	van der Fits and Memelink (2000)
Nornicotine	*Nicotiana tabacum* L.	Gene silencing (RNA interference)	CYP82E4	Decreased content	Gavilano et al. (2006)
			Nicotine demethylase	Decreased content	Lewis et al. (2008)
Heavy metals (lead, cadmium)	*Arabidopsis thaliana* (L.) Heynh.	Heterologous expression	Pb(II)/Cd(II)/Zn(II)-transporting ATPase (ZntA)	Improved heavy metal resistance	Lee et al. (2003)
Caffeine	*Coffea arabica* L.	Gene silencing (RNA interference)	*Theobromine synthase* (CaXMT1, CaMXMT1) and *caffeine synthase* (CaDXMT1)	Decreased caffeine content	Ogita et al. (2003)
p-Hydroxybenzylglucosinolates	*Arabidopsis thaliana* (L.) Heynh.	Heterologous expression	CYP79A1	Production of novel compound	Bak et al. (1999)
Laudanine	*Eschscholzia californica* Cham.	Gene silencing (RNA interference)	*Berberine bridge forming enzyme* (BBE)	Production of novel compound	Fujii et al. (2007)

(continued)

Table 3.4 (continued)

Metabolite	Plant	Strategy	Biosynthetic gene(s)	Outcome	References
Valencene	*Nicotiana benthamiana* Domin	Inhibiting competitive pathway (through gene silencing/RNAi)	*5-Epi-aristolochene synthase* (EAS) and *squalene synthase*	Increased content	Cankar et al. (2014)
Morphine	*Papaver bracteatum* Lindl.	Overexpression of pathway gene	*Codeinone reductase*	Increased content	Sharafi et al. (2013)
Rosmarinic acid	*Salvia miltiorrhiza* Bunge	Engineering regulatory mechanism (through overexpression)	R2R3 MYB (transcriptional repressor)	Decreased content	Zhang et al. (2013)
Apocarotenoids	*Medicago truncatula* Gaertn.	Gene silencing (RNA interference)	*1-Deoxy*-d-*xylulose 5-phosphate synthase* 2	Decreased content	Floss et al. (2008)
Lignin	*Populus tomentosa* Carr.	Engineering regulatory mechanism (through overexpression)	PtoMYB216 (R2R3-MYB transcription factor)	Ectopic deposition of lignin	Tian et al. (2013)
Quercetin 3,4′-diglucoside	*Scutellaria baicalensis* Georgi	Engineering regulatory mechanism (through overexpression)	(SbMYB2 or SbMYB7) R2R3-MYB transcription factors	Decreased content	Yuan et al. (2013)
Tanshinone	*Salvia miltiorrhiza* Bunge	Overexpression of pathway gene	*Allene oxide cyclase*	Increased content	Gu et al. (2012)
Camptothecin	*Camptotheca acuminata* Decne.	Engineering regulatory mechanism (through overexpression)	*Allene oxide cyclase*	Increased content	Yan et al. (2012)
Anthraquinones	*Rubia cordifolia* L.	Engineering regulatory mechanism (through overexpression)	AtCPK1 (*Arabidopsis calcium-dependent kinase gene*)	Increased content	Shkryl et al. (2011)

pathways of secondary metabolites. An analysis of regulatory means may facilitate our understanding of how secondary metabolites are produced efficiently. The concept of gene expression from biosynthetic pathways in heterologous organisms has significantly directed the potentiality for combinatorial biosynthesis. The realm of combinatorial biosynthesis is possibly an exciting area for metabolite production. These may present an opportunity for metabolic engineering of secondary metabolism. Moreover, bioinformatics tools for molecular elucidation are available for microbial species; however, this remains a comparatively unknown area in plants. In the future, metabolic engineering and bioinformatics tools can be used to overcome the limited availability bioactive compounds from medicinal plants. We conclude that persistence and efforts in this area will direct to convenient and innovative production of desired bioactive compounds from medicinal plants. A probable substitute to field-grown plants could be to use transformed cells in reactors under controlled conditions to therefore make them eco-friendly. The long-awaited advancement to plant biotechnology will probably be accomplished with the combination of plant cell culture technologies and metabolic engineering.

Acknowledgements The authors wish to thank the University Grant Commission, Special Assistance Programme (SAP), New Delhi, and Maharshi Dayanand University, Rohtak, India, for the necessary support.

References

Afendi FM, Okada T, Yamazaki M, Hirai-Morita A, Nakamura Y, Nakamura K, Ikeda S, Takahashi H, Altaf-Ul-Amin M, Darusman LK, Saito K, Kanaya S (2012) KNApSAcK family databases: integrated metabolite-plant species databases for multifaceted plant research. Plant Cell Physiol 53:1–12. doi:10.1093/pcp/pcr165

Alam P, Abdin MZ (2011) Over-expression of HMG-CoA reductase and amorpha-4,11-diene synthase genes in *Artemisia annua* L. and its influence on artemisinin content. Plant Cell Rep 30:1919–1928. doi:10.1007/s00299-011-1099-6

Allen RS, Millgate AG, Chitty JA, Thisleton J, Miller JA, Fist AJ, Gerlach WL, Larkin PJ (2004) RNAi-mediated replacement of morphine with the nonnarcotic alkaloid reticuline in opium poppy. Nat Biotechnol 22:1559–1566. doi:10.1038/nbt1033

Aquil S, Husaini AM, Abdin MZ, Rather GM (2009) Overexpression of the HMG-CoA reductase gene leads to enhanced artemisinin biosynthesis in transgenic *Artemisia annua* plants. Planta Med 75:1453–1458. doi:10.1055/s-0029-1185775

Bak S, Olsen CE, Petersen BL, Moller BL, Halkier BA (1999) Metabolic engineering of p-hydroxybenzyl-glucosinolate in *Arabidopsis* by expression of the cyanogenic CYP79A1 from *Sorghum bicolor*. Plant J 20:663–671. doi:10.1046/j.1365-313X.1999.00642.x bit.23077

Bourgaud F, Gravot A, Milesi S, Gontier E (2001) Production of plant secondary metabolites: a historical perspective. Plant Sci 161:839–851. doi:10.1016/S0168 9452(01)00490-3

Cai Y, Jia JW, Crock J, Lin ZX, Chen XY, Croteau R (2002) A cDNA clone for betacaryophyllene synthase from *Artemisia annua*. Phytochemistry 61:523–529. doi:10.1016/S0031-9422(02)00265-0

Cankar K, Jongedijk E, Klompmaker M, Majdic T, Mumm R, Bouwmeester H, Bosch D, Beekwilder J (2014) (+)-Valencene production in *Nicotiana benthamiana* is increased by downregulation of competing pathways. Biotechnol J 10:180–189. doi:10.1002/biot.201400288

Chemler JA, Koffas MAG (2008) Metabolic engineering for plant natural product biosynthesis in microbes. Curr Opin Biotechnol 19:597–605. doi:10.1016/j.copbio.2008.10.011

Chen DH, Ye HC, Li GF (2000) Expression of a chimeric farnesyl diphosphate synthase gene in *Artemisia annua* L. transgenic plants via *Agrobacterium tumefaciens* mediated transformation. Plant Sci 155:179–185. doi:10.1016/S0168-9452(00)00217-X

Chen R, Harada Y, Bamba T, Nakazawa Y, Gyokusen K (2012) Overexpression of an isopentenyl diphosphate isomerase gene to enhance trans-polyisoprene production in *Eucommia ulmoides* Oliver. BMC Biotechnol 12:1–12. doi:10.1186/1472-6750-12-78

Chung HH, Sung YC, Shyur LF (2016) Deciphering the biosynthetic pathways of bioactive compounds *In Planta* using omics approaches. In: Tsay H-S et al (eds) Medicinal plants–recent advances in research and development. Springer Science, Singapore, pp 129–165. doi:10.1007/978-981-10-1085-9_5

Corey EJ, Jardine PS, Rohloff JC (1988) Total synthesis of (+/−)-forskolin. J Am Chem Soc 110:3672–3673. doi:10.1021/ja00219a059

Crozier A, Jaganath IB, Clifford MN (2009) Dietary phenolics:chemistry, bioavailability and effects on health. Nat Prod Rep 26:1001–1043. doi:10.1039/b802662a

Davuluri GR, van Tuinen A, Fraser PD, Manfredonia A, Newman R, Burgess D, Brummell DA, King SR, Palys J, Uhlig J, Bramley PM, Pennings HM, Bowler C (2005) Fruit-specific RNAi-mediated suppression of DET1 enhances carotenoid and flavonoid content in tomatoes. Nat Biotechnol 23:890–895. doi:10.1038/nbt1108

Decker G, Wanner G, Zenk MH, Lottspeich F (2000) Characterization of proteins in latex of the opium poppy (*Papaver somniferum*) using two-dimensional gel electrophoresis and microsequencing. Electrophoresis 21:3500–3516. doi:10.1002/1522-2683(20001001

Demain AL, Adrio JL (2008) Contributions of microorganisms to industrial biology. Mol Biotechnol 38:41–55. doi:10.1007/s12033-007-0035-z

Desai PN, Shrivastava N, Padh H (2010) Production of heterologous proteins in plants: strategies for optimal expression. Biotechnol Adv 28:427–435. doi:10.1016/j.biotechadv.2010.01.005

DiCosmo F, Misawa M (1995) Plant cell and tissue culture: alternatives for metabolite production. Biotechnol Adv 13:425–453. doi:10.1016/0734-9750(95)02005-N

Dudareva N, Andersson S, Orlova I, Gatto N, Reichelt M, Rhodes D, Boland W, Gershenzon J (2005) The nonmevalonate pathway supports both monoterpene and sesquiterpene formation in snapdragon flowers. Proc Natl Acad Sci U S A 102:933–938. doi:10.1073/pnas.0407360102

Floss DS, Hause B, Lange PR, Kuster H, Strack D, Walter MH (2008) Knock-down of the MEP pathway isogene 1-deoxy-D-xylulose 5-phosphate synthase 2 inhibits formation of arbuscular mycorrhiza-induced apocarotenoids, and abolishes normal expression of mycorrhiza-specific plant marker genes. Plant J 56:86–100. doi:10.1111/j.1365-313X.2008.03575.x

Fujii N, Inui T, Iwasa K, Morishige T, Sato F (2007) Knockdown of berberine bridge enzyme by RNAi accumulates (S)-reticuline and activates a silent pathway in cultured California poppy cells. Transgenic Res 16:363–375. doi:10.1007/s11248-006-9040-4

Gandhi SG, Mahajan V, Bedi YS (2015) Changing trends in biotechnology of secondary metabolism in medicinal and aromatic plants. Planta 241:303–317. doi:10.1007/s00425-014-2232-x

Garcia SNJ, Vazquez-Cruz MA, Gonzalez RGG, Pacheco IT, Hernandez AC, Perez AAF (2013) Current approaches for enhanced expression of secondary metabolites as bioactive compounds in plants for agronomic and human health purposes—a review. Pol J Food Nutr Sci 63:67–78. doi:10.2478/v10222-012-0072-6

Gavilano LB, Coleman NP, Burnley LE, Bowman ML, Kalengamaliro NE, Hayes A, Bush L, Siminszky B (2006) Genetic engineering of *Nicotiana tabacum* for reduced nornicotine content. J Agric Food Chem 54:9071–9078. doi:10.1021/jf0610458

Gil-Chavez GJ, Villa JA, Ayala-Zavala JF, Heredia JB, Sepulveda D, Yahia EM, G A G´a-A (2013) Technologies for extraction and production of bioactive compounds to be used as nutraceuticals and food ingredients: an overview. Compr Rev Food Sci Food Saf 12:5–23. doi:10.1111/1541-4337.12005

Grotewold E, Chamberlin M, Snook M, Siamed B, Butlerd L, Swensonb J, Maddockb S, Clairb GS, Bowenb B (1998) Engineering secondary metabolism in maize cells by ectopic expression of transcription factors. Plant Cell 10:721–740. doi:10.1105/tpc.10.5.721

Gu XC, Chen JF, Xiao Y, Di P, Xuan HJ, Zhou X, Zhang L, Chen WS (2012) Overexpression of allene oxide cyclase promoted tanshinone/phenolic acid production in *Salvia miltiorrhiza*. Plant Cell Rep 31:2247–2259. doi:10.1007/s00299-012-1334-9

Guaadaoui A, Benaicha S, Elmajdoub N, Bellaoui M, Hamal A (2014) What is a bioactive compound? A combined definition for a preliminary consensus. Int J Nutr Food Sci 3:174–179. doi:10.11648/j.ijnfs.20140303.16

Guterman I, Shalit M, Menda N, Piestun D, Dafny-Yelin M, Shalev G, Bar E, Davydov O, Ovadis M, Emanuel M, Wang J, Adam Z, Pichersky E, Lewinsohn E, Zamir D, Vainstein A, Weiss D (2002) Rose scent: genomics approach to discovering novel floral fragrance-related genes. Plant Cell 14:2325–2338. doi:10.1105/tpc.005207

Haefner B (2003) Drugs from the deep: marine natural products as drug candidates. Drug Discov Today 8:536–544. doi:10.1016/S1359-6446(03)02713-2

Han JY, Wang HY, Choi YE (2014) Production of dammarenediol-II triterpene in a cell suspension culture of transgenic tobacco. Plant Cell Rep 33:225–233. doi:10.1007/s00299-013-1523-1

Harvey AL, Edrada-Ebel R, Quinn RJ (2015) The re-emergence of natural products for drug discovery in the genomics era. Nat Rev Drug Discov 14:111–129. doi:10.1038/nrd4510

Hashimoto S, Sakata S, Sonegawa M, Ikegami S (1988) A total synthesis of (±)-forskolin. J Am Chem Soc 110:3670–3672. doi:10.1021/ja00219a058

Heinstein PF, Chang CJ (1994) Taxol. Annu Rev Plant Physiol Plant Mol Biol 45:663–674. doi:10.1146/annurev.pp.45.060194.003311

Hranueli D, Cullum J, Basrak B, Goldstein P, Long PF (2005) Plasticity of the *Streptomyces* genome evolution and engineering of new antibiotics. Curr Med Chem 12:1697–1704. doi:10.2174/0929867054367176

Hussain MS, Fareed S, Ansari S, Rahman MA, Ahmad IZ, Saeed M (2012) Current approaches toward production of secondary plant metabolites. J Pharm Bioallied Sci 4:10–20. doi:10.4103/0975-7406.92725

Hwang EI, Kaneko M, Ohnishi Y, Horinouchi S (2003) Production of plant-specific flavanones by *Escherichia coli* containing an artificial gene cluster. Appl Environ Microbiol 69:2699–2706. doi:10.1128/AEM.69.5.2699-2706.2003

Julsing MK, Koulman A, Woerdenbag HJ, Quax WJ, Kayser O (2006) Combinatorial biosynthesis of medicinal plant secondary metabolites. Biomol Eng 23:265–279. doi:10.1016/j.bioeng.2006.08.001

Kadomura-Ishikawa Y, Miyawaki K, Noji S, Takahashi A (2013) Phototropin 2 is involved in blue light-induced anthocyanin accumulation in *Fragaria* x *ananassa* fruits. J Plant Res 126:847–857. doi:10.1007/s10265-013-0582-2

Khan S, Rahman L (2016) Pathway modulation of medicinal and aromatic plants through metabolic engineering using *Agrobacterium tumefaciens*. In: Jha S (ed) Transgenesis and secondary metabolism. Reference series in phytochemistry 1-32. Springer International Publishing, Switzerland. doi:10.1007/978-3-319-27490-4_15-1

Kim HJ, Ono E, Morimoto K, Yamagaki T, Okazawa A, Kobayashi A, Satake H (2009) Metabolic engineering of lignan biosynthesis in *Forsythia* cell culture. Plant Cell Physiol 50:2200–2209. doi:10.1093/pcp/pcp156

Lannenpaa M (2014) Heterologous expression of AtMYB12 in kale (*Brassica oleracea* var. *acephala*) leads to high flavonol accumulation. Plant Cell Rep 33:1377–1388. doi:10.1007/s00299-014-1623-6

Lee J, Bae H, Jeong J, Lee JY, Yang YY, Hwang I, Martinoia E, Lee Y (2003) Functional expression of a bacterial heavy metal transporter in *Arabidopsis* enhances resistance to and decreases uptake of heavy metals. Plant Physiol 133:589–596. doi:10.1104/pp.103.021972

Lewis RS, Jack AM, Morris JW, Robert J V, Gavilano LB, Siminszky B, Bush LP, Hayes AJ, Dewey RE (2008) RNA interference (RNAi)-induced suppression of nicotine demethylase activity reduces levels of a key carcinogen in cured tobacco leaves. Plant Biotechnol J 6:346–354. doi:10.1111/j.1467-7652.2008.00324.x

Li H, Flachowsky H, Fischer TC, Hanke M V, Forkmann G, Treutter D, Schwab W, Hoffmann T, Szankowski I (2007) Maize Lc transcription factor enhances biosynthesis of anthocyanins, distinct proanthocyanidins and phenylpropanoids in apple (*Malus domestica* Borkh.) Planta 226:1243–1254. doi:10.1007/s00425-007-0573-4

Li W, Wang B, Wang M, Chen M, Yin JM, Kaleri GM, Zhang RJ, Zuo TN, You X, Yang Q (2014) Cloning and characterization of a potato StAN11 gene involved in anthocyanin biosynthesis regulation. J Integr Plant Biol 56:364–372. doi:10.1111/jipb.12136

Lloyd AM, Walbot V, Davis RW (1992) *Arabidopsis* and *Nicotiana* anthocyanin production activated by maize regulators R and C1. Science 258:1773–1775. doi:10.1126/science.1465611

Lommen WJM, Schenk E, Bouwmeester HJ, Verstappen FWA (2006) Trichome dynamics and artemisinin accumulation during development and senescence of *Artemisia annua* leaves. Planta Med 72:336–345. doi:10.1055/s-2005-916202

Mahmoud SS, Croteau RB (2001) Metabolic engineering of essential oil yield and composition in mint by altering expression of deoxyxylulose phosphate reductoisomerase and menthofuran synthase. Proc Natl Acad Sci U S A 98:8915–8920. doi:10.1073/pnas.141237298

Majumdar S, Garai S, Jha S (2012) Use of cryptogein gene to stimulate the accumulation of *Bacopa* saponins in transgenic *Bacopa monnieri* plants. Plant Cell Rep 31:1899–1909. doi:10.1007/s00299-012-1303-3

Miettinen K, Dong L, Navrot N, Schneider T, Burlat V, Pollier J, Woittiez L, van der Krol S, Lugan R, Ilc T, Verpoorte R, Oksman-Caldentey KM, Martinoia E, Bouwmeester H, Goossens A, Memelink J, Werck-Reichhart D (2014) The seco-iridoid pathway from *Catharanthus roseus*. Nat Commun 5:1–11. doi:10.1038/ncomms4606

Moore BS, Kalaitzis JA, Xiang L (2005) Exploiting marine actinomycete biosynthetic pathways for drug discovery. Antonie Van Leeuwenhoek 87:49–57. doi:10.1007/s10482-004-6541-0

Muir SR, Collins GJ, Robinson S, Hughes S, Bovy A, Ric De Vos CH, van Tunen AJ, Verhoeyen ME (2001) Overexpression of *Petunia* chalcone isomerase in tomato results in fruit containing increased levels of flavonols. Nat Biotechnol 19:470–474. doi:10.1038/88150

Nalawade SM, Tsay HS (2004) *In vitro* propagation of some important Chinese medicinal plants and their sustainable usage. In Vitro Cell Dev Biol-Plant 40:143–154. doi:10.1079/IVP2003504

Ncube B, Van Staden J (2015) Tilting plant metabolism for improved metabolite biosynthesis and enhanced human benefit. Molecules 20:12698–12731. doi:10.3390/molecules200712698

Niu Y, Luo H, Sun C, Yang TJ, Dong L, Huang L, Chen S (2014) Expression profiling of the triterpene saponin biosynthesis genes FPS, SS, SE, and DS in the medicinal plant *Panax notoginseng*. Gene 533:295–303. doi:10.1016/j.gene.2013.09.045

O'Keefe BR (2001) Biologically active proteins from natural product extracts. J Nat Prod 64:1373–1381. doi:10.1021/np0103362

Ogita S, Uefuji H, Yamaguchi Y, Koizumi N, Sano H (2003) Producing decaffeinated coffee plants. Nature 423:823. doi:10.1038/423823a

Oksman-Caldentey K, Saito K (2005) Integrating genomics and metabolomics for engineering plant metabolic pathways. Curr Opin Biotechnol 16:174–179. doi:10.1016/j.copbio.2005.02.007

Oksman-Caldentey KM, Inze D (2004) Plant cell factories in the post-genomic era: new ways to produce designer secondary metabolites. Trends Plant Sci 9:433–440. doi:10.1016/j.tplants.2004.07.006

Paddon CJ, Westfall PJ, Pitera DJ, Benjamin K, Fisher K, McPhee D, Leavell MD, Tai A, Main A, Eng D, Polichuk DR, Teoh KH, Reed DW, Treynor T, Lenihan J, Jiang H, Fleck M, Bajad S, Dang G, Dengrove D, Diola D, Dorin G, Ellens KW, Fickes S, Galazzo J, Gaucher SP, Geistlinger T, Henry R, Hepp M, Horning T, Iqbal T, Kizer L, Lieu B, Melis D, Moss N, Regentin R, Secrest S, Tsuruta H, Vazquez R, Westblade LF, Xu L, Yu M, Zhang Y, Zhao L, Lievense J, Covello PS, Keasling JD, Reiling KK, Renninger NS, Newman JD (2013) High-level semisynthetic production of the potent antimalarial artemisinin. Nature 496:528–532. doi:10.1038/nature12051

Poulter CD, Wiggins PL, Le AT (1981) Farnesylpyrophosphate synthetase. A stepwise mechanism for the 1′-4 condensation reaction. J Am Chem Soc 103:3926–3927. doi:10.1021/ja00403a054

Prather KLJ, Martin CH (2008) *De novo* biosynthetic pathways: rational design of microbial chemical factories. Curr Opin Biotechnol 19:468–474. doi:10.1016/j.copbio.2008.07.009

Rischer H, Oresic M, Seppanen-Laakso T, Katajamaa M, Lammertyn F, Ardiles-Diaz W, Van Montagu MC, Inze D, Oksman-Caldentey KM, Goossens A (2006) Gene-to-metabolite networks for terpenoid indole alkaloid biosynthesis in *Catharanthus roseus* cells. Proc Natl Acad Sci U S A 103:5614–5619. doi:10.1073/pnas.0601027103

Ro DK, Paradise EM, Ouellet M, Fisher KJ, Newman KL, Ndungu JM, Ho KA, Eachus RA, Ham TS, Kirby J, Chang MCY, Withers ST, Shiba Y, Sarpong R, Keasling JD (2006) Production of the antimalarial drug precursor artemisinic acid in engineered yeast. Nature 440:940–943. doi:10.1038/nature04640

Rohmer M (1999) The discovery of a mevalonate-independent pathway for isoprenoid biosynthesis in bacteria, algae and higher plants. Nat Prod Rep 16:565–574. doi:10.1039/a709175c

Ronen G, Carmel-Goren L, Zamir D, Hirschberg J (2000) An alternative pathway to β-carotene formation in plant chromoplasts discovered by map-based cloning of beta and old-gold color mutations in tomato. Proc Natl Acad Sci U S A 97:11102–11107. doi:10.1073/pnas.190177497

Saema S, Rahman LU, Singh R, Niranjan A, Ahmad IZ, Misra P (2015) Ectopic overexpression of WsSGTL1, a sterol glucosyltransferase gene in *Withania somnifera*, promotes growth, enhances glycowithanolide and provides tolerance to abiotic and biotic stresses. Plant Cell Rep 35:195–211. doi:10.1007/s00299-015-1879-5

Schoendorf A, Rithner CD, Williams RM, Croteau RB (2001) Molecular cloning of a cytochrome P450 taxane 10 betahydroxylase cDNA from *Taxus* and functional expression in yeast. Proc Natl Acad Sci U S A 98:1501–1506. doi:10.1073/pnas.98.4.1501

Seo JW, Jeong JH, Shin CG, Lo SC, Han SS, Yu KW, Harada E, Han JY, Choi YE (2005) Overexpression squalene synthase in *Eleutherococcus senticosus* increases phytosterol and triterpene accumulation. Phytochemistry 66:869–877. doi:10.1016/j.phytochem.2005.02.016

Shahidi F (2009) Nutraceuticals and functional foods:whole versus processed foods. Trends Food Sci Technol 20:376–387. doi:10.1016/j.tifs.2008.08.004

Sharafi A, Sohi HH, Mousavi A, Azadi P, Khalifani BH, Razavi K (2013) Metabolic engineering of morphinan alkaloids by over-expression of codeinone reductase in transgenic hairy roots of *Papaver bracteatum*, the Iranian poppy. Biotechnol Lett 35:445–453. doi:10.1007/s10529-012-1080-7

Shen Q, Chen YF, Wang T, Wu SY, Lu X, Zhang L, Zhang FY, Jiang WM, Wang GF, Tang KX (2012) Overexpression of the cytochrome P450 monooxygenase (cyp71av1) and cytochrome P450 reductase (cpr) genes increased artemisinin content in *Artemisia annua* (Asteraceae). Genet Mol Res 11:3298–3309. doi:10.4238/2012.September.12.13

Shkryl YN, Veremeichik GN, Bulgakov P V, Zhuravlev YN (2011) Induction of anthraquinone biosynthesis in *Rubia cordifolia* cells by heterologous expression of a calcium-dependent protein kinase gene. Biotechnol Bioeng 108:1734–1738. doi:10.1002/bit.23077

Srivastava S, Srivastava AK (2007) Hairy root culture for mass production of high-value secondary metabolites. Crit Rev Biotechnol 27:29–43. doi:10.1080/07388550601173918

Steele CL, Chen Y, Dougherty BA, Li W, Hofstead S, Lam KS, Xing Z, Chiang SJ (2005) Purification, cloning, and functional expression of phenylalanine aminomutase: the first committed step in taxol side-chain biosynthesis. Arch Biochem Biophys 438:1–10. doi:10.1016/j.abb.2005.04.012

Suzuki H, Achnine L, Xu R, Matsuda SP, Dixon RA (2002) A genomics approach to the early stages of triterpene saponin biosynthesis in *Medicago truncatula*. Plant J 32:1033–1048. doi:10.1046/j.1365-313X.2002.01497.x

Tian Q, Wang X, Li C, Lu W, Yang L, Jiang Y, Luo K (2013) Functional characterization of the poplar R2R3-MYB transcription factor PtoMYB216 involved in the regulation of lignin biosynthesis during wood formation. PLoS One 8:e76369. doi:10.1371/journal.pone.0076369

Van der Fits L, Memelink J (2000) ORCA3, a jasmonate-responsive transcriptional regulator of plant primary and secondary metabolism. Science 289:295–297. doi:10.1126/science.289.5477.295

Verpoorte R, Contin A, Memelink J (2002) Biotechnology for the production of plant secondary metabolites. Phytochem Rev 1:13–25. doi:10.1023/A:1015871916833

Verpoorte R, Memelink J (2002) Engineering secondary metabolite production in plants. Curr Opin Biotechnol 13:181–187. doi:10.1016/S0958-1669(02)00308-7

Verpoorte R, Van der Heijden R, Ten Hoopen HJG, Memelink J (1999) Metabolic engineering of plant secondary metabolite pathways for the production of fine chemicals. Biotechnol Lett 21:467–479. doi:10.1023/A:1005502632053

Wang Q, Xing S, Pan Q, Yuan F, Zhao J, Tian Y, Chen Y, Wang G, Tang K (2012) Development of efficient *Catharanthus roseus* regeneration and transformation system using *Agrobacterium tumefaciens* and hypocotyls as explants. BMC Biotechnol 12:1–12. doi:10.1186/1472-6750-12-34

Wenzel S, Muller R (2005) Recent developments towards the heterologous expression of complex bacterial natural product biosynthetic pathways. Curr Opin Biotechnol 16:594–606. doi:10.1016/j.copbio.2005.10.001

Whetten RW, Sederoff RR (1992) Phenylalanine ammonia-lyase from loblolly pine: purification of the enzyme and isolation of complementary DNA clones. Plant Physiol 98:380–386. doi:10.1104/pp.98.1.380

Whitmer S, Canel C, Hallard D, Gonçalves C, Verpoorte R (1998) Influence of precursor availability on alkaloid accumulation by transgenic cell line of *Catharanthus roseus*. Plant Physiol 116:853–857. doi:10.1104/pp.116.2.853

Wilkinson B, Micklefield J (2007) Mining and engineering natural-product biosynthetic pathways. Nat Chem Biol 3:379–386. doi:10.1038/nchembio.2007.7

Wong K V, Law BY, Yao XJ, Chen X, Xu SW, Liu L, Leung EL (2016) Advanced research technology for discovery of new effective compounds from Chinese herbal medicine and their molecular targets. Pharmacol Res 111:546–555. doi:10.1016/j.phrs.2016.07.022

Xu J, Ge X, Dolan MC (2011) Towards high-yield production of pharmaceutical proteins with plant cell suspension culture. Biotechnol Adv 29:278–299. doi:10.1016/j.biotechadv.2011.01.002

Yamazaki M, Saito K (2002) Differential display analysis of gene expression in plants. Cell Mol Life Sci 59:1246–1255. doi:10.1007/s00018-002-8503-x

Yan P, Keji J, Juan L, Ying C, Qian W, Ji Y, Xiaofen S, Kexuan T (2012) Effects of over-expression of allene oxide cyclase on camptothecin production by cell cultures of *Camptotheca acuminata*. Afr J Biotechnol 11:6535–6541. doi:10.5897/AJB11.2789

Yang D, Du X, Yang Z, Liang Z, Guo Z, Liu Y (2014) Transcriptomics, proteomics, and metabolomics to reveal mechanisms underlying plant secondary metabolism. Eng Life Sci 14:456–466. doi:10.1002/elsc.201300075

Yuan Y, Liu W, Zhang Q, Xiang L, Liu X, Chen M, Lin Z, Wang Q, Liao Z (2014) Overexpression of artemisinic aldehyde 11(13) reductase gene-enhanced artemisinin and its relative metabolite biosynthesis in transgenic *Artemisia annua* L. Biotechnol Appl Biochem 62:17–23. doi:10.1002/bab.1234

Yuan Y, Wu C, Liu Y, Yang J, Huang L (2013) The *Scutellaria baicalensis* R2R3-MYB transcription factors modulate flavonoid biosynthesis by regulating GA metabolism in transgenic tobacco plants. PLoS One 8:e77275. doi:10.1371/journal.pone.0077275

Zhang L, Ding R, Chai Y, Bonfill M, Moyano E, Oksman-Caldentey KM, Xu T, Pi Y, Wang Z, Zhang H, Kai G, Liao Z, Sun X, Tang K (2004) Engineering tropane biosynthetic pathway in *Hyoscyamus niger* hairy root cultures. Proc Natl Acad Sci U S A 101:6786–6791. doi:10.1073/pnas.0401391101

Zhang S, Ma P, Yang D, Li W, Liang Z, Liu Y, Liu F (2013) Cloning and characterization of a putative R2R3 MYB transcriptional repressor of the rosmarinic acid biosynthetic pathway from *Salvia miltiorrhiza*. PLoS One 8:e73259. doi:10.1371/journal.pone.0073259

Zulak KG, Cornish A, Daskalchuk TE, Deyholos MK, Goodenowe DB (2005) Gene transcript and metabolite profiling of elicitor-induced opium poppy cell cultures reveals the coordinate regulation of primary and secondary metabolism. Planta 225:1085–1106. doi:10.1007/s00425-006-0419-5

Protein Therapeutic: Production, Application, and Future Scenario

4

Renu Chaudhary, Meenakshi Balhara, and Anil Kumar Chhillar

Abstract

The demands for recombinant therapeutic proteins are escalating rapidly over the past two decades because of its effectiveness in treating human diseases that are uncurable. To meet the increasing demand, there is continuous need to enhance the existing expression systems, furthermore building up an attractive strategy to confront the therapeutic protein demands. Nowadays, human cell line has come up as a novel and effective strategy to generate therapeutic proteins since this expression framework possesses the machinery to modify the recombinant proteins to its final form (posttranslational modifications) analogous to those present human proteins. In addition human cell lines also decrease the possible immunogenicity against non-human antigenic determinants. Therefore, the present chapter discusses about how recombinant proteins with therapeutic properties can be in mammalian cells and their application, and finally we will talk how new innovations might add to the further advancement and generation of therapeutic proteins.

Keywords

Mammalian cell line • Expression system • Production • Food and Drug Administration

---

R. Chaudhary • M. Balhara • A.K. Chhillar (✉)
Centre for Biotechnology Maharshi Dayanand University, Rohtak 124001, Haryana, India
e-mail: anil.chhillar@gmail.com

V.C. Kalia, A.K. Saini (eds.), *Metabolic Engineering for Bioactive Compounds*,
DOI 10.1007/978-981-10-5511-9_4

4.1 Introduction

Therapeutic proteins are proteins that are built in the research facility for clinical use. Insulin was the primary restorative macromolecule to be accustomed to treat diabetes in the 1920s. Therapeutic proteins have reformed the treatment of numerous maladies for over decades (Matasci et al. 2008). There are roughly 225 recombinant restorative proteins as of now endorsed for pharmaceutical used by the Food and Drug Administration (FDA), and few hundreds are still in clinical trials. The identification of around 40,000 genes revealed during human genome analysis results in the development of many more therapeutic protein targets in the near future (Lander et al. 2001; Venter et al. 2001; Pennisi 2003). In 2015, 45 new recombinant products were approved and can be found in Table 4.1 (FDA 2015). From 2010 to 2016, the Center for Drug Evaluation and Research (CDER) has an average of 31 novel drugs approved per year (Exhibit 4.1). The business sector for such protein items is critical and is developing at a rate much higher than that acquired for the pharmaceuticals business. It is estimated that the aggregate worldwide

Table 4.1 New drugs approved by FDA in 2015

S. no.	Drug name	Active ingredient	Date of approval	Treatment
1	Savaysa	Edoxaban	1/8/2015	Stroke and blood clots
2	Cosentyx	Secukinumab	1/21/2015	Plaque psoriasis
3	Natpara	Parathyroid hormone	1/23/2015	Hypocalcemia
4	Ibrance	Palbociclib	2/3/2015	Breast cancer
5	Lenvima	Lenvatinib	2/13/2015	Thyroid cancer
6	Farydak	Panobinostat	2/23/2015	Multiple myeloma
7	Avycaz	Ceftazidime-avibactam	2/25/2015	Infections in the abdomen and urinary tract
8	Cresemba capsule injection	Isavuconazonium sulfate	3/6/2015	Invasive aspergillosis and invasive mucormycosis
9	Unituxin	Dinutuximab	3/10/2015	Neuroblastoma
10	Cholbam	Cholic acid	3/17/2015	Gall bladder and peroxisomal disorders
11	Corlanor	Ivabradine	4/15/2015	Heart failure
12	Kybella	Deoxycholic acid	4/29/2015	Submental fat
13	Viberzi	Eluxadoline	5/27/2015	Bowel and diarrhea disorder
14	Kengreal	Cangrelor	6/22/2015	Blood clots
15	Orkambi	Lumacaftor 200 mg/ ivacaftor 125 mg	7/2/2015	Cystic fibrosis
16	Entresto	Sacubitril/valsartan	7/7/2015	Heart failure
17	Rexulti	Brexpiprazole	7/10/2015	Schizophrenia and depressive disorder
18	Praluent	Alirocumab	7/24/2015	High cholesterol
19	Odomzo	Sonidegib	7/24/2015	Basal cell carcinoma
20	Daklinza	Daclatasvir	7/24/2015	Chronic hepatitis C virus (HCV)
21	Addyi	Flibanserin	8/8/2015	Hypoactive sexual desire disorder (HSDD) in women

Table 4.1 (continued)

S. no.	Drug name	Active ingredient	Date of approval	Treatment
22	Repatha	Evolocumab	8/27/2015	High cholesterol
23	Varubi	Rolapitant	9/2/2015	Delayed phase chemotherapy
24	Xuriden	Uridine triacetate	9/4/2015	Hereditary orotic aciduria
25	Vraylar	Cariprazine	9/17/2015	Schizophrenia and bipolar disorder
26	Lonsurf	Trifluridine and tipiracil	9/22/2015	Colorectal cancer
27	Tresiba	Insulin degludec injection	9/25/2015	Diabetes mellitus
28	Aristada	Aripiprazole lauroxil	10/6/2015	Schizophrenia
29	Praxbind	Idarucizumab	10/16/2015	Reverse pradaxa's blood thinning effect
30	Veltassa	Patiromer for oral suspension	10/21/2015	Hyperkalemia
31	Yondelis	Trabectedin	10/23/2015	Sarcomas, liposarcoma and leiomyosarcoma
32	Strensiq	Asfotase alfa	10/23/2015	Hypophosphatasia
33	Nucala	Mepolizumab	11/4/2015	Asthma
34	Genvoya	Elvitegravir, cobicistat, emtricitabine, and tenofovir alafenamide	11/5/2015	HIV-1 and pediatric
35	Cotellic	Cobimetinib	11/10/2015	Melanoma
36	Tagrisso	Osimertinib	11/13/2015	Lung cancer
37	Darzalex	Daratumumab	11/16/2015	Multiple myeloma
38	Ninlaro	Ixazomib	11/20/2015	Multiple myeloma
39	Portrazza	Necitumumab	11/24/2015	Lung cancer
40	Empliciti	Elotuzumab	11/30/2015	Multiple myeloma
41	Kanuma	Sebelipase alfa	12/8/2015	Lysosomal acid lipase deficiency
42	Alecensa	Alectinib	12/11/2015	Lung cancer
43	Bridion	Sugammadex	12/15/2015	Reverse effect of neuromuscular blocking drugs
44	Uptravi	Selexipag	12/22/2015	Pulmonary arterial hypertension
45	Zurampic	Lesinurad	12/22/2015	High blood uric acid level

business sector will be growing at an average annual rate of 15% for over the next 10 years. The overall yearly income for recombinant antibodies and proteins in 2012 was over $110 billion (Rader 2008; Walsh 2010). Production of recombinant therapeutic proteins can be achieved in particular host systems including (microorganism and yeast) transgenic plants, creatures (mammalian and creepy crawly), and plant cells. Microbial expression systems are alluring in light because of minimal effort, high efficiency, and fast usage. In addition, there is no adventitious virus concern to the regulatory authorities (Zhang 2010). Regardless of the points of interest, articulation of huge complex proteins subsequent to posttranslational metabolic apparatus is just accessible in mammalian cells (Butler 2005; Zhu 2012).

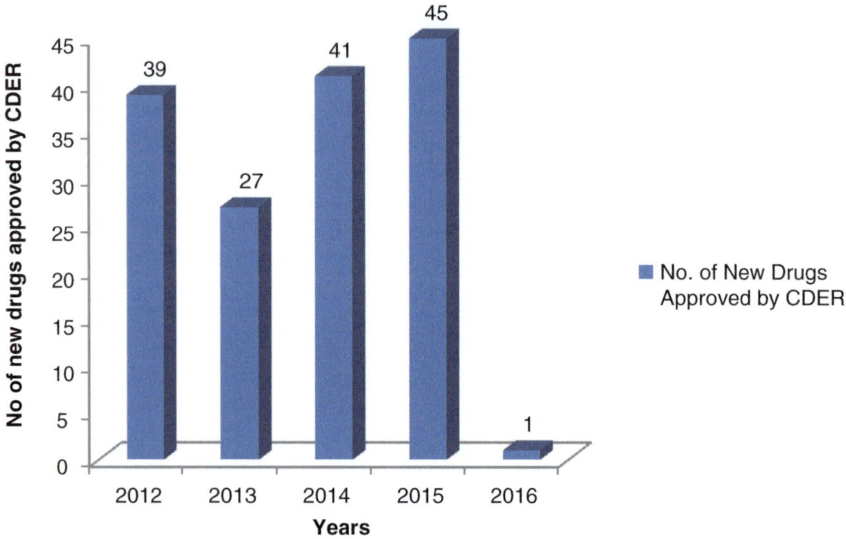

Exhibit 4.1 Number of novel drugs approved by CDER from 2010 to 2016

Plant cells can be considered as a different option for microbial- and animal cell-based systems because they are a less costly method for the production of eukaryotic protein with low contamination and other impurities (Pogue et al. 2010; Zhu 2012). In any case, plant cells don't necessarily deliver the similar three-dimensional structures with indistinguishable sequences, and further posttranslational alteration in them gives unique glycosylation pattern (Swiech et al. 2012). Transgenic animals are additionally an appealing distinct option for therapeutic protein production due to easy generation and high caliber of the proteins. In any case, purification of human proteins from their animal partner is exceptionally troublesome. Also, a percentage of the recombinant proteins may be unsafe to transgenic animals (Houdebine 2009). Recombinant protein creation utilizing animal cells offers numerous focal points over the other referred to expression frameworks. Presently more than 70% of therapeutic proteins are being generated in mammalian cell-based system majorly due to benefits of posttranslational modifications such as glycosylation and sialylation that produce top-notch proteins that are comparable in their biochemical properties to the normally happening individual structures (Matasci et al. 2008; Zhu 2012). This developed interest for superb recombinant therapeutic protein motivates the innovative work of mammalian cell-based assembling frameworks for an excellent upgraded creation field. Using mammalian and human cell lines has risen as another capable option for recombinant restorative protein production since this expression framework is expected to generate therapeutic proteins with diminished potential for immunogenic responses against nonhuman epitopes because of posttranslational alterations more similar to their regular partner (Swiech et al. 2012; Ferrer-Miralles et al. 2015). In the current chapter, we will be discussing therapeutic protein production in human cell lines and their application, and finally we will discuss how new technologies aid in the further development and production of therapeutic proteins.

4.2 Therapeutic Protein Production in Mammalian Cells

Numbers of expression systems are available for the production of therapeutic proteins which includes prokaryotic as well as eukaryotic expression system. Therapeutic protein can be useful only if they are synthesized in their biologically active form with proper folding and posttranslational modifications. The prokaryotic expression systems, for example, *E. coli*, frequently neglect to produce functional three-dimensional structure of eukaryotic proteins. So these proteins can be produced in the eukaryotic expression system, for example, mammalian cells, which facilitate the protein folding crucial for the functionality of recombinant protein. Latest advancements in technologies have altogether upgraded the expression level in mammalian cell lines specifically in relentlessly transfected Chinese hamster ovary (CHO) cells (Wurm 2004; Figueroa et al. 2007). CHO cells have turned out to be standard mammalian host for the generation of therapeutic proteins. Since it offers higher productivity and high capacity of producing secreted proteins, their expression level is much lesser than intracellular proteins (Derouazi et al. 2004; Rosser et al. 2005). Schematic outline of key steps involved during production of therapeutic protein in mammalian cells is shown in Fig. 4.1. First of all, gene of interest (GOI) with marker gene is introduced in vector and transferred in the mammalian cells for expression and thereafter transferred to the cells. Afterward the cells containing transferred GOI are put into specific condition to recapture merely those cells that express firmly incorporated selector genes like glutamine synthetase (GS) or dihydrofolate reductase (DHFR). These two selector genes are most commonly used for the selection. Cells are then cultured in the medium without the proper metabolites, for example, hypoxanthine and thymine on account DHFR determination and glutamine on selection of GS, so that it will prevent a growth of non-transformed cells and only transformed cells survive. Another option for selection is auxotrophy where gene confers imperviousness and is utilized for selection of anti-infection agents. With this methodology, recombinant cells are selected on the basis of suitable antibiotic marker like geneticin or G418, hygromycin-B, puromycin, blasticidin, and zeocin. The utilization of DHFR and GS selection systems is more advantageous than any other selection system because it takes into consideration the amplification of integrated transgenes. Amplification typically brings about the upgradation and accomplishment by exposing selected cells to inhibitors of the selector proteins (methotrexate (MTX) inhibitor of DHFR and methionine sulfoximide (MSX) inhibitor of GS). Thereafter, the selection, transformed cells are placed into another developmental vessel, and cultures are extended for the production of clonal populaces. Each clonal cell populace obtained via the mentioned system is examined for the expression of protein from GOI, and the most economically valuable transformed cells are later analyzed for stable therapeutic protein generation. From these hopeful candidates, cell lines with high productivity and better development are chosen for the production of recombinant therapeutic protein (Matasci et al. 2008).

Tragically, this procedure can be achieved in several months, being the real downside of stable CHO cell lines. Thus, faster and less expensive methodologies for the generation of recombinant proteins are required where numerous proteins

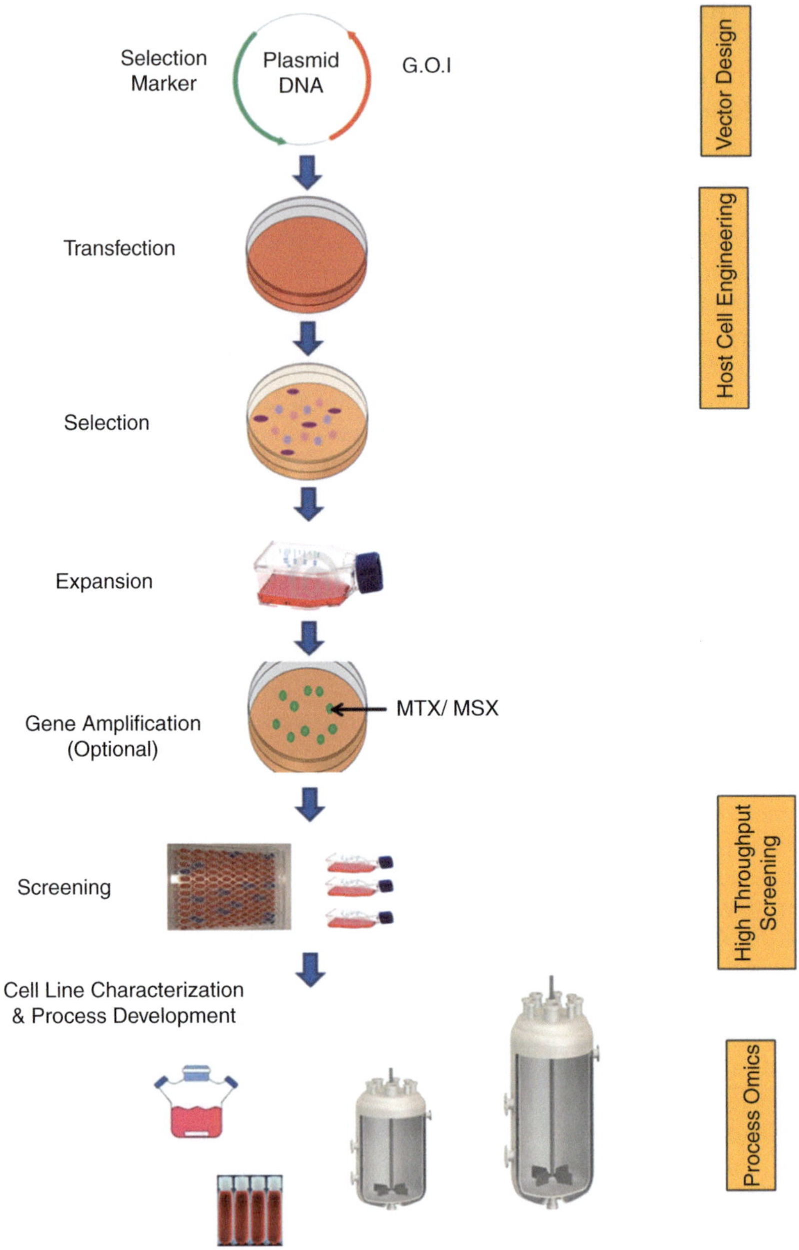

Fig. 4.1 Production of therapeutic protein of interest (Matasci et al. 2008)

must be quickly assessed as potential biotechnological or pharmacological items. Therefore, distinctive human cell lines, for example, child hamster kidney, human embryonic kidney (HEK293), and human inferred (PER.C6) as shown in Table 4.2, turned out to be a better option for the therapeutic protein production (Durocher et al. 2002; Girard et al. 2002; Baldi et al. 2007). Another alternative, for instance,

Table 4.2 Recombinant proteins produced by human cell lines (Picanco-Castro et al. 2013)

Cell line	Recombinant protein	Stable/transient	Culture system	Protein production level	References
HEK 293					
HEK.EBNA	Serine/threonine kinase receptor	Transient	Serum-free suspension culture; WAVE bioreactor 10–20 L	43.5 mg/L	Fischer et al. (2012)
HEK293-6E	TGF-superfamily member	Transient	Serum-free suspension culture; WAVE bioreactor 10–20 L	1.8 mg/L	Fischer et al. (2012)
HEK293-6E	Wnt signaling receptor	Transient	Serum-free suspension culture; shake flask	6 mg/L	Fischer et al. (2012)
HEK293	Endocan	Stable	Serum-free suspension culture; CELLine 1000 classic™ bioreactor; CELLine adhere™ bioreactor	174.3 g/mL (classic) 207.9 g/mL (adhere)	Adam et al. (2008)
HEK293-EBNA	Laminin fragment LG4/5	Stable	Serum-free suspension culture; miniPERM™ classic bioreactor	68 mg/L	Belin and Rousselle (2006)
HEK293 EBNA1	Erythropoietin	Transient	Serum-free suspension culture; batch and fed-batch cultures; 1 L bioreactor	18.1 mg/L (Batch) 33.6 mg/L (Fed-Batch)	Sun et al. (2006)
HEK293E	IgG	Transient	Serum-free suspension culture; ES-W orbital shaker	26 mg/L	Backliwal et al. (2008)
HEK293SF-3F6	BDD-factor VIII	Transient	Serum-free suspension culture; shake flask	400 ng/mL	Swiech et al. (2012)
HEK293	Factor VII	Stable	Serum-containing medium	Total 273 ng/ml Active 132 ng/ml	Wajih et al. (2008)
HEK293H	Factor VIII-Fc	Stable	N.r.	2057 IU/nmol	Dumont et al. (2012)
PER C6					
PER.C6	IgG	Stable	Fed-batch culture	387 mg/L 24 pg/cell/day	Zhu (2012)
PER.C6	α-1-Antitrypsin	Stable	Serum-free suspension culture; fed-batch; shake flask	1.2–2.6 g/L 21.3 pg/cell/day	Ross et al. (2012)

(continued)

Table 4.2 (continued)

Cell line	Recombinant protein	Stable/transient	Culture system	Protein production level	References
PER.C6	IgM	Stable	Serum-free suspension culture; fed-batch; shake flask	>20 pg/cell/day 0.5–2.0 g/L	Tchoudakova et al. (2009)
PER.C6	Oligoclonics™ 3 types of IgG	Stable	Serum-free suspension culture; batch culture; shake flask	232–387 g/mL 24 pg/cell/day	Kruif et al. (2010)
PERC.C6	IgG	Stable	Suspension serum-free medium; roller bottles and hollow fiber	300–500 mg/L (roller bottles) 800–1000 mg/L (hollow fiber)	Jones et al. (2003)
CAP					
CAP-T	IgG1	Transient	Suspension serum-free culture; shake flask	63 mg/L	Fischer et al., (2012)
CAP-T	Serine/threonine kinase receptor	Transient	Suspension serum-free culture; shake flask	180 mg/L	Fischer et al., (2012)
CAP-T	Serine/threonine kinase receptor	Transient	Suspension serum-free culture; shake flask	274 mg/L	Fischer et al., (2012)
CAP-T	TGF-superfamily member	Transient	Suspension serum-free culture; 3 L HDTF	8–10 mg/L	Fischer et al., (2012)
CAP-T	Wnt signaling receptor	Transient	Suspension serum-free culture; shake flask	11.4 mg/L	Fischer et al., (2012)
CAP-T	Cynomolgus cytokine protein	Transient	Suspension serum-free culture; shake flask	5.6 mg/L	Fischer et al., (2012)
CAP-T	BMP antagonist	Transient	Suspension serum-free culture; 3 L HDTF	29.5 mg	Fischer et al., (2012)
CAP	Human alpha-1 antitrypsin	Stable	Serum-free medium, static culture	30 pg/cell/day	Schiedner et al. (2008)

HKB-11					
HKB-11	IgG1	Transient	Serum-free suspension culture; shake flask	28 mg/L	Fischer et al. (2012)
HKB-11	Serine/threonine kinase receptor	Transient	Serum-free suspension culture; WAVE bioreactor	69 mg/L	Fischer et al., (2012)
HKB-11	TGF-superfamily member	Transient	Serum-free suspension culture; WAVE bioreactor	2.7 mg/L	Fischer et al., (2012)
HKB-11	Wnt signaling receptor	Transient	Serum-free suspension culture; WAVE bioreactor	9.3 mg/L	Fischer et al., (2012)
HKB-11	FEG-BDD-factor VIII	Stable	Serum-free medium	11,560 IU/mg	Mei et al. (2010)
HKB-11	BDD-factor VIII	Stable	Suspension culture in chemically defined, proprietary serum-free medium	9.27 IU/mg	Mei et al. (2006)
HKB-11	IL-4	Transient	Serum-free suspension culture; shake flask	30 g/mL	Cho et al. (2002)
HKB-11	IL-4SA	Transient	Serum-containing medium; monolayer cultures	100 ng/mL	Cho et al. (2002)
HKB-11	ICAM-1	Transient	Serum-containing medium; monolayer cultures	1400 ng/e6c/2d	Cho et al. (2002)
HKB-11	IgG (pSH125)	Transient	Serum-free suspension culture; shake flask	90 g/mL	Cho et al. (2002)
HKB-11	IgG (pSA286)	Transient	Serum-free suspension culture; shake flask	30 g/mL	Cho et al. (2002)
HKB-11	FVIII	Transient	Serum-containing medium; monolayer cultures	200 mU/e7c/2d	Cho et al. (2002)
HKB-11	BDD-FVIII	Stable	WAVE system	9050 IU/mg	Tang et al. (2013)

transient gene expression (TGE), is furthermore used where the recombinant cells express the recombinant gene but do not integrate into the host genome. Thus, these cells express for particular period of time from several days to weeks after which the recombinant gene is lost through cell division (Geisse 2009).

4.3 Different Class of Recombinant Therapeutic Protein

Leader et al. (2008) categorized the therapeutic protein based on their restorative application into four totally different groups briefly outlined in Fig. 4.2: Set 1, protein carrying either regulatory or enzymatic traits; Set 2, therapeutic protein with unique targeting trait; Set 3, protein-based vaccines; and Set 4, protein that can be used for diagnostics.

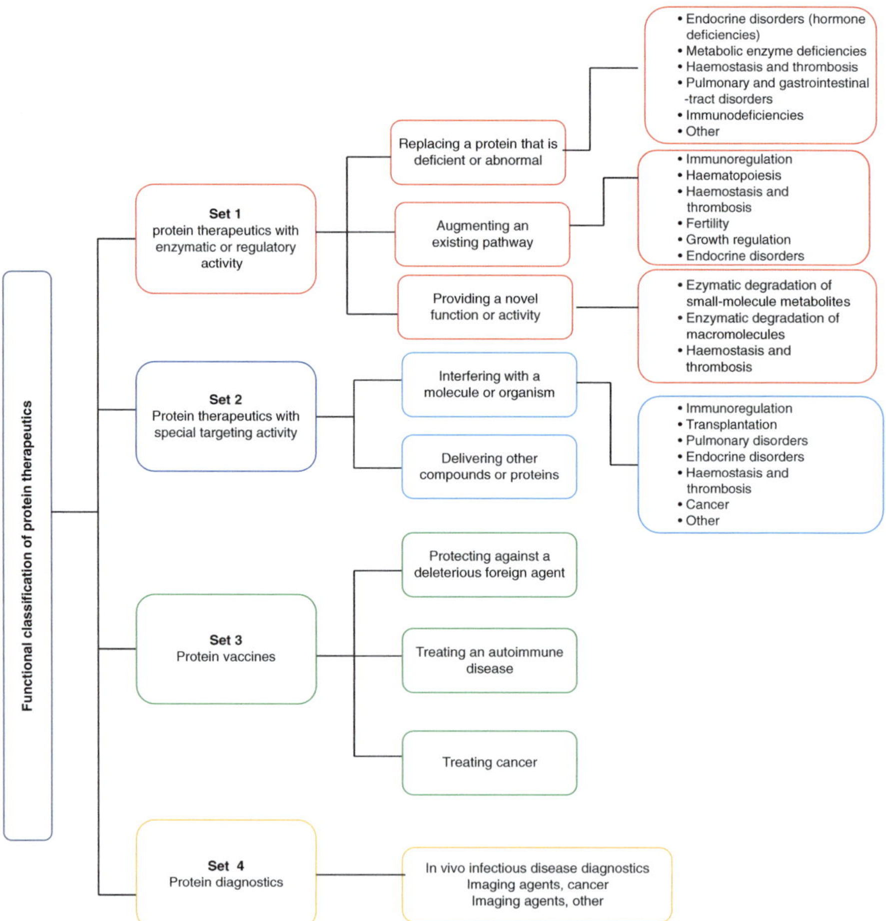

Fig. 4.2 Classification of therapeutic proteins (Akash et al. 2015)

Proteins which are accustomed to treat metabolism-related malformations or endocrine abnormities similar to IL-1Ra for type II diabetes mellitus or diseases akin to hemophilia A, which need coagulation clotting factor VIII, are put into Set 1a (Roth et al. 2001; Akash et al. 2012, 2013). Proteins known to elicit varied hematologic and immune responses for the treatment of viral diseases and persistent deficiency of RBC in the blood are categorized under Set 1b (Corwin et al. 2002; van Zonneveld et al. 2004).

Proteins carrying the remedial traits are used to adjust the physiopathology of structure- and function-related disorders and are encased in Set 1c; few illustrations are neurotoxin types A and B for numerous abnormal muscle tone and recombinant hirudin used for the treatment of thrombocytopenia (Jankovic and Brin 1991; Eriksson et al. 1997). Therapeutic proteins which are characterized as belonging to Set 2a are known to be involved in signaling pathways or in restraining functionality of particles or organism via direct binding. An exemplar is cetuximab, an inhibitor for the receptor for epidermal growth factor, and is employed in treating head and neck cancer (Saltz et al. 2004). Additionally, anakinra (IL-1Ra) that is utilized for treating extreme dynamic atrophic joint pain has been incorporated in Set 2a (Cohen et al. 2004). Compounds that are miraculous in delivering the drugs to patients are encased underneath Set 2b, wherever monoclonal antibody radioimmunotherapy (ibritumomab tiuxetan) and antineoplastic agent (denileukin diftitox) taken as examples are utilized for the treatment of lymphoproliferative disorder and chronic cutaneous T-cell malignant neoplastic disease, correspondingly (Witzig et al. 2002; Ohya and Matsuda 2005). Recombinant/purified proteins used as vaccines are grouped in Set 3.

Hepatitis B surface antigen (HBsAg) used to find out hepatitis B virus (HVB) infection and IgG anti-D (anti-RhD) antibodies used for curing Rh disease in fetus and newborn are few cases of proteins in Set 3 (MacKenzie et al. 2004). Finally, proteins used for diagnostic pathogens and new imaging agent used in detecting and mapping cancers are placed in Set 4 (Sodee et al. 2000; Campos-Neto et al. 2001). In a nutshell, almost every therapeutic protein and antibodies approved by FDA are placed among medical specialty merchandise with impending importance in each field associated with treatments.

4.4 Current Challenges and Future Scenario of Therapeutic Protein

Despite the rapid increases in the knowledge of production and expression system of therapeutic proteins, many basic questions involving absorption, distribution, metabolism, and extraction of recombinant proteins remain to be answered. All therapeutic proteins are certainly capable of inducing humoral- and/or cell-mediated immune response in patients and sometimes associated with antidrug-antibody response which will cause a severe safety issue. On the whole, massive challenges lie ahead for the production and identification of recombinant proteins (Mahmood and Green 2005). Currently existing biophysical and analytical methodologies used for characterization of proteins are incapable of differentiating bioactive and non-bioactive elements of many therapeutic proteins. Analyses of metabolic pathways and evaluation of affinity and/or kinetics of the binding interaction of most of the therapeutic proteins have not yet been done, and these obstacles continue to limit the victorious use of therapeutic protein to treat human diseases.

The administration of therapeutic proteins provokes the immune response to generate endogenous antibodies that alter pharmacodynamics and pharmacokinetics of therapeutic proteins. Evaluation of pharmacodynamics, pharmacokinetic parameters, and toxicology of these proteins is difficult and depends on the number of factors including antibody formation in response to therapeutic protein, metabolites, binding proteins, sequences of protein molecules, and size of therapeutic proteins. The other complicating factors that may change the pharmacokinetics of therapeutic protein are immunogenicity and dosing strategies (dose, route, and frequency of drug administration), but the interaction of therapeutic proteins with immune system aids in building up the pharmacodynamics of therapeutic proteins and may lead to the discovery of new approaches to reduce immunogenicity (Vugmeyster et al. 2012).

In the near future, high demand for therapeutic protein in medical field will be overcome by the advancement of protein engineering tools and the use of combinatorial chemistry and phage-display technologies which will generate an oversized repertoire of ligands specific for nearly any protein. It's doubtless that these ligands are going to be exaggeratedly employed in future applications for large-scale recombinant protein purification and creation of proteins with completely novel activities (Carter 2011). Therapeutic proteins have modified the face of recent drugs in the past decades and still continue to produce novel and efficient therapeutics for various human diseases starting from melanoma to physiological conditions.

4.5 Perspective

The development of recombinant technology within the 1970s revolutionized the face of the existing medicine and initiated effective therapies for a number of unmanageable diseases. Currently, the Food and Drug Administration approved more or less 225 therapeutic proteins up to 2015 (including interferons, natural enzymes, hormones, monoclonal antibodies, vaccines, and cell therapies). The present and future prospectives of these therapies are huge; besides recombinant protein not only offers treatment for particular diseases but also may be utilized in combination with drugs to produce additive or synergic effect. With recent advances in technological approaches, such as fusion and platform, technologies have heralded a new horizon of innovative therapeutic approaches and bring us closer to the likelihood of a practical cure for varied diseases. Therefore, the future success of protein therapeutic will depend on discovery and production of novel therapeutic protein for hemorrhagic fever (Ebola) and other viruses which will be there for years to come.

Acknowledgment R.C. is supported by University Research Scholarship of M. D. University, Rohtak, India.

References

Adam E, Sarrazin S, Landolfi C, Motte V, Lortat-Jacob H, Lassalle P, Delehedde M (2008) Efficient long-term and high-yielded production of a recombinant proteoglycan in eukaryotic HEK293 cells using a membrane-based bioreactor. Biochem Biophys Res Commun 369:297–302. doi:10.1016/j.bbrc.2008.01.141

Akash MSH, Rehman K, Chen S (2013) IL-1Ra and its delivery strategies: inserting the association in perspective. Pharm Res 30:2951–2966. doi:10.1007/s11095-013-1118-0

Akash MSH, Shen Q, Rehman K, TARIQ M, Chen S (2015) Development of therapeutic proteins: advances and challenges. Turk J Biol 39:1–16. doi:10.3906/biy-1411-8

Akash MSH, Shen Q, Rehman K, Chen S (2012) Interleukin-1 receptor antagonist: a new therapy for type 2 diabetes mellitus. J Pharm Sci 101:1647–1658. doi:10.3906/biy-1411-8

Backliwal G, Hildinger M, Hasija V, Wurm FM (2008) High-density transfection with HEK-293 cells allows doubling of transient titers and removes need for a priori DNA complex formation with PEI. Biotechnol Bioeng 99:721–727. doi:10.1002/bit.21596

Baldi L, Hacker DL, Adam M, Wurm FM (2007) Recombinant protein production by large scale transient gene expression in mammalian cells: state of the art and future perspectives. Biotechnol Lett 29:677–684. doi:10.1007/s10529-006-9297

Belin V, Rousselle P (2006) Production of a recombinantly expressed laminin fragment by HEK293-EBNA cells cultured in suspension in a dialysis-based bioreactor. Prot Expr Purif 48:43–48. doi:10.1016/j.pep.2006.02.014

Butler M (2005) Animal cell cultures: recent achievements and perspectives in the production of biopharmaceuticals. Appl Microbiol Biotechnol 68:283–291. doi:10.1007/s00253-005-1980-8

Campos-Neto A, Rodrigues-Junior V, Pedral-Sampaio DB, Netto EM, Ovendale PJ, Coler RN, Skeiky YA, Badaro R, Reed SG (2001) Evaluation of DPPD, a single recombinant Mycobacterium tuberculosis protein as an alternative antigen for the Mantoux test. Tuberculosis (Edinb) 81:353–358. doi:10.1054/tube.2001.0311

Carter PJ (2011) Introduction to current and future protein therapeutics: a protein engineering perspective. Exp Cell Res 17:1261–1269. doi:10.1016/j.yexcr.2011.02.013

Cho MS, Yee H, Chan S (2002) Establishment of a human somatic hybrid cell line for recombinant protein production. J Biomed Sci 9:631–638. doi:10.1159/000067294

Cohen SB, Moreland LW, Cush JJ, Greenwald MW, Block S, Shergy WJ, Hanrahan PS, Kraishi MM, Patel A, Sun G, Bear MB (2004) A multicentre, double blind, randomised, placebo controlled trial of anakinra (Kineret), a recombinant interleukin 1 receptor antagonist, in patients with rheumatoid arthritis treated with background methotrexate. Ann Rheum Dis 63:1062–1068. doi:10.1136/ard.2003.016014

Corwin HL, Gettinger A, Pearl RG, Fink MP, Levy MM, Shapiro MJ, Corwin MJ, Colton T (2002) Efficacy of recombinant human erythropoietin in critically Ill patients: a randomized controlled trial. JAMA 288:2827–2835. doi:10.1001/jama.288.22.2827

Derouazi M, Girard P, Van Tilborgh F, Iglesias K, Muller N, Bertschinger M, Wurm FM (2004) Serum-free large-scale transient transfection of CHO cells. Biotechnol Bioeng 87:537–545. doi:10.1002/bit.20161

Dumont JA, Liu T, Low SC, Zhang X, Kamphaus G, Sakorafas P, Fraley C, Drager D, Reidy T, McCue J, Franck HWG, Merricks EP, Nichols TC, Bitonti AJ, Pierce GF, Jiang H (2012) Prolonged activity of a recombinant factor VIII-Fc fusion protein in hemophilia A mice and dogs. Blood 119:3024–3030. doi:10.1182/blood-2011-08-367813

Durocher Y, Perret S, Kamen A (2002) High level and high-throughput recombinant protein production by transient transfection of suspension-growing human 293-EBNA1 cells. Nucleic Acids Res 30:E9. doi:10.1093/nar/30.2.e9

Eriksson BI, Wille-Jorgensen P, Kalebo P, Mouret P, Rosencher N, Bosch P, Baur M, Ekman S, Bach D, Lindbratt S, Close P (1997) A comparison of recombinant hirudin with a low-molecular weight heparin to prevent thromboembolic complications after total hip replacement. N Engl J Med 337:1329–1335. doi:10.1056/NEJM199711063371901

Ferrer-Miralles N, Saccardo P, Corchero JL, Xu Z, García-Fruitós E (2015) General introduction: recombinant protein production and purification of insoluble proteins. Methods Mol Biol 1258:1–24. doi:10.1007/978-1-4939-2205-5_1

Figueroa B Jr, Ailor E, Osborne D, Hardwick JM, Reff M, Betenbaugh MJ (2007) Enhanced cell culture performance using inducible anti-apoptotic genes E1B-19K and Aven in the production of a monoclonal antibody with Chinese hamster ovary cells. Biotechnol Bioeng 97:877–892. doi:10.1002/bit.21222

Fischer S, Charara N, Gerber A, Wolfel J, Schiedner G, Voedisch B, Geisse S (2012) Transient recombinant protein expression in a human amniocytecell line: the CAP-T1 cell system. Biotechnol Bioeng 109:2250–2261. doi:10.1002/bit.24514

FDA (2015) Novel drug approvals for 2015. http://www.fda.gov/Drugs/DevelopmentApprovalProcess/DrugInnovation/ucm430302

Geisse S (2009) Reflections on more than 10 years of TGE approaches. Protein Expr Purif 64:99–107. doi:10.1016/j.pep.2008.10.017

Girard P, Derouazi M, Baumgartner G, Bourgeois M, Jordan M, Jacko B, Wurm FM (2002) 100-liter transient transfection. Cytotechnology 38:15–21. doi:10.1023/A:1021173124640

Houdebine LM (2009) Production of pharmaceutical proteins by transgenic animals. Comp Immunol Microbiol Infect Dis 32:107–121. doi:10.1016/j.cimid.2007.11.005

Jankovic J, Brin MF (1991) Therapeutic uses of botulinum toxin. N Engl J Med 324:1186–1194. doi:10.1056/NEJM199104253241707

Jones D, Kroos N, Anema R, Montfort B, Vooys A, Kraats S, Helm E, Smits S, Schouten J, Brouwer K, Lagerwerf F, Berkel P, Opstelten D, Logtenberg T, Bout A (2003) High-level expression of recombinant IgG in the human cell line PER C6. Biotechnol Prog 19:163–168. doi:10.1021/bp025574h

Kruif J, Kramer A, Nijhuis R, Zande V, Blanken R, Clements C, Visser T, Keehnen R, Hartog M, Throsby M, Logtenberg T (2010) Generation of stable cell clones expressing mixtures of human antibodies. Biotechnol Bioeng 106:741–750. doi:10.1002/bit.22763

Lander ES, Linton LM, Birren B, Nusbaum C et al (2001) Initial sequencing and analysis of the human genome. Nature 409:860–921. doi:10.1038/35057062

Leader B, Baca QJ, Golan DE (2008) Protein therapeutics: a summary and pharmacological classification. Nat Rev Drug Discov 7:21–39. doi:10.1038/nrd2399

MacKenzie IZ, Bichler J, Mason GC, Lunan CB, Stewart P, Al-Azzawi F, De Bono M, Watson N, Andresen I (2004) Efficacy and safety of a new, chromatographically purified rhesus (D) immunoglobulin. Eur J Obstet Gynecol Reprod Biol 117:154–161. doi:10.1016/j.ejogrb.2004.03.009

Mahmood I, Green MD (2005) Pharmacokinetic and pharmacodynamic considerations in the development of therapeutic proteins. Clin Pharmacokinet 44:331–347. doi:10.2165/00003088-200544040-00001

Matasci M, Hacker DL, Baldi L, Wurm FM (2008) Recombinant therapeutic protein production in cultivated mammalian cells: current status and future prospects. Drug Discov Today Technol 5:e37–e42. doi:10.1016/j.ddtec.2008.12.003

Mei B, Chen Y, Chen J, Pan CQ, Murphy JE (2006) Expression of human coagulation factor VIII in a human hybrid cell line HKB11. Mol Biotechnol 34:165–178. doi:10.1385/MB:34:2:165

Mei B, Pan C, Jiang H, Tjandra H, Strauss J, Chen Y, Liu T, Zhang X, Severs J, Newgren J, Chen J, JM G, Subramanyam B, Fournel MA, Pierce GF, Murphy JE (2010) Rational design of a fully active, long-acting PEGylated factor VIII for hemophilia A treatment. Blood 116:270–279. doi:10.1182/blood-2009-11-254755

Ohya S, Matsuda T (2005) Poly (N-isopropylacrylamide) (PNIPAM)—grafted gelatin as thermo responsive three-dimensional artificial extracellular matrix: molecular and formulation parameters vs. cell proliferation potential. J Biomater Sci Polym Ed 16:809–827. doi:10.1163/1568562054255736

Pennisi E (2003) Bioinformatics Gene counters struggle to get the right answers. Science 301:1040–1041. doi:10.1126/science.301.5636.1040

Picanco-Castro V, Biaggio RT, Cova DT, Swiech K (2013) Production of recombinant therapeutic protein in human cells: current achievements and future perspectives. Protein Pept Lett 20:1373–1381. doi:10.2174/09298665201213111213032 2

Pogue GP, Vojdani F, Palmer KE, Hiatt E, Hume S, Phelps J, Long L, Bohorova N, Kim D, Pauly M, Velasco J, Whaley K, Zeitlin L, Garger SJ, White E, Bai Y, Haydon H, Bratcher B (2010) Production of pharmaceutical-grade recombinant aprotinin and a monoclonal antibody product using plant-based transient expression systems. Plant Biotechnol J 8:638–654. doi:10.1111/j.1467-7652.2009.00495

Rader RA (2008) Redefining biopharmaceutical. Nat Biotechnol 26:743–751. doi:10.1038/nbt0708-743

Ross D, Brown T, Harper R, Pamarthi M, Nixon J, Bromirski J, Li CM, Ghali R, Xie H, Medvedeff G, Li H, Scuderi P, Arora V, Hunt J, Barnett T (2012) Production and characterization of a novel human recombinant alpha-1-antitrypsin in PER.C6 cells. J Biotechnol 162:262–723. doi:10.1016/j.jbiotec.2012.09.018

Rosser MP, Xia W, Hartsell S, McCaman M, Zhu Y, Wang S, Harvey S, Bringmann P, Cobb RR (2005) Transient transfection of CHO-K1-S using serum-free medium in suspension: a rapid mammalian protein expression system. Protein Expr Purif 40:237–243. doi:10.1016/j.pep.2004.07.015

Roth DA, Kessler CM, Pasi KJ, Rup B, Courter SG, Tubridy KL (2001) Human recombinant factor IX: safety and efficacy studies in hemophilia B patients previously treated with plasma-derived factor IX concentrates. Blood 98:3600–3606. doi:http://dx.doi.org/10.1182/blood-2002-07-2080

Saltz LB, Meropol NJ, Loehrer PJ Sr, Needle MN, Kopit J, Mayer RJ (2004) Phase II trial of cetuximab in patients with refractory colorectal cancer that expresses the epidermal growth factor receptor. J Clin Oncol 22:1201–1208. doi:10.1200/JCO.2004.10.182

Schiedner G, Hertel S, Bialek C, Kewes H, Waschütza G, Volpers C (2008) Efficient and reproducible generation of high expressing stable human cell lines without need for antibiotic selection. BMC Biotechnol 8:13. doi:10.1186/1472-6750-8-13

Sodee DB, Malguria N, Faulhaber P, Resnick MI, Albert J, Bakale G (2000) Multicenter ProstaScint imaging findings in 2154 patients with prostate cancer. The Prosta Scint Imaging Centers. Urology 56:988–993. doi:10.1016/S0090-4295(00)00824-4

Sun X, Goh PE, Wong KTK, Mori T, Yap MGS (2006) Enhancement of transient gene expression by fed-batch culture of HEK 293 EBNA1 cells in suspension. Biotechnol Lett 28:843–848. doi:10.1007/s10529-006-9010-1

Swiech K, Picanço-Castro V, Covas DT (2012) Human cells: new platform for recombinant therapeutic protein production. Protein Expr Purif 84:147–153. doi:10.1016/j.pep.2012.04.023

Tang L, Leong L, Sim D, Ho E, JM G, Schneider D, Feldman RI, Monteclaro F, Jiang H, Murphy JE (2013) von Willebrand factor contributes to longer half-life of PEGylated factor VIII in vivo. Haemophilia 19:539–545. doi:10.1111/hae.12116

Tchoudakova A, Hensel F, Murillo A, Eng B, Foley M, Smith L, Schoenen F, Hildebrand A, Kelter AR, Ilag LL, Vollmers HP, Brandlein S, McIninch J, Chon J, Lee G, Cacciuttolo M (2009) High level expression of functional human IgMs in human PER.C6 cells. MAbs 1:163–171. doi:10.4161/mabs.1.2.7945

Van Zonneveld M, Honkoop P, Hansen BE, Niesters HG, Darwish Murad S, de Man RA, Schalm SW, Janssen HL (2004) Long term follow-up of alpha-interferon treatment of patients with chronic hepatitis B. Hepatology 39:804–810. doi:10.1002/hep.20128

Venter JC, Adams MD, Myers EW, Li PW et al (2001) The sequence of the human genome. Science 291:1304–1351. doi:10.1126/science.1058040

Vugmeyster Y, Xu X, Theil FP, Khawli LA, Leach MW (2012) Pharmacokinetics and toxicology of therapeutic proteins: advances and challenges world. J Biol Chem 26(3):73–92. doi:10.4331/wjbc.v3.i4.73

Wajih N, Owen J, Wallin R (2008) Enhanced functional recombinant factor VII production by HEK 293 cells stably transfected with VKORC1 where the gamma-carboxylase inhibitor calumenin is stably suppressed by shrna transfection. Thromb Res 122:405–410. doi:10.1016/j.thromres.2007.11.002

Walsh G (2010) Biopharmaceutical benchmarks. Nat Biotechnol 28:917–924. doi:10.1038/nbt0910-917

Witzig TE, Gordon LI, Cabanillas F, Czuczman MS, Emmanouilides C, Joyce R, Pohlman BL, Bartlett NL, Wiseman GA, Padre N, Grillo-López AJ, Multani P, White CA (2002) Randomized controlled trial of yttrium-90-labeled ibritumomab tiuxetan radioimmunotherapy versus rituximab immunotherapy for patients with relapsed or refractory low-grade, follicular, or transformed B-cell non-Hodgkin's lymphoma. J Clin Oncol 20:2453–2463. doi:10.1200/JCO.2002.11.076

Wurm FM (2004) Production of recombinant protein therapeutics in cultivated mammalian cells. Nat Biotechnol 22:1393–1398. doi:10.1038/nbt1026

Zhang J (2010) In: Richard HB, Davies JE, Demain AL (eds) Manual of industrial microbiology and biotechnology. ASM Press, Washington, DC, pp 157–178. doi:10.1128/9781555816827

Zhu J (2012) Mammalian cell protein expression for biopharmaceutical production. Biotechnol Adv 30:1158–1170. doi:10.1016/j.biotechadv.2011.08.022

Part II

Metabolic Engineering in Plants

Engineering in Plant Genome Using *Agrobacterium*: Progress and Future

5

Anshu Alok, Shivam Sharma, Jitesh Kumar, Subodh Verma, and Hemant Sood

Abstract

Plant genetic and metabolic engineering has emerged as an important area in plant science for crop improvement programmes. Efficient and stable genetic transformation is a great need for successful engineering in plants. Various significant improvements have been made for an efficient genetic transformation method in plants. However, *Agrobacterium*-mediated genetic transformation of plants is the most widely used among all known transformation methods. In the past few decades, progress in *Agrobacterium*-mediated transformation has led to the emergence of various transgenic technologies, including overexpression of recombinant protein in plants, knockdown of targeted gene through RNAi approach, targeted genome editing, and generation of marker-free transgenic plants. Various binary and super binary vectors have been developed in past few decades, which overcome hurdle of genetic transformation in higher and recalcitrant plant species. In addition, the use of different strains of *Agrobacterium* and other parameters during optimization of plant transformation protocol also enhanced genetic transformation efficiency. This recent advancement of genetic engineering tools has boosted agricultural biotechnology and overcome the

A. Alok • J. Kumar
Department of Biotechnology (DBT), National Agri-Food Biotechnology Institute, Sector-81 (Knowledge City), S.A.S. Nagar, Mohali, Punjab, India

S. Sharma
Jaypee University of Information Technology, Solan, Himachal Pradesh, India

S. Verma
National Institute of Plant Genome Research, Aruna Asaf Ali Marg, New Delhi, India

H. Sood (✉)
Jaypee University of Information Technology, Solan, Himachal Pradesh, India

Department of Biotechnology and Bioinformatics, Jaypee University of Information Technology, Solan, Himachal Pradesh, India
e-mail: hemant.sood@juit.ac.in

V.C. Kalia, A.K. Saini (eds.), *Metabolic Engineering for Bioactive Compounds*, DOI 10.1007/978-981-10-5511-9_5

limitation of conventional plant breeding methods. The recently developed genome-editing tool based on "CRISPR/Cas-mediated genome editing" is being applied to various plant species including crop and woody plant. Using this tool we can generate "marker- or transgene-free" genetically modified plants, which are therefore more acceptable for field release. The use of different genetic engineering tools in the area of agricultural and industrial biotechnology leads to production of improved crop variety and economically important metabolites.

Keywords
Agrobacterium • Binary vectors • Plant transformation • Gene stacking • VIGS • Rnai • CRISPR/Cas

5.1 Introduction

Genetic engineering is an essential tool for functional genomics and proteomics studies of plants. The term genetic engineering is commonly used when genome of any living organism is modified using recombinant DNA technology. Genetic engineering has emerged rapidly when Paul Berg combined *E. coli* genome with the genes of a bacteriophage and SV40 virus in 1972 to create the first recombinant DNA molecule (Nemudryi et al. 2014). Since then, genetic engineering has gained tremendous success and is being applied to various areas including plant science. Transformation is one of the most important techniques for plant genetic engineering. Significant progress has been made in the past few decades to develop a novel and efficient transformation methods in plants. However, *Agrobacterium* and biolistic-mediated transformation are most popular among all the methods of DNA delivery being used for various plant transformations (Barampuram and Zhang 2011; Zalewski et al. 2012). But, the main disadvantage of biolistic mediated method is low transformation efficiency and multiple gene insertion. However, an *Agrobacterium*-mediated plant transformation is more acceptable, because of low cost and single-gene integration into plant genome (Zalewski et al. 2012; Alok et al. 2016; Bhati et al. 2016).

Agrobacterium species belongs to gram-negative bacterium and is mostly found in the soil. It causes a plant disease symptom "crown gall" and has the ability to transfer and integrate a new genetic material into the plant cell (Pitzschke and Hirt 2010; Gelvin 2012). Due to this unique property, *Agrobacterium* is known to be a "nature-genetic engineer." The natural ability of an *Agrobacterium* species to alter the plant genomic content was the foundation of plant transformation. Wild-type *Agrobacterium* causes either "tumor" because of Ti (tumor-inducing) plasmid or "hairy root" due to Ri (root-inducing) plasmids of *A. tumefaciens* and *A. rhizogenes*, respectively. Both Ti and Ri plasmids are larger in size and contain a T-DNA (transferred DNA) which consists of oncogenic and opine catabolism genes. The size of T-DNA within its natural plasmids may be large adequate to encode ten genes. For example, pTiC58 consists of approx. 23 kbp large T-DNA region (Gelvin 2003). The oncogenic genes encode different enzymes which are involved in biosynthesis of auxins and cytokines. The opine are produced by condensation of amino acid and sugars and excreted by crown gall cells, which is further utilized by bacterium as carbon and nitrogen sources (Vladimirov et al. 2015). The T-DNA is also bordered by 25-bp repeats called the left border (LB) and right border (RB). Besides these genes, virulence (vir) genes are also important for T-DNA transfer, and it is positioned outside the T-DNA region within plasmid, and few genes are located on chromosomal DNA (Zupan et al. 2000). The vir genes are structured in the form of operons on the Ti plasmid (virA, virB, virC, virD, virE, virF, virG, and virH) and bacterial chromosome (chvA, chvB, and chvF) (Barampuram and Zhang 2011). The eight operons of vir region encode their respective proteins which regulate the processing of T-DNA transfer and integration (Subramoni et al. 2014). Gene transfer during *Agrobacterium*-mediated plant cell transformation involves five essential steps: (1) activation of virulence genes, (2) formation of T-DNA complex, (3) transfer of T-DNA from *Agrobacterium* to the plant cell nucleus, (4) integration of T-DNA into the plant genome, and (5) expression of T-DNA genes (Pitzschke and Hirt 2010; Gelvin 2012). T-DNA complex and several vir proteins are exported into the plant cell by a VirB/D4 type IV secretion system, and finally a T-DNA is integrated into the chromosome of plant (Christie 2004).

Using recombinant DNA technology, the wild-type Ti or Ri plasmid has been modified and used frequently for plant transformation. Wild-type plasmids which have deleted tumor-inducing genes are called disarmed Ti or Ri plasmid, and the strain which harbors these plasmids is known as disarmed strain (Hellens et al. 2015). Typical binary vectors consist of modified T-DNA region flanked by 25-bp border sequences. These modified T-DNA regions consist of multiple cloning site (MCS) and appropriate selectable marker gene. For example, binary vector pCambia1301 consists of a pUC18-derived MCS having lacZ alpha and *hpt*II gene as plant selectable marker (http://www.cambia.org).

Agrobacterium-mediated genetic transformation of plants became more popular, when the first successful transgenics of petunia and tobacco were recovered in 1983 (Fraley et al. 1983). After various modification in strains and binary vectors, protocols were optimized that made it most popular and an important technique in the area of plant genetic engineering. *A. tumefaciens* can be used to overexpress any gene of interest (gain of function) or knock down/out (loss of function). In the past several decades, numerous reports have been published to overexpress different gene or multiple genes under the control of different promoter in different plant species (Dafny-Yelin and Tzfira 2007). The RNAi-mediated gene suppression technology is widely known to scientific community due to serendipity in their research work by two groups (Napoli et al. 1990; van der Krol et al. 1990). They were attempting to make petunia flower more dark purple by introducing a transgene which overproduces the chalcone synthase enzyme, but surprisingly they found some flower turned white and some purple pigmented instead of dark purple. This type of phenotype occurred due to silencing of a transgene and its endogenous homologs, and thus, Napoli et al. introduce a new term "co-suppression" (Napoli et al. 1990). In RNAi mechanism, small molecules known as interfering RNA suppress a targeted gene very effectively. Various binary vectors are available to induce RNAi in different plant species, and using these vectors, we can generate abiotic and biotic tolerant plants by knocking down any gene of interest (Miki 2004; Mann et al. 2012). However, knockdown of any gene of interest through RNAi has some disadvantages such as unpredictable off-target effects and temporary gene function inhibition (Gaj et al. 2013).

To overcome this problem, genome-editing technologies and tools that emerged in the past few decades will permanently and precisely knock out any targeted gene from complex genome of plant species. Zinc-finger nucleases (ZFNs), transcription activator-like effector nucleases (TALENs), and clustered regularly interspaced short palindromic repeats/Cas (CRISPR/Cas) are three advanced tools for genome editing. These tools induce double-strand break at specific sites in the plant genome, which might be repaired by nonhomologous end-joining break (NHEJ) or homology-directed repair (HDR) that results in gene mutations at the target site (Marton et al. 2010; Upadhyay et al. 2013; Xiong et al. 2015). ZFNs and TALENs are based on the protein-guided recognition mechanism, and it requires modular assembly of its components on binary vector which is time-consuming and tedious. Among these tools, CRISPR/Cas-mediated genome editing is on the boom due to simplicity and efficiency of engineering. Another advantage is easy construction of CRISPR/Cas binary vector to introduce multiple gene disruptions simultaneously using a single vector. Therefore, this technique is being widely adopted nowadays by the plant research community.

A. rhizogenes plays an important role in plant genetic engineering because it induces "hairy roots" which are the source for secondary metabolite production. Hairy root induction by *A. rhizogenes* into host cell mimics a natural engineering, which is based on the genetic transfer of wild or modified Ri T-DNA (Makhzoum et al. 2013). *A. rhizogenes*-mediated genetic transformation of plants and hairy root induction of various important economical and medicinal plants species are in demand. Scale-up of hairy root culture using different techniques is frequently used for scale production of valuable secondary metabolites and recombinant proteins (Sivakumar et al. 2010; Pala et al. 2016).

5.2 Agrobacterium Strains and Genotypes

Agrobacterium has been categorized into a different species depending upon the symptom of plant disease and host range. For example, *A. tumefaciens* and *A. rhizogenes* cause crown gall and hairy root disease, respectively, whereas *A. radiobacter* is an avirulent species (Gelvin 2003; Chandra 2012). There are different classifications of *Agrobacterium* reported considering different parameters. However mostly *A. tumefaciens* and *A. rhizogenes* are used in plant genetic engineering. For an efficient and stable genetic transformation of any plant species for engineering in plant genome, the most important is selection of strains of *Agrobacterium*. For example, most of monocots such as wheat, rice, maize, barley, etc. showed high transformation efficiency using AGL1 and LBA4404 strain as compared to other strains (Ishida et al. 2007; He et al. 2010). The most commonly used strains of *A. tumefaciens* are listed in Table 5.1. The molecular and genetic basis of the host range of *Agrobacterium*

Table 5.1 List of *Agrobacterium tumefaciens* strains being used for plant transformation

Bacterial strain	Chromosomal background	Ti plasmid	Opine classification[a]	Antibiotic resistance[b]	References
AGL0	C58	Disarmed pTiBo542	Succ	Rif	Lazo et al. (1991)
AGL1	C58	Disarmed pTiBo542	Succ	Rif, Car	Lazo et al. (1991)
A136	C58	Cured	Nopa	Rif, Nal, Carb	Watson et al. (1975)
C58C1	Cured/ disarmed?	Disarmed pTiC58	Nopa	Rif	Deblaere et al. (1985)
C58-Z707		Disarmed pTiC58	Nopa	Rif	Hepburn et al. (1985)
EHA101	C58	Disarmed pTiBo542	Nopa	Rif, Kan	Hood et al. (1986)
EHA105	C58	Disarmed pTiBo542	Succ	Rif, Kan	Hood et al. (1993)
GV3101	C58		Nopa	Rif,Carb	Holsters et al. (1980)
GV3101::pMP90	C58	Disarmed pTiC58	Nopa	Rif, Gen	Koncz and Schell (1986)
LBA4404	Ach5	Disarmed pTiAch5	Octo	Rif, Spec, Strep	Hoekema et al. (1983)
NT1(pKPSF2)	C58	Disarmed pTichry5		Ery	Palanichelvam et al. (2000)
GV2260	C58	Disarmed pTiB6S3	Octo	Carb	McBride and Summerfelt (1990)
GV3100	C58	Cured	Nopa	Carb	Holsters et al. (1980)
GV3850	C58	Disarmed pTiC58	Nopa	Rif, Carb	Zambryski et al. (1983)

[a]*Agro* agrocinopine, *Octo* octopine, *Succ* succinamopine, *Leuc* leucinopine, *Nopa* nopaline
[b]*Car* carbenicillin, *Kan* kanamycin, *Rif* rifampicin, *Gen* gentamicin, *Ery* erythromycin

strain remains unclear for genetic transformation among different plant species. The capability of *Agrobacterium* strain to successfully transform plant cells is dependent upon their genetic makeup (Jones et al. 2005). Earlier, it was thought to be only Ti plasmid is key genetic determinant for host range rather than chromosomal DNA (Gelvin 2003). Besides genetic makeup, the host range of *Agrobacterium* is controlled by various parameters within bacterium and plant. Host range may be enhanced by placing particular Ti plasmids with assured bacterial chromosomal backgrounds. For example, when plasmid pTiBo542 mobilized in Bo542 strain of *A. tumefaciens*, it was inadequate for genetic transformation of legume plants. Conversely, while pTiBo542 positioned with C58 chromosomal background, it showed powerful virulence to soybeans and other legumes (Hood et al. 1987).

5.3 Binary Vectors and Its Properties

Binary vector is an essential tool for stable or transient genetic transformation. The term binary vector literally refers to the entire system that consists of two replicons, i.e., T-DNA and genes for virulence (Komari et al. 2006). For example, "AGL1-containing pCambia1301" means pCambia1301 consists of T-DNA region, whereas pTibo542 consists of vir gene. The modified Ti plasmid having vir gene is also known as "helper plasmid" because it is necessary for plant genetic transformation via binary vector (Gelvin 2003). The plasmid that carries the T-DNA region is frequently called a binary vector, which we use in this chapter. Binary vectors which carry both T-DNA region and vir operon are called as super binary vector (Komari et al. 2006). For example, a super binary vector "pTOK233" consists of vir B, vir C, and vir G genes which were cloned from pTiBo542 into backbone of pGA472 (Hiei et al. 1994). A typical binary vector consists of (a) T-DNA region flaking with right and left border, (b) multiple cloning sites (MCS), (c) bacterial selectable marker gene, (d) origin of replication for *E. coli* and *Agrobacterium*, and (e) plasmid mobilization and function gene.

The binary vectors are of different types, but it can be broadly classified into following types depending upon their function in plant genetic engineering (Table 5.2):

1. Expression vectors (expression of any gene of interest)
2. RNAi vectors (suppression of mRNA of gene of interest)
3. CRISPR/Cas vectors (editing or mutation of gene of interest)

5.3.1 Expression Vectors

The overexpression vectors are used to express or overexpress gene of interest into the plants. Examples of such vectors are pCambia1301, pBI101, pRI101AN, etc. The binary vector consists of unique multiple cloning sites inside the T-DNA region and is used for plant overexpression vector (Fig. 5.1). Since these vectors are used for delivery of transgenes into plants, they should be efficient in plant transformation, easy for transgene cloning, widely available for researchers, small in size, and high in copy number (Komari et al. 2006). These vectors can be also used to express gene of interest in targeted tissue by regulating them with tissue-specific promoter in plants. Different components of binary vectors reside on T-DNA region and vector backbone.

T-DNA region: The transferable region of binary vector flanking between right and left border is known as T-DNA. This is integrated into the plant genome and mainly consists of T-DNA borders, MCS, plant selectable marker, and reporter gene.

Vector backbone: This part of any binary vectors is necessary for its replication and mobilization. It consists of bacterial selectable marker gene, plasmid replication functions, and plasmid mobilization functions.

Table 5.2 List of different binary vectors for plant transformation

Binary vector	Size (kb), Mob	Bacterial selection	Replication origin Agro, *E. coli*	Plant selection gene	References
Expression vectors					
pBIN19	11.7, Yes	Kan	pRK2, pRK2	Kan	Bevan (1984)
pBI121	14.7, Yes	Kan	pRK2, pRK2	Kan	Koncz and Schell (1986)
pPZP111	8.9, Yes	Chloramphenicol	pVS1, ColE1	Kan	Hajdukiewicz et al. (1994)
pPCV001	9.2, Yes	Ampicillin	pRK2, ColE1	Kan	Koncz et al. (1989)
pGreen0029	4.6, No	Kanamycin	pSA, pUC	Kan	Hellens et al. (2000)
pOCA18	24.3, Yes	Tet	pRK290, pRK2	Kan absent	Olszewski et al. (1988)
pSoup	9.3, No	Tet	IncR/pSa, pRK2	Absent	Hellens et al. (2000)
pCB301	5.0, No	Kanamycin	mini-RK2	Bar	Xiang et al. (1999)
Suppression vectors					
pRNAi-GG	15.7, Yes	Kan	pRK2, pRK2	Kan	Yan et al. (2012)
pIPKb008	15.8, Yes	Chloramphenicol	pVS1, ColE1	Hyg	Himmelbach et al. (2007)
pANIC 8A	19.7, Yes	Kanamycin	pVS1, ColE1	Hyg	Mann et al. (2012)
pFGC5941	11.4, Yes	Kanamycin	pVS1, ColE1	Bar	http://www.chromDB.org
Gene-editing vectors					
pRGEB31	15.0, Yes	Kanamycin	pVS1, ColE1	Hyg	Xie and Yang (2013)
pHSN401	12.7, No	Kanamycin	pSA, pUC	Hyg	Xing et al. (2014)
pBSN401	12.5, No	Kanamycin	pSA, pUC	Bar	Xing et al. (2014)
pKSN401	12.5, No	Kanamycin	pSA, pUC	Kanamycin	Xing et al. (2014)
pHSE401	16.6, Yes	Kanamycin	pVS1, ColE1	Hyg	Xing et al. (2014)

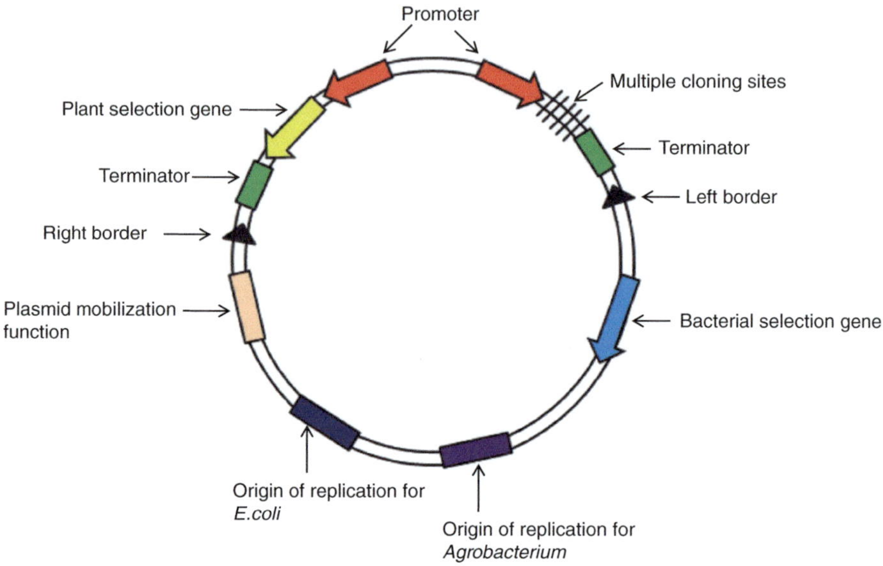

Fig. 5.1 A typical map of plant expression vector

5.3.2 RNAi Vectors

RNA interference (RNAi) vectors are very useful tool for functional genomic analysis of plants. This binary vector has been successfully used to downregulate the levels of targeted genes, enabling loss-of-function studies in living plant cells (Dafny-Yelin et al. 2007). A typical RNAi vectors consist a hairpin RNA expression cassettes on binary vectors. These cassettes consist of a promoter and two different MCS attached by an intron as linker (Fig. 5.2), for example, pMCG161, pFGC5941, etc. RNA silencing is a mechanism that downregulates the expression level of target gene by either transcriptional or posttranscriptional gene silencing (Agrawal et al. 2003; Bhati et al. 2016). RNAi vectors and its applications open a new avenue in the area of functional genomics. The main advantage of these vectors is that it can be used to explore any gene function using either transient or stable transformation of plants.

5.3.3 CRISPR/Cas Vectors

CRISPR/Cas vector consists of two components, i.e., guide RNA which directs double-strand cleavage of the target DNA and Cas9 nuclease (Fig. 5.3). Both components should be regulated and expressed by an efficient promoter within the plant cell where editing is to be done. The construction of CRISPR/Cas vectors is based on the backbone of plant transformation vector that contains essential components for replication. Examples of such vectors are pRGEB31, pRGE31, pBUN411, etc. (Xie and Yang 2013; Xing et al. 2014). Recently, various vectors have been constructed that

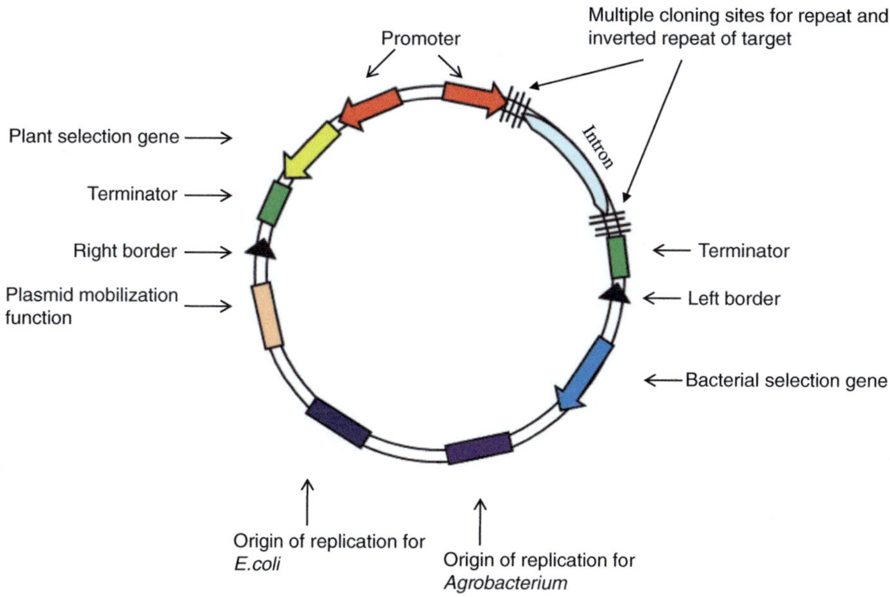

Fig. 5.2 A typical map of plant RNAi vector

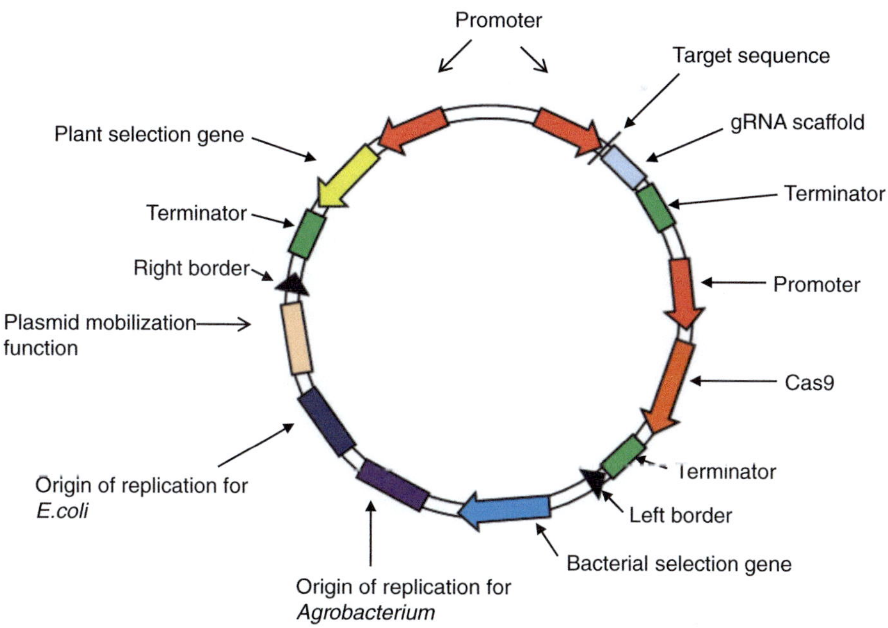

Fig. 5.3 A typical map of plant CRISPR/Cas vector

differs in promoter and cloning of target. By adding more than two or more guide RNA, this vector can be used to target two or more genes within same gene to delete large fragment of that gene. These vectors facilitate a development of a variety of mutant with multiple targeted mutations in plant system (Xing et al. 2014).

5.4 Plant Transformation Process via *Agrobacterium*

Plant genetic transformation via *Agrobacterium* is a highly complex and evolved mechanism. The process is dependent upon the genetic makeup of *Agrobacterium*, host plant, and various other parameters which can influence this (Ziemienowicz 2014). The main genetic arrangement of *Agrobacterium* required T-DNA, vir gene, and three chromosomal virulence loci chvA, chvB, and pscA for successful plant transformation (Stachel et al. 1986; Reuhs et al. 1997). Different methods and technique are frequently used for genetic transformation of plant species via *Agrobacterium*, but it can be classified as tissue culture-dependent technique and tissue culture-independent technique (Bent 2000).

5.4.1 Tissue Culture-Dependent *Agrobacterium*-Mediated Transformations

Tissue culture-dependent genetic transformation has been widely used for the fast production of uniform, virus-free, and good quality of transgenic plants. In this technique, the initial explants such as leaf, stem, immature and mature embryos, and internode and embryonic calli are generally transformed depending upon regeneration capacity of that targeted plant. Various dicot plants such as tobacco (Gallois and Marinho 1995), potato (Cruz-Mendívil et al. 2011), tomato (Sun et al. 2015), and *Artemisia* (Alok et al. 2016) and few monocot plants such as rice (Karthikeyan et al. 2011) have the capability to regenerate a whole plant directly from the transformed leaf tissue. But various monocots such as wheat, barley, banana, etc. are recalcitrant and difficult to regenerate directly from leaf tissues, but in this case suspension cell transformation is very useful (Remakanthan et al. 2014). The suspension cell is single cells that arise from embryonic calli, and it has the capability to regenerate into an individual plant. The suspension cell transformation is very efficient for genetic transformation of recalcitrant crops as demonstrated in wheat (Weir et al. 2001), barley (Wu et al. 1998), banana (Tripathi et al. 2015), etc.

5.4.2 Tissue Culture-Independent *Agrobacterium*-Mediated Transformations

To reduce the cost of expensive laboratory facilities and time consumption, some researchers avoid tissue culture during *Agrobacterium*-mediated transformation. In this method we generally target germ cell or embryos which can directly regenerate into a whole plantlet. The most frequently used method is floral dip and *in planta* transformation of seed (Risacher et al. 2009; Martins et al. 2015). In the early decades, the seeds were transformed by *Agrobacterium* and then directly subjected for appropriated antibiotic selection (Feldmann and David Marks 1987). The main disadvantage of this method is that it gives mosaic expression of transgene and has very low transformation efficiency. Floral dip method is a genetic transformation approach where the inflorescence of some plant species at appropriate stage is dipped into an *Agrobacterium* solution. Initially the vacuum infiltration was used successfully during floral dip of *Arabidopsis* (Bechtold and Pelletier 1997). Various modifications were made in the protocol, and different parameters were optimized for an efficient floral dip-mediated transformation (Clough and Bent 1998).

5.4.3 Factors Affecting Genetic Transformation of Plants

During the process of genetic transformation, transfer and integration of gene of interest along with T-DNA into genome of host plant are regulated by various factors such as type of explants, bacterial strains (Table 5.1), type of binary vectors, genotype of the plant, composition of culture medium, physical damage, cocultivation temperature and duration, addition surfactant in media, pH of medium, etc. (Egnin et al. 1998; Sood et al. 2011; Tiwari and Tuli 2012; Alok et al. 2016). *Agrobacterium*-mediated transformations were done in groundnut and showed that transformation frequencies are related to cocultivation duration and temperature, various antioxidants and their concentrations, bacterial strains, and explant type (Tiwari and Tuli 2012). In vitro manipulation of explants such as wounding, sonication, desiccation, and osmotic treatment was shown to enhance the transfer process of T-DNA and subsequent recovery of transgenic plants. However, in vitro manipulation of explants is largely dependent upon plant species. For example, sucrose treatment of explants increased T-DNA delivery in case of rice, whereas desiccation improved T-DNA delivery and stable transformation of some species such as sugarcane, maize, wheat, and soybean (Ziemienowicz 2014).

5.5 Engineering in Plants for Gain of Gene Function

Plant genetic engineering has reduced the time required for the improvement of crop varieties in comparison to traditional method such as plant breeding. In addition, it overcomes genetic barriers such as pollen compatibility with the pistil. In this approach, gene responsible for a useful trait is identified and cloned into a binary vector under a strong promoter. The selected gene of interest may be from endogenous (cisgene) or exogenous (transgene) source. For example, R1 gene, responsible for late-blight resistance from wild potato, is cloned and inserted into cultivated potato and is known as cisgenesis, and plants are called cisgenic. The Bt gene which encodes toxic proteins from *Bacillus thuringiensis* inserted into cotton for pest-resistant plant is known as transgenesis (Hou et al. 2014). The cloned cisgene or transgene in binary vector is finally integrated into the plant genome after successful genetic transformation. The single or multiple gene of interest can be transferred by both *A. tumefaciens* and *A. rhizogene* (Tiwari et al. 2008; Pandey et al. 2014b, a; Alok et al. 2016; Pala et al. 2016). The gene of interest is assembled to develop an expression cassette, which consists of promoter and terminator for an efficient expression of transgene. Different promoters such as constitutive, seed specific, and root specific are frequently used depending upon the need of transgenic. For example, CaMV35S promoter is a constitutive promoter; ACC oxidase and ACO synthase promoter is fruit-specific promoter; and *Arabidopsis* PP2C-like promoter is abscisic acid-inducible promoter (Bhalothia et al. 2013, 2016; Dutt et al. 2014). The stacking of different traits or genes in plants is another strategy to engineer metabolic pathway for crop improvement.

5.5.1 Gene Stacking Using *Agrobacterium*-Mediated Co-transformation

Multiple genes can be simultaneously into same plants through co-transformation strategy. Co-transformation via *Agrobacterium* has been extensively applied for genetic manipulation of plant genome to produce better crops with enhanced nutritional and medicinal value or stress-tolerant plants. For example, transgenic rice resistant against the sheath blight disease was developed after stacking of two genes, i.e., rice chitinase gene for disruption of fungal cell wall along with tobacco osmotin gene to target cell membrane (Ramana Rao et al. 2011). Another noteworthy application of this approach was used to produce "golden rice." The "golden rice" with enriched provitamin A (β-carotene) was developed by co-transforming phytoene synthase and lycopene β-cyclase genes from *Narcissus pseudonarcissus* and phytoene desaturase gene from *Erwinia uredovora* (Beyer et al. 2002). However, co-transformation is used to generate marker-free plants. The gene of interest along with plant selectable marker can be transformed either using single or two independent binary vectors. The plant transformation can be proceeded by putting these both vectors in one or two *Agrobacterium* strains (McKnight et al. 1987; Daley et al. 1998). Sugita et al. (2000) used co-transformation strategy and found marker-free plants when they used the same selectable marker for sequential *Agrobacterium*-mediated transformation.

5.5.2 Gene Stacking by Multiple Expression Cassettes Linked Together

Stacking multiple genes of interest in single plant transformation vectors has an advantage over simultaneously co-transformation using two or more vectors. For example, high amount of artemisinin was produced in tobacco by expressing required multiple genes in this pathway (Fuentes et al. 2016). During transformation T-DNA will transfer and integrate randomly into the plant genome; therefore, to avoid integration of T-DNA at multiple sites, it is better to link inside a single T-DNA (Dafny-Yelin and Tzfira 2007). Designing and assembling multigene cassette into single plant transformation vectors are very lengthy process due to limitation of cloning site (Dafny-Yelin and Tzfira 2007). Due to this difficulty, different plant transformation vectors with rare restriction sites came to exist (Goderis et al. 2002). Type IIS restriction enzymes such as BsaI and XcmI cleave DNA outside of their recognition site and produce 3′ or 5′ overhangs (Emami et al. 2013). Gateway recombination cloning system required an entry and destination vector which utilizes λ phage integrase. Later this system was modified for stacking multigene and known as MultiSite Gateway vectors which utilize two entry vectors with different recombinases (Chen et al. 2006).

5.6 Engineering for Loss of Gene Function in Plants

5.6.1 Virus-Induced Gene Silencing (VIGS)

VIGS has come into sight as techniques which are very fast to explore unknown gene function in recent years. VIGS vector is prepared by cloning a 200–1300 bp coding sequence of target gene into the genome of a virus (Gilchrist and Haughn 2010). *Agrobacterium*-mediated transient transformation is performed further with the help of binary vector containing this viral genome with target sequence into a plant. In plant viral genome, replication occurs which resulted the formation of dsRNA (double-stranded RNA) intermediate along with target gene sequence. Dicer-like enzymes chop this intermediate RNA into small interfering RNA (siRNA) and finally activating siRNA silencing pathway. If the gene of interest/target gene is similar to host plant, the siRNAs would target the host mRNA also, and finally loss of the function in the encoded protein will occur (Lu 2003). For example, VIGS vectors are based on the tobacco rattle virus, and apple latent spherical virus is most frequent in use due to its wide host range (Igarashi et al. 2009). VIGS has been used in different plant species using monocot and dicot plant-specific VIGS vectors. The main limitation of this technique is host range of the prepared viral vector and transient effect. Other disadvantage of this is it cannot be used for gene having one or more homologous sequences (Gilchrist and Haughn 2010).

5.6.2 RNA-Mediated Interference (RNAi)

RNAi is a natural mechanism which leads to posttranscriptional gene silencing (PTGS) triggered by dsRNA to prevent the expression of targeted gene. During this, dsRNA is processed by enzyme dicer to convert the small interfering RNAs (siRNAs). The antisense strand of ~22-nucleotide short siRNAs become specific to endonuclease-protein complex (RNA-induced silencing complex or RISC), which then targets the homologous RNA and degrades it at specific site. Finally, knockdown of protein expression occurs (Zamore et al. 2000).

There are several methods used to activate RNAi pathway in plants such as RNAi vectors, in vitro dicing, and synthetic molecules (artificial microRNAs) (Agrawal et al. 2003). Among all these methods, artificial microRNAs (amiRNA) are growing fast and most popular in use. In this procedure, a miRNA gene carrying a 21 bp insert complementary to the target gene is transformed into the plant where it is processed by the dicer pathway to produce the small-silencing RNA molecules which degrade endogenous transcript of plant. An efficient suppression of transcript level depends upon the use of promoter in RNAi vector specific to monocot and dicot plant. For example, pMCG161 is most frequently used for wheat, rice, banana, maize, etc. (Gasparis et al. 2011; Bhati et al. 2016). Modified version of this RNAi vectors involves the production of interfering RNAs using promoters that are temporally or spatially specific or that are inducible by some exogenous factor (Estornell et al. 2009). The advantages of RNAi are that transcripts of multiple genes from a family can be silenced by a single construct. A disadvantage of the RNAi approach is that few genes are resistant to silencing by exogenous RNA, possibly due to sequence or structural features of these genes. In addition, transcripts of endogenous genes which are similar in sequence to the targeted gene may be downregulated inadvertently, and this is known as "off-target" silencing (Gilchrist and Haughn 2010).

5.6.3 Targeted Genome Modification Using Site-Specific Nucleases

Target-specific genome editing using site-specific nucleases includes ZFNs, TALENs, and CRISPR/Cas tool. These tools first create DNA double-strand breaks at target site which activate plant DNA repair mechanism to join the DNA break. The broken DNA is joined by either homologous recombination or nonhomologous end joining (Chen and Gao 2014). The main advantages of these tools are that we can generate transgene-free genetically modified plants which are not concerned to regulatory issues (Zhang et al. 2016).

The Cys2-His2 zinc finger (ZF) domain is a DNA-binding motif and found in eukaryotes. Each ZF consists of approx. 30 amino acids that fold into a ββα configuration which is stabilized by a zinc ion. The ZF can recognize and bind to a specific 3-bp DNA sequence by inserting an α-helix into the major groove of the DNA double helix (Pabo et al. 2001). ZFNs are generated by fusing an artificial DNA-binding domain consisting of a tandem array of ZFs to the nonspecific DNA cleavage domain of the FokI restriction endonuclease (Durai et al. 2005). A typical ZF DNA-binding domain is composed of two arrays of three or four individual ZFs each that can bind to a 9- or 12-bp target sequence. The cleavage domain functions typically as a dimer, and the pair of ZFAs is designed to bind sequences within a distance of 5–7 nucleotides from each other; the result is an enzyme capable of targeting a unique DNA sequence and able to induce targeted DSBs (Durai et al. 2005).

TALENs are artificially engineered nuclease created by fusing a TAL effector DNA-binding domain to a DNA cleavage domain that cleaves target DNA in a nonspecific manner (Joung and Sander 2013). TALENs operate in a similar way ZFNs does, but each domain in TALENs recognizes a single base instead of DNA triplets. TALEs are naturally occurring protein from pathogenic bacteria of genus *Xanthomonas* and are 33–34 amino acid tandem repeats of motifs that contain two variable positions which determine the specificity of nucleic acid recognition (Deng et al. 2014).

CRISPR/Cas9 system has emerged as a potential tool in comparison to ZFNs and TALENs for targeted genome editing (Upadhyay et al. 2013; Xiong et al. 2015; Zhang et al. 2016). The CRISPR/Cas system has been adopted from bacterial RNA-guided immune system against invading foreign DNA (Wiedenheft et al. 2012). The CRISPR/Cas is categorized into three different types, but among them type II is the simplest one (Bhaya et al. 2011). CRISPR/Cas module requires a CRISPR RNA (crRNA) to recognize target DNA and a Cas9 nuclease to create a double-stranded break. The only requirement of CRISPR/Cas tool is a protospacer adjacent motif (PAM) sequence, i.e., NGG downstream of target (Jinek et al. 2012). CRISPR/Cas has enormous potential for crop improvement for a high-quality product due to its ease of engineering and is affordable. The tools discussed above are being applied for understanding the function of unknown gene toward improving quality, resistance to pathogen, engineering pathways, etc.

5.7 Perspectives

Genetic transformation of plants has been achieved by numerous transformation techniques such as electroporation, PEG-mediated protoplast transformation, and nanoparticle-mediated and biolistic method in the past three decades. However, mostly *Agrobacterium*-mediated transformations are now regular and frequently used in various labs due to low cost and ease of availability. Recent advancement in plant biotechnology explored different genetic and physical factors which influence T-DNA transfer and integration by *Agrobacterium*. Regardless of these progresses, *Agrobacterium*-mediated genetic transformation of various plants species is a challenge due to its highly recalcitrant nature (Barampuram and Zhang 2011). The transformation and regeneration efficiency of these recalcitrant plant species can be increased by optimizing various factors affecting *Agrobacterium*-mediated transformation. The use of selectable marker gene during genetic transformation and selection is not acceptable due to safety regulation; therefore, marker-free transgenic plants will be more acceptable in the future. Hairy root induction and culture of some plant species by *A. rhizogenes* are very useful in production of secondary metabolite and recombinant proteins (i.e., molecular farming). It is also useful in phytoremediation and healthcare industry (Makhzoum et al. 2013; Pala et al. 2016). *Agrobacterium*-mediated chloroplast transformation is a technology where gene of interest is integrated in chloroplast genome and frequently used for molecular farming (Fuentes et al. 2016). RNAi-mediated gene suppression approach has become a highly valuable tool for functional genomics of plants. These approaches have been used for crop improvement, knockdown targeted gene, and generation of biotic (insect/pest/pathogen) and abiotic (heat/drought/salt) resistance transgenic plants. However, RNAi technology has some limitation, and therefore genome editing tools emerged. ZFN-, TALEN-, and CRISPR/Cas-mediated technology are used to edit targeted gene permanently. But, among these, CRISPR/Cas is very recent and efficient tool for genome editing. Advancements in genetic engineering have also led to the development of gene editing and gene correction via CRISPR/Cas. Marker-free genetically modified (GM) plants can be achieved easily through CRISPR/Cas technology, and hence these will be more acceptable in the future (Zhang et al. 2016). Large DNA sequences, including multiple genes, could be introduced into plant genome or mutated by these overexpression and knockout tools via *Agrobacterium*. Therefore, advances in plant genetic engineering would afford a solution for production of improved crop and medicinal plant species to meet the needs of increasing world population. Thus, it is expected that in the near future, GM plants with minimal genomic modifications can be developed and released.

Acknowledgments The authors thank Executive Director, National Agri-Food Biotechnology Institute (NABI), Director, National Institute Of Plant Genome Research (NIPGR), and Jaypee University of Information Technology (JUIT) for facilitating the research work. A.A. and J.K. acknowledge CSIR, for providing fellowship. S.V. is thankful to DBT for granting fellowship.

References

Agrawal N, Dasaradhi PVN, Mohmmed A, Malhotra P, Bhatnagar RK, Mukherjee SK (2003) RNA interference: biology, mechanism, and applications. Microbiol Mol Biol Rev 67:657–685. doi:10.1128/MMBR.67.4.657-685.2003

Alok A, Shukla V, Pala Z et al (2016) *In vitro* regeneration and optimization of factors affecting agrobacterium mediated transformation in Artemisia Pallens, an important medicinal plant. Physiol Mol Biol Plants 22:261–269. doi:10.1007/s12298-016-0353-3

Barampuram S, Zhang ZJ (2011) Recent advances in plant transformation. Methods Mol Biol 701:1–35. doi:10.1007/978-1-61737-957-4_1

Bechtold N, Pelletier G (1997) In planta *Agrobacterium* mediated transformation of adult *Arabidopsis thaliana* plants by vacuum infiltration. Methods Mol Biol 82:259–266. doi:10.1385/0-89603-391-0:259

Bent AF (2000) Arabidopsis in planta transformation uses, mechanisms, and prospects for transformation of other species. Plant Physiol 124:1540–1547. doi:10.1104/pp.124.4.1540

Bevan M (1984) Binary *Agrobacterium* vectors for plant transformation. Nucleic Acids Res 12:8711–8721

Beyer P, Al-Babili S, Ye X, Lucca P, Schaub P, Welsch R, Potrykus I (2002) Golden rice: introducing the beta-carotene biosynthesis pathway into rice endosperm by genetic engineering to defeat vitamin A deficiency. J Nutr 132:506S–5510

Bhalothia P, Alok A, Mehrotra S, Mehrotra R (2013) AACA element negatively regulates expression of protein phosphatase 2C (*PP2C*) like promoter in *Arabidopsis thaliana*. Am J Plant Sci 4:549–554. doi:10.4236/ajps.2013.43071

Bhalothia P, Sangwan C, Alok A, Mehrotra S, Mehrotra R (2016) PP2C-like promoter and its deletion variants are induced by ABA but not by MeJA and SA in *Arabidopsis thaliana*. Front Plant Sci 7:547. doi:10.3389/fpls.2016.00547

Bhati KK, Alok A, Kumar A, Kaur J, Tiwari S, Pandey AK (2016) Silencing of ABCC13 transporter in wheat reveals its involvement in grain development, phytic acid accumulation and lateral root formation. J Exp Bot 67:4379–4389. doi:10.1093/jxb/erw224

Bhaya D, Davison M, Barrangou R (2011) CRISPR-Cas systems in bacteria and archaea: versatile small RNAs for adaptive defense and regulation. Annu Rev Genet 45:273–297. doi:10.1146/annurev-genet-110410-132430

Chandra S (2012) Natural plant genetic engineer *Agrobacterium rhizogenes*: role of T-DNA in plant secondary metabolism. Biotechnol Lett 34:407–415. doi:10.1007/s10529-011-0785-3

Chen K, Gao C (2014) Targeted genome modification technologies and their applications in crop improvements. Plant Cell Rep 33:575–583. doi:10.1007/s00299-013-1539-6

Chen Q-J, Zhou H-M, Chen J, Wang X-C (2006) A gateway-based platform for multigene plant transformation. Plant Mol Biol 62:927–936. doi:10.1007/s11103-006-9065-3

Christie PJ (2004) Type IV secretion: the *Agrobacterium* VirB/D4 and related conjugation systems. Biochim Biophys Acta 1694:219–234. doi:http://dx.doi.org/10.1016/j.bbamcr.2004.02.013

Clough SJ, Bent AF (1998) Floral dip: a simplified method for *Agrobacterium*-mediated transformation of *Arabidopsis thaliana*. Plant J 16:735–743. doi:10.1046/j.1365-313x.1998.00343.x

Cruz-Mendívil A, Rivera-López J, Germán-Báez LJ, López-Meyer M, Hernández-Verdugo S, López-Valenzuela JA, Reyes-Moreno C, Valdez-Ortiz A (2011) A simple and efficient protocol for plant regeneration and genetic transformation of tomato cv. micro-tom from leaf explants. Hortscience 46:1655–1660

Dafny-Yelin M, Chung S-M, Frankman EL, Tzfira T (2007) pSAT RNA interference vectors: a modular series for multiple gene down-regulation in plants. Plant Physiol 145:1272–1281. doi:10.1104/pp.107.106062

Dafny-Yelin M, Tzfira T (2007) Delivery of multiple transgenes to plant cells. Plant Physiol 145:1118–1128. doi:10.1104/pp.107.106104

Daley M, Knauf VC, Summerfelt KR, Turner JC (1998) Co-transformation with one *Agrobacterium tumefaciens* strain containing two binary plasmids as a method for producing marker-free transgenic plants. Plant Cell Rep 17:489–496. doi:10.1007/s002990050430

Deblaere R, Bytebier B, De Greve H, Deboeck F, Schell J, Van Montagu M, Leemans J (1985) Efficient octopine Ti plasmid-derived vectors for *Agrobacterium*-mediated gene transfer to plants. Nucleic Acids Res 13:4777–4788

Deng D, Yan C, Wu J, Pan X, Yan N (2014) Revisiting the TALE repeat. Protein Cell 5:297–306. doi:10.1007/s13238-014-0035-2

Durai S, Mani M, Kandavelou K, Wu J, Porteus MH, Chandrasegaran S (2005) Zinc finger nucleases: custom-designed molecular scissors for genome engineering of plant and mammalian cells. Nucleic Acids Res 33:5978–5990. doi:10.1093/nar/gki912

Dutt M, Dhekney SA, Soriano L, Kandel R, Grosser JW (2014) Temporal and spatial control of gene expression in horticultural crops. Hortic Res 1:14047

Egnin M, Mora A, Prakash C (1998) Factors enhancing *Agrobacterium tumefaciens*-mediated gene transfer in peanut (*Arachis hypogaea* L.) In Vitro Cell Dev Biol Plant 34:310–318. doi:10.1007/BF02822740

Emami S, Yee M, Dinneny JR (2013) A robust family of golden gate *Agrobacterium* vectors for plant synthetic biology. Front Plant Sci 4:339. doi:10.3389/fpls.2013.00339

Estornell LH, Orzáez D, López-Peña L, Pineda B, Antón MT, Moreno V, Granell A (2009) A multisite gateway-based toolkit for targeted gene expression and hairpin RNA silencing in tomato fruits. Plant Biotechnol J 7:298–309. doi:10.1111/j.1467-7652.2009.00402.x

Feldmann K, David Marks M (1987) *Agrobacterium*-mediated transformation of germinating seeds of *Arabidopsis thaliana*: a non-tissue culture approach. Mol Gen Genet 208:1–9. doi:10.1007/BF00330414

Fuentes P, Zhou F, Erban A, Karcher D, Kopka J, Bock R (2016) A new synthetic biology approach allows transfer of an entire metabolic pathway from a medicinal plant to a biomass crop. elife 5:pii: e13664. doi:10.7554/eLife.13664

Fraley RT, Rogers SG, Horsch RB et al (1983) Expression of bacterial genes in plant cells. Proc Natl Acad Sci U S A 80:4803–4807

Gaj T, Gersbach CA, Barbas CF (2013) ZFN, TALEN and CRISPR/Cas-based methods for genome engineering. Trends Biotechnol 31:397–405. doi:10.1016/j.tibtech.2013.04.004

Gallois P, Marinho P (1995) Leaf disk transformation using *Agrobacterium tumefaciens*-expression of heterologous genes in tobacco. Methods Mol Biol 49:39–48. doi:10.1385/0-89603-321-X:39

Gasparis S, Orczyk W, Zalewski W, Nadolska-Orczyk A (2011) The RNA-mediated silencing of one of the pin genes in allohexaploid wheat simultaneously decreases the expression of the other, and increases grain hardness. J Exp Bot 62:4025–4036. doi:10.1093/jxb/err103

Gelvin SB (2012) Traversing the cell: *Agrobacterium* T-DNA's journey to the host genome. Front Plant Sci 3:52. doi:10.3389/fpls.2012.00052

Gelvin SB (2003) *Agrobacterium*-mediated plant transformation: the biology behind the "Gene-jockeying" tool. Microbiol Mol Biol Rev 67:16–37. doi:10.1128/MMBR.67.1.16-37.2003

Gilchrist E, Haughn G (2010) Reverse genetics techniques: engineering loss and gain of gene function in plants. Brief Funct Genomics 9:103–110. doi:10.1093/bfgp/elp059

Goderis IWM, De Bolle MC, François IJA et al (2002) A set of modular plant transformation vectors allowing flexible insertion of up to six expression units. Plant Mol Biol 50:17–27. doi:10.1023/A:1016052416053

Hajdukiewicz P, Svab Z, Maliga P (1994) The small, versatile pPZP family of *Agrobacterium* binary vectors for plant transformation. Plant Mol Biol 25:989–994. doi:10.1007/BF00014672

He Y, Jones HD, Chen S, Chen XM, Wang DW, Li KX, Wang DS, Xia LQ (2010) *Agrobacterium*-mediated transformation of durum wheat (*Triticum turgidum* L. var. durum cv Stewart) with improved efficiency. J Exp Bot 61:1567–1581. doi:10.1093/jxb/erq035

Hellens R, Edwards EA, Leyland N, Bean S, Mullineaux PM (2000) pGreen: a versatile and flexible binary Ti vector for *Agrobacterium*-mediated plant transformation. Plant Mol Biol 42:819–832. doi:10.1023/A:1006496308160

Hellens R, Mullineaux P, Klee H (2015) Technical focus: a guide to *Agrobacterium* binary Ti vectors. Trend Plant Sci 5:446–451. doi:10.1016/S1360-1385(00)01740-4

Hepburn AG, White J, Pearson L, Maunders MJ, Clarke LE, Prescott AG, Blundy KS (1985) The use of pNJ5000 as an intermediate vector for the genetic manipulation of *Agrobacterium* Ti-plasmids. J Gen Microbiol 131:2961–2969. doi:10.1099/00221287-131-11-2961

Himmelbach A, Zierold U, Hensel G, Riechen J, Douchkov D, Schweizer P, Kumlehn J (2007) A set of modular binary vectors for transformation of cereals. Plant Physiol 145:1192–1200. doi:10.1104/pp.107.111575

Hoekema A, Hirsch PR, Hooykaas PJJ, Schilperoort RA (1983) A binary plant vector strategy based on separation of vir- and T-region of the *Agrobacterium tumefaciens* Ti-plasmid. Nature 303:179–180. doi:10.1038/303179a0

Holsters M, Silva B, Van Vliet F, Genetello C, De Block M, Dhaese P, Depicker A, Inzé D, Engler G, Villarroel R et al (1980) The functional organization of the nopaline *A. tumefaciens* plasmid pTiC58. Plasmid 3:212–230. doi:10.1016/0147-619X(80)90110-9

Hood E, Gelvin S, Melchers L, Hoekema A (1993) New *Agrobacterium* helper plasmids for gene transfer to plants. Transgenic Res 2:208–218. doi:10.1007/BF01977351

Hood EE, Fraley RT, Chilton M-D (1987) Virulence of *Agrobacterium tumefaciens* strain A281 on legumes. Plant Physiol 83:529–534. doi:10.1104/pp.83.3.529

Hood EE, Helmer GL, Fraley RT, Chilton MD (1986) The hypervirulence of *Agrobacterium tumefaciens* A281 is encoded in a region of pTiBo542 outside of T-DNA. J Bacteriol 168:1291–1301

Hou H, Atlihan N, Z-X L (2014) New biotechnology enhances the application of cisgenesis in plant breeding. Front Plant Sci 5:389. doi:10.3389/fpls.2014.00389

Igarashi A, Yamagata K, Sugai T, Takahashi Y, Sugawara E, Tamura A, Yaegashi H, Yamagishi N, Takahashi T, Isogai M, Takahashi H, Yoshikawa N (2009) Apple latent spherical virus vectors for reliable and effective virus-induced gene silencing among a broad range of plants including tobacco, tomato, *Arabidopsis thaliana*, cucurbits, and legumes. Virology 386:407–416. doi:10.1016/j.virol.2009.01.039

Ishida Y, Hiei Y, Komari T (2007) *Agrobacterium*-mediated transformation of maize. Nat Protoc 2:1614–1621. doi:10.1038/nprot.2007.241

Jinek M, Chylinski K, Fonfara I, Hauer M, Doudna JA, Charpentier E (2012) A programmable dual-RNA guided DNA endonuclease in adaptive bacterial immunity. Science 337:816–822. doi:10.1126/science.1225829

Jones HD, Doherty A, Wu H (2005) Review of methodologies and a protocol for the *Agrobacterium*-mediated transformation of wheat. Plant Methods 1:5. doi:10.1186/1746-4811-1-5

Joung JK, Sander JD (2013) TALENs: a widely applicable technology for targeted genome editing. Nat Rev Mol Cell Biol 14:49–55. doi:10.1038/nrm3486

Karthikeyan A, Pandian SK, Ramesh M (2011) *Agrobacterium*-mediated transformation of leaf base derived callus tissues of popular indica rice (*Oryza sativa* L. sub sp. indica cv. ADT 43). Plant Sci 181:258–268. doi:10.1016/j.plantsci.2011.05.011

Hiei Y, Ohta S, Komari T, Kumashiro T (1994) Efficient transformation of rice (*Oryza sativa* L.) mediated by *Agrobacterium* and sequence analysis of the boundaries of the T-DNA. Plant J 6:271–282. doi:10.1046/j.1365-313X.1994.6020271.x

Komari T, Takakura Y, Ueki J, Komari T, Kato N, Ishida Y, Hiei Y (2006) Binary vectors and super-binary vectors. Methods Mol Biol 343:15–42. doi:10.1104/pp.107.105734

Koncz C, Martini N, Mayerhofer R, Koncz-Kalman Z, Körber H, Redei GP, Schell J (1989) High-frequency T-DNA-mediated gene tagging in plants. Proc Natl Acad Sci U S A 86:8467–8471

Koncz C, Schell J (1986) The promoter of TL-DNA gene 5 controls the tissue-specific expression of chimaeric genes carried by a novel type of *Agrobacterium* binary vector. Mol Gen Genet 204:383–396. doi:10.1007/BF00331014

Lazo GR, Stein PA, Ludwig RA (1991) A DNA transformation-competent *Arabidopsis* genomic library in *Agrobacterium*. Nat Biotechnol 9:963–967. doi:10.1038/nbt1091-963

Lu R (2003) Virus-induced gene silencing in plants. Methods 30:296–303. doi:10.1016/S1046-2023(03)00037-9

Makhzoum A, Sharma P, Bernards M, Trémouillaux-Guiller J (2013) Hairy roots: an ideal platform for transgenic plant production and other promising applications. In: Gang DR (ed) Phytochemicals, plant growth, and the environment. Springer, New York, pp 95–142. doi:10.1007/978-1-4614-4066-6_6

Mann DG, Fayette PR, Abercrombie LL, Abercrombie LL, King ZR, Mazarei M, Halter MC, Poovaiah CR, Baxter H, Shen H, Dixon RA, Parrott WA, Neal Stewart CJ (2012) Gateway-compatible vectors for high-throughput gene functional analysis in switchgrass (*Panicum virgatum* L.) and other monocot species. Plant Biotechnol J 10:226–236. doi:10.1111/j.1467-7652.2011.00658.x

Martins PK, Nakayama TJ, Ribeiro AP, da Cunha BA, Nepomuceno AL, Harmon FG, Kobayashi AK, Molinari HB (2015) Setaria viridis floral-dip: a simple and rapid Agrobacterium-mediated transformation method. Biotechnology Reports 6:61–3

Marton I, Zuker A, Shklarman E, Zeevi V, Tovkach A, Roffe S, Ovadis M, Tzfira T, Vainstein A (2010) Nontransgenic genome modification in plant cells. Plant Physiol 154:1079–1087. doi:10.1104/pp.110.164806

McBride K, Summerfelt K (1990) Improved binary vectors for *Agrobacterium*-mediated plant transformation. Plant Mol Biol 14:269–276. doi:10.1007/BF00018567

McKnight T, Lillis M, Simpson R (1987) Segregation of genes transferred to one plant cell from two separate *Agrobacterium* strains. Plant Mol Biol 8:439–445. doi:10.1007/BF00017989

Miki D (2004) Simple RNAi vectors for stable and transient suppression of gene function in rice. Plant Cell Physiol 45:490–495. doi:10.1093/pcp/pch048

Napoli C, Lemieux C, Jorgensen R (1990) Introduction of a chimeric chalcone synthase gene into petunia results in reversible co-suppression of homologous genes in trans. Plant Cell 2:279–289. doi:10.1105/tpc.2.4.279

Nemudryi AA, Valetdinova KR, Medvedev SP, Zakian SM (2014) TALEN and CRISPR/Cas genome editing systems: tools of discovery. Acta Nat 6:19–40

Olszewski NE, Martin FB, Ausubel FM (1988) Specialized binary vector for plant transformation: expression of the *Arabidopsis thaliana* AHAS gene in *Nicotiana tabacum*. Nucleic Acids Res 16:10765–10782. doi:10.1093/nar/16.22.10765

Pabo CO, Peisach E, Grant RA (2001) Design and selection of novel Cys2His2 zinc finger proteins. Annu Rev Biochem 70:313–340. doi:10.1146/annurev.biochem.70.1.313

Pala Z, Shukla V, Alok A, Kudale S, Desai N (2016) Enhanced production of an anti-malarial compound artesunate by hairy root cultures and phytochemical analysis of Artemisia pallens Wall. 3 Biotech 6(2):182

Palanichelvam K, Oger P, Clough SJ et al (2000) A second T-region of the soybean-supervirulent chrysopine-type Ti plasmid pTiChry5, and construction of a fully disarmed vir helper plasmid. Mol Plant-Microbe Interact 13:1081–1091. doi:10.1094/MPMI.2000.13.10.1081

Pandey A, Misra P, Bhambhani S, Bhatia C, Trivedi PK (2014a) Expression of *Arabidopsis* MYB transcription factor, *AtMYB111*, in tobacco requires light to modulate flavonol content. Sci Rep 4:5018. doi:10.1038/srep05018

Pandey A, Misra P, Khan MP, Swarnkar G, Tewari MC, Bhambhani S, Trivedi R, Chattopadhyay N, Trivedi PK (2014b) Co-expression of *Arabidopsis* transcription factor, AtMYB12, and soybean isoflavone synthase, GmIFS1, genes in tobacco leads to enhanced biosynthesis of isoflavones and flavonols resulting in osteoprotective activity. Plant Biotechnol J 12:69–80. doi:10.1111/pbi.12118

Pitzschke A, Hirt H (2010) New insights into an old story: *Agrobacterium*-induced tumour formation in plants by plant transformation. EMBO J 29:1021–1032. doi:10.1038/emboj.2010.8

Ramana Rao MV, Parameswari C, Sripriya R, Veluthambi K (2011) Transgene stacking and marker elimination in transgenic rice by sequential *Agrobacterium*-mediated co-transformation with the same selectable marker gene. Plant Cell Rep 30:1241–1252. doi:10.1007/s00299-011-1033-y

Risacher T, Craze M, Bowden S, Paul W, Barsby T (2009) Highly efficient Agrobacterium-mediated transformation of wheat via in planta inoculation. Methods Mol Biol 478:115–124. doi:10.1007/978-1-59745-379-0_7

Reuhs BL, Kim JS, Matthysse AG (1997) Attachment of *Agrobacterium tumefaciens* to carrot cells and Arabidopsis wound sites is correlated with the presence of a cell-associated, acidic polysaccharide. J Bacteriol 179:5372–5379

Remakanthan A, Menon TG, Soniya E V (2014) Somatic embryogenesis in banana (Musa acuminata AAA cv. Grand Naine): effect of explant and culture conditions. In Vitro Cell Dev Biol Plant 50:127–136. doi:10.1007/s11627-013-9546-4

Sivakumar G, Liu C, Towler MJ, Weathers PJ (2010) Biomass production of hairy roots of *Artemisia annua* and *Arachis hypogaea* in a scaled-up mist bioreactor. Biotechnol Bioeng 107:802–813. doi:10.1002/bit.22892

Sood P, Bhattacharya A, Sood A (2011) Problems and possibilities of monocot transformation. Biol Plant 55:1–15. doi:10.1007/s10535-011-0001-2

Stachel SE, Nester EW, Zambryski PC (1986) A plant cell factor induces *Agrobacterium tumefaciens* vir gene expression. Proc Natl Acad Sci U S A 83:379–383

Subramoni S, Nathoo N, Klimov E, Yuan Z-C (2014) *Agrobacterium tumefaciens* responses to plant-derived signaling molecules. Front Plant Sci 5:322. doi:10.3389/fpls.2014.00322

Sugita K, Kasahara T, Matsunaga E, Ebinuma H (2000) A transformation vector for the production of marker-free transgenic plants containing a single copy transgene at high frequency. Plant J 22(5):461–9

Sun S, Kang X-P, Xing X-J, X-Y X, Cheng J, Zheng S-W, Xing G-M (2015) *Agrobacterium-mediated* transformation of tomato (*Lycopersicon esculentum* L. cv. Hezuo 908) with improved efficiency. Biotechnol Biotechnol Equip 29:861–868. doi:10.1080/13102818.2015.1056753

Tiwari S, Mishra D, Singh A et al (2008) Expression of a synthetic cry1EC gene for resistance against *Spodoptera litura* in transgenic peanut (*Arachis hypogaea* L.) Plant Cell Rep 27:1017–1025. doi:10.1007/s00299-008-0525-x

Tiwari S, Tuli R (2012) Optimization of factors for efficient recovery of transgenic peanut (*Arachis hypogaea* L.) Plant Cell Tissue Organ Cult 109:111–121. doi:10.1007/s11240-011-0079-4

Tripathi JN, Oduor R, Tripathi L (2015) A high-throughput regeneration and transformation platform for production of genetically modified banana. Front Plant Sci 6:1025. doi:10.3389/fpls.2015.01025

Upadhyay SK, Kumar J, Alok A, Tuli R (2013) RNA-guided genome editing for target gene mutations in wheat. G3 (Bethesda, MD) 3:2233–2238. doi:10.1534/g3.113.008847

Van der Krol AR, Mur LA, Beld M, Mol JN, Stuitje AR (1990) Flavonoid genes in petunia: addition of a limited number of gene copies may lead to a suppression of gene expression. Plant Cell 2:291–299. doi:10.1105/tpc.2.4.291

Vladimirov IA, Matveeva T V, Lutova LA (2015) Opine biosynthesis and catabolism genes of *Agrobacterium tumefaciens and Agrobacterium rhizogenes*. Russ J Genet 51:121–129. doi:10.1134/S1022795415020167

Watson B, Currier TC, Gordon MP et al (1975) Plasmid required for virulence of *Agrobacterium tumefaciens*. J Bacteriol 123:255–264

Weir B, Gu X, Wang M et al (2001) *Agrobacterium tumefaciens*-mediated transformation of wheat using suspension cells as a model system and green fluorescent protein as a visual marker. Funct Plant Biol 28:807–818. doi:10.1071/PP99211

Wiedenheft B, Sternberg SH, Doudna JA (2012) RNA-guided genetic silencing systems in bacteria and archaea. Nature 482:331–338. doi:10.1038/nature10886

Wu H, McCormac A, Elliott M, Chen D-F (1998) *Agrobacterium*-mediated stable transformation of cell suspension cultures of barley (*Hordeum vulgare*). Plant Cell Tissue Organ Cult 54:161–171. doi:10.1023/A:1006106916542

Xiang C, Han P, Lutziger I et al (1999) A mini binary vector series for plant transformation. Plant Mol Biol 40:711–717. doi:10.1023/A:1006201910593

Xie K, Yang Y (2013) RNA-guided genome editing in plants using a CRISPR–Cas System. Mol Plant 6:1975–1983. doi:10.1093/mp/sst119

Xing HL, Dong L, Wang ZP, Zhang HY, Han CY, Liu B, Wang XC, Chen Q (2014) A CRISPR/Cas9 toolkit for multiplex genome editing in plants. BMC Plant Biol 14:327. doi:10.1186/s12870-014-0327-y

Xiong JS, Ding J, Li Y (2015) Genome-editing technologies and their potential application in horticultural crop breeding. Hortic Res 2:15019. doi:10.1038/hortres.2015.19

Yan P, Shen W, Gao X, Li X, Zhou P, Duan J (2012) High-throughput construction of intron-containing hairpin RNA vectors for RNAi in plants. PLoS One 7:e38186. doi:10.1371/journal.pone.0038186

Zalewski W, Orczyk W, Gasparis S, Nadolska-Orczyk A (2012) HvCKX2 gene silencing by biolistic or *Agrobacterium*-mediated transformation in barley leads to different phenotypes. BMC Plant Biol 12:206. doi:10.1186/1471-2229-12-206

Zambryski P, Joos H, Genetello C et al (1983) Ti plasmid vector for the introduction of DNA into plant cells without alteration of their normal regeneration capacity. EMBO J 2:2143–2150

Zamore PD, Tuschl T, Sharp PA, Bartel DP (2000) RNAi: double-stranded RNA directs the ATP-dependent cleavage of mRNA at 21 to 23 nucleotide intervals. Cell 101:25–33. doi:10.1016/S0092-8674(00)80620-0

Ziemienowicz A (2014) *Agrobacterium*-mediated plant transformation: factors, applications and recent advances. Biocatal Agric Biotechnol 3:95–102. doi:http://dx.doi.org/10.1016/j.bcab.2013.10.004

Zhang Y, Liang Z, Zong Y, Wang Y, Liu J, Chen K, Qiu JL, Gao C (2016) Efficient and transgene-free genome editing in wheat through transient expression of CRISPR/Cas9 DNA or RNA. Nat Commun 7:12617. doi:10.1038/ncomms12617

Zupan J, Muth TR, Draper O, Zambryski P (2000) The transfer of DNA from *Agrobacterium tumefaciens* into plants: a feast of fundamental insights. Plant J 23:11–28. doi:10.1046/j.1365-313x.2000.00808.x

Gaining Insight into Plant Responses to Beneficial and Pathogenic Microorganisms Using Metabolomic and Transcriptomic Approaches

6

Fernando Matías Romero, María Marina, Fernando Luis Pieckenstain, Franco Ruben Rossi, María Elisa Gonzalez, Paulina Vignatti, and Andrés Gárriz

Abstract

Plants are constantly interacting with microorganisms. Many of them have the potential to cause disease, while many other may establish beneficial interactions where plants enhance their ability to incorporate important nutrients and improve disease resistance. During these interactions, plants must regulate the expression of thousands of genes, which ultimately triggers distinct hormonal signaling pathways and affects the concentration of numerous metabolites. Transcriptomics and metabolomics have played a pivotal role in identifying the genes and metabolites involved in such responses, which has given crucial hints to refine our current strategies for plant protection and crop yield improvement. However, there is still a gap on our knowledge on many features that distinguish the interplay between plants and microorganisms. This chapter initially discusses the contributions of these high-throughput technologies to the understanding of this field of research and ends with future prospects in the search for interaction-specific biomarker genes and metabolites.

Keywords

Transcriptomics • Metabolomics • Plant-microobe interactions

F.M. Romero • M. Marina • F.L. Pieckenstain • F.R. Rossi
M.E. Gonzalez • P. Vignatti • A. Gárriz (✉)
Laboratorio de Interacciones Planta-microorganismo, Instituto de Investigaciones Biotecnológicas-Instituto Tecnológico de Chascomús/Universidad Nacional de General San Martín-Consejo Nacional de Investigaciones Científicas y Técnicas (IIB-INTECH/UNSAM-CONICET), Intendente Marino KM 8.200, Chascomús, Buenos Aires (CP 7130), Argentina
e-mail: garriz@intech.gov.ar

V.C. Kalia, A.K. Saini (eds.), *Metabolic Engineering for Bioactive Compounds*,
DOI 10.1007/978-981-10-5511-9_6

6.1 Introduction

The complexity inherent to plant-microorganism interactions is utterly stunning. These biological associations may present beneficial effects for the plant, such as those established with plant growth-promoting microorganisms (PMPMs), or rather act in detriment of plant fitness (e.g., pathogens) (Asai and Shirasu 2015). The study of this field has enormous implications in the development of the so-called green biotechnology, the combination of technological approaches tailored to improve crop production and quality.

As a consequence of the close contact with a multitude of microorganisms, plants have developed effective strategies to prevent the invasion by potential pathogens (Bernoux et al. 2011; Asai and Shirasu 2015). A first line of defense is exerted by a set of preformed chemical and physical barriers. For instance, the cuticle and cell wall thickness are crucial for plant defense, as well as the shape and size of the plant stomata, which may avoid the entrance of some microorganisms using these natural openings to invade plant tissues. In turn, many low molecular weight compounds collectively known as phytoanticipins and proteins classified as defensins are constantly synthesized to inhibit the growth and development of microorganisms (Tam et al. 2015; Pedras and Yaya 2015). In addition, plants have developed a bunch of surveillance systems to detect the presence of pathogens in the early stages of the infection and consequently trigger additional defensive responses. One of these systems requires the action of pattern recognition receptors (PRR), which are localized at the cell surface and detect foreign molecules associated with the pathogen, or pathogen-associated molecular patterns (PAMPs), and compounds derived from the degradation of self-structures, or damage-associated molecular patterns (DAMPs). As PAMPs and DAMPs are quite conserved among microbes, the defenses triggered following PAMP and DAMP detection (known as PTI for PAMP-triggered immunity) usually repel a broad spectrum of pathogens. In turn, many pathogens are able to overcome PTI by secreting effectors, some of which are injected directly into host plant cells and contribute to attenuate or completely block these initial plant defense responses. However, during evolution plants have developed various cytoplasmic proteins that detect the presence (or the activity) of effectors, an event that triggers the so-called effector-triggered immunity (ETI). This response simply reinforces previously activated defenses and also induces the activation of newly ones. Contrary to the multigenic nature of PTI, ETI requires the presence of one (or a few) particular gene in the plant and the microorganism, and it sometimes involves the activation of the hypersensitive response (HR), a type of programmed cell death in the host cells surrounding the infection site aimed to restrict the spreading of the disease. The evolution of microbial effectors and plant receptors is under constant selection pressure, which has originated disease cycles characterized by the appearance of novel pathogens overcoming ETI and the consequent resurgence of plant genotypes able to recognize and fight back against them (Dodds and Rathjen 2010). Several reviews have covered in detailed the functionality of plant immunity mechanisms and the forces driving plant defense evolution (Anderson et al. 2010; Dodds and Rathjen 2010; Bernoux et al. 2011; Asai and Shirasu 2015).

Both PTI and ETI involve the activation of an intricate network of signaling pathways where a large number of genes must be regulated in a very precise manner. Gene regulation leads to a tight control on the concentration and activity of metabolites and proteins. The advent of the so-called "omics" platforms has fostered a great progress in the identification of the most important components underlying successful plant defense responses. The term "omics" informally refers to different high-throughput techniques intended to characterize and quantify the entire pool of a kind of molecules in individual cells, tissues, or organisms under a given set of conditions. For instance, transcriptomic approaches provide information about overall gene expression levels simultaneously, while proteomics and metabolomics do it for proteins and metabolites, respectively. The analytical challenge associated to any of these approaches was one of the main setbacks that researchers faced not so long ago, as they usually required complicated, time-consuming, and expensive methods. Fortunately, this issue has been partially solved (and further advances should be expected) with the development of reliable and robust techniques that allow researchers to get an instantaneous snapshot of the entire collection of mRNAs, proteins, and metabolites in a relatively easy and inexpensive manner. Because they generate an enormous amount of data, perhaps the main challenge nowadays is our ability to analyze and interpret the results from these studies. In addition, as it is quite evident now that a single "omics" approach cannot decipher by itself the complexity pertaining cell physiology, the integration of multidimensional "omics" data has been proposed as an essential procedure (Zhang et al. 2010).

In this chapter, we first made a concise description of the most relevant techniques used in "omics" studies and later reviewed the general conclusions from transcriptomic and metabolomic works focused to unravel the molecular events occurring during the interactions between plants and microbes. Because of length restrictions, we will describe mostly those transcriptomic and metabolomic studies conducted to explore plant responses. However, readers should be aware that "omics" studies focused on microorganisms are also critical to have a fair comprehension of the interplay between both partners in the interaction.

6.1.1 Transcriptomic Platforms

The expression level of thousands of genes has been mostly assessed using two different methodologies, DNA microarrays (also known as DNA chips) and RNA-Seq technologies.

There are many protocols for DNA microarray-based studies. They basically require a first step of mRNA extraction from two or more samples under comparison, which must be then converted to cDNA by reverse transcriptase polymerase chain reaction (RT-PCR). Later, each sample is labeled using two distinct fluorochromes and mixed together previous to a hybridization step against individual DNA sequences spotted to a solid surface. Finally, the excess of non-hybridized cDNA is washed off and fluorescence determined by laser scanning. Relative fluorescence of each fluorochrome indicates whether the gene is up- or downregulated in the experimental sample with respect to the control (Schena et al. 1995). Depending on the platform being used, the DNA probes spotted to the solid surface may be double-stranded or 16–20 bp oligonucleotides (Lodha and Basak 2012). Importantly, the construction of the DNA chip requires the pre-existing knowledge of the genomic sequence of the organism under study.

Even though microarrays are at the top among the most used platforms for transcriptomic studies, the considerable improvements achieved in sequencing technologies in the last years have led to the development of novel sequence-based approaches. These methods have emerged as the dominant platforms and have revolutionized the transcriptional landscape. The most widely used of these new methodologies is RNA-Seq. In this method, isolated RNA is used to construct a double-stranded cDNA library, and each cDNA is later individually sequenced (Wang et al. 2009). There are several methods for massive sequencing, all requiring a first step of in vitro cloning and amplification of the individual cDNA strands (Qian et al. 2014). Reads can be counted, which allows to infer the level of expression of a defined gene. The main obstacle with RNA-Seq is the fact that sequencing produces millions of short sequences ranging from 25 to 450 bp, which constitute a serious bioinformatic challenge. Nevertheless, the advent of powerful software in the last years has accelerated the mapping and assembling of these sequences, thus speeding up data management and interpretation.

6.1.2 Metabolomics

Metabolomics is the combination of techniques that monitor the metabolome, that is, the pool of small organic molecules defining a biological sample. In a standard metabolomic procedure, metabolites are solvent extracted from biological samples and then detected and quantified using different chemical detection procedures. In this trend, mass spectrometry (MS) is a powerful tool for studying metabolites due to its sensitivity and flexibility for detection of different classes of molecules. Besides, in order to enhance the capability to unravel the complexity of biological extracts, MS has been coupled to chromatography, a combination known as a "metabolomic platform." The most widely used platforms are liquid chromatography-mass spectrometry (LC-MS) and gas chromatography-mass spectrometry (GC-MS). Although there are other platforms, like Fourier transform ion cyclotron resonance mass spectrometry (FTICR-MS) or nuclear magnetic resonance spectroscopy (NMR), they have some disadvantages that make them inappropriate to profile complex mixture of compounds (Kopka et al. 2004).

The great amount of data generated with MS-associated techniques makes the annotation of the detected molecular features the bottleneck in metabolomic studies (Wishart 2011). This has given rise to the development of two workflows: targeted and nontargeted metabolomics. Readers should be aware that nontargeted metabolomics is also referred as "untargeted," "unbiased," or "global" metabolomics (Heuberger et al. 2014). Targeted metabolomics is focused on the analysis of a particular set of metabolites, thus requiring the knowledge of the compounds of interest in advance. There are different ways to target the metabolomic profile to a certain group of molecules, such as the optimization of the extraction protocol and the use of solvents with different polarities and pH to maximize recovery of the desired compounds. This reduction in the complexity of the sample allows a better calibration of the spectrometer and significantly raises the sensitivity in the detection of a particular group of metabolites. In addition, it gathers data of metabolites with known masses and retention times and uses data analysis procedures that are relatively simple. Data normalization is conducted employing isotope-labeled internal standards for each compound of interest or a class of compounds. The main disadvantages of targeted workflows are the demand for authentic standards as well as the time, labor, and costs inherent to the optimization of these methods. On the other hand, nontargeted workflows are designed to globally profile all detectable metabolites in the sample. This is perhaps the main advantage of this approach, as it is able to detect not only the same metabolites than in a targeted design but it can also collect novel chemical identities. Therefore, in this case, the extraction and detection procedures must be optimized to include all different classes of metabolites. One of the main issues in nontargeted analyses is the identification and annotation of unknown compounds. This is relatively difficult as every molecule ionized in MS prior detection leads to several mass signals. One way to solve this problem is to cluster and cure this redundant data by computational procedures. Once the spectrometric data are cured, molecular features are annotated as metabolites by comparing data against different spectral databases. In those cases where a perfect match is not found, partial matches may involve molecules with similar structures.

In summary, targeted workflows require significant efforts to optimize the experiment, and they result in only the identification of a defined group of compounds. However, they produce less-complex data and allow more confident statistical analyses. On the other hand, nontargeted metabolomic workflows might lead to the discovery of novel compounds by using relatively simple analytical procedures; even so they require higher efforts in the analysis and interpretation of the resulting data.

6.2 High-Throughput Analysis of Plant-Microbe Interactions

The technologies described above have made an enormous contribution to biological research and accelerated in an impressive way our knowledge of the molecular events that govern plant-microbe interactions. They are becoming gradually less expensive, require less effort, and may be performed in almost any organism. Perhaps the main concern in future studies will be the interpretation of the extensive amount of data generated as well as the integration of results coming from different “omics” technologies. It is important to understand that plant responses may vary from a situation where a few cells are engaged in the interaction (as occurs generally with pathogens unable to overcome the first passive lines of defense) to a very different situation where cells are seriously affected (as in the case of compatible pathogens) (see Fig. 6.1). In addition, the responses change considerably over time, and samples may include a mosaic of cells

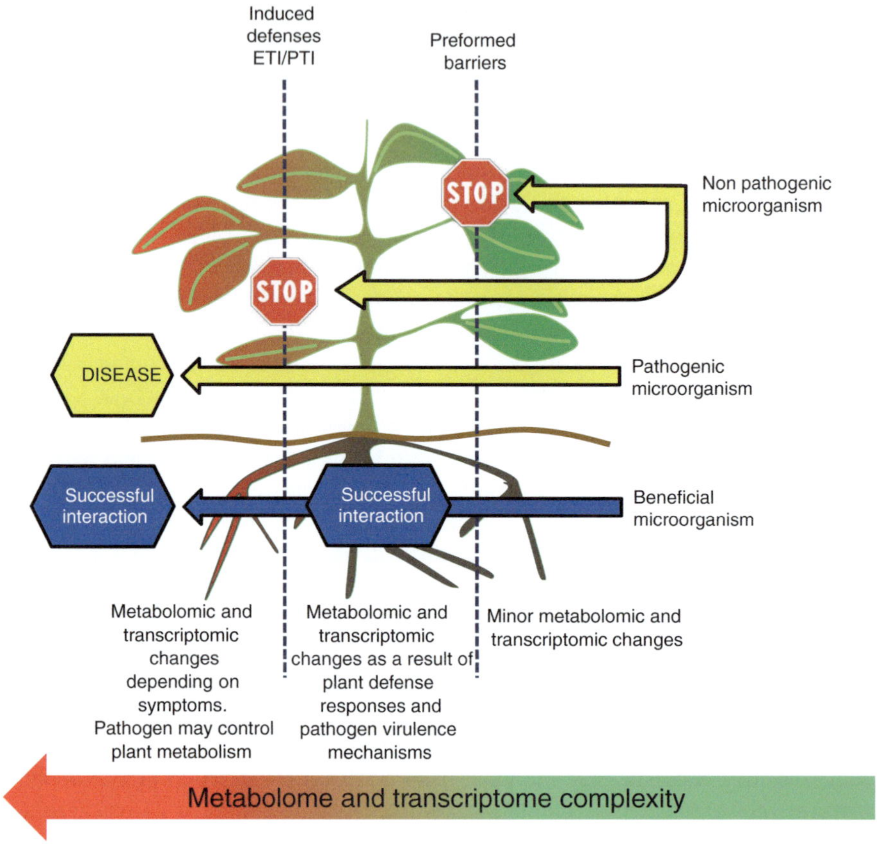

Fig. 6.1 Scheme depicting molecular modifications that could occur during the interactions between plants and microorganisms. Nonpathogenic microorganisms fail to overcome the first line of plant preformed barriers or induce responses associated to PTI and ETI. In turn, pathogens are able to attenuate or block these defensive mechanisms and establish the disease. Even though beneficial microorganisms evade preformed barriers, they may not necessarily induce full additional responses as those triggered by other invading microorganisms. Transcriptomic and metabolomic complexity is expected to grow proportionally to the perturbation on plant cell physiology

with different physiological status and belong to different organisms. Therefore, when interpreting omics data, researchers should consider the temporal evolution of plant-microbe interactions and take special care on the type of samples under analysis.

6.2.1 Transcriptomic Analysis of Plant Responses During Their Interactions with Microbes

6.2.1.1 Pathogenic Microorganisms

Currently, we have a very good idea of the main hallmarks of the transcriptomic landscape in plant responses to pathogens. However, it is rather problematic to identify common patterns of responses among all the plant species studied. This is so because the specific responses depend not only on the plant species but also on other factors as the interacting microorganisms, the type of interaction, and the plant organs involved. This limits our capability to extrapolate the results obtained from a particular plant-microbe interaction to another. Nevertheless, in spite of these setbacks, we are still able to look at the entire picture and make some broad conclusions that may be considered when working with related plant species. In this trend, a bunch of plant responses regulated at the transcription level might be expected to occur during any type of interaction, such as the production of antimicrobial compounds and the activation of upstream gene regulatory factors and common signaling cascades. There are, of course, appreciable quantitative as well as qualitative differences among all these responses, which ultimately decide the fate of the interaction.

Global transcriptomic analysis made a substantial contribution to tear down a long-standing paradigm on plant-pathogen interactions, the assumption that the defense responses mediated by salicylic acid (SA) protect the plant against biotrophic microorganisms (those colonizing and obtaining nutrients from living tissues), whereas those responses induced by hormones like jasmonic acid (JA) and ethylene (Et) operate against necrotrophic microorganisms (which kill plant cells and feed on them) (Glazebrook 2005). It was also proposed that both sets of activation mechanisms act in an antagonistic mode. However, gene expression profiling demonstrated that hormonal cross talk is not such a simple mechanism and exposed a very different scenario, in which significant overlapping occurs between both signaling networks (De Vos et al. 2005; Salzman et al. 2005; Garg et al. 2012; Okamoto et al. 2012). The work by Garg et al. (2012) gives a good example of this statement. These authors evaluated global gene transcription after treating rice seedlings with SA, JA, Et, or other phytohormones such as auxins, cytokinins, and abscisic acid (ABA). The range of genes being altered by these hormones ranged from a maximum of 3635 (ABA treatment) to a minimum of 183 (in seedlings treated with the ethylene precursor 1-aminocyclopropane-1-carboxylic acid, ACC), indicating that at least in their experimental conditions, the effects of plant hormones on gene regulation were significantly different. Nevertheless, 28% of these regulated genes were responsive to two or more treatments; most of them showed a similar response (upregulated or downregulated in all treatments), while a few showed contrasting responses (upregulated in a treatment and downregulated in others). The extension of the overlapping response was quite high in some cases, demonstrating that hormones may play crucial roles in the same response. For instance, 91% and 81% of the genes regulated by auxin were also responsive to SA and ABA, respectively. Importantly, the authors also showed a significant overlapping between the SA and JA responses. It should be noted that the similarities between these pathways might be even higher since hormones can regulate the same cellular processes by nonoverlapping transcriptional responses. Similar conclusions were drawn from global transcriptional studies in other model systems (Salzman et al. 2005; Nemhauser et al. 2006; Goda et al. 2008; Okamoto et al. 2012; Wang et al. 2013). Based on this, it is clear now that hormones work in a cooperative manner in the activation of a subset of defense mechanisms but may antagonize the activation of other responses. Plants must then interpret the combination of hormone-induced pathways to respond properly to pathogen attack. Moreover, transcriptomics is helping to uncover the link between novel molecules and some of defense mechanisms activated by plant hormones. For instance, microarray analysis of *Arabidopsis* lines with different levels of the polyamine spermine demonstrated that this compound regulates at the transcription level a broad spectrum of genes associated to biotic and abiotic stress (Gonzalez et al. 2011). Likewise, it was shown that potato plants treated with β-aminobutyric acid (BABA) are resistant against *Phytophthora infestans*, which might be related to its positive effect on the expression of PR proteins and phytoalexin biosynthesis (Bengtsson et al. 2014).

Global gene expression analysis also showed that plants shift their primary metabolism to a higher demand for energy and biosynthetic rate during defense (Scheideler et al. 2002). This seemed to contradict the observation that photosynthesis and chlorophyll biosynthesis are downregulated upon contact with pathogens (Bilgin et al. 2010), which was thought to preserve hard-earned metabolites to build up defense responses. However, more recent studies showed that plants facing the attack of a microorganism upregulate main primary metabolic pathways, such as those involved in the synthesis and degradation of carbohydrates, amino acids, and lipids (Rojas et al. 2014). By these means, plants are able to generate important energetic molecules and, by the same token, other signaling and active compounds that drive the defense responses. In particular, the induction of genes involved in glycolysis, the pentose phosphate and the tricarboxylic acid pathways, electron transport, and ATP biosynthesis are thought to provide energy and favor downstream defense processes as radical oxygen species (ROS) generation and expression of PR proteins. Collectively, these works suggest that energy is acquired from an increase in the respiratory metabolism and a higher import of hexoses by cells engaged in the defense against the pathogen (Essmann et al. 2008; Proels and Huckelhoven 2014; Rojas et al. 2014). Even though transcriptomics is giving a good hint of the function of the plant metabolism in plant-pathogen interactions, a deeper exploration remains to be done in order to assess the main impact of a single metabolite in this process. The shift on the primary metabolism occurs mainly during incompatible interactions, where the disease is successfully suppressed. By the contrary, the patterns of global gene expression might be considerably different in compatible interactions, where it results from the activation of plant responses to pathogen perception as well as the action of virulence mechanisms triggered by the pathogen (as virulence factors are secreted by biotrophs to modulate plant metabolism in its own favor or by necrotrophs to affect plant cell fitness).

For many microorganisms, to exert a tight control of the plant metabolism may be crucial to support the infection process (Okmen and Doehlemann 2014). For instance, it has been demonstrated by microarray experiments that the expression of plant heat-shock proteins (HSPs) is induced during viral infections (Whitham et al. 2003; Senthil et al. 2005) and that HSPs are important to ensure proper synthesis of viral proteins (Qanungo et al. 2004). In the same line of evidence, viral particles of BNYVV (*Beet necrotic yellow vein virus*) containing the single-stranded RNA known as RNA4 modify the expression of several genes from *Nicotiana benthamiana* involved in RNA silencing, ubiquitin-proteasome pathway, cellulose synthesis, and gibberellin metabolism, which may contribute to virus multiplication in the infected leaves (Fan et al. 2014). Similarly, *Plasmodiophora brassicae* causes the upregulation of brassinosteroid synthesis and signal perception genes in *Arabidopsis*, and it has been demonstrated that this process is essential for symptom development (Schuller et al. 2014). Biotrophic pathogens also provoke the attenuation of plant defense responses. Thus, microarrays have demonstrated that the establishment of successful biotrophy by the maize fungal pathogen *Ustilago maydis* leads to a suppression of several defense responses triggered during the initial phase of the interaction, such as PR protein expression and phytoalexin synthesis (Doehlemann et al. 2008). The ability of the pathogen to attenuate the initial plant responses and avoid subsequent massive changes in gene expression is a key element for complete virulence (Truman et al. 2006; van de Mortel et al. 2007).

Transcriptomics has been an attractive approach to compare plant responses to compatible and incompatible pathogens. The initial thought was that the main cellular functions responsible for pathogen recognition and control could be distinguished by cutting off all those genes being exclusively regulated during the incompatible interactions. However, it was clear from the beginning that both compatible and incompatible pathogens roughly induce the same patterns of cellular functions and that the outcome of the interactions depends rather on complex quantitative and temporal differences of common transcribed genes or a different set of genes contributing to the same functional category (Tao et al. 2003; Zimmerli et al. 2004; Zou et al. 2005; Rinaldi et al. 2007; Baebler et al. 2009; Zellerhoff et al. 2010; Wichmann et al. 2011; Bagnaresi et al. 2012; Bai et al. 2013; Bordenave et al. 2013; Chen et al. 2013). For instance, Bai et al. (2013) demonstrated that the transcriptional responses in roots of banana cultivars with different susceptibilities to *Fusarium oxysporum* were quite alike. However, the resistant cultivar reprogrammed a greater set of genes in a faster manner against the fungus. Most of these genes were related to biotic stress, such as PR proteins, cell wall metabolism, and PAMP receptors. In turn, the susceptible cultivar over-expressed genes associated to HR and senescence, suggesting that these processes favor fungal colonization. In turn, very similar functional categories were demonstrated to be induced in susceptible and resistant ecotypes of *Solanum lycopersicum* and *Solanum tuberosum* against the *Tomato yellow leaf curl virus* (TYLCV) and the *Potato virus Y* (NTN), respectively, but they differed in the expression levels of a group of defense genes. Thus, susceptible cultivars showed a general gene shutting-off when challenged by the virus, whereas resistant cultivars were able to induce PR genes, WRKY transcriptional factors, protein kinases, and enzymes involved in the secondary metabolism (Baebler et al. 2009; Chen et al. 2013). Interestingly, transcriptomic studies also suggest that resistance may be associated to higher basal expression of important genes rather than their upregulation after pathogen perception. In this trend, several defense genes were found to be highly expressed in non-inoculated leaves of the ecotype Gifu B-129 of *Lotus japonicus* when compared to the ecotype MG-20, showing moderately resistant and susceptible phenotypes against *Pseudomonas syringae* pv. tomato DC3000, respectively. Importantly, several of these genes were upregulated in MG-20 after bacterial infiltration (Bordenave et al. 2013). Comparable results seem to explain at least in part the resistance of the rice variety CL 161 to *Burkholderia glumae* (Magbanua et al. 2014).

6.2.1.2 Beneficial Microbes

Plants may be beneficiated from interactions established with microorganisms including different kinds of fungi, bacteria, and yeasts. Because of their effect on the plant hosts, these microorganisms are commonly known as plant growth-promoting microorganisms (PGPM). Mechanisms of growth promotion by well-studied PGPMs include phosphate solubilization, production of phytohormones, nitrogen fixation (performed by diazotrophic microorganisms), and production of siderophores to facilitate the transport of ferric iron into plant cells (Glick 1995; Bloemberg and Lugtenberg 2001). PGPMs can also stimulate plant growth by preventing the deleterious effects of phytopathogenic microorganisms. In this case, they act as biological control agents by direct antagonistic effects on pathogenic organisms or indirectly by priming plant defense responses, a state that allows the plant to react more rapidly to attacking pathogens (Bloemberg and Lugtenberg 2001; Haas and Keel 2003; Ryu et al. 2004; Balmer et al. 2015). In the light of public concern about the use of agrochemicals and the need to find alternative methods for increasing plant yield and protection against pathogenic microorganisms, the abovementioned features of PGPMs give them an unlimited potential for agronomic use. The best-known PGPMs are microorganisms able to colonize the rhizosphere and, in some cases, invade root interior and establish endophytic populations (Kloepper et al. 1999). In addition, some PGPMs are capable to pass the root endodermis barrier and reach the vascular system, from where they subsequently colonize other organs of the plant (Kobayashi and Palumbo 2000; Lundberg et al. 2012; Romero et al. 2014). Even though they usually enter plants through roots, endophytic communities can also be originated from the phyllosphere (aboveground portions of plants), the anthosphere (flowers), and the spermosphere (seeds) (Hallmann et al. 1997).

Among the different PGPMs, bacteria are probably the most studied microorganisms of this group. In the last years, many works have been published describing the isolation and characterization of several plant growth-promoting bacteria (PGPB) from different environments and hosts (Luna et al. 2012; Abraham et al. 2013; Rashid et al. 2012; Huang et al. 2013; Romero et al. 2015). However, the molecular mechanisms underlying the beneficial effects provoked by most of these microorganisms are not always easy to be established. Many attempts have been made to address this issue by using transcriptomic approaches in model plants and microorganisms. van de Mortel et al. (2012) analyzed the transcriptomic changes induced by the rhizobacterium *Pseudomonas fluorescens* SS101 (Pf.SS101) in *Arabidopsis*. Seed treatment with Pf.SS101 resulted in a total of 1179 and 920 differentially expressed genes in roots and leaves, respectively. Functional classification of the 556 genes upregulated in roots showed a large group belonging to functional categories related to defense, such as general stress response, root morphogenesis, metal ion transport, secondary metabolism (mainly those from the glucosinolate biosynthesis), and SA signaling. A similar trend was observed in leaves, where a large proportion of the responsive transcripts were involved in SA signaling and secondary metabolism (phenylpropanoids, flavonoids, glucosinolates, and camalexin biosynthesis pathways). In turn, downregulated genes in these tissues were associated to hormone signaling and abiotic factors such as temperature, desiccation, and oxidative stress. Importantly, many of the upregulated genes related to defense in roots and leaves of inoculated plants were previously described to participate in PTI activation (Thilmony et al. 2006). In a similar study, Wang et al. (2005) analyzed the transcriptional changes induced by *P. fluorescens* strain FPT9601-T5 in *Arabidopsis* shoots. The authors found that 95 genes were upregulated during the infection, whereas 105 were downregulated. The majority of these genes belong to functional categories associated to metabolism, transcription, cellular communication/signal transduction mechanism, cell rescue, development, and disease resistance (PR proteins, WRKY transcription factors, secondary metabolism, and stress response proteins such as glutathione S-transferase and peroxidase). Moreover, among the upregulated genes, these authors found many transcripts that could be responsible for plant-growth promotion, such as auxin-induced genes as well as genes involved in the metabolism of C compounds and carbohydrates. Auxin-induced genes were also induced in *Arabidopsis* plants inoculated with *P. chlororaphis*, along with many transcripts related to disease resistance such as the PR proteins, the broad-spectrum mildew resistance gene RPW, and the defense-associated transcription factor WRKY18 (Cho et al. 2013). Similarly, genes involved in stress response, oxidative burst, response to auxin, and wounding were induced in this plant species following the inoculation with *P. thivervalensis* MLG45, whereas many transcripts related to photosynthesis and chloroplast function were downregulated (Cartieaux et al. 2003). An additional study focused on the transcriptomic changes induced in the aerial tissues of *Arabidopsis* by root-colonizing bacteria was done by Verhagen et al. (2004) using *P. fluorescens* strain WCS417r, which is known to induce systemic resistance against different pathogens. Interestingly, these authors found that bacterial inoculation provoked the upregulation of a set of genes in leaves involved in the JA and Et signaling. These observations are in agreement with previous findings demonstrating the requirement of these plant hormones for full expression of induced systemic resistance (ISR) (Pieterse et al. 1998, 2000).

This trend of defense activation/auxin pathway induction is not only restricted to infections by *Pseudomonas* strains, since it was also described to occur following inoculation with *Bacillus subtilis* (Lakshmanan et al. 2013). Interestingly, this work also showed that "transcription" is among the most represented functional categories in the downregulated group. This category is comprised of members from two major transcription factor families, ERF (ethylene response factor)/AP2 and MYB, which have been widely associated to defense, indicating that some of the signaling pathways related to plant defense are suppressed during the interaction of plants with endophytic bacteria. The work by Wang et al. (2005) also identifies four nodulin-like genes among those activated after bacterial inoculation. As nodulin proteins play crucial roles during symbiotic nitrogen fixation with rhizobia, it was proposed that these PGPBs and rhizobia might share common signaling pathways.

The upregulation of genes involved in auxin biosynthesis was also observed during the interaction between canola and *Pseudomonas putida* UW4 (Stearns et al. 2012) and *Arabidopsis*/*Burkholderia phytofirmans* PsJN (Poupin et al. 2013). In the last system, the authors also reported the upregulation of different genes involved in the biosynthesis and signaling effects of other phytohormones such as gibberellins, SA, JA, and Et. Auxin-responsive genes, such as a putative auxin-regulated protein, a xyloglucan endo-1,4-β-D-glucanase precursor, and the heat-/auxin-/ethylene-/wounding-induced small protein, were also induced in *Arabidopsis* by the PGPMs *P. thivervalensis* MLG45 and *P. fluorescens* FPT9601-T5 (Cartieaux et al. 2003; Wang et al. 2005). Noteworthy, *P. fluorescens* WCS417r, *P. fluorescens* FPT9601-T5, and *Pseudomonas chlororaphis* O6 colonization led to downregulation of *Arabidopsis* genes involved in the Et signaling pathway (Verhagen et al. 2004; Wang et al. 2005; Cho et al. 2013). Interestingly, it was shown that during the interaction between *Arabidopsis* and *Azospirillum brasilense* Sp245, genes from the GH3 family (GH3.2, GH3.3, GH3.4, GH3.5, and GH3.12) were upregulated. Members of the GH3 family encode for enzymes that conjugate indole acetic acid to amino acids, an important function in auxin homeostasis (Spaepen et al. 2014). In this work, plants were also inoculated with a mutant strain of *A. brasilense* unable to produce IAA, which provoked just a milder induction of GH3.3 and GH3.4. These results suggest that the induction in the auxin-response genes during beneficial interactions could be, at least in part, due to the production of IAA by microorganisms.

The symbiotic interactions between nitrogen-fixing rhizobia and legume plants require the coordinated expression of numerous genes. The activation of the plant transcriptional machinery after plant-rhizobia mutual recognition leads to the formation of root nodules, specialized structures for nitrogen fixation and assimilation (Popp and Ott 2011). Microarray and RNA-Seq approaches have been used to profile gene expression during nodule development. Moreau et al. (2011) and Maunoury et al. (2010) used a series of *Sinorhizobium meliloti* mutants, as well as mutant lines of the model legume *Medicago truncatula* with impaired symbiotic properties, in order to analyze transcriptional reprogramming during different nodule developmental stages. A similar approach was used by Hogslund et al. (2009) to study the interaction between *Mesorhizobium loti* and *Lotus japonicus*. By these means, these works were able to distinguish different groups of genes depending on their expression patterns. The first phase of nodule development was usually characterized by the repression of many defense-associated genes, particularly those belonging to the phenylpropanoid pathway, suggesting that plant defense is switched off from the beginning. The rest of the process is accompanied by a transient induction of genes regulating cell cycle progression and protein synthesis followed by the activation of genes from the secretory pathway. It is thought that the last groups of genes are in charge of nodule organogenesis and maintenance, respectively. This process was also explored in depth by Roux et al. (2014), using RNA-Seq coupled to laser microdissection of nodule regions. An important finding of these authors was that some of the key genes controlling the root apical meristem seem to participate also in the nodule meristem region, suggesting that both processes share common regulatory pathways.

Plants also establish interactions with nonpathogenic beneficial fungi. For instance, *Piriformospora indica* has a broad spectrum of mutualistic symbiosis with plants. Schäfer et al. (2009) analyzed the transcriptomic changes induced by the colonization of *P. indica* in barley roots at different stages of the infection, from the development of the fungus on the root surface to penetration and root cell colonization. They observed that the number of genes differentially expressed increased with the progress of the colonization process. The most represented functional categories were defense/stress response, signaling, secondary metabolism, transport, transcription, and protein biosynthesis. Most of the upregulated transcripts coded for many proteins are involved in regulation of cell death, and genes are involved in the biosynthesis of phytoalexins, carotenoids, and gibberellins. In addition, as observed with plant-beneficial bacteria interactions, colonization by *P. indica* induces the expression of genes involved in auxin biosynthesis and signaling. Lahrmann et al. (2015) compared the response to *P. indica* with that induced by the related fungus *Sebacina vermifera*. This work showed that both responses were quite similar, particularly in the timing of induction of genes responsible for the biosynthesis of secondary metabolites such as glucosinolates, phytoalexins, and triterpenoids. At the same time, key regulators in the signaling pathways of SA, JA, and Et were also upregulated in both systems. Interestingly, these modifications in gene expression were accompanied by an increase in the concentration of JA in inoculated plants and a decrease of SA along with the accumulation of its catabolic products 2,3- and 2,5-dihydroxybenzoic acid (2,3-DHBA and 2,5-DHBA, respectively). As two genes for the conversion of SA to 2,3-DHBA and 2,5-DHBA were among those induced by the fungi, it was suggested that these symbionts do not repress the production of SA but induce the SA catabolism by which plants regulate the concentrations of the phytohormone (Lahrmann et al. 2015). Similarly, the endophytic colonization of barley by *P. indica* or the nematophagous fungus *Pochonia chlamydosporia* also provokes the upregulation of genes involved in the metabolism of auxins, Et, and JA (Larriba et al. 2015).

Fungi from the genera *Trichoderma* are broadly used as biocontrol agents. Mathys et al. (2012) analyzed the transcriptional regulation during the induction of ISR in *Arabidopsis* by *T. hamatum* and compared it to the defense responses elicited by the nechrotrophic pathogen *Botrytis cinerea*. Both responses were found to overlap in the induction of HR-related genes, as well as different components of the response to chitin, and intermediate compounds of the SA and ABA pathways. On the other hand, the main differences were found in the induction of negative regulators of the defense mechanisms, which were upregulated only by *T. hamatum*. These results differ considerably to those obtained using *T. harzianum*, another species of this genus. In this case, the colonization of *Arabidopsis* plants downregulated the SA signaling pathway, while typical markers for the JA defense cascade remained unaffected (Morán-Diez et al. 2012). The basis explaining such differences has not yet been established.

Arbuscular mycorrhizal fungi (AMF) are obligate biotrophs that colonize inter- and intracellularly the root cortex of host plants forming specialized structures called arbuscules and assist plants in obtaining important nutrients from the soil (Parniske 2008). This interaction also provokes profound changes in the transcriptomic profile of plants. Thus, tomato roots inoculated with two different species of AMF, *Glomus mosseae* and *Glomus intraradices*, showed the upregulation of defense genes encoding a chitinase, glutathione S-transferase, β-1,3-glucanase, patatin, β-D-xylosidase, pathogenesis PR10-like protein, and DXS-2, a key enzyme of the mevalonate-independent pathway of carotenoid biosynthesis (López-Ráez et al. 2010). Interestingly, JA- and oxylipin-related genes were induced in mycorrhizal plants, but no other changes were observed in genes related to ABA, SA, or Et signaling. A large group of genes induced in *L. japonicus* during the colonization by the AMF *Gigaspora margarita* were related to membrane synthesis and cell wall architecture (Guether et al. 2009). This observation provides a strong support for the hypothesis that the plant host plays an active role in the restructuration of roots to facilitate the infection process. A recent microarray analysis showed that exudates from *G. margarita* spores are able to induce a strong response in roots of *L. japonicus*. This response includes an increased expression of many defense genes at the onset of the treatment, which are later on downregulated to let the infection proceed normally (Giovannetti et al. 2015).

6.3 Changes of the Plant Metabolome During Plant-Microbe Interactions

Metabolomics lets us detect and quantify the set of metabolites that define any time point in the interaction between plant and microbes. As metabolites represent the final step of all the regulatory networks triggered during these interactions, the analysis of metabolomic profiles brings our conclusions closer to phenotype (Feussner and Polle 2015). One of the main benefits of this approach resides in the fact that in contrast to transcriptomics and proteomics, it does not require previous knowledge on the genome and transcriptome of the studied organisms. This means that metabolomic approaches can be easily focused in any plant, regardless of the availability of genomic information.

6.3.1 Pathogenic Interactions

Even though in a lesser extent than transcriptomics, metabolic profiling has been used to investigate the global responses of plants against pathogenic microorganisms. In this trend, Ward et al. (2010) evaluated the changes in the metabolic profiles of *Arabidopsis* plants infected with virulent and avirulent strains of *P. syringae*. These authors found that the responses to these strains differ in the changes associated to the abundance of some specific metabolites. These changes involved molecules such as amino acids and nitrogenous compounds, as well as carbohydrates and

ROS modulators. Interestingly, many of these differences were observed in metabolites with very low abundance, suggesting that the outcome of the interaction relies on subtle but very important changes in the defense metabolome. Perhaps the most significant reconfiguration was observed in the levels of aromatic amino acids. In agreement with this, it has been shown that the metabolism of tryptophan is dramatically activated in *Arabidopsis* after treatment with a bacterial lipopolysaccharide (which functions as a MAMP and elicits plant defense responses), a mechanism supposed to fuel the synthesis of the antimicrobial phytoalexin camalexin (Beets et al. 2012). Additional works should be addressed to understand the real impact of these metabolites in the outcome of the interactions. Phytoalexins belonging to the sesquiterpene alkaloid family were also shown to accumulate in potato sprouts infected by the fungus *Rhizoctonia solani* (Aliferis and Jabaji 2012), a phenomenon that could help to delay pathogen multiplication. This work also demonstrated that potato plants respond to the pathogen with the production of a vast amount of other defense molecules with antimicrobial activity, such as oxidized fatty acids and organic acids, including oxalic acid and ferulic acid.

Some studies focused on the effect of pathogen effectors and other known defense elicitors. For instance, rice plants exposed to chitin accumulate diterpenoid phytoalexins, which was demonstrated to be activated by a signaling pathway regulated by the MAPK kinase OsMKK4 (Kishi-Kaboshi et al. 2010). In turn, the effect of the priming agent BABA in *Arabidopsis* plants was also analyzed, which was shown to induce the accumulation of indole-3-carboxylic acid, whereas other indolic compounds remained unaltered (Gamir et al. 2012). It was proposed that the production of this intermediate compound accelerates callose deposition and downregulates the production of phytoalexins. Similar studies were conducted in tobacco cells exposed to the plant stress-derived metabolite isonitrosoacetophenone and the fungal steroid ergosterol (Madala et al. 2013; Tugizimana et al. 2014), which provoked different variations in the concentration of important components of the secondary metabolism (terpenoids, coumarins, lignin, phenylpropanoids, and flavonoids) and hormones. All of these investigations represent an invaluable contribution to the understanding of the molecular components that orchestrate plant defense responses. Metabolomic profiling was essential to understand the functionality of a chorismate mutase enzyme secreted by *U. maydis* during the biotrophic phase of infection of maize plants. Interestingly, this protein is incorporated by plant cells, spreads to adjacent cells, and perturbs their metabolism, contributing to attenuate plant defense by diverting the synthesis of SA to the production of phenylpropanoids such as coumaroyl and caffeoylquinate, syringin, and lignan (Djamei et al. 2011). Likewise, a metabolomic analysis has recently shown that the effector protein Tin2 from *U. maydis* also promotes plant colonization by activating anthocyanin synthesis at the expenses of other important plant defense compounds (Tanaka et al. 2014). These results demonstrated that a fine-tuned regulation of the host metabolome must be coordinated by the pathogen for a successful infection. This is also evident in the interaction between wheat and the hemibiotrophic pathogen *Zymoseptoria tritici* (Rudd et al. 2015). This fungus grows initially in the extracellular space of plant tissues and shows a long symptomless biotrophic phase but then develops a fast necrotrophic phase and then kills host cells. Metabolic profiling conducted in colonized plants has shown that during the initial phase of the infection, the host physiology remains unaltered, and there is no activation of plant defense responses. A comparison of these results with those from studies performed on in vitro cultures of the fungus suggests that the microorganism supports its growth during this stage at the expense of the oxidation of stored lipids and fatty acids. This sub-utilization of plant-derived nutrients coincides with the induction in the expression of fungal genes that presumably suppress plant defense responses. However, this phase is followed by a notable switch in plant and fungal physiology, during which the fungus downregulates the expression of effectors and activates cell wall-degrading enzymes. Accordingly, plant defense signaling is induced leading to host cell death.

With the aim to explore the molecular basis explaining plant resistance, metabolomics has been used to compare the time-course changes in metabolite concentration between resistant and susceptible lines. For instance, metabolomic studies demonstrated that many metabolites are modulated in resistant sunflower in response to the fungus *Sclerotinia sclerotiorum* and in resistant barley and wheat after inoculation with *Fusarium graminearum* (Jobic et al. 2007; Bollina et al. 2010; Peluffo et al. 2010; Gunnaiah et al. 2012; Kumaraswamy et al. 2012). Interestingly, these works demonstrate that a considerable proportion of the resistance is explained by the accumulation of pre-existing metabolites, such as flavonoids and phenolic acids. In addition, many other metabolites were shown to be produced in response to the pathogen, such as different sugars, organic acids, amino acids, terpenoids, and phenylpropanoids. Interestingly, infected barley plants presented elevated levels of a glycosylated form of the fungal virulence factor deoxynivalenol (DON). As it has considerable effects on the plant's secondary metabolism, conjugation of this factor toward an inactive form was proposed to be a proper detoxifying strategy (Bollina et al. 2010). In this trend, the amount of phenylpropanoids in barley is reduced in plants infected by the wild-type strain of *F. graminearum* but not in plants infected by a hypovirulent strain unable to produce DON and other trichothecenes. Moreover, the quantity of resistance-related metabolites in barley-resistant lines is higher when inoculated with the hypovirulent strain (Kumaraswamy et al. 2012). Interestingly, the resistance mechanism against *F. graminearum* in wheat seems to differ considerably. Gunnaiah et al. (2012) tried to discern the bases for resistance in this plant species making a nice integration between metabolomic and proteomic data. They found that many metabolites from the phenylpropanoid pathway were only present in the resistant line (or induced proportionally in a higher manner compared to a near-isogenic susceptible line), what was associated to the expression of some of the most important enzymes of this metabolic pathway. Finally, these authors showed that the plant cell wall was strengthened by the deposition of metabolites of this metabolic pathway. Similarly, metabolic profiling demonstrated that soluble phenylpropanoids such as sinapoyl glucosides, coniferin, syringin, and lignans are accumulated in *Arabidopsis* during the early stages of the infection by the fungus *Verticillium longisporum* (Konig et al. 2014). The participation of these metabolites in plant defense was verified by using mutant lines in this pathway. Thus, it was shown that symptoms were incremented in a mutant line unable to produce sinapoyl esters, whereas a coniferin-accumulating transgenic line was more tolerant.

Recently, Warth et al. (2015) evaluated the changes in the plant metabolome after DON treatment in six genotypes of wheat with different levels of resistance to *F. graminearum*. They found that treatment with the fungal factor produces a considerable reduction in the concentration of intermediates from glycolysis and other sugars as well as the amino acids alanine and serine. By the contrary, aromatic amino acids and biologic amines associated to plant defense were incremented. However, as this phenomenon occurs in all tested genotypes, this work was unable to associate metabolic modifications to resistance.

Other works support the idea that amino acid metabolism may have a considerable role in plant defense. In this trend, GC-MS analysis demonstrated that priming for defense in *Arabidopsis* after root infection by *Trichoderma asperelloides* seems to be related to the activation of the amino acid metabolism in leaves (Brotman et al. 2012). Moreover, it was shown that pipecolic acid, one of the intermediate compounds of the lysine catabolic pathway, is accumulated in many plant species in response to pathogen infection. Interestingly, MS analyses demonstrated that exogenous application of pipecolic acid in *Arabidopsis* is sufficient to promote common plant defense responses such as SA and camalexin biosynthesis (Navarova et al. 2012). Recent studies also identified metabolites such as indole acetic acid, indole-3-carboxaldehyde, and camalexin as key components of the defense signaling against the necrotrophic fungus *Plectosphaerella cucumerina* in *Arabidopsis* (Gamir et al. 2014).

Samples used for omics studies in plant-microbe interactions usually consist of a mix of infected and non-infected cells, an issue that may mask actual changes in metabolite abundance and lead to erroneous interpretations of the results obtained. Even though this issue can be overcome in transcriptomic and proteomic studies with the availability of detailed genomic information of the interacting organisms, it is one of the major challenges in plant metabolomics. This is because it is very hard (if not impossible) to distinguish whether the metabolites are derived from the plant or the microorganism. Allwood et al. (2010) tackled this problem by using a plant cell-pathogen co-culture-based approach, where *Arabidopsis* cells were separated from *P. syringae* cells via differential filtering and centrifugation. Afterwards, Fourier transform infrared spectroscopy (FTIR) was conducted to analyze metabolic profiles in both organisms. Even though separation procedures are not adaptable to all plant-microbe interaction and despite the fact that these culture conditions only partially represent the real situation existing *in planta*, science will certainly benefit from studies on the interface between plants and their microbial pathogens/symbionts. For instance, the plant apoplast (the extracellular space including plant cell wall) comprises one of the first compartments where host-pathogen interactions occur. This compartment suffers remarkable changes during the pathogenic process and offers the means for delivering proteins and molecules with important roles on virulence or defense. However, just a few attempts have been made to analyze the metabolic status of the apoplastic compartment. In this trend, Floerl et al. (2012) identified the modulation in the abundance of hundreds of defense-associated metabolites in the apoplast of *Arabidopsis* in response to *V. longisporum*, such as glycosides of SA, lignans, dihydroxybenzoic acid, and oxylipins.

6.3.2 Beneficial

The changes provoked by beneficial microorganisms on the plant metabolome have not been studied in depth so far. There are, however, a few works describing the modifications in the content of primary and secondary metabolites. For instance, the metabolome of poplar plants inoculated with an endophytic strain of the endophyte *Paenibacillus* sp. was analyzed by GC-MS, which identified 11 metabolites that were consistently modified by inoculation (Scherling et al. 2009). Most of these metabolites decreased after infection, including organic acids (malate, succinate, fumarate, and citrate), amino acids (phenylalanine and oxoproline), and sugar phosphates (fructose-6-P). On the other hand, asparagine, urea, and threitol were increased in inoculated plants. These results indicate that the presence of *Paenibacillus* sp. induces the downregulation of central metabolic pathways of the plant while increases amino acid turnover. A comparable response was shown to occur in legumes during their associations with rhizobia (Prell and Poole 2006). The metabolome of *Arabidopsis* plants after inoculation with the rhizobacterium *P. fluorescens* SS101 was analyzed by an untargeted LC-MS-based approach (van de Mortel et al. 2012). In this work, the authors found 46 and 13 differentially modified metabolites in roots and leaves, respectively. The metabolites that were increased in both plant organs were mainly the glucosinolates, phytoalexin scopoletin glucoside, D-gluconate, and indole-3-carboxylic acid β-D-glucopyranosyl ester. These changes in metabolite content in Pf.SS101-inoculated plants were accompanied by the upregulation of the genes involved in their synthesis. In turn, the maize response to *P. putida* KT2440 was characterized by the accumulation of several phospholipids, particularly diacylglycerophosphocholine (Planchamp et al. 2015), whereas the benzoxazinone identified as 2-hydroxy-4,7-dimethoxy-1,4-benzoxazin-3-one glucoside was reduced considerably. These results are in agreement with previous studies analyzing the changes in secondary metabolites in maize plants under greenhouse conditions inoculated with several microorganisms, such as different strains of the PGPB *Azospirillum* spp. In this case, changes in the content of benzoxazinoids and phenolic compounds were reported to occur in a strain-dependent manner (Walker et al. 2011). Similar results were also observed in maize plants under field conditions using different combinations of the inoculants *Azospirillum lipoferum* CRT1, *P. fluorescens* F113, and *G. intraradices* JJ291 (Walker et al. 2012). More work should be conducted in order to understand the implication of these metabolites in the interaction of plants with beneficial microbes.

6.4 Perspectives

Bit by bit, modern science has made a remarkable progress in the understanding on how plant and microbes communicate with each other. Thus, a broad spectrum of the repertory of molecules sensed by plants to recognize microorganisms has been discovered. We should add to this list thousands of genes activated and chemicals synthesized afterwards to hinder pathogen attack or facilitate the infection by beneficial microorganisms, as well as myriad virulence components produced by microorganisms to clear a path through plant defensive responses. In addition, our knowledge on the regulatory mechanisms that governs the production of these molecules is growing exponentially. Undoubtedly, a large part of the progress in this field is due to the development of reliable omics approaches. Even though we have only reviewed transcriptomics and metabolomics in this chapter, proteomics has made a tremendous contribution to this research area. Table 6.1 shows a brief summary of some of the most relevant proteomic works.

Table 6.1 Summary of some relevant works studying the changes of the plant proteome during plant-microbe interactions

Plant	Interaction	Comparison or treatment	References
Arabidopsis	Pathogenic	*P. syringae* pv. tomato DC3000	Jones et al. (2004)
Arabidopsis	Pathogenic	*P. syringae* pv. tomato DC3000	Jones et al. (2006)
Lycopersicon hirsutum	Pathogenic	*Clavibacter michiganensis* subsp. michiganensis	Coaker et al. (2004)
Oryza sativa	Pathogenic	Compatible and incompatible races of *Xanthomonas oryzae* pv. oryzae	Mahmood et al. (2006)
Oryza sativa	Pathogenic	*Xanthomonas oryzae* pv. oryzae infection	Chen et al. (2007)
Nicotiana occidentalis	Pathogenic	'*Candidatus Phytoplasma mali*' strain AT	Luge et al. (2014)
Actinidia chinensis	Pathogenic	*P. syringae* pv. actinidiae	Petriccione et al. (2013)
Actinidia deliciosa	Pathogenic	*P. syringae* pv. actinidiae	Petriccione et al. (2014)
Glycine max [L.] Merr. cv. Akishirome	Symbiotic	Nodule mitochondria vs. root mitochondria	Hoa et al. (2004)
Glycine max cv. Stevens	Symbiotic	Soybean peribacteroid membrane-specific proteins	Panter et al. (2000)
Glycine max [L.] Merr, cv. Williams 82	Symbiotic	Inoculated vs. uninoculated (*Bradyrhizobium japonicum*) root hairs	Wan et al. (2005)
Melilotus alba	Symbiotic	Nodule vs. root	Natera et al. (2000)
Medicago truncatula	Symbiotic	Ethylene-insensitive mutant vs. wild-type after *Sinorhizobium*	Prayitno et al. (2006)
Medicago truncatula	Symbiotic	*Sinorhizobium meliloti*	Schenkluhn et al. (2010)
Oryza sativa	Endophytic	*Azoarcus* sp. strain BH72	Miché et al. (2006)
Vigna unguiculata	Symbiotic	Root hairs inoculated with wild-type *Rhizobium* sp. NGR234 vs. hair-deformation minus mutants	Krause and Broughton (1992)
Arabidopsis	Symbiotic	Volatiles organic compounds from the rhizobacterium *Bacillus subtilis* GB03	Kwon et al. (2010)

Nevertheless, the entire interplay between these organisms is still far from being fully elucidated. For instance, the roles played by numerous regulated genes and metabolites overproduced during plant immune responses are unknown. There is also a lack of information in the regulatory mechanisms that operate on important genes, the relationship between the activation of known defense mechanisms with other metabolic pathways, and the responses elicited in a vast amount of non-model plant species. At last, we only have a glimpse of the contribution of metabolites produced by microorganisms in the outcome of their interactions with plants.

More advances in this direction will be prompted by the advent of new tools able to analyze these components right in the interface between plants and microbes, while they are establishing associations as it occurs in nature. This must be accompanied by a deeper exploration of the defense mechanisms activated in model and non-model plants as well as the virulence activity of different strains of microorganisms. This goal may be addressed by a combination of different omics approaches, which will lead in the future to a thorough comprehension of plant immunity and provide us with the key to improve plant protection and productivity.

Acknowledgments This work was supported by grants of *Agencia Nacional de Promoción Científica y Tecnológica* (ANPCyT) (PICT 2014-3286, 2012-1716, and 2013-0477) and *Consejo Nacional de Investigaciones Científicas y Técnicas* (CONICET) (PIP 114-201101-00367). A.G., M.M., M.E.G., and F.L.P. are members of the Research Career of *Consejo Nacional de Investigaciones Científicas y Técnicas* (CONICET). FMR and FRR are postdoctoral fellows of CONICET. P.V. is a fellow of ANPCyT.

References

Abraham A, Philip S, Kuruvilla Jacob C, Jayachandran K (2013) Novel bacterial endophytes from *Hevea brasiliensis* as biocontrol agent against *Phytophthora* leaf fall disease. BioControl 58:675–684. doi:10.1007/s10526-013-9516-0

Aliferis KA, Jabaji S (2012) FT-ICR/MS and GC-EI/MS metabolomics networking unravels global potato sprout's responses to *Rhizoctonia solani* infection. PLoS One 7:e42576. doi:10.1371/journal.pone.0042576

Allwood JW, Clarke A, Goodacre R, Mur LA (2010) Dual metabolomics: a novel approach to understanding plant-pathogen interactions. Phytochemistry 71:590–597. doi:10.1016/j.phytochem.2010.01.006

Anderson JP, Gleason CA, Foley RC, Thrall PH, Burdon JB, Singh KB (2010) Plants versus pathogens: an evolutionary arms race. Funct Plant Biol 37:499–512. doi:10.1071/FP09304

Asai S, Shirasu K (2015) Plant cells under siege: plant immune system versus pathogen effectors. Curr Opin Plant Biol 28:1–8. doi:10.1016/j.pbi.2015.08.008

Baebler S, Krecic-Stres H, Rotter A, Kogovsek P, Cankar K, Kok EJ, Gruden K, Kovac M, Zel J, Pompe-Novak M, Ravnikar M (2009) PVY(NTN) elicits a diverse gene expression response in different potato genotypes in the first 12 h after inoculation. Mol Plant Pathol 10:263–275. doi:10.1111/j.1364-3703.2008.00530.x

Bagnaresi P, Biselli C, Orru L, Urso S, Crispino L, Abbruscato P, Piffanelli P, Lupotto E, Cattivelli L, Vale G (2012) Comparative transcriptome profiling of the early response to *Magnaporthe oryzae* in durable resistant vs susceptible rice (*Oryza sativa* L.) genotypes. PLoS One 7:e51609. doi:10.1371/journal.pone.0051609

Bai TT, Xie WB, Zhou PP, ZL W, Xiao WC, Zhou L, Sun J, Ruan XL, Li HP (2013) Transcriptome and expression profile analysis of highly resistant and susceptible banana roots challenged with

Fusarium oxysporum f. sp. *cubense* tropical race 4. PLoS One 8:e73945. doi:10.1371/journal.pone.0073945
Balmer A, Pastor V, Gamir J, Flors V, Mauch-Mani B (2015) The 'prime-ome': towards a holistic approach to priming. Trends Plant Sci 20:443–452. doi:10.1016/j.tplants.2015.04.002
Beets CA, Huang JC, Madala NE, Dubery I (2012) Activation of camalexin biosynthesis in *Arabidopsis thaliana* in response to perception of bacterial lipopolysaccharides: a gene-to-metabolite study. Planta 236:261–272. doi:10.1007/s00425-012-1606-1
Bengtsson T, Weighill D, Proux-Wera E, Levander F, Resjo S, Burra DD, Moushib LI, Hedley PE, Liljeroth E, Jacobson D, Alexandersson E, Andreasson E (2014) Proteomics and transcriptomics of the BABA-induced resistance response in potato using a novel functional annotation approach. BMC Genomics 15:315–334. doi:10.1186/1471-2164-15-315
Bernoux M, Ellis JG, Dodds PN (2011) New insights in plant immunity signaling activation. Curr Opin Plant Biol 14:512–518. doi:10.1016/j.pbi.2011.05.005
Bilgin DD, Zavala JA, Zhu J, Clough SJ, Ort DR, DeLucia EH (2010) Biotic stress globally downregulates photosynthesis genes. Plant Cell Environ 33:1597–1613. doi:10.1111/j.1365-3040.2010.02167.x
Bloemberg GV, Lugtenberg BJJ (2001) Molecular basis of plant growth promotion and biocontrol by rhizobacteria. Curr Opin Plant Biol 4:343–350. doi:10.1016/S1369-5266(00)00183-7
Bollina V, Kumaraswamy GK, Kushalappa AC, Choo TM, Dion Y, Rioux S, Faubert D, Hamzehzarghani H (2010) Mass spectrometry-based metabolomics application to identify quantitative resistance-related metabolites in barley against Fusarium head blight. Mol Plant Pathol 11:769–782. doi:10.1111/j.1364-3703.2010.00643.x
Bordenave CD, Escaray FJ, Menendez AB, Serna E, Carrasco P, Ruiz OA, Garriz A (2013) Defense responses in two ecotypes of *Lotus japonicus* against non-pathogenic *Pseudomonas syringae*. PLoS One 8:e83199. doi:10.1371/journal.pone.0083199
Brotman Y, Lisec J, Meret M, Chet I, Willmitzer L, Viterbo A (2012) Transcript and metabolite analysis of the Trichoderma-induced systemic resistance response to *Pseudomonas syringae* in *Arabidopsis thaliana*. Microbiology 158:139–146. doi:10.1099/mic.0.052621-0
Cartieaux F, Thibaud M-C, Zimmerli L, Lessard P, Sarrobert C, David P, Gerbaud A, Robaglia C, Somerville S, Nussaume L (2003) Transcriptome analysis of *Arabidopsis* colonized by a plant-growth promoting rhizobacterium reveals a general effect on disease resistance. Plant J 36:177–188. doi:10.1046/j.1365-313X.2003.01867.x
Chen F, Yuan Y, Li Q, He Z (2007) Proteomic analysis of rice plasma membrane reveals proteins involved in early defense response to bacterial blight. Proteomics 7:1529–1539. doi:10.1002/pmic.200500765
Chen T, Lv Y, Zhao T, Li N, Yang Y, Yu W, He X, Liu T, Zhang B (2013) Comparative transcriptome profiling of a resistant vs. susceptible tomato (*Solanum lycopersicum*) cultivar in response to infection by tomato yellow leaf curl virus. PLoS One 8:e80816. doi:10.1371/journal.pone.0080816
Cho S-M, Kang BR, Kim YC (2013) Transcriptome analysis of induced systemic drought tolerance elicited by *Pseudomonas chlororaphis* O6 in *Arabidopsis thaliana*. Plant Pathol J 29:209–220. doi:10.5423/PPJ.SI.07.2012.0103
Coaker GL, Willard B, Kinter M, Stockinger EJ, Francis DM (2004) Proteomic analysis of resistance mediated by Rcm 2.0 and Rcm 5.1, two loci controlling resistance to bacterial canker of tomato. Mol Plant-Microbe Interact 17:1019–1028. doi:10.1094/MPMI.2004.17.9.1019
De Vos M, Van Oosten VR, Van Poecke RM, Van Pelt JA, Pozo MJ, Mueller MJ, Buchala AJ, Metraux JP, Van Loon LC, Dicke M, Pieterse CM (2005) Signal signature and transcriptome changes of *Arabidopsis* during pathogen and insect attack. Mol Plant-Microbe Interact 18:923–937. doi:10.1094/MPMI-18-0923
Djamei A, Schipper K, Rabe F, Ghosh A, Vincon V, Kahnt J, Osorio S, Tohge T, Fernie AR, Feussner I, Feussner K, Meinicke P, Stierhof YD, Schwarz H, Macek B, Mann M, Kahmann R (2011) Metabolic priming by a secreted fungal effector. Nature 478:395–398. doi:10.1038/nature10454
Dodds PN, Rathjen JP (2010) Plant immunity: towards an integrated view of plant-pathogen interactions. Nat Rev Genet 11:539–548. doi:10.1038/nrg2812

Doehlemann G, Wahl R, Horst RJ, Voll LM, Usadel B, Poree F, Stitt M, Pons-Kuhnemann J, Sonnewald U, Kahmann R, Kamper J (2008) Reprogramming a maize plant: transcriptional and metabolic changes induced by the fungal biotroph *Ustilago maydis*. Plant J 56:181–195. doi:10.1111/j.1365-313X.2008.03590.x

Essmann J, Schmitz-Thom I, Schon H, Sonnewald S, Weis E, Scharte J (2008) RNA interference-mediated repression of cell wall invertase impairs defense in source leaves of tobacco. Plant Physiol 147:1288–1299. doi:10.1104/pp.108.121418

Fan H, Sun H, Wang Y, Zhang Y, Wang X, Li D, Yu J, Han C (2014) Deep sequencing-based transcriptome profiling reveals comprehensive insights into the responses of *Nicotiana benthamiana* to beet necrotic yellow vein virus infections containing or lacking RNA4. PLoS One 9:e85284. doi:10.1371/journal.pone.0085284

Feussner I, Polle A (2015) What the transcriptome does not tell—proteomics and metabolomics are closer to the plants' patho-phenotype. Curr Opin Plant Biol 26:26–31. doi:10.1016/j.pbi.2015.05.023

Floerl S, Majcherczyk A, Possienke M, Feussner K, Tappe H, Gatz C, Feussner I, Kues U, Polle A (2012) *Verticillium longisporum* infection affects the leaf apoplastic proteome, metabolome, and cell wall properties in *Arabidopsis thaliana*. PLoS One 7:e31435. doi:10.1371/journal.pone.0031435

Gamir J, Pastor V, Cerezo M, Flors V (2012) Identification of indole-3-carboxylic acid as mediator of priming against *Plectosphaerella cucumerina*. Plant Physiol Biochem 61:169–179. doi:10.1016/j.plaphy.2012.10.004

Gamir J, Pastor V, Kaever A, Cerezo M, Flors V (2014) Targeting novel chemical and constitutive primed metabolites against *Plectosphaerella cucumerina*. Plant J 78:227–240. doi:10.1111/tpj.12465

Garg R, Tyagi AK, Jain M (2012) Microarray analysis reveals overlapping and specific transcriptional responses to different plant hormones in rice. Plant Signal Behav 7:951–956. doi:10.4161/psb.20910

Giovannetti M, Mari A, Novero M, Bonfante P (2015) Early *Lotus japonicus* root transcriptomic responses to symbiotic and pathogenic fungal exudates. Front Plant Sci 6:480. doi:10.3389/fpls.2015.00480

Glazebrook J (2005) Contrasting mechanisms of defense against biotrophic and necrotrophic pathogens. Annu Rev Phytopathol 43:205–227. doi:10.1146/annurev.phyto.43.040204.135923

Glick BR (1995) The enhancement of plant growth by free-living bacteria. Can J Microbiol 41:109–117. doi:10.1139/m95-015

Goda H, Sasaki E, Akiyama K, Maruyama-Nakashita A, Nakabayashi K, Li W, Ogawa M, Yamauchi Y, Preston J, Aoki K, Kiba T, Takatsuto S, Fujioka S, Asami T, Nakano T, Kato H, Mizuno T, Sakakibara H, Yamaguchi S, Nambara E, Kamiya Y, Takahashi H, Hirai MY, Sakurai T, Shinozaki K, Saito K, Yoshida S, Shimada Y (2008) The AtGenExpress hormone and chemical treatment data set: experimental design, data evaluation, model data analysis and data access. Plant J 55:526–542. doi:10.1111/j.0960-7412.2008.03510.x

Gonzalez ME, Marco F, Minguet EG, Carrasco-Sorli P, Blázquez MA, Carbonell J, Ruiz OA, Pieckenstain FL (2011) Perturbation of spermine synthase gene expression and transcript profiling provide new insights on the role of the tetraamine spermine in *Arabidopsis* defense against *Pseudomonas viridiflava*. Plant Physiol 156:2266–2277. doi:10.1104/pp.110.171413

Guether M, Balestrini R, Hannah M, He J, Udvardi MK, Bonfante P (2009) Genome-wide reprogramming of regulatory networks, transport, cell wall and membrane biogenesis during arbuscular mycorrhizal symbiosis in *Lotus japonicus*. New Phytol 182:200–212. doi:10.1111/j.1469-8137.2008.02725.x

Gunnaiah R, Kushalappa AC, Duggavathi R, Fox S, Somers DJ (2012) Integrated metabolo-proteomic approach to decipher the mechanisms by which wheat QTL (Fhb1) contributes to resistance against *Fusarium graminearum*. PLoS One 7:e40695. doi:10.1371/journal.pone.0040695

Haas D, Keel C (2003) Regulation of antibiotic production in root-colonizing *Peudomonas* spp. and relevance for biological control of plant disease. Annu Rev Phytopathol 41:117–153. doi:10.1146/annurev.phyto.41.052002.095656

Hallmann J, Quadt-Hallmann A, Mahaffee WF, Kloepper JW (1997) Bacterial endophytes in agricultural crops. Can J Microbiol 43:895–914. doi:10.1139/m97-131

Heuberger AL, Robison FM, Lyons SM, Broeckling CD, Prenni JE (2014) Evaluating plant immunity using mass spectrometry-based metabolomics workflows. Front Plant Sci 5:291. doi:10.3389/fpls.2014.00291

Hoa LT-P, Nomura M, Kajiwara H, Day DA, Tajima S (2004) Proteomic analysis on symbiotic differentiation of mitochondria in soybean nodules. Plant Cell Physiol 45:300–308. doi:10.1093/pcp/pch035

Hogslund N, Radutoiu S, Krusell L, Voroshilova V, Hannah MA, Goffard N, Sanchez DH, Lippold F, Ott T, Sato S, Tabata S, Liboriussen P, Lohmann GV, Schauser L, Weiller GF, Udvardi MK, Stougaard J (2009) Dissection of symbiosis and organ development by integrated transcriptome analysis of *Lotus japonicus* mutant and wild-type plants. PLoS One 4:e6556. doi:10.1371/journal.pone.0006556

Huang J, Wei Z, Tan S, Mei X, Yin S, Shen Q, Xu Y (2013) The rhizosphere soil of diseased tomato plants as a source for novel microorganisms to control bacterial wilt. Appl Soil Ecol 72:79–84. doi:10.1016/j.apsoil.2013.05.017

Jobic C, Boisson AM, Gout E, Rascle C, Fevre M, Cotton P, Bligny R (2007) Metabolic processes and carbon nutrient exchanges between host and pathogen sustain the disease development during sunflower infection by *Sclerotinia sclerotiorum*. Planta 226:251–265. doi:10.1007/s00425-006-0470-2

Jones AM, Thomas V, Bennett MH, Mansfield J, Grant M (2006) Modifications to the *Arabidopsis* defense proteome occur prior to significant transcriptional change in response to inoculation with *Pseudomonas syringae*. Plant Physiol 142:1603–1620. doi:10.1104/pp.106.086231

Jones AME, Thomas V, Truman B, Lilley K, Mansfield J, Grant M (2004) Specific changes in the *Arabidopsis* proteome in response to bacterial challenge: differentiating basal and R-gene mediated resistance. Phytochemistry 65:1805–1816. doi:10.1016/j.phytochem.2004.04.005

Kishi-Kaboshi M, Okada K, Kurimoto L, Murakami S, Umezawa T, Shibuya N, Yamane H, Miyao A, Takatsuji H, Takahashi A, Hirochika H (2010) A rice fungal MAMP-responsive MAPK cascade regulates metabolic flow to antimicrobial metabolite synthesis. Plant J 63:599–612. doi:10.1111/j.1365-313X.2010.04264.x

Kloepper JW, Rodríguez-Ubana R, Zehnder GW, Murphy JF, Sikora E, Fernández C (1999) Plant root-bacterial interactions in biological control of soilborne diseases and potential extension to systemic and foliar diseases. Australas Plant Pathol 28:21–26. doi:10.1071/ap99003

Kobayashi D, Palumbo J (2000) Bacterial endophytes and their effects on plants and uses in agriculture. In: Bacon CW, White J (eds) Microbial endophytes. CRC, Boca Raton, pp 199–233

Konig S, Feussner K, Kaever A, Landesfeind M, Thurow C, Karlovsky P, Gatz C, Polle A, Feussner I (2014) Soluble phenylpropanoids are involved in the defense response of *Arabidopsis* against *Verticillium longisporum*. New Phytol 202:823–837. doi:10.1111/nph.12709

Kopka J, Fernie A, Weckwerth W, Gibon Y, Stitt M (2004) Metabolite profiling in plant biology: platforms and destinations. Genome Biol 5:109. doi:10.1186/gb-2004-5-6-109

Krause A, Broughton WJ (1992) Proteins associated with root-hair deformation and nodule initiation in *Vigna unguiculata*. Mol Plant-Microbe Interact 5:96–103

Kumaraswamy GK, Kushalappa AC, Choo TM, Dion Y, Rioux S (2012) Differential metabolic response of barley genotypes, varying in resistance, to trichothecene-producing and -nonproducing (tri5−) isolates of *Fusarium graminearum*. Plant Pathol 61:509–521. doi:10.1111/j.1365-3059.2011.02528.x

Kwon YS, Ryu C-M, Lee S, Park HB, Han KS, Lee JH, Lee K, Chung WS, Jeong M-J, Kim HK, Bae D-W (2010) Proteome analysis of *Arabidopsis* seedlings exposed to bacterial volatiles. Planta 232:1355–1370. doi:10.1007/s00425-010-1259-x

Lahrmann U, Strehmel N, Langen G, Frerigmann H, Leson L, Ding Y, Scheel D, Herklotz S, Hilbert M, Zuccaro A (2015) Mutualistic root endophytism is not associated with the reduction of saprotrophic traits and requires a noncompromised plant innate immunity. New Phytol 207:841–857. doi:10.1111/nph.13411

Lakshmanan V, Castaneda R, Rudrappa T, Bais H (2013) Root transcriptome analysis of *Arabidopsis thaliana* exposed to beneficial *Bacillus subtilis* FB17 rhizobacteria revealed genes for bacterial recruitment and plant defense independent of malate efflux. Planta 238:657–668. doi:10.1007/s00425-013-1920-2

Larriba E, Jaime MLA, Nislow C, Martín-Nieto J, Lopez-Llorca L (2015) Endophytic colonization of barley (*Hordeum vulgare*) roots by the nematophagous fungus *Pochonia chlamydosporia* reveals plant growth promotion and a general defense and stress transcriptomic response. J Plant Res 128:665–678. doi:10.1007/s10265-015-0731-x

Lodha TD, Basak J (2012) Plant-pathogen interactions: what microarray tells about it? Mol Biotechnol 50:87–97. doi:10.1007/s12033-011-9418-2

López-Ráez JA, Verhage A, Fernández I, García JM, Azcón-Aguilar C, Flors V, Pozo MJ (2010) Hormonal and transcriptional profiles highlight common and differential host responses to arbuscular mycorrhizal fungi and the regulation of the oxylipin pathway. J Exp Bot 61:2589–2601. doi:10.1093/jxb/erq089

Luge T, Kube M, Freiwald A, Meierhofer D, Seemüller E, Sauer S (2014) Transcriptomics assisted proteomic analysis of *Nicotiana occidentalis* infected by *Candidatus Phytoplasma mali* strain AT. Proteomics 14:1882–1889. doi:10.1002/pmic.201300551

Luna MF, Aprea J, Crespo JM, Boiardi JL (2012) Colonization and yield promotion of tomato by *Gluconacetobacter diazotrophicus*. Appl Soil Ecol 61:225–229. doi:10.1016/j.apsoil.2011.09.002

Lundberg DS, Lebeis SL, Paredes SH, Yourstone S, Gehring J, Malfatti S, Tremblay J, Engelbrektson A, Kunin V, TGd R, Edgar RC, Eickhorst T, Ley RE, Hugenholtz P, Tringe SG, Dangl JL (2012) Defining the core *Arabidopsis thaliana* root microbiome. Nature 488:86–90. doi:10.1038/nature11237

Madala NE, Steenkamp PA, Piater LA, Dubery IA (2013) Metabolomic analysis of isonitrosoacetophenone-induced perturbations in phenolic metabolism of *Nicotiana tabacum* cells. Phytochemistry 94:82–90. doi:10.1016/j.phytochem.2013.05.010

Magbanua ZV, Arick M II, Buza T, Hsu CY, Showmaker KC, Chouvarine P, Deng P, Peterson DG, Lu S (2014) Transcriptomic dissection of the rice-*Burkholderia glumae* interaction. BMC Genomics 15:755. doi:10.1186/1471-2164-15-755

Mahmood T, Jan A, Kakishima M, Komatsu S (2006) Proteomic analysis of bacterial-blight defense-responsive proteins in rice leaf blades. Proteomics 6:6053–6065. doi:10.1002/pmic.200600470

Mathys J, De Cremer K, Timmermans P, Van Kerkhove S, Lievens B, Vanhaecke M, Cammue B, De Coninck B (2012) Genome-wide characterization of ISR induced in *Arabidopsis thaliana* by *Trichoderma hamatum* T382 against *Botrytis cinerea* infection. Front Plant Sci 3:108. doi:10.3389/fpls.2012.00108

Maunoury N, Redondo-Nieto M, Bourcy M, Van de Velde W, Alunni B, Laporte P, Durand P, Agier N, Marisa L, Vaubert D, Delacroix H, Duc G, Ratet P, Aggerbeck L, Kondorosi E, Mergaert P (2010) Differentiation of symbiotic cells and endosymbionts in *Medicago truncatula* nodulation are coupled to two transcriptome-switches. PLoS One 5:e9519. doi:10.1371/journal.pone.0009519

Miché L, Battistoni F, Gemmer S, Belghazi M, Reinhold-Hurek B (2006) Upregulation of jasmonate-inducible defense proteins and differential colonization of roots of *Oryza sativa* cultivars with the endophyte *Azoarcus* sp. Mol Plant-Microbe Interact 19:502–511. doi:10.1094/MPMI-19-0502

Morán-Diez E, Rubio B, Domínguez S, Hermosa R, Monte E, Nicolás C (2012) Transcriptomic response of *Arabidopsis thaliana* after 24 h incubation with the biocontrol fungus *Trichoderma harzianum*. J Plant Physiol 169:614–620. doi:10.1016/j.jplph.2011.12.016

Moreau S, Verdenaud M, Ott T, Letort S, de Billy F, Niebel A, Gouzy J, de Carvalho-Niebel F, Gamas P (2011) Transcription reprogramming during root nodule development in *Medicago truncatula*. PLoS One 6:e16463. doi:10.1371/journal.pone.0016463

Natera SHA, Guerreiro N, Djordjevic MA (2000) Proteome analysis of differentially displayed proteins as a tool for the investigation of symbiosis. Mol Plant-Microbe Interact 13:995–1009. doi:10.1094/MPMI.2000.13.9.995

Navarova H, Bernsdorff F, Doring AC, Zeier J (2012) Pipecolic acid, an endogenous mediator of defense amplification and priming, is a critical regulator of inducible plant immunity. Plant Cell 24:5123–5141. doi:10.1105/tpc.112.103564

Nemhauser JL, Hong F, Chory J (2006) Different plant hormones regulate similar processes through largely nonoverlapping transcriptional responses. Cell 126:467–475. doi:10.1016/j.cell.2006.05.050

Okamoto M, Tsuboi Y, Goda H, Yoshizumi T, Shimada Y, Hirayama T (2012) Multiple hormone treatment revealed novel cooperative relationships between abscisic acid and biotic stress hormones in cultured cells. Plant Biotechnol 29:19–34. doi:10.5511/plantbiotechnology.11.1130a

Okmen B, Doehlemann G (2014) Inside plant: biotrophic strategies to modulate host immunity and metabolism. Curr Opin Plant Biol 20:19–25. doi:10.1016/j.pbi.2014.03.011

Panter S, Thomson R, de Bruxelles G, Laver D, Trevaskis B, Udvardi M (2000) Identification with proteomics of novel proteins associated with the peribacteroid membrane of soybean root nodules. Mol Plant-Microbe Interact 13:325–333. doi:10.1094/MPMI.2000.13.3.325

Parniske M (2008) Arbuscular mycorrhiza: the mother of plant root endosymbioses. Nat Rev Microbiol 6:763–775. doi:10.1038/nrmicro1987

Pedras MS, Yaya EE (2015) Plant chemical defenses: are all constitutive antimicrobial metabolites phytoanticipins? Nat Prod Commun 10:209–218

Peluffo L, Lia V, Troglia C, Maringolo C, Norma P, Escande A, Esteban Hopp H, Lytovchenko A, Fernie AR, Heinz R, Carrari F (2010) Metabolic profiles of sunflower genotypes with contrasting response to *Sclerotinia sclerotiorum* infection. Phytochemistry 71:70–80. doi:10.1016/j.phytochem.2009.09.018

Petriccione M, Di Cecco I, Arena S, Scaloni A, Scortichini M (2013) Proteomic changes in *Actinidia chinensis* shoot during systemic infection with a pandemic *Pseudomonas syringae* pv. *actinidiae* strain. J Proteome 78:461–476. doi:10.1016/j.jprot.2012.10.014

Petriccione M, Salzano AM, Di Cecco I, Scaloni A, Scortichini M (2014) Proteomic analysis of the *Actinidia deliciosa* leaf apoplast during biotrophic colonization by *Pseudomonas syringae* pv. actinidiae. J Proteome 101:43–62. doi:10.1016/j.jprot.2014.01.030

Pieterse CMJ, Van Pelt JA, Ton J, Parchmann S, Mueller MJ, Buchala AJ, Métraux J-P, Van Loon LC (2000) Rhizobacteria-mediated induced systemic resistance (ISR) in *Arabidopsis* requires sensitivity to jasmonate and ethylene but is not accompanied by an increase in their production. Physiol Mol Plant Pathol 57:123–134. doi:10.1006/pmpp.2000.0291

Pieterse CMJ, van Wees SCM, van Pelt JA, Knoester M, Laan R, Gerrits H, Weisbeek PJ, van Loon LC (1998) A novel signaling pathway controlling induced systemic resistance in *Arabidopsis*. Plant Cell 10:1571–1580

Planchamp C, Glauser G, Mauch-Mani B (2015) Root inoculation with *Pseudomonas putida* KT2440 induces transcriptional and metabolic changes and systemic resistance in maize plants. Front Plant Sci 5:719. doi:10.3389/fpls.2014.00719

Popp C, Ott T (2011) Regulation of signal transduction and bacterial infection during root nodule symbiosis. Curr Opin Plant Biol 14:458–467. doi:10.1016/j.pbi.2011.03.016

Poupin MJ, Timmermann T, Vega A, Zuniga A, Gonzalez B (2013) Effects of the plant growth-promoting bacterium *Burkholderia phytofirmans* PsJN throughout the life cycle of *Arabidopsis thaliana*. PLoS One 8:e69435. doi:10.1371/journal.pone.0069435

Prayitno J, Imin N, Rolfe BG, Mathesius U (2006) Identification of ethylene-mediated protein changes during nodulation in *Medicago truncatula* using proteome analysis. J Proteome Res 5:3084–3095. doi:10.1021/pr0602646

Prell J, Poole P (2006) Metabolic changes of rhizobia in legume nodules. Trends Microbiol 14:161–168. doi:10.1016/j.tim.2006.02.005

Proels RK, Huckelhoven R (2014) Cell-wall invertases, key enzymes in the modulation of plant metabolism during defence responses. Mol Plant Pathol 15:858–864. doi:10.1111/mpp.12139

Qanungo KR, Shaji D, Mathur M, Banerjee AK (2004) Two RNA polymerase complexes from vesicular stomatitis virus-infected cells that carry out transcription and replication of genome RNA. Proc Natl Acad Sci U S A 101:5952–5957. doi:10.1073/pnas.0401449101

Qian X, Ba Y, Zhuang Q, Zhong G (2014) RNA-Seq technology and its application in fish transcriptomics. OMICS 18:98–110. doi:10.1089/omi.2013.0110

Rashid S, Charles TC, Glick BR (2012) Isolation and characterization of new plant growth-promoting bacterial endophytes. Appl Soil Ecol 61:217–224. doi:10.1016/j.apsoil.2011.09.011

Rinaldi C, Kohler A, Frey P, Duchaussoy F, Ningre N, Couloux A, Wincker P, Le Thiec D, Fluch S, Martin F, Duplessis S (2007) Transcript profiling of poplar leaves upon infection with compatible and incompatible strains of the foliar rust *Melampsora larici-populina*. Plant Physiol 144:347–366. doi:10.1104/pp.106.094987

Rojas CM, Senthil-Kumar M, Tzin V, Mysore KS (2014) Regulation of primary plant metabolism during plant-pathogen interactions and its contribution to plant defense. Front Plant Sci 5:17. doi:10.3389/fpls.2014.00017

Romero FM, Marina M, Pieckenstain FL (2014) The communities of tomato (*Solanum lycopersicum* L.) leaf endophytic bacteria, analyzed by 16S-ribosomal RNA gene pyrosequencing. FEMS Microbiol Lett 351:187–194. doi:10.1111/1574-6968.12377

Romero FM, Marina M, Pieckenstain FL (2015) Novel components of leaf bacterial communities of field-grown tomato plants and their potential for plant growth promotion and biocontrol of tomato diseases. Res Microbiol 167:222–233. doi:10.1016/j.resmic.2015.11.001

Roux B, Rodde N, Jardinaud MF, Timmers T, Sauviac L, Cottret L, Carrere S, Sallet E, Courcelle E, Moreau S, Debelle F, Capela D, de Carvalho-Niebel F, Gouzy J, Bruand C, Gamas P (2014) An integrated analysis of plant and bacterial gene expression in symbiotic root nodules using laser-capture microdissection coupled to RNA sequencing. Plant J 77:817–837. doi:10.1111/tpj.12442

Rudd JJ, Kanyuka K, Hassani-Pak K, Derbyshire M, Andongabo A, Devonshire J, Lysenko A, Saqi M, Desai NM, Powers SJ, Hooper J, Ambroso L, Bharti A, Farmer A, Hammond-Kosack KE, Dietrich RA, Courbot M (2015) Transcriptome and metabolite profiling of the infection cycle of *Zymoseptoria tritici* on wheat reveals a biphasic interaction with plant immunity involving differential pathogen chromosomal contributions and a variation on the hemibiotrophic lifestyle definition. Plant Physiol 167:1158–1185. doi:10.1104/pp.114.255927

Ryu C-M, Murphy JF, Mysore KS, Kloepper JW (2004) Plant growth-promoting rhizobacteria systemically protect *Arabidopsis thaliana* against *Cucumber mosaic virus* by a salicylic acid and NPR1-independent and jasmonic acid-dependent signaling pathway. Plant J 39:381–392. doi:10.1111/j.1365-313X.2004.02142.x

Salzman RA, Brady JA, Finlayson SA, Buchanan CD, Summer EJ, Sun F, Klein PE, Klein RR, Pratt LH, Cordonnier-Pratt MM, Mullet JE (2005) Transcriptional profiling of sorghum induced by methyl jasmonate, salicylic acid, and aminocyclopropane carboxylic acid reveals cooperative regulation and novel gene responses. Plant Physiol 138:352–368. doi:10.1104/pp.104.058206

Schäfer P, Pfiffi S, Voll LM, Zajic D, Chandler PM, Waller F, Scholz U, Pons-Kühnemann J, Sonnewald S, Sonnewald U, Kogel K-H (2009) Manipulation of plant innate immunity and gibberellin as factor of compatibility in the mutualistic association of barley roots with *Piriformospora indica*. Plant J 59:461–474. doi:10.1111/j.1365-313X.2009.03887.x

Scheideler M, Schlaich NL, Fellenberg K, Beissbarth T, Hauser NC, Vingron M, Slusarenko AJ, Hoheisel JD (2002) Monitoring the switch from housekeeping to pathogen defense metabolism in *Arabidopsis thaliana* using cDNA arrays. J Biol Chem 277:10555–10561. doi:10.1074/jbc.M104863200

Schena M, Shalon D, Davis RW, Brown PO (1995) Quantitative monitoring of gene expression patterns with a complementary DNA microarray. Science 270:467–470

Schenkluhn L, Hohnjec N, Niehaus K, Schmitz U, Colditz F (2010) Differential gel electrophoresis (DIGE) to quantitatively monitor early symbiosis- and pathogenesis-induced changes of the *Medicago truncatula* root proteome. J Proteome 73:753–768. doi:10.1016/j.jprot.2009.10.009

Scherling C, Ulrich K, Ewald D, Weckwerth W (2009) A metabolic signature of the beneficial interaction of the endophyte *Paenibacillus* sp. isolate and *in vitro*–grown poplar plants revealed by metabolomics. Mol Plant-Microbe Interact 22:1032–1037. doi:10.1094/mpmi-22-8-1032

Schuller A, Kehr J, Ludwig-Muller J (2014) Laser microdissection coupled to transcriptional profiling of Arabidopsis roots inoculated by *Plasmodiophora brassicae* indicates a role for brassinosteroids in clubroot formation. Plant Cell Physiol 55:392–411. doi:10.1093/pcp/pct174

Senthil G, Liu H, Puram VG, Clark A, Stromberg A, Goodin MM (2005) Specific and common changes in *Nicotiana benthamiana* gene expression in response to infection by enveloped viruses. J Gen Virol 86:2615–2625. doi:10.1099/vir.0.81043-0

Spaepen S, Bossuyt S, Engelen K, Marchal K, Vanderleyden J (2014) Phenotypical and molecular responses of *Arabidopsis thaliana* roots as a result of inoculation with the auxin-producing bacterium *Azospirillum brasilense*. New Phytol 201:850–861. doi:10.1111/nph.12590

Stearns JC, Woody OZ, McConkey BJ, Glick BR (2012) Effects of bacterial ACC deaminase on *Brassica napus* gene expression. Mol Plant-Microbe Interact 25:668–676. doi:10.1094/mpmi-08-11-0213

Tam JP, Wang S, Wong KH, Tan WL (2015) Antimicrobial peptides from plants. Pharmaceuticals (Basel) 8:711–757. doi:10.3390/ph8040711

Tanaka S, Brefort T, Neidig N, Djamei A, Kahnt J, Vermerris W, Koenig S, Feussner K, Feussner I, Kahmann R (2014) A secreted *Ustilago maydis* effector promotes virulence by targeting anthocyanin biosynthesis in maize. elife 3:e01355. doi:10.7554/eLife.01355

Tao Y, Xie Z, Chen W, Glazebrook J, Chang HS, Han B, Zhu T, Zou G, Katagiri F (2003) Quantitative nature of *Arabidopsis* responses during compatible and incompatible interactions with the bacterial pathogen *Pseudomonas syringae*. Plant Cell 15:317–330. doi:10.1105/tpc.007591

Thilmony R, Underwood W, He SY (2006) Genome-wide transcriptional analysis of the *Arabidopsis thaliana* interaction with the plant pathogen *Pseudomonas syringae* pv. Tomato DC3000 and the human pathogen *Escherichia coli* O157:H7. Plant J 46:34–53. doi:10.1111/j.1365-313X.2006.02725.x

Truman W, de Zabala MT, Grant M (2006) Type III effectors orchestrate a complex interplay between transcriptional networks to modify basal defence responses during pathogenesis and resistance. Plant J 46:14–33. doi:10.1111/j.1365-313X.2006.02672.x

Tugizimana F, Steenkamp PA, Piater LA, Dubery IA (2014) Multi-platform metabolomic analyses of ergosterol-induced dynamic changes in *Nicotiana tabacum* cells. PLoS One 9:e87846. doi:10.1371/journal.pone.0087846

van de Mortel JE, de Vos RCH, Dekkers E, Pineda A, Guillod L, Bouwmeester K, van Loon JJA, Dicke M, Raaijmakers JM (2012) Metabolic and transcriptomic changes induced in *Arabidopsis* by the rhizobacterium *Pseudomonas fluorescens* SS101. Plant Physiol 160:2173–2188. doi:10.1104/pp.112.207324

van de Mortel M, Recknor JC, Graham MA, Nettleton D, Dittman JD, Nelson RT, Godoy CV, Abdelnoor RV, Almeida AM, Baum TJ, Whitham SA (2007) Distinct biphasic mRNA changes in response to Asian soybean rust infection. Mol Plant-Microbe Interact 20:887–899. doi:10.1094/MPMI-20-8-0887

Verhagen BWM, Glazebrook J, Zhu T, Chang H-S, van Loon LC, Pieterse CMJ (2004) The transcriptome of rhizobacteria-induced systemic resistance in *Arabidopsis*. Mol Plant-Microbe Interact 17:895–908. doi:10.1094/mpmi.2004.17.8.895

Walker V, Bertrand C, Bellvert F, Moënne-Loccoz Y, Bally R, Comte G (2011) Host plant secondary metabolite profiling shows a complex, strain-dependent response of maize to plant growth-promoting rhizobacteria of the genus *Azospirillum*. New Phytol 189:494–506. doi:10.1111/j.1469-8137.2010.03484.x

Walker V, Couillerot O, Von Felten A, Bellvert F, Jansa J, Maurhofer M, Bally R, Moënne-Loccoz Y, Comte G (2012) Variation of secondary metabolite levels in maize seedling roots induced by inoculation with *Azospirillum, Pseudomonas* and *Glomus* consortium under field conditions. Plant Soil 356:151–163. doi:10.1007/s11104-011-0960-2

Wan J, Torres M, Ganapathy A, Thelen J, DaGue BB, Mooney B, Xu D, Stacey G (2005) Proteomic analysis of soybean root hairs after infection by *Bradyrhizobium japonicum*. Mol Plant-Microbe Interact 18:458–467. doi:10.1094/MPMI-18-0458

Wang Y, Ohara Y, Nakayashiki H, Tosa Y, Mayama S (2005) Microarray analysis of the gene expression profile induced by the endophytic plant growth-promoting rhizobacteria, *Pseudomonas fluorescens* FPT9601-T5 in *Arabidopsis*. Mol Plant-Microbe Interact 18:385–396. doi:10.1094/MPMI-18-0385

Wang Y, Tao X, Tang XM, Xiao L, Sun JL, Yan XF, Li D, Deng HY, Ma XR (2013) Comparative transcriptome analysis of tomato (*Solanum lycopersicum*) in response to exogenous abscisic acid. BMC Genomics 14:841. doi:10.1186/1471-2164-14-841

Wang Z, Gerstein M, Snyder M (2009) RNA-Seq: a revolutionary tool for transcriptomics. Nat Rev Genet 10:57–63. doi:10.1038/nrg2484

Ward JL, Forcat S, Beckmann M, Bennett M, Miller SJ, Baker JM, Hawkins ND, Vermeer CP, Lu C, Lin W, Truman WM, Beale MH, Draper J, Mansfield JW, Grant M (2010) The metabolic transition during disease following infection of *Arabidopsis thaliana* by *Pseudomonas syringae* pv. Tomato. Plant J 63:443–457. doi:10.1111/j.1365-313X.2010.04254.x

Warth B, Parich A, Bueschl C, Schoefbeck D, Neumann NK, Kluger B, Schuster K, Krska R, Adam G, Lemmens M, Schuhmacher R (2015) GC-MS based targeted metabolic profiling identifies changes in the wheat metabolome following deoxynivalenol treatment. Metabolomics 11:722–738. doi:10.1007/s11306-014-0731-1

Whitham SA, Quan S, Chang HS, Cooper B, Estes B, Zhu T, Wang X, Hou YM (2003) Diverse RNA viruses elicit the expression of common sets of genes in susceptible *Arabidopsis thaliana* plants. Plant J 33:271–283. doi:10.1046/j.1365-313X.2003.01625.x

Wichmann F, Asp T, Widmer F, Kolliker R (2011) Transcriptional responses of Italian ryegrass during interaction with *Xanthomonas translucens* pv. *graminis* reveal novel candidate genes for bacterial wilt resistance. Theor Appl Genet 122:567–579. doi:10.1007/s00122-010-1470-y

Wishart DS (2011) Advances in metabolite identification. Bioanalysis 3:1769–1782. doi:10.4155/bio.11.155

Zellerhoff N, Himmelbach A, Dong W, Bieri S, Schaffrath U, Schweizer P (2010) Nonhost resistance of barley to different fungal pathogens is associated with largely distinct, quantitative transcriptional responses. Plant Physiol 152:2053–2066. doi:10.1104/pp.109.151829

Zhang W, Li F, Nie L (2010) Integrating multiple 'omics' analysis for microbial biology: application and methodologies. Microbiology 156:287–301. doi:10.1099/mic.0.034793-0

Zimmerli L, Stein M, Lipka V, Schulze-Lefert P, Somerville S (2004) Host and non-host pathogens elicit different jasmonate/ethylene responses in *Arabidopsis*. Plant J 40:633–646. doi:10.1111/j.1365-313X.2004.02236.x

Zou J, Rodriguez-Zas S, Aldea M, Li M, Zhu J, Gonzalez DO, Vodkin LO, DeLucia E, Clough SJ (2005) Expression profiling soybean response to *Pseudomonas syringae* reveals new defense-related genes and rapid HR-specific downregulation of photosynthesis. Mol Plant-Microbe Interact 18:1161–1174. doi:10.1094/MPMI-18-1161

Part III

Metabolic Engineering in Microbes

Cellulases: Industrial Workhorse in Bioenergy Sector

7

Tanveer Bilal Pirzadah, Bisma Malik, Reiaz Ul Rehman, and Manoj Kumar

Abstract

There exists a diverse array of microorganisms in the environments which secretes cellulase enzyme. Cellulases are multienzyme proteins acting on cellulose to convert them into smaller sugar components such as glucose. These enzymes are widely used in many industrial applications such as processing cotton, recycling paper and additivities of feed. The challenges which the world faces currently are concerned with rising oil prices and global warming, and in this context, the cellulases have gained importance as the property of cellulases in cellulose degradation has played an important role in sustainable biofuel production. However, high cost of cellulases is one of the biggest hindrances for commercialization of cellulosic biofuel technology and for bringing out the cost-effective technology; major research is focused on enhancing enzyme efficiency and in reducing capital costs.

Keywords

Biofuel • Cellulases • Lignocellulose biomass • Synergy

T.B. Pirzadah • B. Malik • R.U. Rehman (✉)
Department of Bioresources, University of Kashmir,
Srinagar 190006, Jammu & Kashmir, India
e-mail: reiazrehman@yahoo.co.in

M. Kumar (✉)
Amity Institute of Microbial Technology, Amity University,
Noida 201303, Uttar Pradesh, India
e-mail: manojjnu@gmail.com

V.C. Kalia, A.K. Saini (eds.), *Metabolic Engineering for Bioactive Compounds*,
DOI 10.1007/978-981-10-5511-9_7

7.1 Introduction

The living standards in Middle East, China and India are changing which might result in enhanced world energy demands by one third up to 2035. In 2010, the majority (81%) of the energy demand was fulfilled by non-renewable sources such as coal, oil and natural gas (IEA 2012). The ever-increasing demand of the energy has led to over-exploitation of fossil fuels that directly led to the deleterious effects on environment because of the generation of greenhouse gases which contribute to global warming. The phenomenon of global warming is a severe threat to environment around the globe because of its ill effects on ecosystem. It has been reported that climate change led to the extinction of large number of species (Thomas et al. 2004). Besides, the non-renewable energy sources are unevenly distributed leading to a negative impact on nations dependent on foreign oil, coal and natural gas. In order to combat the problems associated with fossil energy resources, scientists all over the world are focusing towards green energy approaches (IEA 2012). Cellulosic biomass being the abundant polymer on earth is considered potential bioresources for the producing clean and green fuel (Naik et al. 2010; Himmel et al. 2007). This includes residues from agriculture and forest residues, solid municipal waste and bioenergy crops. Lignocellulosic bioethanol is considered as second-generation biofuels composed of three different types of polymers, viz. cellulose, hemicellulose and lignin. However, recalcitrant nature of lignocellulosic biomass hinders its use in the biofuel industry, and thus the ongoing research is focused towards novel approaches for the efficient utilization of biomass feedstock (Himmel et al. 2007; Ghimire et al. 2015; Yang et al. 2016). The bioconversion of lignocellulosic waste to value-added chemicals has led to extensive studies on microorganisms for searching candidates capable of producing efficient cell wall-degrading enzymes. Among various hydrolytic enzymes which form microbes, cellulases are a widely studied multienzyme complex protein which converts cellulose to glucose. In the industry, *Trichoderma* is the leading source of cellulases and hemicellulases used for depolymerizing cellulosic biomass to simple sugars (Pirzadah et al. 2014).

The cellulases have become indispensable enzymes for converting simple sugars from biomass to liquid fuels (Adsul et al. 2011). Among the cellulase-producing microorganisms, the majority of them produce complete cellulase enzymes acting on native cellulose synergistically (Saggi and Dey 2016; Wilson 2008) besides producing protein byproducts that also help in cellulose degradation (Wang et al. 2011). However, the fungal cellulase enzymes lack β-glucosidase activity which reduces its hydrolysis efficiency (Lynd et al. 2002). Moreover, another constraint in the production of cellulosic bioethanol is the high cost of cellulose, and it has been estimated that for the production of 1 gal of ethanol, approximately 100 g of cellulase is required for biomass saccharification (Zhu et al. 2009). In order to reduce the cellulase consumption, the elevation of volumetric productivity, reduction of capital cost and consolidated bioprocessing (CBP) have been introduced in a single step by assimilating cellulase production, cellulose hydrolysis and ethanol fermentation (Lynd et al. 2008). Recent research has been focusing on reducing cellulase cost by using structural techniques for enhancing their properties. Besides, novel genetic engineering approaches have been employed to determine the genome sequencing of cellulolytic microorganisms and then coupled with transcriptomic and proteomic studies of these microbes which have led to recognition of novel proteins that play vital roles in degradation of cellulose (Zhou et al. 2010). The fungal GH-61 proteins essentially act as novel enzymes which enhance the function of commercial cellulases resulting in the reduction of total protein load and thus reducing the cost. The oxidoreductase enzymes (carrying C1 activity) are the focus of current research because they are highly efficient in targeting crystalline cellulose and allow easy access to cellulase enzymes. The main function of cellobiose dehydrogenase is to interact with GH-61 proteins, which provides a supportive role to this puzzling enzyme. Remarkable progress has been made in cellulase research and needs further attention to make this technology cost-effective.

7.2 Lignocellulosic Biomass: Major Sustainable Biofuel Reserve

In nature, lignocellulosic biomass is in abundance and is considered as a sustainable and largest biomass reserve for ethanol production. Approximately 7.5×10^{10} tons/year of cellulose is manufactured by photosynthesis (Monserrate et al. 2001). These biomass reserves are usually classified into four types such as agricultural waste, forest residues, MSW and other organic residues. Lignocellulosic feedstock is generally comprised of cellulose, hemicellulose and lignin. Cellulose is one of the abundant biopolymers on earth constituting approximately about 40–50% of the lignocellulosic biomass. Chemically, it is composed of repeating D-glucose units (glucose or glucan units) bonded by β-1,4 linkages (Lynd et al. 2002), and the smallest repetitive unit in cellulose is referred as cellobiose (composed of two glucose units, disaccharide) (Varrot et al. 2003). The degrees of polymerization (DPs) of cellulose chains in primary plant cell walls vary from 5000 to 7500 glucose monomer units, while the DP in wood and cotton cellulose is about 10,000 and 15,000, respectively (O'Sullivan 1997). In cellulose, glucose chains are firmly bounded by van der Waals forces and H linkages (intramolecular and intermolecular) (Gan et al. 2003) into crystalline structures termed as elementary fibril (composed of about 40 glucan chains), about 40 Å wide, 30 Å tick and 100 Å long (Bidlack et al. 1992). Generally cellulose exists in distinct forms such as amorphous and crystalline form, and the region that is eminently disordered (low in crystallinity) is termed as amorphous cellulose. Besides, the natural cellulose also contains least ordered amorphous or paracrystalline regions (Teeri 1997). Hemicellulose is considered as the second largest abundant organic polymer on the earth constituting about 20–30% (Gilbert and Hazlewood 1993). For converting lignocellulose to fuels and chemicals, the utilization of hemicellulose as a byproduct is essential to make overall economics of processing wood into chemicals feasible. This biopolymer is heterogeneous polysaccharide located in the matrix between lignin and cellulose and is composed of both linear and branched polysaccharides (D-xylose, D-mannose, L-fucose, D-glucuronic acid, D-galactose, D-glucose, L-arabinose and L-rhamnose) that may be acetylated or methylated. Majority of the hemicellulase possesses 2–6 of the above sugar units (Biely 1993). However, the DP ranges from 70 to 200, but it can vary among species (Fengel and Wegener 1983). Lignin is the third aromatic biopolymer constituting about 15–20% of the lignocellulosic biomass. It is rigid and tough in nature providing strength to plants, and importantly it acts as a primary line of defence against pathogens preventing plants from infections. Chemically it is composed of *p*-hydroxyphenyl, guaiacyl and/or syringyl monomers linked in three dimensions, and these monomers have different methoxylation patterns in the aromatic ring (Douglas 1996). Lignin acts as a physical barrier to cellulases and thus is considered as one of the major hindrances in cellulosic biofuel technology (Mansfield et al. 1999). Besides, the irreversible binding of cellulases to lignin influences the amount of enzymes needed for the hydrolysis and the recovery of the enzymes after the hydrolysis (Lu et al. 2002). Thus, for the production of cellulosic ethanol, lignin is degraded by means of various chemical and biological techniques. Normally the lignocellulosic biomass shows recalcitrance to microbial action, but the suitable pretreatment brings about disruption in lignin structure in biomass resulting in the increased accessibility of enzymes and thus increasing the biodegradation rate of the biomass (Lynd et al. 2002).

7.3 Cellulases: Green Workhorse for Cellulose Degradation

Cellulases are O-glucoside hydrolase enzymes (GH-EC 3.2.1) that comprise of independently folded and structurally and functionally discrete units, known as modules or domains (Henrissat et al. 1998). Generally, cellulases are composed of one carbohydrate-binding module (CBM) and one catalytic domain (CD) usually joined to the CD by a glycosylate linker (30–44 amino acids). GH is generally classified as cellulases families on the basis of similarity in amino acid sequences (Percival-Zhang et al. 2006). To date, there occur about 125 families, out of which 15 belong to cellulases and 64 families constitute the cellulose-binding domains (CBDs). It has been revealed that the main function of CBD is to provide physical contact of the enzyme to the cellulose polymer besides enhancing its effective concentration. Moreover, it helps the enzyme to spend more time near the proximity of the substrate. These CBDs are usually distributed within 49 families varying from few (30–40 amino acids in family 1) peptides to large domains (200 residues in families 11 and 17). There are few reports that cellulase enzymes without CBDs (possessing only CD) have lower affinity towards cellulose (Karlsson et al. 2002). As cellulase is a multienzyme complex protein, these are usually classified into three categories: endoglucanases, exoglucanases and β-glucosidases.

7.3.1 Endoglucanases or Endo-1,4-β-D-Glucan Glucanohydrolases (EC 3.2.1.4)

Endoglucanases are the hydrolytic enzymes that usually act randomly on the amorphous regions (β-1,4 glycosidic bonds) of the cellulose polymer and carry out indiscriminate nicks, thus producing oligosaccharides (Boisset et al. 2000). Therefore, this class of cellulase enzyme is mainly responsible for reducing the DP of the cellulosic biomass, increasing the specific surface area of the substrate for exocellulase activity and thus making it susceptible to exoglucanase. However, both endoglucanases and exoglucanases exhibit product (glucose/cellobiose) inhibition. Holtzapple et al. (1990) reported that cellobiose exhibits stronger inhibitory effect than glucose. The inhibitory activity of these byproducts depends upon numerous parameters such as enzyme concentration, nature of substrate, binding efficiency and substrate surface area (Gruno et al. 2004).

7.3.2 Exoglucanases or Cellobiohydrolases (CBH) (1,4-β-D-Glucan Cellobiohydrolase; EC 3.2.1.91)

This class of enzyme acts on the reducing and non-reducing ends of the cellulose chain, thus generating a disaccharide molecule, namely, cellobiose, which in turn is acted upon by β-glucosidases. It constitutes about 40–70% of the total cellulase system and possesses a unique ability to depolymerize microcrystalline cellulose (Teeri 1997). CBH exhibits specificity on the ends of cellulose, for instance, *T. reesei* contains two CBH units such as CBHI and CBHII which, respectively, acts on reducing and non-reducing ends (Teeri et al. 1998). Besides, crystalline cellulose which is usually resistant to endoglucanases is degraded by exoglucanases (Zhang and Lynd 2004). However, for efficient hydrolysis of crystalline cellulose, both enzymes (EG and CBH) are needed (Henrissat and Bairoch 1996).

7.3.3 β-Glucosidase or β-D-Glucoside Glucohydrolases (EC 3.2.1.21)

β-Glucosidase enzymes hydrolyse the soluble disaccharide (cellobiose) as well as short-chain (soluble) cello-oligosaccharides to glucose subunits. This reaction is catalysed in the aqueous phase rather than on the surface of the insoluble cellulose particles, which is the case for the 'real' cellulases (EGs and CBHs); besides, it avoids cellobiose inhibition (Zhang and Lynd 2004). For instance, *T. reesei* BG exhibits product (glucose) inhibition (Chen et al. 1992), while *Aspergillus* is more glucose tolerant (Decker et al. 2000). In cellulose hydrolysis, BG is considered as a rate limiting step that is why commercial cellulase is usually supplemented with BG enzyme (Tolan and Foody 1999). The molecular weight of these enzymes varies from 35 to 640 kDa; and they can be monomeric or exist as homo-oligomers as is the case of BG of the yeast *Rhodotorula minuta* (Onishi and Tanaka 1996).

7.4 Synergism in Cellulase Enzyme

Synergistic action is a natural phenomenon in cellulase enzyme for efficient degradation of cellulosic biomass. This phenomenon involves cooperative action between various classes of cellulase enzymes such as EG with CBH, CBH with CBH, EG with EG (e.g. *Gloeophyllum sepiarium* and *G. trabeum* (Mansfield et al. 1998)),

CBH or EG with BG (Woodward 1991), CD with CBD (e.g. CenA of *Cellulomonas fimi*) (Din et al. 1995) or two CDs (Zverlov et al. 1999), cellulose-enzyme-microbe synergism (Schwarz 2001) and spatial synergism for cellulase complexes (i.e. the cellulosome of *C. thermocellum*) (Lynd et al. 2002). However, such synergistic action depends upon various factors like source of cellulase enzyme, features of substrate, concentration and composition of enzyme, cellulase-cellulose affinity, component stereospecificity and enzyme-substrate ratio. The endo-exo synergy is known between set of enzymes such as EG and CBH wherein the former presents the free ends from the cellulose chain to the latter. This synergy helps the CBH to act progressively on reducing as well as non-reducing cellulose chains. Further, the synergy between another set, i.e. CBH and BG, results in the formation of cellobiose as the final product which in turn is acted upon by BG to produce glucose. Although more information has been generated regarding the cellulase action and dynamics of cellulase-cellulose interaction, the crystalline cellulose biodegradation is still an inactive process because of the substrate insolubility as well as poor enzyme accessibility. For solving this problem, scientists optimized the ratio between the cellulolytic enzymes and the enzyme blend: 60:20:20 (CBHI/CBHII/EGI) was found to be optimum for crystalline cellulose saccharification. In this case, for eradicating the cellobiose inhibition, a saturated level of BG was also incorporated (Baker et al. 1998). The main challenge in cellulase enzyme research is to decode the synergistic action between individual constituents (Walker and Wilson 1991). For determining the degree of synergism between various constituents of cellulase enzyme, it becomes indispensable that every component be purified to homogeneity. However, it has been reported that cellulase being complex enzyme is very difficult to break down into its constituents (e.g. *T. reesei* cellulase) (Sprey and Lambert 1983). Moreover, other classes of cellulase enzyme such as EG are very difficult to purify to homogeneity.

7.5 Current Progress in Understanding Enzymatic Action and Enzyme Design

Enzyme technology is recognized as green technology for bioethanol production. Generally three steps are involved in the hydrolysis of cellulosic biomass which comprise of adsorption of enzymes onto cellulose surface, the cellulose biodegradation and desorption of cellulase (Sun and Cheng 2002). As cellulose is a complex structure composed of tightly packed polysaccharide chains, it needs certain supplementary factors that would make the substrate more amenable to hydrolysis (Horn et al. 2012). Besides these cell wall-degrading enzymes that act precisely on polysaccharides, the cellulolytic organisms also secrete certain proteins that aid in cellulose modification and thus enhance its hydrolysis process, and these include ‘expansins’, ‘swollenins’ and ‘loosenins’. These accessory proteins interact with the cellulosic biomass substrate and modify its structure by lengthening or slippage of its constituents and thus increase its surface area so that it is easily accessible to glycosyl hydrolases (Sweeney and Xu 2012). Currently, during transcriptional analysis of *T. reesei*, few cellulose-induced proteins (CIP1/CIP2) exhibit certain synergistic activity with both swollenin and GH61 and are considered as an essential player to break down the

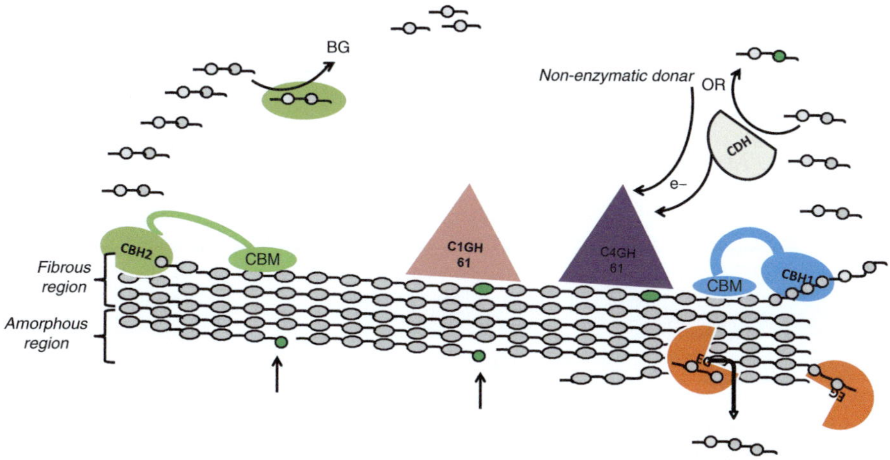

Fig. 7.1 Degradation of cellulose by fungal enzymes in a synergistic manner. *BG* β-Glucisidase, *EG* endoglucanase, *CBH* cellobiohydrolase, *CDH* cellobiose dehydrogenase, *CBM* carbohydrate-binding module

linkage present in the hemicellulose-lignin matrix (Scott et al. 2009). Certain saprophytic microorganisms possess the ability to degrade cellulosic biomass by using surface-linked enzyme complexes known as cellulosomes. These are actually clusters of numerous hemicellulase and cellulase proteins bonded together by a particular linkage between 'dockerin' domains on the enzyme and 'cohesins' linked to structural scaffoldings on surface of microbes (Kumar et al. 2008). Currently genetic engineering plays a vital role to develop hybrid scaffolding molecules that encompass cohesins with distinct binding specificity from various microorganisms in order to develop efficient cellulose-degrading enzymes (Wilson 2009). Morais et al. (2010) reported that the synergistic action between purified xylanases and cellulases produced by thermophilic bacterium (*Thermobifida fusca*) displayed on engineered cellulosomes was found to retain optimum activity when wheat straw was used as a substrate source than the analogous free enzymes. Besides catalytic core, some enzymes contain non-catalytic domains (dockerins and carbohydrate-binding module (CBM)) that play an important role to enhance its activity as it grips the enzyme to desired substrate or onto scaffolding to agglomerate cellulosome and thus disrupt the microfibrils of crystalline cellulose (Sweeney and Xu 2012) (Fig. 7.1). Currently, novel proteins such as bacterial proteins CBM33 (family 33 CBM) in the carbohydrate-active enzyme (CAZy) database and GH61 (fungal proteins belong to family 61 glycoside hydrolases) have been discovered that possess the ability to catalyse oxidative cleavage of polysaccharides (Horn et al. 2012). These novel proteins (copper-dependent monooxygenases) usually produce nicks in the polysaccharide chains by acting on the surfaces of insoluble substrates and thus enhancing their accessibility to cellulolytic enzymes (Horn et al. 2012). Nowadays, cellulases are the main industrial workhorses in biofuel sector, and therefore reduction in the cost of cellulases is major issue. Therefore, current research should be focused to develop novel cellulases with high activity to reduce enzyme requirement in order to hydrolyse cellulosic biomass. Improvement in cellulase should be done by using advanced

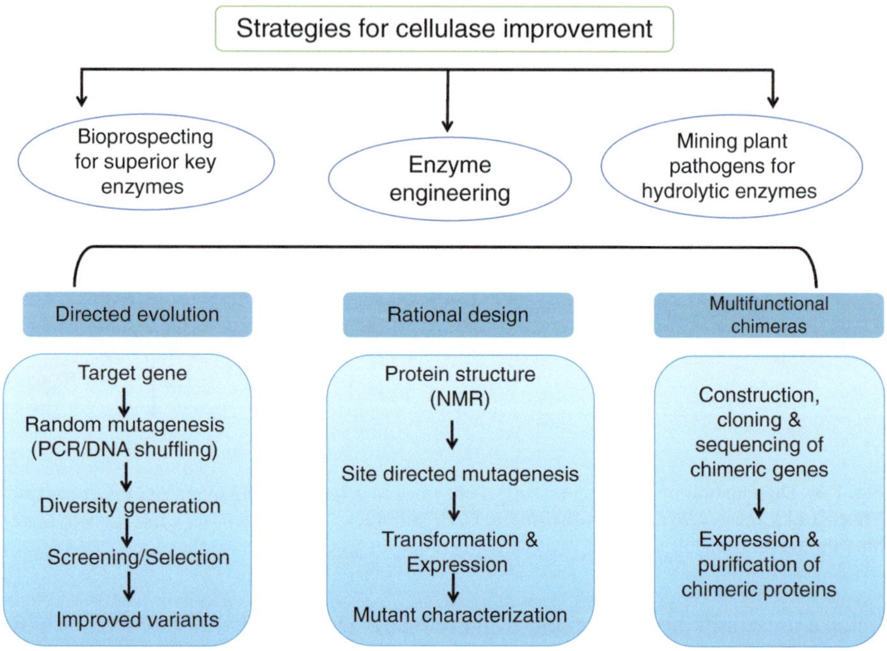

Fig. 7.2 Routes to advancement in cellulase enzyme technology

engineering approaches and is depicted in Fig. 7.2. The main routes involve bioprospection of novel microorganisms with higher cellulase activity and mining plant pathogens for cellulolytic enzymes; besides, it also involves cellulase engineering to enhance the innate diversity of enzymes.

7.6 Perspective

Lignocellulosic biomass is an excellent candidate for bioenergy production because it is abundant, cost-effective and renewable energy resource. Biofuel is a promising alternative to current non-renewable energy resources as it can be employed in automobiles with less modification upon recent technologies and has considerable potential to boost sustainability and decrease GHG. While shifting from fossil fuel to a biofuel-based economy, it would require major advancements in biofuel production, such as (a) bioprospection of specific microorganisms that are highly efficient to degrade biomass (C5 and C6 sugars), (b) high-resistant microorganisms to avoid end product inhibition and (c) ability to decompose lignin content so as to reduce the pretreatment cost. Further, the key issues in cellulosic ethanol technology are to enhance the enzyme activity and reduce its production cost. For this, genetic engineering plays a vital role to improve the efficiency of microorganisms for increase yield as well as to minimize cost of production. It has also been suggested that the use of engineered material (substrate with high sugar content) integrating with latest conversion technologies could make the ethanol cost-effective. The cellulase enzymes will generally find its way for utilization in various industrial processes, and the demand for the enzymes with high stability, potency and specificity will be also growing rapidly. So, cellulase enzyme will be the most stirring technology of today's times gaining the worldwide attention.

References

Adsul MG, Singhvi MS, Gaikaiwari SA, Gokhale DV (2011) Development of biocatalysts for production of commodity chemicals from lignocellulosic biomass. Bioresour Technol 102:4304–4312. doi:10.1016/j.biortech.2011.01.002

Baker JO, Ehrman CI, Adney WS, Thomas SR, Himmel ME (1998) Hydrolysis of cellulose using ternary mixtures of purified celluloses. Appl Biochem Biotechnol 72:395–403. doi:10.1007/BF02920154

Bidlack J, Malone M, Benson R (1992) Molecular structure and component integration of secondary cell wall in plants. Proc Okla Acad Sci 72:51–56

Biely P (1993) Biochemical aspects of the production of microbial hemicellulases. In: Coughlan MP, Hazlewood GP (eds) Hemicellulose and Hemicellulases. Portland Press, Cambridge, pp 29–51

Boisset C, Fraschini C, Schulein M, Henrissat B, Chanzy H (2000) Imaging the enzymatic digestion of bacterial cellulose ribbons reveals endo character of the cellobiohydrolase Cel6A from *Humicolainsolens* and its mode of synergy with cellobiohydrolase Cel7A. Appl Environ Microbiol 66:1444–1452. doi:10.1128/AEM.66.4.1444-1452

Chen H, Hayn M, Esterbauer H (1992) Purification and characterization of two extracellular beta-glucosidases from *Trichoderma reesei.* Biochem Biophys Acta 1121:54–60. doi:10.1016/0167-4838(92)90336-C

Decker CH, Visser J, Schreier P (2000) β-glucosidases from five black *Aspergillus* species: study of their physico-chemical and biocatalytic properties. J Agric Food Chem 48:4929–4936. doi:10.1021/jf000434d

Din N, Coutinho JB, Gilkes NR et al (1995) Interactions of cellulases from *Cellulomonas fimi* with cellulose. Prog Biotechnol 10:261–270. doi:10.1016/S0921-0423(06)80109-7

Douglas CJ (1996) Phenylpropanoid metabolism and lignin biosynthesis: from weeds to trees. Trends Plant Sci 1:171–178. doi:10.1016/1360-1385(96)10019-4

Fengel D, Wegener G (1983) Wood: chemistry, ultrastructure and reactions. Walter de Gruyter& Co, Berlin, p 613

Gan Q, Allen SJ, Taylor G (2003) Kinetic dynamics in heterogeneous enzymatic hydrolysis of cellulose: an overview, an experimental study and mathematical modeling. Process Biochem 38:1003–1018. doi:10.1016/S0032-9592(02)00220-0

Ghimire A, Frunzo L, Pirozzi F, Trably E, Escudie R, Lens PN, Esposito G (2015) A review on dark fermentative biohydrogen production from organic biomass: process parameters and use of by-products. Appl Energy 144:73–95. doi:10.1016/j.apenergy.2015.01.045

Gilbert HJ, Hazlewood GP (1993) Bacterial cellulases and xylanases. J Gen Microbiol 139:187–194. doi:10.1099/00221287-139-2-187

Gruno M, Valjamae P, Pettersen G, Johansson G (2004) Inhibition of the *Trichoderma reesei* cellulases by cellobiose is strongly dependent on the nature of the substrate. Biotechnol Bioeng 86:503–511. doi:10.1002/bit.10838

Henrissat B, Bairoch A (1996) Updating the sequence-based classification of glycosyl hydrolases. Biochem J 316:695–696. doi:10.1042/bj3160695

Henrissat B, Teeri TT, Warren RAJ (1998) A scheme for designating enzymes that hydrolyze the polysaccharides in the cell walls of plants. FEBS Lett 425:352–354. doi:10.1016/S0014-5793(98)00265-8

Himmel ME, Ding SY, Johnson DK, Adney WS, Nimlos MR, Brady JW, Foust TD (2007) Biomass recalcitrance: engineering plants and enzymes for biofuels production. Science 315:804–807. doi:10.1126/science.1137016

Holtzapple M, Cognata M, Shu Y, Hendrickson C (1990) Inhibition of *Trichoderma reesei* cellulase by sugars and solvents. Biotechnol Bioeng 36:275–287

Horn SJ, Vaaje-Kolstad G, Westereng B, Eijsink VGH (2012) Novel enzymes for the degradation of cellulose. Biotechnol Biofuels 5:45. doi:10.1186/1754-6834-5-45

International Energy Agency (2012) World energy outlook. OECD/IEA, Paris

Karlsson J, Siika-aho M, Tenkanen M, Tjerneld F (2002) Enzymatic properties of the low molecular mass endoglucanases Cel12A (EG III) and Cel45A (EG V) of *Trichoderma reesei*. J Biotechnol 99:63–78. doi:10.1016/S0168-1656(02)00156-6

Kumar R, Singh S, Singh OV (2008) Bioconversion of lignocellulosic biomass: biochemical and molecular perspective. Ind J Microbiol Biotechnol 35:377–391. doi:10.1007/s10295-008-0327-8

Lu Y, Yang B, Gregg D, Saddler JN, Mansfield SD (2002) Cellulase adsorption and an evaluation of enzyme recycle during hydrolysis of steam-exploded softwood residues. Appl Biochem Biotechnol 98-100:641–654. doi:10.1385/ABAB:98-100:1-9:641

Lynd LR, Laser MS, Bransby D, Dale BE, Davison B, Hamilton R, Himmel M, Keller M, McMillan JD, Sheehan J, Wyman CE (2008) How biotech can transform biofuels. Nat Biotechnol 26:169–172. doi:10.1038/nbt0208-169

Lynd LR, Weimer PJ, van Zyl WH, Pretorius IS (2002) Microbial cellulose utilization: fundamentals and biotechnology. Microbiol Mol Biol Rev 66:506–577. doi:10.1128/MMBR.66.3.506-577

Mansfield S, Saddler J, Gübitz G (1998) Characterization of endoglucanases from the brown rot fungi *Gloeophyllum sepiarium* and *Gloeophyllum trabeum*. Enzyme Microb Technol 23(1–2):133–140. doi:10.1016/S0141-0229(98)00033-7

Mansfield SD, Mooney C, Saddler JN (1999) Substrate and enzyme characteristics that limit cellulose hydrolysis. Biotechnol Proc 15:804–816. doi:10.1021/bp9900864

Monserrate E, Leschine SB, Canale-Parola E (2001) Clostridium hungatei sp. Nov. a mesophilic, N2-fixing cellulolytic bacterium isolated from soil. Int J Syst Evol Microbiol 51:123–132

Morais S, Barak Y, Caspi J, Hadar Y, Lamed R, Shoham Y, Wilson DB, Bayer EA (2010) Cellulase-xylanase synergy in designer cellulosomes for enhanced degradation of a complex cellulosic substrate. mBio 1:00285–00210. doi:10.1128/mBio.00285-10

Naik SN, Goud VV, Rout PK, Dalai AK (2010) Production of first and second generation biofuels: a comprehensive review. Renew Sust Energy Rev 14:578–597. doi:10.1016/j.rser.2009.10.003

O'Sullivan AC (1997) Cellulose: the structure slowly unravels. Cellulose 4:173–207. doi:10.1023/A:1018431705579

Onishi N, Tanaka T (1996) Purification and properties of a galacto- and gluco-oligosaccharide-producing beta-glycosidase from *Rhodotorula minuta* IFO879. J Ferment Bioeng 82:439–443

Percival-Zhang YH, Himmel ME, Mielenz JR (2006) Outlook for cellulase improvement: screening and selection strategies. Biotechnol Adv 24:452–481. doi:10.1016/j.biotechadv.2006.03.003

Pirzadah T, Garg S, Singh J, Vyas A, Kumar M, Gaur N, Bala M, Rehman R, Varma A, Kumar V, Kumar M (2014) Characterization of Actinomycetes and Trichoderma spp. for cellulase production utilizing crude substrates by response surface methodology. Spring 3:622. doi:10.1186/2193-1801-3-622

Saggi SK, Dey P (2016) An overview of simultaneous saccharification and fermentation of starchy and lignocellulosic biomass for bio-ethanol production. Biofuels. doi:10.1080/17597269.2016.1193837

Schwarz WH (2001) The cellulosome and cellulose degradation by anaerobic bacteria. Appl Microbiol Biotechnol 56:634–649. doi:10.1007/s002530100710

Scott BR, Hill C, Tomashek J, Liu C (2009) Enzymatic hydrolysis of lignocellulosic feedstocks using accessory enzymes. United States Patent Application-0061484

Sprey B, Lambert C (1983) Titration curves of cellulases from *Trichoderma reesei*: demonstration of a cellulase xylanase glucosidase containing complex. FEMS Microbiol Lett 18:217–222. doi:10.1111/j.1574-6968.1983.tb00481.x

Sun Y, Cheng J (2002) Hydrolysis of lignocellulosic materials for ethanol production. Bioresour Technol 83:1–11. doi:10.1016/S0960-8524(01)00212-7

Sweeney MD, Xu F (2012) Biomass converting enzymes as industrial biocatalysts for fuels and chemicals. Recent developments. Catalysts 2:244–263. doi:10.3390/catal2020244

Teeri T (1997) Crystalline cellulose degradation: new insight into the function of cellobiohydrolases. Trends Biotechnol 15:160–167. doi:10.1016/S0167-7799(97)01032-9

Teeri T, Koivula A, Linder M, Wohlfahrt G, Divne C, Jones T (1998) *Trichoderma reesei* cellobiohydrolases: why so efficient on crystalline cellulose? Biochem Soc Trans 26:173–178. doi:10.1042/bst0260173

Thomas CD, Cameron A, Green RE, Bakkenes M, Beaumont LJ, Collingham YC, Erasmus BFN, Ferreira De Siqueira M, Grainger A, Hannah L, Hughes L, Huntley B, Van Jaarsveld AS, Midgley GF, Miles L, Ortega-Huerta MA, Peterson AT, Phillips OL, Williams SE (2004) Extinction risk from climate change. Nature 427:145–148. doi:10.1038/nature02121

Tolan JS, Foody B (1999) Cellulases from submerged fermentation. Adv Biochem Eng Biotechnol 65:41–67. doi:10.1007/3-540-49194-5_3

Varrot A, Frandsen TP, von Ossowski I, Boyer V, Cottaz S, Driguez H, Sculein M, Davies GJ (2003) Structural basis for ligand binding and processivity in cellobiohydrolase Cel6A from *Humicola insolens*. Structure 11:855–864. doi:10.1016/S0969-2126(03)00124-2

Walker L, Wilson D (1991) Enzymatic hydrolysis of cellulose: an overview. Bioresour Technol 36:3–14. doi:10.1016/0960-8524(91)90095-2

Wang Y, Tang R, Tao J, Gao G, Wang X, Mu Y, Feng Y (2011) Quantitative investigation of non-hydrolytic disruptive activity on crystalline cellulose and application to recombinant swollenin. Appl Microbiol Biotechnol 91:1353–1363. doi:10.1007/s00253-011-3421-1

Wilson DB (2008) Three microbial strategies for plant cell wall degradation. Ann N Y Acad Sci 1125:289–297. doi:10.1196/annals.1419.026

Wilson DB (2009) Cellulases and biofuels. Curr Opin Biotechnol 20:295–299. doi:10.1016/j.copbio.2009.05.007

Woodward J (1991) Synergism in cellulase systems. Bioresour Technol 36:67–75. doi:10.1016/0960-8524(91)90100-X

Yang H, Wei H, Ma G, Antunes MS, Vogt S, Cox J, Zhang X, Liu X, Bu L, Gleber SC, Carpita NC (2016) Cell wall targeted in planta iron accumulation enhances biomass conversion and seed iron concentration in Arabidopsis and rice. Plant Biotechnol J 14:1998–2000. doi:10.1111/pbi.12557

Zhang YHP, Lynd LR (2004) Toward an aggregated understanding of enzymatic hydrolysis of cellulose: non-complexed cellulase systems. Biotechnol Bioeng 88:797–824. doi:10.1002/bit.20282 ·

Zhou F, Chen H, Xu Y (2010) GASdb: a large-scale and comparative exploration database of glycosyl hydrolysis systems. BMC Microbiol 10:69. doi:10.1186/1471-2180-10-69

Zhu Z, Sathitsuksanoh N, Vinzant T, Schell DJ, McMillan JD, Zhang YHP (2009) Comparative study of corn stover pretreated by dilute acid and cellulose solvent-based lignocellulose fractionation: enzymatic hydrolysis, supramolecular structure, and substrate accessibility. Biotechnol Bioeng 103:715–724. doi:10.1002/bit.22307

Zverlov VV, Volkov IY, Lunina NA, Velikodvorskaya GA (1999) Enzymes of thermophilic anaerobic bacteria hydrolyzing cellulose, xylan, and other b-glucans. J Mol Biol 33:89–95

Part IV

Metabolic Engineering in Yeast

Engineering *Saccharomyces cerevisiae* for C5 Fermentation: A Step Towards Second-Generation Biofuel Production

8

Farnaz Yusuf and Naseem A. Gaur

Abstract

Recently, biofuels are receiving increased attention because of the rising cost of fossil fuels and global warming. Huge attention is being paid to the development of technologies for second-generation biofuel production as it could be the long-term promising alternative to fossil fuels. Lignocellulosic material is the most abundant renewable source of biomass on earth and has been widely considered for bio-alcohol production. An optimum ethanol yield from lignocellulosic material is critical for successful industrial ethanol production and depends on fermentation of both C5 and C6 sugars present in the biomass. Xylose is the second highest sugar after glucose in the lignocellulosic biomass therefore, it is necessary to ferment xylose along with glucose. *Saccharomyces cerevisiae* is the most preferred microorganism for industrial ethanol production and has several advantages over other microbes, including higher ethanol productivity and high ethanol and pretreatment inhibitor tolerance. However, *S. cerevisiae* lacks the functionally active xylose-fermenting pathway. Therefore, genetic engineering of *S. cerevisiae* to transport and ferment xylose in a glucose/xylose mix has been an intense area of research. Over the past few decades, efforts are being made to construct recombinant yeast strains capable of fermenting xylose along with glucose. This chapter highlights various genetic engineering approaches carried out for developing C5 fermenting *S. cerevisiae* strains.

Keywords

Biofuel • Lignocellulosic biomass • Fermentation • *S. cerevisiae* • Xylose

F. Yusuf • N.A. Gaur (✉)
Yeast Biofuel Group, International Centre for Genetic Engineering and Biotechnology (ICGEB), Aruna Asaf Ali Road, New Delhi 110067, India
e-mail: naseem@icgeb.res.in

V.C. Kalia, A.K. Saini (eds.), *Metabolic Engineering for Bioactive Compounds*,
DOI 10.1007/978-981-10-5511-9_8

8.1 Introduction

Ever-increasing fossil fuel demand and price along with the growing concern about global warming due to increased greenhouse gas emissions has refocused the attention to second-generation biofuel production. Due to their sustainability and low impact on environment and health, fossil fuels are being replaced with biofuels. It is believed that replacing fossil fuels with biofuel could help in decreasing the carbon dioxide emissions nearly by 60–90% (Wang et al. 2007). Worldwide, cellulosic ethanol technology is one of the most important technologies for the second-generation biofuel production (Stephanopoulos 2007; Chu and Majumdar 2012). In the last one decade, biofuel production has increased fivefold (Lennartsson et al. 2014). In the USA and Brazil, bioethanol is currently being used as a blending component with petroleum.

The most abundant feedstock material available for second-generation bioethanol production includes agricultural wastes like crop straws, husks, corn stover and bagasse, forestry residues, waste paper, short rotation woody crops, and industrial and municipal wastes (Kuhad et al. 2011; Jouzani and Taherzadeh 2015). Lignocellulosic biomass consists of cellulose, hemicellulose, and lignin. Cellulose is the most abundant polymer in biomass (40–60% of the total dry weight) and consists of linear subunits of the glucose linked by β-1-4-glycosidic bond. Hemicellulose is the second most abundant component of the biomass (20–40%) and consists of C5 and C6 sugar including glucose, xylose, mannose, arabinose, etc. It is primarily composed of xylan linkages including α-xylose (90%) and L-arabinose (10%) (Scheller and Ulvskov 2010). The degree of branching and the xylan composition vary with the nature and the source of raw materials. Lignin is the third major component of lignocellulosic biomass, and it is responsible for the structural rigidity to plant cells (Galbe and Zacchi 2007). Although the percentage of each component varies among different types of biomass, cellulose remains the most abundant fraction followed by hemicellulose (Saha 2003) (Table 8.1).

Conversion of lignocellulosic material to biofuels involves three major sequential steps: (a) physicochemical and chemical pretreatment of the lignocellulosic biomass, (b) hydrolysis of celluloses and hemicelluloses to sugars by cellulases, and

Table 8.1 Potential lignocellulosic biomass sources and composition (Anwar et al. 2014)

Raw material	Hemicelluloses (%)	Cellulose (%)	Lignin (%)
Agricultural wastes	25–50	37–50	5–15
Grasses	35–50	25–40	–
Softwood	25–29	40–45	30–60
Hardwood	25–40	45–47	20–25
Switch grass	30–35	40–45	12
Newspaper	25–40	40–55	18–30
Corncob	34.7	31.7	20.3
Wheat straw	21.2	38.2	23.4
Rice straw	24	37	14

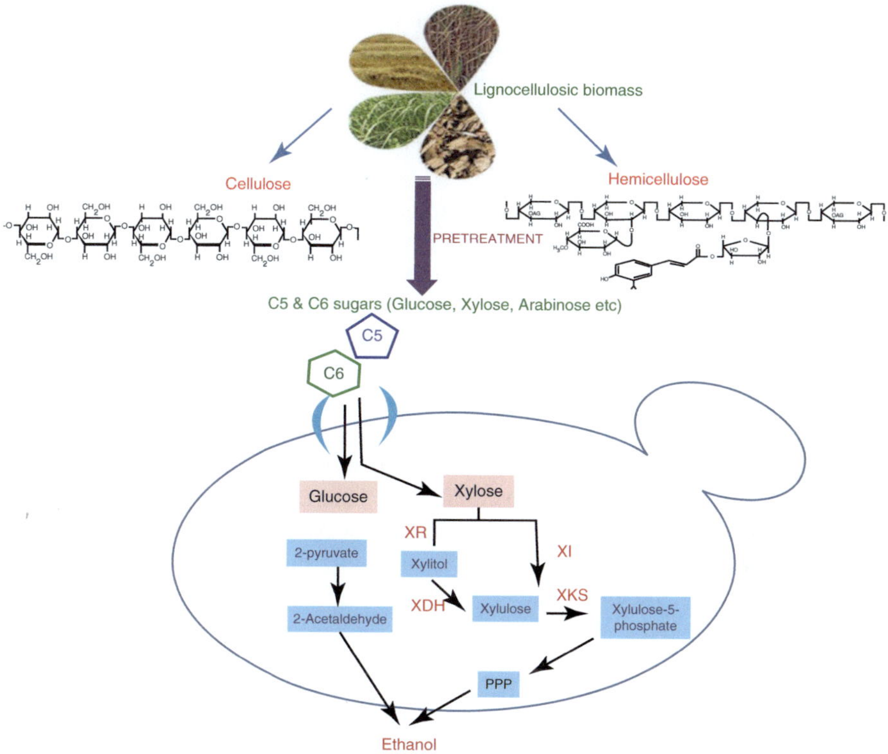

Fig. 8.1 Bioconversion of lignocellulosic biomass to biofuel by recombinant yeast: schematic representation of lignocellulosic biomass pretreatment and fermentation of C5 and C6 sugars to bioethanol by recombinant *Saccharomyces cerevisiae*

(c) fermentation of released sugars to ethanol or higher alcohols. The number of pretreatment technologies, like chemical pretreatment (alkali, acid, ozonolysis, organosolv, and ionic liquids), physical pretreatment (microwave and extrusion, grinding, and milling), physicochemical pretreatment (liquid hot water, steam explosion, ammonia fiber explosion, CO_2 explosion, and wet oxidation), and biological pretreatment (using microorganisms), has been developed (Palmqvist and Hahn-Hagerdal 2000; Isroi et al. 2011; Sarkar et al. 2012; Mood et al. 2013, 2014; Rajendran and Taherzadeh 2014). Pretreatment of lignocellulosic material for biofuel production attributes to about 18–20% of the total cost. Pretreatment helps to overcome the chemical and physical obstacles present in lignocellulosic biomass in order to enhance enzyme accessibility and thereby increase the fermentable sugar yields (Zhang et al. 2015). Xylose is the second highest sugar in the lignocellulosic biomass comprising up to 35% of the total carbohydrates (Hahn-Hagerdal et al. 2006; van Maris et al. 2007; Zhou et al. 2012; Kim et al. 2013; Latimer et al. 2014) therefore, efficient utilization of xylose along with glucose has to be ensured for economically feasible production of lignocellulosic biofuels (Fig. 8.1). The use of lignocellulosic biomass for biofuel generation is complex not only because the polysaccharides are more stable but also due to the fact that the pentose sugars are not fermented by *Saccharomyces cerevisiae*.

Since native xylose-utilizing organisms exist, they either lack well-developed genetic tools for host engineering or exhibit low product and inhibitor tolerances. Therefore, emphasis is given to develop *S. cerevisiae* host platforms with more efficient xylose utilization and fermentation. *S. cerevisiae* is the most preferred microbe for fermentation due to its high ethanol productivity, high tolerance to low pH, and robust growth (Chu and Lee 2007; Hector et al. 2011; Li et al. 2015; Kavscek et al. 2015). In addition, *S. cerevisiae* serves as an excellent model organism with its genome sequenced; its genetics is well studied and arrays of genetic engineering tools are available (Kavscek et al. 2015). However, it cannot metabolize xylose, has lower tolerance to pretreatment inhibitors, and does not produce cellulases. These limitations increase the cost and largely decrease the efficiency of ethanol production. To overcome this, various biotechnological approaches are being currently used, including the development of xylose-fermenting strains of *S. cerevisiae*, engineering of enzymes that break down cellulose and hemicellulose into simple sugars, and use of yeast species that naturally ferment xylose. The commercialization of cellulose-to-ethanol conversion remains challenging due to various limitations in process technology and microbial physiology.

8.2 Xylose-Assimilating Microbes and Pathways

Xylose-fermenting microorganisms include bacteria, filamentous fungi, and yeast species (Skoog and Hahn-Hagerdal 1988). Among the bacteria, *Aerobacter*, *Aeromonas*, *Bacillus*, *Bacteroides*, *Erwinia*, *Klebsiella*, *Clostridium*, and *Thermoanaerobacter* (Kuhad et al. 2011) are known to naturally ferment xylose. However, lower ethanol tolerance and secondary metabolite production during fermentation discouraged the use of these bacteria in second-generation industrial ethanol production. Fungal xylose-metabolizing genera include *Fusarium*, *Piromyces* (Kuyper et al. 2005; van Maris et al. 2007), *Orpinomyces* (Madhavan et al. 2009), *Aspergillus*, *Monilia*, *Neurospora*, *Mucor*, *Polyporus*, *Paecilomyces*, and *Rhizopus* (Kuhad et al. 2011). They often exhibit high yields but suffer from several drawbacks like low ethanol productivity, long fermentation period, and formation of large amounts of by-products. The most widely studied xylose-fermenting yeasts include *Candida shehatae* (Preez et al. 1986), *Pachysolen tannophilus* (Schneider et al. 1983), and *Pichia stipitis* (Slininger et al. 1990). *P. stipitis* and *C. shehatae* are the best native xylose fermenters producing 0.51 g ethanol/g xylose. In bacteria and some anaerobic fungi, xylose is metabolized by XI pathway which involves xylose isomerase enzyme. However, in yeast and aerobic fungi, xylose is metabolized by XR/XDH pathway. This pathway is more complex and involves reduction and oxidation reactions with the help of NAD (P) H and NAD $(P)^+$ cofactors (Fig. 8.2). Interestingly, *S. cerevisiae* cannot utilize and ferment xylose to ethanol despite of the fact that the genes encoding the oxido-reductive pathway enzymes are present in its genome (Batt et al. 1986).

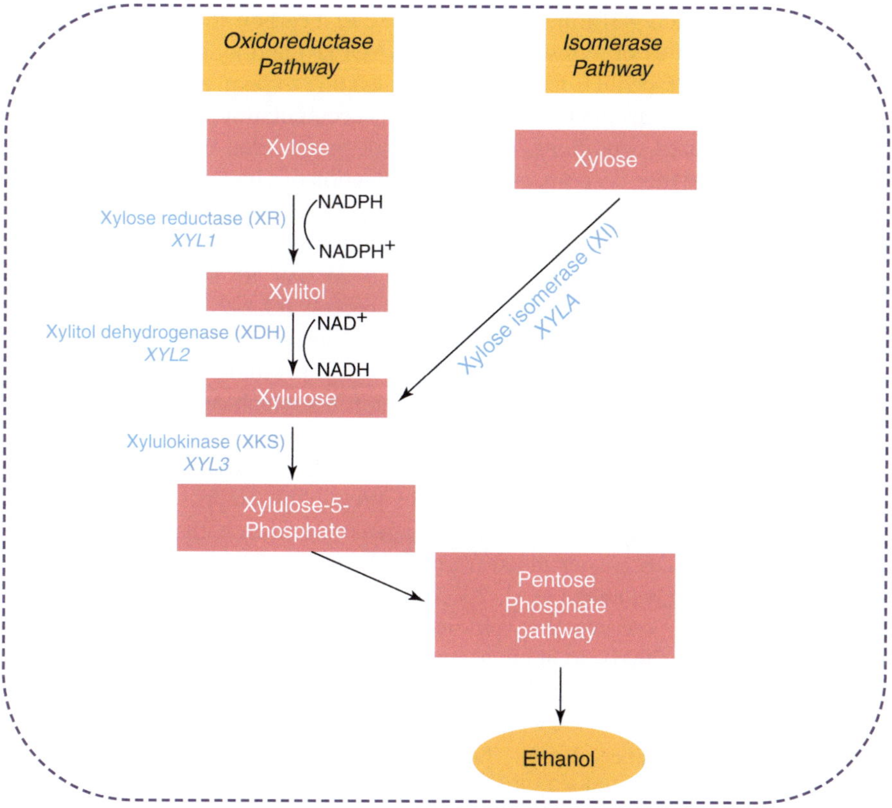

Fig. 8.2 Two different xylose-metabolizing pathways: the xylose metabolic pathway found in bacteria and fungi

8.3 Engineering *Saccharomyces cerevisiae* for Xylose Fermentation

Metabolic engineering is widely used and a powerful tool for genetic manipulation and construction of heterologous pathway in a desired host organism. *S. cerevisiae* is not capable of assimilating xylose naturally; therefore, developing strains capable of efficiently fermenting xylose along with glucose is a major challenge. The following approaches have been used for developing xylose-fermenting yeast strains.

8.3.1 Metabolic Engineering Using XR/XDH Pathway

The construction of oxidoreductase pathway in *S. cerevisiae* was first carried out by Kotter. The heterologous expression of the XR and XDH genes derived from the xylose-utilizing yeast *Pichia stipitis* was carried out in *S. cerevisiae* (Kotter et al. 1990). Surprisingly, the recombinant *S. cerevisiae* strain showed very low growth and poor fermentation performance using xylose as a carbon source. Difference in coenzyme specificities of xylose reductase ($NADPH^+$) and xylitol dehydrogenase ($NADP^+$) caused intracellular redox imbalance leading to poor growth and fermentation (Ostergaard et al. 2000). This results in accumulation of NADH because under oxygen-limited conditions, NADH cannot be recycled through respiration. This cofactor imbalance is compensated by production of glycerol, but the assimilation of xylose leads to the accumulation of xylitol in the cell. To resolve this issue, one approach is to alter the affinity of the protein for a substrate (K_m) without significantly reducing the turnover number (K_{cat}) so that its specificity (K_{cat}/K_m) increases while maintaining a high catalytic efficiency. It has been shown that protein engineering of *P. stipitis XR* gene by changing one residue in the putative phosphate pocket that binds NADPH changes the cofactor affinity and concentration (Watanabe et al. 2007). Mutation was carried out in XR which decreased the affinity for NADPH than the wild-type enzyme, and it was found that expressing such enzyme shifted the utilization of cofactor from NADPH to NADH, thereby increasing the ethanol yield with decreased xylitol production (Jeppsson et al. 2006). Similarly by using multiple site-directed mutagenesis approach, the cofactor preference of *P. stipitis* XDH was changed from NAD^+ to $NADP^+$ resulting in improved ethanol production (Watanabe et al. 2005). Similarly, a mutant of *Gluconobacter oxydans* XDH was constructed, and it was able to use $NADP^+$ exclusively (Ehrensberger et al. 2006). In another approach, cofactor imbalance problem was resolved by increasing the affinity of XDH to $NADP^+$ by introducing the $NADP^+$ recognition sequence from *Thermoanaerobium brockii* to the *XYL2* gene of *P. stipitis* (Metzger and Hollenberg 1995). There was increase in specific activity of xylose reductase and ethanol yield by 25% when the deletion of NADPH-dependent glutamate dehydrogenase was carried out along with the overexpression of an NADH-dependent glutamate dehydrogenase as it resulted in the shift of ammonia assimilation from NADPH to NADH dependent (Grotkjaer et al. 2005). It was observed that expression of xylulokinase gene from *P. stipitis* (*XYL3*) helps in decreasing the xylitol production in addition to creating a complete heterologous pathway (Bettiga et al. 2008).

8.3.2 Metabolic Engineering Using Xylose Isomerase Pathway

The isomerase pathway involves the metal ion-dependent enzyme xylose isomerase (XI), encoded by the *XYLA* gene which catalyzes the isomerization of xylose to xylulose in bacteria and some anaerobic fungi (Jeffries and Jin 2000; Van Maris et al. 2007). The XI pathway bypasses the cofactor requirements and hence eliminates the cofactor imbalance created in XR/XDH pathway. A number of bacterial *XYLA* genes coding for XIs have been heterogeneously expressed in *S. cerevisiae*, including *XYLA* from *Escherichia coli* (Sarthy et al. 1987), *Bacillus subtilis* or *Actinoplanes missouriensis* (Amore et al. 1989), *Clostridium thermosulfurogenes* (Moes et al. 1996), *Thermus thermophilus* (Walfridsson et al. 1996), and

Streptomyces rubiginosus (Gardonyi and Hahn-Hagerdal 2003). Since most of the XIs expressed in yeast are native to bacteria, they could not be expressed fully with optimum enzyme activity. This could be due to posttranslational modifications, improper protein folding, and inter- and intramolecular disulfide bridge formation (Gardonyi and Hahn-Hagerdal 2003). The first attempt to express the prokaryotic gene *XYLA* in *S. cerevisiae* was from *Thermus thermophilus*, and the recombinant gene was active at 85 °C. However, the recombinant strain could not convert xylose to ethanol efficiently due to high temperature and optimum pH requirements (Walfridsson et al. 1996). The problem of functional expression of xylose isomerase in yeast was resolved by expression of xylose isomerase gene from anaerobic rumen fungus *Piromyces* (Kuyper et al. 2003, 2005; van Maris et al. 2007) and *Orpinomyces* (Madhavan et al. 2009). Bacterial XI gene from anaerobic bacterium *Clostridium phytofermentans* was reported to be more robust to xylitol tolerance than the enzyme from the *Piromyces* strain (Brat et al. 2009). This new enzyme showed low sequence similarities to the xylose isomerase of *Thermus thermophilus* and *Piromyces* sp. As isomerase-based pathway does not require any cofactor, therefore no by-product is necessarily produced to compensate for cofactor imbalance. It is believed that heterologous expression of xylose isomerase pathway in yeasts provides an opportunity to bypass the redox cofactor imbalance problems associated with oxido-reductase pathway and would help in enhancing the ethanol yield.

8.3.3 Overexpression of Xylulokinase (Xyl3) Gene

It was observed that recombinant yeast strains expressing xylose reductase and xylitol dehydrogenase genes from *Pichia stipitis* poorly ferment xylose to ethanol. One of the possible reasons is the low activity of endogenous xylulokinase which is the next downstream enzyme in the xylose metabolic pathway (Jin and Jeffries 2003; Jeffries 2006). Recombinant *S. cerevisiae* overexpressing the *Pichia stipitis* xylose reductase and xylitol dehydrogenase enzymes usually accumulates xylitol as the primary by-product from xylose fermentations suggesting that XKS is a metabolic roadblock to further metabolize xylose (Hallborn et al. 1991; Takuma et al. 1991). The endogenous xylulokinase gene was first successfully overexpressed in a strain which also expressed *P. stipitis* XR and XDH. This strain was able to carry out the fermentation of xylose as well as xylose and glucose co-fermentation (Ho et al. 1998). Similarly, in other study, it was found that the overexpression of endogenous xylulokinase (XKS) in *S. cerevisiae* along with the *Pichia stipitis* XR and XDH helps in efficient conversion of xylose to ethanol (Toivari et al. 2001). These studies demonstrated that native XKS activity was not sufficient for xylulose fermentation, and hence its overexpression was required for improved ethanol yields. It was found that overexpression of XKS gene in recombinant xylose utilizing *S. cerevisiae* produces ethanol from xylose (Eliasson et al. 2000). Later, it was found that only fine-tuned overexpression of XKS in *S. cerevisiae* led to improved ethanol production (Jin et al. 2003).

8.3.4 Deletion of *GRE3* Gene

Xylose isomerase pathway bypasses cofactor requirements; therefore, it was widely believed that the heterologous expression of xylose isomerase in *S. cerevisiae* would significantly improve ethanol yield. Interestingly, most of the xylose isomerase-based pathways reported in *S. cerevisiae* often suffer from low ethanol productivity, low xylose consumption rates, poor cell growth, and high xylitol production (Karhumaa et al. 2007; Lee et al. 2012). This could be explained by the fact that the *S. cerevisiae* have aldolase reductase encoded by *GRE3* gene which causes the reduction of xylose to xylitol, thereby hampering the overall ethanol yield and growth of recombinant strains (Traff et al. 2001). Xylitol has a negative effect on xylose isomerase (XI) gene; it strongly inhibits the activity of xylose isomerase. Therefore, deletion of *GRE3* gene is believed to improve the xylose assimilation in xylose isomerase pathway expressing *S. cerevisiae* strains (Traff et al. 2001). Since aldolase reductase belongs to stress-induced proteins, its deletion results in consequences of poor growth. Deletion of *GRE3* gene reduces the growth by 30% (Lönn et al. 2003), thereby limiting its deletion in industrial strains (Hahn-Hägerdal et al. 2007).

8.3.5 Modifications in Downstream Pentose Phosphate Pathway

Transketolase (*TKL1*) and transaldolase (*TAL1*) are one of the two enzymes of the downstream pentose phosphate pathway which are considered rate limiting for efficient fermentation of xylose. Transketolase catalyzes conversion of xylose-5-phosphate to sedoheptulose-7-phosphate and glyceraldehyde-3-phosphate, and transaldolase converts sedoheptulose-7-phosphate to fructose-6-phosphate and erythrose-4-phosphate (Senac and Hahn-Hagerdal 1991). It was observed that the *Pichia stipitis TAL1* gene seems to be more advantageous when overexpressed along with endogenous *TAL1* in *S. cerevisiae* (Walfridsson et al. 1995). Deletion of either *TAL1* or *TKL1* led to decreased growth and ethanol production from xylose, demonstrating the essential nature of these enzymes in xylose fermentation (Matsushika et al. 2012). Later it was found that the overexpression of other two enzymes of pentose phosphate pathway, i.e., ribulose-5-phosphate 4-epimerase (*RPE1*) and ribulokinase (*RKI1*) along with *TKL1* and *TAL1*, is important in efficient xylose fermentation (Johansson and Hahn-Hagerdal 2002; Matsushika et al. 2012).

Transcriptomics played an important role for identification of the target genes required for improved xylose fermentation in recombinant strains. Cells grown in xylose showed increased mRNA levels of pentose phosphate pathway genes including *TAL1*, *TKL1*, *RPE1*, and *RKI1*. Interestingly, many genes such as galactose metabolic pathway enzymes which are not directly involved in xylose metabolism were also overexpressed indicating that xylose metabolism is controlled by the expression of a variety of genes that involves complex genetic and biochemical interactions (Kricka et al. 2015).

8.3.6 Improvement in Transport of Pentose Sugars

Native xylose-metabolizing yeasts possess both xylose-specific and xylose-non-specific transport systems (Leandro et al. 2006). However, *Saccharomyces cerevisiae* lack the xylose-specific transporters. Therefore, xylose is taken inside the cell through glucose transporters encoded by the *HXT* gene family (Hamacher et al. 2002). The hexose transporters in *S. cerevisiae* include Hxt7p, Hxt5p, Hxt4p, Hxt2p, Hxt1p, and Gal2p (Saloheimo et al. 2007). In a mixed sugar medium, glucose is preferentially transported into the cells as xylose has lower affinity for the hexose transporters. This problem has been addressed by the overexpression of various native hexose transporters which increase the xylose transport into the cells; however, ethanol production was not significantly increased (Young et al. 2011; Tanino et al. 2012). It has been found that native *S. cerevisiae* expresses Hxt5p and Hxt7p in large amounts when xylose is used as a carbon source; thus, these two HXT transporters were thought to be important for transport of xylose. Another approach involves the introduction of genes encoding xylose transporters into *S. cerevisiae.* Interestingly, overexpression of genes such as *GXF1/GXS1* from *Candida intermedia*, *Trxlt1* from *T. reesei*, *SUT1* from *P. stipitis*, and *At5g17010* from *Arabidopsis thaliana* has increased xylose uptake in mixed sugar fermentations and has shown improved ethanol productivity (Leandro et al. 2006; Saloheimo et al. 2007; Runquist et al. 2010; Kricka et al. 2015). Adaptive evolutionary approach has also helped in improving the xylose uptake. It was found that when an engineered yeast strain expressing heterologous xylose utilization pathway was grown using xylose as the sole carbon source, the strain showed significant growth in xylose. The evolved strain was found to have a single amino acid change in the hexose transporter *HXT7* (F79S) coding sequence to be responsible for the evolved phenotype (Apel et al. 2015).

8.3.7 Improvement in Inhibitor Tolerance

Pretreatment of lignocellulosic biomass results in production of various toxic compounds including weak acids (formic acid, acetic acid, and levulinic acids), furfurals, 5-hydroxymethyl furfural (5-HMF), and phenolic compounds. These compounds inhibit cell growth and metabolism and lead to reduced ethanol yield and productivity, making industrial ethanol production cost expensive. Weak acids affect cellular growth and ethanol yield through diffusion across the plasma membrane and alter cytosolic pH (Kricka et al. 2015) that lead to increase in cellular ATP concentrations and cause DNA damage. The inhibitory effect of acetic acid was addressed by overexpression of acetylating acetaldehyde dehydrogenase; it reduces the acetate concentration in the medium and leads to improved cell growth and ethanol yield (Wei et al. 2013). Furfurals and 5-HMF are formed when pentose and hexose sugars are degraded at high temperature and pressure. Furfurals cause cellular redox imbalance that leads to reduced XR/XDH enzyme activity (Modig et al. 2002; Ask et al. 2013). Furthermore, it was demonstrated that Furfural and 5-HMF reduce cell growth and metabolism by producing oxidative stress. Indeed, it was shown that overexpression of certain genes involved in oxidative stress tolerance such as *MSN2* (Sasano et al. 2012), *GSH1* (Ask et al. 2013), and *FLO1* (Westman et al. 2014) improved cell growth and ethanol productivity under furfural and 5-HMF stresses. Phenolic compounds are by-product of lignin degradation, and they reduce xylose metabolism in recombinant yeast expressing xylose-utilizing genes (Wang et al. 2014). It was observed that these inhibitors result in accumulation of various pentose phosphate pathway intermediates (Hasunuma et al. 2011). Interestingly, expression of laccase gene of *Trametes versicolor*, a white rot fungus under constitutive *PGK1* promoter in *S. cerevisiae*, enhanced resistance to phenolic inhibitors (Larsson et al. 2001). Moreover, transcriptome analysis in the presence of pretreatment inhibitors revealed that expression of genes involved in carbohydrate metabolism was reduced along with the genes involved in basic transcriptional and translational control of gene regulation, indicating that these inhibitors exert pleotropic effect on cell metabolism and growth (Bajwa et al. 2013; Li and Yuan 2010). In another study, it was demonstrated that overexpression of alcohol dehydrogenase genes *ADH6* and *ADH7* and a mutated *ADH1* led to increased growth, ethanol yields, and productivity in the presence of pretreatment inhibitors (Ishii et al. 2013; Kricka et al. 2015). Since *ADH6* (NADH) and *ADH7* (NADPH) have different cofactor requirements, therefore the overexpression of these enzymes leads to redox imbalances (Ask et al. 2013). Likewise, the effect of inhibitors can also be overcome by encapsulation of recombinant cells or by using highly flocculent strains (Kricka et al. 2015). Other than genetic engineering, approaches like adaptive evolution and mutagenesis can be used to adapt the yeast strains to various inhibitors (Demeke et al. 2013; Koppram et al. 2012).

8.3.8 Global Transcription Machinery Engineering

Global transcription machinery engineering (gTME) is a novel method to reprogram gene transcription in order to evoke cellular phenotypes. The main proposition of this approach is that gene networks and cellular metabolism of cell can be reprogrammed by introducing dominant mutant alleles of generic transcription-related proteins. This method allows the engineering of transcription factors to develop cellular systems for improved pentose fermentation. Several phenotypes like pentose utilization, high sugar and ethanol tolerance, and other regulatory issues can be improved by this approach (Young et al. 2010; Moysés et al. 2016). Mutagenesis of the transcription factor Spt15p has been carried out using gTME method leading to dominant mutations that helped in improving the ethanol production (Alper et al. 2006).

8.4 Conclusion and Future Perspective

Significant research progress has been made in the past decades making second-generation biofuel production an economically successful industrial endeavor. Rapid progress has been made by metabolic engineering methods guided by genomic and transcriptomic studies. The extensive genetic engineering approach made simultaneous hexose and pentose fermentation feasible in yeast, although pentose and hexose metabolic pathways are fundamentally different. Despite huge efforts, the complexities of cellular metabolism and unforeseen subtle interactions between cellular metabolites may limit our ability to bioengineer yeasts for efficient bioethanol production. Bottlenecks, such as reduced enzyme activities at fermentation temperatures, sugar uptake, and effects of inhibitors on yeast cell growth and metabolism, continue to provide challenges for bioengineering efficient xylose-fermenting yeast. More emphasis on *Saccharomyces cerevisiae* genome and regulatory structure must be given to understand the effect of biological complexity on a pathway. Multidisciplinary approaches like pathway engineering, evolutionary engineering, directed evolution, transcriptomics, metabolomics, and combinatorial genetics can be used to understand the cellular complexity. The main future challenge is to transfer the metabolic engineering strategies from laboratory strain to the industrial production strains.

References

Alper H, Moxley J, Nevoigt E, Fink GR, Stephanopoulos G (2006) Engineering yeast transcription machinery for improved ethanol tolerance and production. Science 314:1565–1568. doi:10.1126/science.1131969

Amore R, Wilhelm M, Hollenberg CP (1989) The fermentation of xylose-an analysis of the expression of *bacillus* and *actinoplanes* xylose isomerase genes in yeast. Appl Microbiol Biotechnol 30:351–357. doi:10.1007/BF00296623

Anwar Z, Gulfraz M, Irshad M (2014) Agro-industrial lignocellulosic biomass a key to unlock the future bio-energy: a brief review. J Radiat Res Appl Sci 7:163–173. doi:10.1016/j.jrras.2014.02.003

Apel AR, Ouellet M, Szmidt-Middleton H, Keasling JD, Mukhopadhyay A (2015) Evolved hexose transporter enhances xylose uptake and glucose/xylose co-utilization in *Saccharomyces cerevisiae*. Sci Rep 6:19512. doi:10.1038/srep19512

Ask M, Bettiga M, Mapelli V, Olsson L (2013) The influence of HMF and furfural on redox-balance and energy-state of xylose-utilizing *Saccharomyces cerevisiae*. Biotechnol Biofuels 6:22. doi:10.1186/1754-6834-6-22

Bajwa PK, Ho CY, Chan CK, Martin VJ, Trevors JT, Lee H (2013) Transcriptional profiling of *Saccharomyces cerevisiae* T2 cells upon exposure to hardwood spent sulphite liquor: comparison to acetic acid, furfural and hydroxymethylfurfural. Antonie Van Leeuwenhoek 103:1281–1295. doi:10.1007/s10482-013-9909-1

Batt CA, Caryallo S, Easson DD Jr, Akedo M, Sinskey AJ (1986) Direct evidence for a xylose metabolic pathway in *Saccharomyces cerevisiae*. Biotechnol Bioeng 28:549–553. doi:10.1002/bit.260280411

Bettiga M, Hahn-Hagerdal B, Gorwa-Grauslund MF (2008) Comparing the xylose reductase/xylitol dehydrogenase and xylose isomerase pathways in arabinose and xylose fermenting *Saccharomyces cerevisiae* strains. Biotechnol Biofuels 1:16. doi:10.1186/1754-6834-1-16

Brat D, Boles E, Wiedemann B (2009) Functional Expression of a Bacterial Xylose Isomerase in *Saccharomyces cerevisiae*. Appl Environ Microbiol 75:2304–2311. doi:10.1128/AEM.02522-08

Chu BC, Lee H (2007) Genetic improvement of *Saccharomyces cerevisiae* for xylose fermentation. Biotechnol Adv 25:425–441. doi:10.1016/j.biotechadv.2007.04.001

Chu S, Majumdar A (2012) Opportunities and challenges for a sustainable energy future. Nature 488:294–303. doi:10.1038/nature11475

Demeke MM, Dietz H, Li Y, Foulquie-Moreno MR, Mutturi S, Deprez S, Den Abt T, Bonini BM, Liden G, Dumortier F, Verplaetse A, Boles E, Thevelein JM (2013) Development of a D-xylose fermenting and inhibitor tolerant industrial *Saccharomyces cerevisiae* strain with high performance in lignocellulose hydrolysates using metabolic and evolutionary engineering. Biotechnol Biofuels 6:89. doi:10.1186/1754-6834-6-89

Ehrensberger AH, Elling RA, Wilson DK (2006) Structure-guided engineering of xylitol dehydrogenase cosubstrate specificity. Structure 14:567–575. doi:10.1016/j.str.2005.11.016

Eliasson A, Christensson C, Wahlbom CF, Hahn-Hägerdal B (2000) Anaerobic xylose fermentation by recombinant *Saccharomyces cerevisiae* carrying XYL1, XYL2, and XKS1 in mineral medium chemostat cultures. Appl Environ Microbiol 66:3381–3386. doi:10.1128/AEM.66.8.3381-3386.2000

Galbe M, Zacchi G (2007) Pretreatment of lignocellulosic materials for efficient bioethanol production. Adv Biochem Eng Biotechnol 108:41–65. doi:10.1007/10_2007_070

Gardonyi M, Hahn-Hagerdal B (2003) The *Streptomyces rubiginosus* xylose isomerase is misfolded when expressed in *Saccharomyces cerevisiae*. Enzym Microb Technol 32:252–259. doi:10.1016/S0141-0229(02)00285-5

Grotkjaer T, Christakopoulos P, Nielsen J, Olsson L (2005) Comparative metabolic network analysis of two xylose fermenting recombinant *Saccharomyces cerevisiae* strains. Metab Eng 7:437–444. doi:10.1016/j.ymben.2005.07.003

Hahn-Hagerdal B, Galbe M, Gorwa-Grauslund MF, Liden G, Zacchi G (2006) Bioethanol-the fuel of tomorrow from the residues of today. Trends Biotechnol 24:549–556. doi:10.1016/j.tibtech.2006.10.004

Hahn-Hägerdal B, Karhumaa K, Jeppsson M, Gorwa-Grauslund MF (2007) Metabolic engineering for pentose utilization in *Saccharomyces cerevisiae*. Adv Biochem Eng Biotechnol 108:147–177. doi:10.1007/10_2007_062

Hallborn J, Walfridsson M, Airaksinen U, Ojamo H, Hahn-Hägerdal B, Penttila M, Kerasnen S (1991) Xylitol production by recombinant *Saccharomyces cerevisiae*. Nat Biotechnol 9:1090–1095. doi:10.1038/nbt1191-1090

Hamacher T, Becker J, Gardonyi M, Hahn-Hagerdal B, Boles E (2002) Characterization of the xylose-transporting properties of yeast hexose transporters and their influence on xylose utilization. Microbiology 148:2783–2788. doi:10.1099/00221287-148-9-2783

Hasunuma T, Sanda T, Yamada R, Yoshimura K, Ishii J, Kondo A (2011) Metabolic pathway engineering based on metabolomics confers acetic and formic acid tolerance to a recombinant xylose-fermenting strain of *Saccharomyces cerevisiae*. Microb Cell Factories 10:2. doi:10.1186/1475-2859-10-2

Hector RE, Dien BS, Cotta MA, Qureshi N (2011) Engineering industrial *Saccharomyces cerevisiae* strains for xylose fermentation and comparison for switchgrass conversion. J Ind Microbiol Biotechnol 38:1193–1202. doi:10.1007/s10295-010-0896-1

Ho NWY, Chen Z, Brainard AP (1998) Genetically engineered *Saccharomyces* yeast capable of effective co fermentation of glucose and xylose. Appl Environ Microbiol 64:1852–1859

Ishii J, Yoshimura K, Hasunuma T, Kondo A (2013) Reduction of furan derivatives by overexpressing NADH-dependent Adh1 improves ethanol fermentation using xylose as sole carbon source with *Saccharomyces cerevisiae* harboring XR-XDH pathway. Appl Microbiol Biotechnol 97:2597–2607. doi:10.1007/s00253-012-4376-6

Isroi MR, Syamsiah S, Niklasson C, Cahyanto MN, Lundquist K, Taherzadeh MJ (2011) Biological pretreatment of lignocelluloses with white-rot fungi and its applications: a review. BioResource 6:5224–5259

Jeffries TW (2006) Engineering yeasts for xylose metabolism. Curr Opin Biotechnol 17:320–326. doi:10.1016/j.copbio.2006.05.008

Jeffries TW, Jin YS (2000) Ethanol and thermotolerance in the bioconversion of xylose by yeasts. Adv Appl Microbiol 47:221–268. doi:10.1186/s13068-015-0227-x

Jeppsson M, Bengtsson O, Franke K, Lee H, Hahn-Hagerdal B, Gorwa-Grauslund MF (2006) The expression of a *Pichia stipitis* xylose reductase mutant with higher K_M for NADPH increases ethanol production from xylose in recombinant *Saccharomyces cerevisiae*. Biotechnol Bioeng 93:665–673. doi:10.1002/bit.20737

Jin YS, Jeffries TW (2003) Changing flux of xylose metabolites by altering expression of xylose reductase and xylitol dehydrogenase in recombinant *Saccharomyces cerevisiae*. Appl Biochem Biotechnol 106:277–285

Jin YS, Ni H, Laplaza JM, Jeffries TW (2003) Optimal growth and ethanol production from xylose by recombinant *Saccharomyces cerevisiae* require moderate D-xylulokinase activity. Appl Environ Microbiol 69:495–503. doi:10.1128/AEM.69.1.495-503.2003

Johansson B, Hahn-Hagerdal B (2002) Overproduction of pentose phosphate pathway enzymes using a new CRE-loxP expression vector for repeated genomic integration in *Saccharomyces cerevisiae*. Yeast 19:225–231. doi:10.1002/yea.833

Jouzani GS, Taherzadeh MJ (2015) Advances in consolidated bioprocessing systems for bioethanol and butanol production from biomass: a comprehensive review. Biofuel Res J 5:152–195. doi:10.18331/BRJ2015.2.1.4

Karhumaa K, Garcia Sanchez R, Hahn-Hagerdal B, Gorwa-Grauslund MF (2007) Comparison of the xylose reductase-xylitol dehydrogenase and the xylose isomerase pathways for xylose fermentation by recombinant *Saccharomyces cerevisiae*. Microb Cell Factories 6:5. doi:10.1186/1475-2859-6-5

Kavscek M, Stražar M, Curk T, Natter K, Petrovič U (2015) Yeast as a cell factory: current state and perspectives. Microb Cell Fact 14:94. doi:10.1186/s12934-015-0281-x

Kim B, Du J, Eriksen DT, Zhao H (2013) Combinatorial design of a highly efficient xylose-utilizing pathway in *Saccharomyces cerevisiae* for the production of cellulosic biofuels. Appl Environ Microbiol 79:931–941. doi:10.1128/AEM.02736-12

Koppram R, Albers E, Olsson L (2012) Evolutionary engineering strategies to enhance tolerance of xylose utilizing recombinant yeast to inhibitors derived from spruce biomass. Biotechnol Biofuels 5:32. doi:10.1186/1754-6834-5-32

Kotter P, Amore R, Hollenberg CP, Ciriacy M (1990) Isolation and characterization of the *Pichia stipitis* xylitol dehydrogenase gene, XYL2, and construction of a xylose-utilizing *Saccharomyces cerevisiae* transformant. Curr Genet 18:493–500

Kricka W, Fitzpatrick J, Bond U (2015) Challenges for the production of bioethanol from biomass using recombinant yeasts. Adv Appl Microbiol 92:89–125. doi:10.1016/bs.aambs.2015.02.003

Kuhad RC, Gupta R, Khasa YP, Singh A, Perciv Zhang YH (2011) Bioethanol production from pentose sugars: current status and future prospects. Renew Sust Energ Rev 15:4950–4962. doi:10.1016/j.rser.2011.07.058

Kuyper M, Harhangi HR, Stave AK, Winkler AA, Jetten MS, de laat WT, Den Ridder JJ, Op den Camp HJ, van Dijken J, Pronk JT (2003) High level functional expression of a fungal xylose isomerase: the key to efficient ethanolic fermentation of xylose by *Saccharomyces cerevisiae*. FEMS Yeast Res 4:69–78. doi:http://dx.doi.org/10.1016/S1567-1356(03)00141-7

Kuyper M, Hartog MM, Toirkens MJ, Almering MJ, Winkler AA, van Dijken JP, Pronk JT (2005) Metabolic engineering of a xylose isomerase expressing *Saccharomyces cerevisiae* strain for rapid anaerobic xylose fermentation. FEMS Yeast Res 5:399–409. doi:http://dx.doi.org/10.1016/j.femsyr.2004.09.010

Larsson S, Cassland P, Jonsson LJ (2001) Development of a *Saccharomyces cerevisiae* strain with enhanced resistance to phenolic fermentation inhibitors in lignocellulose hydrolysates by heterologous expression of laccase. Appl Environ Microbiol 67:1163–1170. doi:10.1128/AEM.67.3.1163-1170.2001

Latimer LN, Lee ME, Medina-Cleghorn D, Kohnz RA, Nomura DK, Dueber JE (2014) Employing a combinatorial expression approach to characterize xylose utilization in *Saccharomyces cerevisiae*. Metab Eng 25:20–29. doi:10.1016/j.ymben.2014.06.002

Leandro MJ, Goncalves P, Spencer-Martins I (2006) Two glucose/xylose transporter genes from the yeast *Candida intermedia*: first molecular characterization of a yeast xylose-H^+ symporter. Biochem J 395:543–549. doi:10.1042/BJ20051465

Lee SM, Jellison T, Alper HS (2012) Directed evolution of xylose isomerase for improved xylose catabolism and fermentation in the yeast *Saccharomyces cerevisiae*. Appl Environ Microbiol 78:5708–5716. doi:10.1128/AEM.01419-12

Lennartsson PR, Erlandsson P, Taherzadeh MJ (2014) Integration of the first and second generation bioethanol processes and the importance of by-products. Bioresour Technol 165:3–8. doi:10.1016/j.biortech.2014.01.127

Li BZ, Yuan YJ (2010) Transcriptome shifts in response to furfural and acetic acid in *Saccharomyces cerevisiae*. Appl Microbiol Biotechnol 86:1915–1924. doi:10.1007/s00253-010-2518-2

Li C, Knierim B, Manisseri C, Arora R, Scheller HV, Auer M, Vogel KP, Simmons BA, Singh S (2010) Comparison of dilute acid and ionic liquid pretreatment of switchgrass: biomass recalcitrance, delignification and enzymatic saccharification. Bioresour Technol 101:4900–4906. doi:10.1016/j.biortech.2009.10.066

Li H, Wu M, Xu L, Hou J, Guo T, Bao X, Shen Y (2015) Evaluation of industrial *Saccharomyces cerevisiae* strains as the chassis cell for second-generation bioethanol production. Microb Biotechnol 8:266–274. doi:10.1111/1751-7915.12245

Lönn A, Träff-Bjerre KL, Cordero Otero RR, van Zyl WH, Hahn-Hägerdal B (2003) Xylose isomerase activity influences xylose fermentation with recombinant *Saccharomyces cerevisiae* strains expressing mutated xylA from *Thermus thermophilus*. Enzym Microb Technol 32:567–573. doi:10.1016/S0141-0229(03)00024-3

Madhavan A, Tamalampudi S, Ushida K, Kanai D, Katahira S, Srivastava A, Fukuda H, Bisaria VS, Kondo A (2009) Xylose isomerase from polycentric fungus *Orpinomyces*: gene sequencing, cloning, and expression in *Saccharomyces cerevisiae* for bioconversion of xylose to ethanol. Appl Microbiol Biotechnol 82:1067–1078. doi:10.1007/s00253-008-1794-6

Matsushika A, Goshima T, Fujii T, Inoue H, Sawayama S, Yano S (2012) Characterization of non-oxidative transaldolase and transketolase enzymes in the pentose phosphate pathway with regard to xylose utilization by recombinant *Saccharomyces cerevisiae*. Enzym Microb Technol 51:16–25. doi:10.1016/j.enzmictec.2012.03.008

Metzger MH, Hollenberg CP (1995) Amino acid substitutions in the yeast *Pichia stipitis* xylitol dehydrogenase coenzyme binding domain affect the coenzyme specificity. Eur J Biochem 228:50–54. doi:10.1111/j.1432-1033.1995.0050o

Modig T, Liden G, Taherzadeh MJ (2002) Inhibition effects of furfural on alcohol dehydrogenase, aldehyde dehydrogenase and pyruvate dehydrogenase. Biochem J 363:769–776. doi:10.1042/bj3630769

Moes CJ, Pretorius IS, vanZyl WH (1996) Cloning and expression of the *Clostridium thermosulfurogenes* D-xylose isomerase gene (xylA) in *Saccharomyces cerevisiae*. Biotechnol Lett 18:269–274

Mood SH, Golfeshan AH, Tabatabaei M, Abbasalizadeh S, Ardjmand M, Jouzani GS (2014) Comparison of different ionic liquids pretreatment for corn stover enzymatic saccharification. Prep Biochem Biotechnol 44:451–463. doi:10.1080/10826068.2013.833112

Mood SH, Golfeshan AH, Tabatabaei M, Jouzani GS, Najafi GH, Gholami M, Ardjmand M (2013) Lignocellulosic biomass to bioethanol, a comprehensive review with a focus on pretreatment. Renew Sust Energ Rev 27:77–93. doi:10.1016/j.rser.2013.06.033

Moysés DN, Reis VCB, Moreira de Almeida JR, Pepe de Moraes LM, Torres FAG (2016) Xylose fermentation by *Saccharomyces cerevisiae*: challenges and prospects. Int J Mol Sci 17:207. doi:10.3390/ijms17030207

Ostergaard S, Olsson L, Nielsen J (2000) Metabolic engineering of *Saccharomyces cerevisiae*. Microbiol Mol Biol Rev 64:34–50

Palmqvist E, Hahn-Hagerdal B (2000) Fermentation of lignocellulosic hydrolysates. I. Inhibition and detoxification. Bioresour Technol 74:17–24. doi:10.1016/S0960-8524(99)00160-1

Preez JCD, Bosch M, Prior BA (1986) The fermentation of hexoses and pentoses sugars by *Candida shehatae* and *Pichia stipitis*. Appl Microbiol Biotechnol 23:228–233. doi:10.1007/BF00261920

Rajendran K, Taherzadeh MJ (2014) Pretreatment of lignocellulosic materials. In: Bisaria V, Kondo A (eds) Bioprocessing of renewable resources to commodity bioproducts. Wiley, USA, pp 43–76. ISBN: 978-1118175835

Runquist D, Hahn-Hagerdal B, Radstrom P (2010) Comparison of heterologous xylose transporters in recombinant *Saccharomyces cerevisiae*. Biotechnol Biofuels 3:5. doi:10.1186/1754-6834-3-5

Saha BC (2003) Hemicellulose bioconversion. J Ind Microbiol Biotechnol 30:279–291. doi:10.1007/s10295-003-0049-x

Saloheimo A, Rauta J, Stasyk OV, Sibirny AA, Penttila M, Ruohonen L (2007) Xylose transport studies with xylose-utilizing *Saccharomyces cerevisiae* strains expressing heterologous and homologous permeases. Appl Microbiol Biotechnol 74:1041–1052. doi:10.1007/s00253-006-0747-1

Sarkar N, Ghosh SK, Bannerjee S, Aikat K (2012) Bioethanol production from agricultural wastes: an overview. Renew Energy 37:19–27. doi:10.1016/j.renene.2011.06.045

Sarthy AV, McConaughy BL, Lobo Z, Sundstrom JA, Furlong CE, Hall BD (1987) Expression of the *Escherichia coli* xylose isomerase gene in *Saccharomyces cerevisiae*. Appl Environ Microbiol 53:1996–2000

Sasano Y, Watanabe D, Ukibe K, Inai T, Ohtsu I, Shimoi H, Tagaki H (2012) Overexpression of the yeast transcription activator Msn2 confers furfural resistance and increases the initial fermentation rate in ethanol production. J Biosci Bioeng 113:451–455. doi:10.1016/j.jbiosc.2011.11.017

Scheller HV, Ulvskov P (2010) Hemicelluloses. Annu Rev Plant Biol 61:263–289. doi:10.1146/annurev-arplant-042809-112315

Schneider H, Maleszka R, Neirinck L, Veliky A, Wang PY, Chan YK (1983) Ethanol production from D-xylose and several other carbohydrate by *Pachysolen tannophilus* and other yeasts. Adv Biochem Eng Biotechnol 27:57–71

Senac T, Hahn-Hagerdal B (1991) Effects of increased transaldolase activity on d-xylulose and d-glucose metabolism in *Saccharomyces cerevisiae* cell extracts. Appl Environ Microbiol 57:1701–1706

Skoog K, Hahn-Hagerdal B (1988) Xylose fermentation. Enzym Microb Technol 10:66–80

Slininger PJ, Bothast RJ, Ladisch MR, Okos MR (1990) Optimum pH and temperature conditions for xylose fermentation by *Pichia stipitis*. Biotechnol Bioeng 35:727–731

Stephanopoulos G (2007) Challenges in engineering microbes for biofuels production. Science 315:801–804. doi:10.1126/science.1139612

Takuma S, Nakashima N, Tantirungkij M, Kinoshita S, Okada H, Seki T, Yoshida T (1991) Isolation of xylose reductase gene of *Pichia stipitis* and its expression in *Saccharomyces cerevisiae*. Appl Biochem Biotechnol 28–29:327–340. doi:10.1007/BF02922612

Tanino T, Ito T, Ogino C, Ohmura N, Ohshima T, Kondo A (2012) Sugar consumption and ethanol fermentation by transporter-overexpressed xylose-metabolizing *Saccharomyces cerevisiae* harboring a xylose isomerase pathway. J Biosci Bioeng 114:209–211. doi:10.1016/j.jbiosc.2012.03.004

Toivari MH, Aristidou A, Ruohonen L, Penttila M (2001) Conversion of xylose to ethanol by recombinant *Saccharomyces cerevisiae*: importance of Xylulokinase (XKS1) and oxygen availability. Metab Eng 3:236–249. doi:10.1006/mben.2000.0191

Traff KL, Cordero RRO, van Zyl WH, Hahn-Hagerdal B (2001) Deletion of the *GRE3* aldose reductase Gene and its influence on xylose metabolism in recombinant strains of *Saccharomyces cerevisiae* expressing the *xylA* and *XKS1* genes. Appl Environ Microbiol 67:5668–5674. doi:10.1128/AEM.67.12.5668–5674.2001

van Maris AJ, Winkler AA, Kuyper M, de Laat WT, van Dijken JP, Pronk JT (2007) Development of efficient xylose fermentation in *Saccharomyces cerevisiae*: xylose isomerase as a key component. Adv Biochem Eng Biotechnol 108:179–204. doi:10.1007/10_2007_057

Walfridsson M, Bao X, Anderlund M, Lilius G, Bülow L, Hahn-Hägerdal B (1996) Ethanolic fermentation of xylose with *Saccharomyces cerevisiae* harboring the *Thermus thermophilus xylA gene*, which expresses an active xylose (glucose) isomerase. Appl Environ Microbiol 62:4648–4651

Walfridsson MJ, Hallborn M, Penttilä M, Keränen S, Hahn-Hägerdal B (1995) Xylose-metabolizing *Saccharomyces cerevisiae* strains overexpressing the TKL1 and TAL1 genes encoding the pentose phosphate pathway enzymes transketolase and transaldolase. Appl Environ Microbiol 61:4184–4190

Wang M, Wu M, Huo H (2007) Life cycle energy and greenhouse gas emission impacts of different corn ethanol plant types. Environ Res Lett 2:1–13. doi:10.1088/1748-9326/2/2/024001

Wang X, Jin M, Balan V, Jones AD, Li X, Li BZ, Dale BE, Yuan YJ (2014) Comparative metabolic profiling revealed limitations in xylose-fermenting yeast during co-fermentation of glucose and xylose in the presence of inhibitors. Biotechnol Bioeng 111:152–164

Watanabe S, Abu Saleh A, Pack SP, Annaluru N, Kodaki T, Makino K (2007) Ethanol production from xylose by recombinant *Saccharomyces cerevisiae* expressing protein-engineered NADH-preferring xylose reductase from *Pichia stipitis*. Microbiology 153:3044–3054. doi:10.1099/mic.0.2007/007856-0

Watanabe S, Kodaki T, Makino K (2005) Complete reversal of coenzyme specificity of xylitol dehydrogenase and increase of thermostability by the introduction of structural zinc. J Biol Chem 280:10340–10349. doi:10.1074/jbc.M409443200

Wei N, Quarterman J, Kim SR, Cate JH, Jin YS (2013) Enhanced biofuel production through coupled acetic acid and xylose consumption by engineered yeast. Nat Commun 4:2580. doi:10.1038/ncomms3580

Westman JO, Bonander N, Taherzadeh MJ, Franzen CJ (2014) Improved sugar co-utilisation by encapsulation of a recombinant *Saccharomyces cerevisiae* strain in alginate chitosan capsules. Biotechnol Biofuels 7:102. doi:10.1186/1754-6834-7-102

Young E, Lee SM, Alper H (2010) Optimizing pentose utilization in yeast: the need for novel tools and approaches. Biotechnol Biofuels 3:24. doi:10.1186/1754-6834-3-24

Young E, Poucher A, Comer A, Bailey A, Alper H (2011) Functional survey for heterologous sugar transport proteins, using *Saccharomyces cerevisiae* as a host. Appl Environ Microbiol 77:3311–3319. doi:10.1128/AEM.02651-10

Zhang Y, Li Q, Su J, Lin Y, Huang Z, Lu Y, Sun G, Yang M, Huang A, Hu H, Zhu Y (2015) A green and efficient technology for the degradation of cellulosic materials: structure changes and enhanced enzymatic hydrolysis of natural cellulose pretreated by synergistic interaction of mechanical activation and metal salt. Bioresour Technol 177:176–181. doi:10.1016/j.biortech.2014.11.085

Zhou H, Cheng J, Wang BL, Fink GR, Stephanopoulos G (2012) Xylose isomerase overexpression along with engineering of the pentose phosphate pathway and evolutionary engineering enable rapid xylose utilization and ethanol production by *Saccharomyces cerevisiae*. Metab Eng 14:611–622. doi:10.1016/j.ymben.2012.07.011

Engineering Yeast as Cellular Factory

9

Poonam, Ritu Ghildiyal, Gopal Singh Bisht, and Rahul Shrivastava

Abstract

Yeast is used as an attractive host for metabolic engineering. It is widely used in industries for the production of enzymes, fine chemicals, alcoholic beverages, baker's yeast, pharmaceutical products, biofuels, recombinant proteins, etc. Yeast can be grown easily as it requires simple growth factors such as carbon, vitamins, and salts which makes it economic for large-scale production. Among different types of yeast like *Pichia pastoris*, *Hansenula polymorpha*, *Yarrowia lipolytica*, *Kluyveromyces lactis*, and *Saccharomyces cerevisiae*, *S. cerevisiae* is most commonly used. Yeast can be genetically modified to produce chemicals (glycerol, propanediol, organic acids, etc.) and fatty acid derivatives (fatty acid ester, fatty alkanes, and fatty alcohols). Yeast engineering for the production of biofuel has made the production of biofuel affordable and eco-friendly. Yeast has the ability to perform posttranslational modification and, therefore, is widely employed for producing human recombinant proteins. Generally modifications of enzyme activity are employed in yeast engineering to obtain desired product. Yeast engineering can be achieved by either introducing a new pathway in yeast or by altering the native pathway. In this chapter we have presented general considerations along with different strategies developed for the synthesis of various products using engineered yeast.

Keywords

Metabolic engineering • Yeast • *S. cerevisiae* • Cell factory • System biology • Synthetic biology

Poonam • R. Ghildiyal • G.S. Bisht • R. Shrivastava (✉)
Department of Biotechnology and Bioinformatics, Jaypee University of Information Technology, Waknaghat, Solan 173 234, Himachal Pradesh, India
e-mail: juit.rahul@gmail.com; rahulmicro@gmail.com

V.C. Kalia, A.K. Saini (eds.), *Metabolic Engineering for Bioactive Compounds*, DOI 10.1007/978-981-10-5511-9_9

9.1 Introduction

Metabolic engineering is an applied science that involves the development of living cells as factories, either by optimization of existing biochemical pathway or introduction of new pathway components by the use of genetic engineering (Tyo et al. 2007). Metabolic engineering is commonly applied in bacteria, plant, and yeast to generate cell factories which may lead to production of various chemicals of commercial importance. Success of industrial biotechnology depends upon the efficient development of such cell factory designs. Microbial cellular metabolism is a very complex process that produces a lot of organic compounds for growth of the cell, and it is very difficult to produce these products in vitro by providing the same physical conditions outside the cell. Therefore, industrial biotechnology is focused on the creation of efficient, self-regenerating, microbial factories to produce chemicals, fuels, and other industrial products.

Metabolic engineering surfaced as a research field in the early 1990s. It is concerned with the refinement of cell factories, i.e., either to develop novel cell factories or improve the existing ones (Bailey 1991; Stephanopoulos and Vallino 1991; Tyo et al. 2007). It is an applied science that uses genetic engineering to have more efficient and developed cell factories with enhanced properties (Kim et al. 2012). The major attractions of this field are the many products of commercial interest that have natural scarcity and their complicated synthetic methods which make them infeasible for manufacturing at a commercial level and enable them to be produced in ways that are cost-effective and environmentally friendly using metabolic engineering approaches (Krivoruchko et al. 2011). This can be done by manipulating the natural producers of the respective compounds or by transferring the involved synthetic routes into an appropriate cell factory that can increase the productivity of these compounds.

Though metabolic engineering uses genetic engineering but is different from it in terms of the modern analytical tools, it uses for target identification suitable for genetic alterations, along with application of available mathematical systems to carry out in silico formulation of living cell factories. Thus, it can be perceived as a cyclic practice (Gombert et al. 2001), where analysis of cell factory is followed by recognizing a suitable target (the design stage). Subsequently, the target is experimentally executed, and the created strain is analyzed again. Therefore, like systems biology, it involves a constant iteration amid design and experimental work.

9.2 Development Steps for Metabolically Engineered Yeast Cell Factories

There are five important decision points for designing yeast cell factories. The first decision focuses on the kind of product to be produced. This influences the selection of chassis to be utilized for fermentation of the desired product. On the basis of factors such as host cell biosynthetic competence, intermediates and product, substrate's effect on the strain, and convenience of fermentation process, the chassis is selected (Keasling 2010). Validating whether the selected chassis can naturally make the preferred product is the aim of the second decision point. If yes, one can advance to step three, but if it is not the case, pathway engineering approaches are used for initiating product formation. In certain cases, the selected chassis can produce the product, but the yield is low, necessitating incorporation of a heterologous pathway for enhanced productivity. Ro and coworkers were able to engineer strains of *S. cerevisiae* to produce high titers of artemisinic acid (antimalarial drug precursor) (Ro et al. 2006). Recognition of genes which can be suitably modified to produce the desired product in high titers is the focus of the third point. These genes can be categorized as positive effect genes and negative effect genes based on their effect on the product formation. Negative effect genes include genes that lead to deterioration of product and genes responsible for enzymes that contend for intermediates. On the other hand, positive effect genes consist of the biosynthetic genes, genes that respond to stress, and genes responsible for substrate utilization (Klein-Marcuschamer et al. 2010; Prather and Martin 2008). The penultimate decision point is how to be in command of the quality and quantity of the target gene(s). Protein engineering can be employed to manage enzyme activity levels thus managing the quality of the protein required, whereas to manage the quantity of target gene(s), gene expression levels need to be controlled. This can be done by controlling gene copy number and modifications of the promoter at transcriptional or translational level. The final decision point aims to attain maximum yield by making certain final and fine adjustments to the biosynthetic network of desired product (Klein-Marcuschamer et al. 2010). The development steps are discussed below under three broad phases.

9.2.1 Step I: Design

Two metabolic engineering strategies can be used to execute this phase:

1. Rational—using genetic engineering to create a microorganism that produces the required products (Lee et al. 2011; Tavares et al. 2011)
2. Inverse strategy—which involves detailed analysis of the desired phenotype to assist in preliminary selection of cellular systems followed by comparative study to determine the genetic variations among the chosen systems and, finally, construction of genetically manipulated strains enriched with identified target genes

A substitute to metabolic engineering approaches is adaptive evolution. Adaptive evolution strategies have major advantage in comparison to metabolic engineering as there is no requirement to comprehend the molecular mechanisms involved or leading to the formation of a desirable phenotype of the organism. Strains with a preferred phenotype can be created by using either direct evolution (recombination) or indirect evolution (random mutagenesis) (Sauer 2001). The notion behind the phenomenon of indirect adaptive evolution is that the microorganisms have a tendency to become accustomed to unnatural conditions (can be in the form of intracellular or extracellular stimuli) and evolve their inherent characteristics accordingly. Random genetic mutations may accompany these evolution processes, and favorable mutations can be selected if the selection pressure is applied in a right direction. As opposed to indirect adaptive evolution, directed evolution attempts to imitate natural evolution in vitro. Directed evolution involves methods or procedures (like DNA recombination) to introduce genetic variations in the organism followed by the selection of the desired superior strain (Liu et al. 2011; Tyo et al. 2011).

9.2.2 Step II: Construction

9.2.2.1 Target Gene(s) Selection

Sources for the target gene(s) can be endogenous, i.e., gene from the host cell; exogenous, i.e., gene from a different organism; or synthetic, i.e., gene is wholly designed or resultant of mutagenesis (natural or artificial). They can be manipulated by deletion, overexpression, or attenuation.

9.2.2.2 Regulation of Target Gene Expression

Quantitative Control

Most basic approach is to regulate copy number of the gene, thus regulating the expression level of the corresponding gene. Removing negative effect target genes or integrating target genes with positive effect on expression of the desired gene into the chromosome is one way to accomplish this. Other ways for regulating the expression levels are as follows: (1) alteration of promoters of target genes, (2) modification of 5′-UTRs and 3′-UTRs, and (3) generation of riboregulators. In the first approach, optimal promoters are identified or designed using either endogenous (Partow et al. 2010; Tochigi et al. 2010) or artificially constructed promoters (Jeppsson et al. 2003; Nevoigt et al. 2006; Hartner et al. 2008) using promoter libraries. In the alternative approach, particular RNA-binding proteins interact with the 5′-UTRs and 3′-UTRs by binding to the nucleotide motifs located in these regions, or sequence components found in the UTRs interact with a riboregulator, i.e., noncoding RNAs that are specifically complementary to abovementioned sequences. Short interfering RNAs (siRNAs), ribozymes (i.e., RNA enzymes), microRNA (miRNA), and antisense RNA (Nasr et al. 1995) are examples of riboregulators, which have been found to be proficient in the regulation of expression at the genetic level (Isaacs et al. 2006).

Qualitative Control

By means of protein engineering, mutations can be incorporated into the target protein. This may lead to superior thermostability, avoidance of substrate or product inhibition, resistance to organic solvents, and better enzyme activity (Luetz et al. 2008). There are two strategies while performing protein engineering (Tang and Zhao 2009):

1. Using site-directed mutagenesis for rational designing—on-hand information in relation to the structure and kinetic characteristics of the target enzyme is the basis for this approach (Saven 2011). For example, site-directed mutagenesis in lactate dehydrogenase gene of the bacterium *Lactobacillus plantarum* enhanced production of lactate in *S. cerevisiae.*
2. Using random mutagenesis (e.g., by PCR that is susceptible to errors or by shuffling of DNA) to accomplish directed evolution (Labrou 2010).

9.2.2.3 Network Fine-Tuning

This is a critical step particularly when dealing with parallel expression of multiple genes. Multiple gene promoter shuffling (MGPS) approach is used in *S. cerevisiae* when multiple genes need to be expressed together and optimally. It consists of selecting a promoter from a promoter library followed by the union of chosen promoters to various target genes, and finally a mixture of promoter gene pairs is created to achieve optimal expression (Lu and Jeffries 2007). Internal sites for ribosomal entry have also been demonstrated to permit concurrent expression of multiple genes in *S. cerevisiae* (Seino et al. 2005; Xia and Holcik 2009). Changing the spatial configuration of pathway enzymes is another strategy for redirecting fluxes in the direction of desired product. It can avoid the loss of intermediates by degeneration or diffusion (Albertsen et al. 2011).

9.2.3 Step III: Analysis

Tools of systems biology, for example, omics tools (gen**omics**, transcript**omics, **prote**omics,** metabol**omics**, and flux**omics**), are important for the analysis of engineered microorganisms. A brief overview of these technologies is discussed in the following section.

9.2.3.1 Genomics

Genomics is a discipline where bioinformatics, genetics, and high-throughput analytical approaches are combined to study the genes of a cell. It involves sequencing the DNA and analysis of function and structure of genomes. Whole-genome sequencing was used to identify mutations (insertions, deletions, and single-nucleotide polymorphisms (SNPs)) when yeast was grown under glucose-limiting conditions. Finding of mutations that take place in yeast during evolution experiments is a typical application of genomics (Kvitek and Sherlock 2011). These results led to the discovery of mutations individually accountable for adaptations in yeast.

9.2.3.2 Transcriptomics

Transcriptomics involves analysis of the transcriptome (complete set of RNA transcripts in a specified cell population under specific circumstances), and it utilizes high-throughput methods for quantification of mRNA levels. Examples of such methods include SAGE (Velculescu et al. 1995), RNA tag sequencing, cross-linking immunoprecipitation (CLIP), and whole-RNA sequencing. Wolf et al. used CLIP method coupled with Illumina sequencing platform to analyze role of an RNA-binding protein, Khd1 in yeast (Wolf et al. 2010). Methods such as CLIP are used to study RNA and protein interactions which may influence cellular regulation and phenotype of the organism.

9.2.3.3 Proteomics

Proteomics is the study of structure and function of proteins. As regulatory mechanisms such as degradation of proteins and posttranslational modifications lead to protein levels that are not complementary or related to transcript levels (Foss et al. 2011; Olivares-Hernández et al. 2011; Straub 2011), transcriptomics alone is not sufficient for the understanding of cell factories. Proteomics is helpful in cases where modifications engineered to the cell may have hinder the performance itself leading to unwanted outcomes of the engineering done (Redding-Johanson et al. 2011). In one study, transcriptome and proteomics along with metabolome analysis were used to construct a comprehensive network map for Snf1, a protein kinase (Usaite et al. 2008).

9.2.3.4 Metabolomics

Metabolomics involves the analysis of metabolite profiles, i.e., a quantitative study of all metabolites (intracellular and extracellular) and the changes they undergo under specific environmental deviations. This knowledge helps in identification of bottlenecks in the cellular metabolic network. An example of metabolomics application in yeast is the illustration of intracellular aerobic glucose metabolism (Christen and Sauer 2011).

9.2.3.5 Fluxomics

Fluxomics is a discipline where rates of metabolic reactions are inferred that leads to the complete description of metabolic networks, and consequently, the phenotype of the cell factory can be judged. Flux balance analysis (FBA) is one of the models used to study the metabolic fluxes and is used to discover those metabolic pathways that show a possibility for improved product formation. An example of its application in yeast is the identification of an important enzyme that enhances the yield of bioethanol (Bro et al. 2006).

Currently, biological cell factories are being improved by an integrated approach utilizing knowledge and outcomes from synthetic biology and systems biology, leading to refinement of the metabolic engineering strategies (Fig. 9.1); an introduction of these approaches is being provided. Knowledge obtained from different approaches is exploited based on iterative design-build-test-analyze (DBTA) cycles. As the data is gathered, the frequency of iterations required to attain the required design will reduce, and this will lead to a radical cutback of the cost of development of cell factories (Li and Borodina 2014).

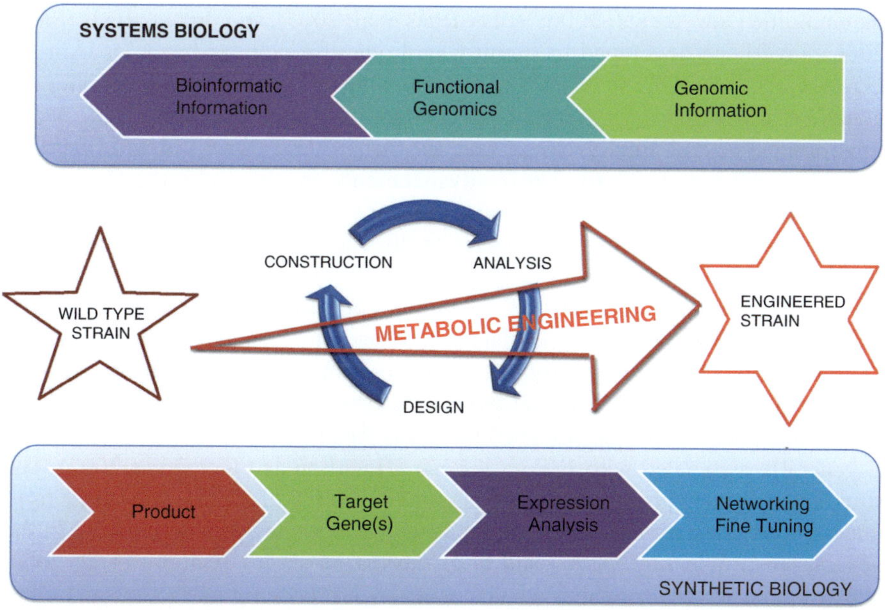

Fig. 9.1 Schematic representation of the integrative approach followed to develop superior yeast cell factories

9.3 Alternative Approaches for Superior Yeast Cell Factories

9.3.1 System Level Approach for Yeast Engineering

Systems biology is a fundamental science that attempts to comprehend the intricacies and functions of the biological systems by mapping relations between cellular constituents (Barrett et al. 2006). The principal objective of systems biology is to gain novel insights by quantitatively describing the biological systems and its cellular processes. This description is generally in the form of mathematical models (Nielsen and Jewett 2008). The models can be employed to envisage the behavior of the biological system under different states. Through apprehension of the different processes and understanding of the relationships between these processes, there is a possibility to get universal models for protein secretion. This can be utilized to devise secretion pathways and, thus, lead to better cell factories for effective manufacturing of recombinant proteins (Graf et al. 2009). In addition, the advances in the domain of systems biology are useful for making metabolites through metabolic engineering.

Various functional genomics tools like analysis of transcriptome (Lashkari et al. 1997), proteome (Zhu et al. 2001), interactome (Lee et al. 2002), metabolome (Jewett et al. 2006; Villas-Boâs et al. 2005), and flux analysis (Sauer 2006) have been developed to monitor and quantify cellular behavior using *S. cerevisiae* as a vehicle, to allow a more extensive understanding of regulation of genetic components in this organism and the quantitative description of its cellular processes, and, finally, to envisage the "whys" and "hows" of cell functioning (Mustacchi et al. 2006). These advancements are driven by the fact that *S. cerevisiae* is being utilized as a model system for understanding the molecular procedures of complex human diseases or disorders and serves as a vital role as a cellular factory.

In addition to traditional uses of *S. cerevisiae* for production of beer, wine, etc., it is also employed for the commercial production of drugs (e.g., human insulin) and vaccines. It is also used for production of bioethanol and organic acids, such as lactic acid (Porro et al. 1999) and pyruvic acid (van Maris et al. 2003). Another potential application of yeast as a cell factory is the production of antibodies, which is generally associated with *Pichia pastoris* (Hamilton and Gerngross 2007). The simplicity of genetic engineering of yeast and the ability of *S. cerevisiae* proteins to undergo posttranslational modifications have led to a considerable increase in the generation of *S. cerevisiae* N- and O-glycosylation strains (Chigira et al. 2008; Nakayama et al. 1992), which would facilitate in producing antibodies and other humanized glycoprotein production at lesser costs.

Thus, the abovementioned applications exemplify the strength of an approach for decreasing R&D time and bringing more products to the market. Still, the development of such yeast cell factories is often not up to the expectations due to many bottlenecks, such as contending metabolic pathways, transcription factor regulation, or product toxicity. To deal with such problems, a more advanced and modern approach is required where systems biology along with synthetic biology and metabolic engineering can synthetically construct whole systems.

9.3.2 Synthetic Biology

Introduced in the early twenty-first century (de Jong et al. 2014), synthetic biology can be defined as "the design and development of novel biological systems that do not exist in nature" (Nielsen 2014). It is anticipated to refine the models of the cell factories, and as stated by the report of NEST High-Level Group, it will facilitate designing in a coherent and methodical way. Synthetic biology works at numerous levels of biological organization. At the DNA level, it operates through development made in gene synthesis techniques and the utilization of codon optimization for most favorable gene expression (Gustafsson et al. 2004; Tian et al. 2009). At the protein level, it tries to impregnate novel functions into proteins and make unnatural amino acids (UAAs) that increase the catalytic potential (Grünberg and Serrano 2010, Voloshchuk and Montclare 2010). At the pathway level, it seeks to design new regulatory systems, recreate biological pathways, or assemble totally novel pathways from biological constituents (Martin et al. 2009; Tyo et al. 2010).

9.4 Yeast as an Ideal Cell Factory

The first approach for the development of cell factory is to select a potential host. There are mainly two strategies for selecting a host as cell factory. One of these two strategies involves the product yield of the host, tolerance of host to the product, and effect of environmental stress on the host (pH, temperature, salt, etc.). Second strategy includes the use of well-studied host and optimization or modulation of the host for the desired product overproduction and required environmental conditions (Kavšček et al. 2015). Metabolic engineering plays an important role in this optimization.

Metabolic engineering can be applied on plant cell, bacterial cell, multicellular fungi, and yeast. Among these hosts, yeast *S. cerevisiae* acts as an attractive host because:

- It can be easily cultured using cheaper media resources consisting only of carbon source, vitamins, and salts.
- Large-scale fermentation can be easily applied on yeast cells.
- Residual cell mass can act as feed for animals if the chemicals of the fermentation products are nontoxic.
- Fermentation using yeast cells can be carried out at acidic pH, hence reducing chances of bacterial contamination.
- Tools for genetic manipulations are well established in yeast.
- *S. cerevisiae* is an organism "generally regarded as safe (GRAS)."

In addition to aforementioned reasons, there is one more attribute of this yeast that makes it a preferred choice as a striking cell factory over other eukaryotes—its DNA repair system, which leads to proficient homologous recombination between instilled donor DNA and the corresponding genomic target (David and Siewers 2014).

S. cerevisiae is commonly used for fermentation procedures leading to various commercial products by using a common carbon source. *S. cerevisiae* is called as conventional yeast, whereas the yeast other than *S. cerevisiae* such as *Pichia pastoris*, *Hansenula polymorpha*, *Yarrowia lipolytica*, and *Kluyveromyces lactis* are called as nonconventional yeast. Nonconventional yeast doesn't require high sugar media and can be used as an alternative for production of various products.

Production of cell factory requires discovery and synthesis of new biochemical pathway for production of desired products and optimization of enzymatic activity of the pathway and the host cell for the desired biochemical production. Metabolic engineering of a cell is a result of advances in system biology, synthetic biology, and omics technology. Synthetic biology is the addition of new parts and functions in the cell, and these additions can be at DNA level (synthetic genes, codon optimization), protein level (new protein structure, unnatural amino acids), pathway level (new and unnatural pathway, new regulatory modules), and organism level (synthetic organism). System biology involves in silico genome-scale modeling, whereas omics technology consists of analysis of whole genome, high-throughput protein analysis, transcript analysis, and intracellular metabolite analysis (Krivoruchko et al. 2011).

Yeast has been associated with the preparation of alcoholic beverages, baker's yeast, pharmaceutical products, biofuels, and recombinant proteins. Pyruvate, xylitol, carotenoids, succinic acid, and malic acid can also be produced by metabolically engineered yeast. Pharmaceutical products such as flavonoids and artemisinin have been produced by yeast more efficiently as compared to other prokaryotic hosts. Yeast engineering for the manufacture of biofuel has replaced the use of fossil fuels that makes the production of biofuel affordable and eco-friendly. Yeast has been involved with the production of chemicals that may range from commodity chemicals (1,2-propanediol) to fine chemicals (resveratrol).

Glycoengineering of yeast provides a large number of pharmaceutical products for the production of human specific glycans. Yeast cell wall is rigid consisting of outer mannoprotein layer and inner glucan layer. Cellular surface of yeast can also be used for the expression of heterologous proteins and immobilization-related applications. The yeast cell surface can serve as a new target, when target substrates are not accessible to the cellular machinery.

9.5 Production of Fine and Bulk Chemicals by Yeast

Yeast can be engineered to produce a variety of chemicals such as sugar alcohols, organic acids, propanediol, L-glycerol 3-phosphate (L-G3P), ergosterol, isoprenoids, and other steroids. Yeast can be exploited for the production of chemicals either through fermentation, the biological method of product formation due to the complex metabolism by an organism, or enzymation, the conversion of complex substances into simpler forms by the use of biocatalyst or enzyme (Faber 2011). Fermentation process that is used for the production of chemicals should be economically viable, which depends upon the yield and titer of the organism used for the fermentation. Fermentation-based as well as enzymation-based chemical production by yeast makes use of metabolic engineering for economically viable product formation. Metabolic engineering finds its application in the production of heterologous protein, modifications leading to substrate range expansion of the host, discovery of pathways for synthesis of novel products, and degeneration of xenobiotics. It is also useful in the improvement of processes and yield and in the elimination or reduction of the amount of by-product formation (Nielsen 2001).

9.5.1 Glycerol

Fermentation of ethanol by *S. cerevisiae* produces glycerol as a major by-product. Glycerol is used for numerous industrial applications such as the nitroglycerin production, manufacture of drugs, cosmetics, toothpastes, etc. It has also been used as a central raw material in chemical industries such as in synthesis of 1,3-propanediol.

Different microbes including yeasts (particularly osmotolerant yeasts) naturally produce large amounts of glycerol as compared to the addition of sulfite and alkali during fermentation process for glycerol overproduction (Wang et al. 2001). Expanding demand of glycerol has led to significant role of metabolic engineering for overproduction of glycerol.

Synthesis of glycerol involves the reduction of dihydroxyacetone phosphate (DHAP) to glycerol 3-phosphate. This step is concurrent with the oxidation of NADH by NAD^+-dependent glycerol 3-phosphate dehydrogenase (Gpd p) (Albertyn et al. 1992). Following which, glycerol 3-phosphatase enzyme (Gpp p) dephosphorylates glycerol 3-phosphate to glycerol (Norbeck et al. 1996). Yeast cells produce glycerol to enable it to maintain a cytosolic redox state for glycolytic catabolism, under hypoxic and glucose-repressing growth conditions (Albers et al. 1996).

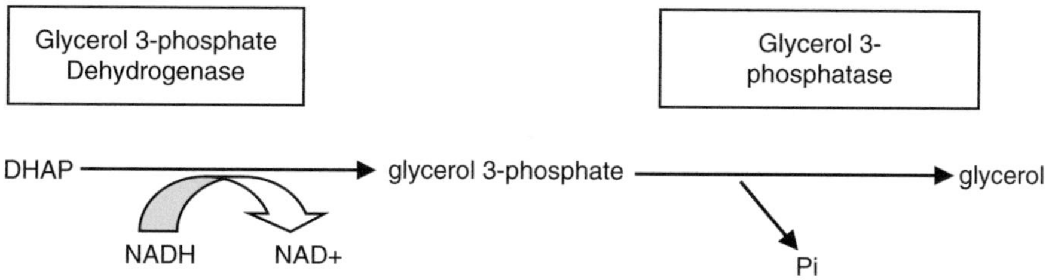

Glycerol can be produced by metabolic engineering of yeast, and some of such strategies are:

- Overexpression of the enzymes encoding isoenzymes of NAD^+-dependent glycerol-3-phosphate dehydrogenase, i.e., *GPD1* or the *GPD2* gene.
- Deletion of gene responsible for acetaldehyde dehydrogenase enzyme.
- Removal of pyruvate decarboxylase gene.
- Prevention of accumulation of DHAP that may affect the growth of yeast. It can be achieved by the removal of *NDE1*and *NDE2* isogenes and *GUT2*. *NDE1* and *NDE2* isogenes encode for external NADH dehydrogenase, while *GUT2* encodes FAD^+-dependent GPD. These different enzymes cause reoxidation of cytosolic NADH inside the mitochondria. Hence, nonavailability of these enzymes increases the accessibility of cytosolic NADH which reduces the accumulation of DHAP pool.

The major limitation in glycerol overproduction using *S. cerevisiae* is that only 50% of the glucose is utilized for NADH-dependent glycerol production and the other 50% is indispensable for the generation of NADH via glyceraldehyde 3-phosphate dehydrogenase. Diversion of glycolytic flux was possible using a *pdc1*Δ *pdc5*Δ *pdc6*Δ *nde1*Δ *nde2*Δ *gut2*Δ mutant strain combined with *GPD2* overexpression, with simultaneous overexpression of *FDH1*, thus managing to produce a glycerol yield of 1.08 mol per mole of glucose (Geertman et al. 2006).

9.5.2 Propanediol

1,2-Propanediol finds multiple applications in the domains of cosmetics, unsaturated polyester resins, pharmaceutical products, antifreeze agent, liquid laundry detergents, and in removal of snow and ice (Cameron et al. 1998). There is a growing demand for polyester polypropylene terephthalate, which is the most common thermoplastic polymer resin used in carpeting, textiles, and other industries. Hence, 1,3-propanediol is considered as a special chemical which is used for the production of polyester polypropylene terephthalate polymer. Synthesis of 1,2-propanediaol and 1,3-propanediol can be done via different pathways using glycolytic intermediate DHAP. Yeast does not have any enzyme which is responsible for the conversion of acetol to 1,2-propanediol. Therefore, *mgs* and *gldA* genes from *E. coli* were integrated into the genome of for such biotransformation to be possible by yeast cells, *mgs* encodes for methylglyoxal synthase and *gldA* for glycerol dehydrogenase. These two genes were integrated under the regulation of the *CUP1* promoter into *S. cerevisiae* genome. A major demerit of the strategy was that the desired product was produced intracellularly (Lee and DaSilva 2006).

1,3-Propanediol cannot be produced naturally by the fermentation of sugars. Some microorganisms have the ability for producing 1,3-propanediol from glycerol by fermentation; however, such microbes fail to produce glycerol from DHAP due to the absence of enzyme glycerol-3-posphate dehydrogenase. Baker's yeast can produce glycerol, but it does not have ability to convert glycerol to 1,3-propanediol. Therefore, metabolic engineering is the only way to produce 1,3-propanediol through fermentation by any single organism. Expression of genes responsible for the conversion of glycerol to 1,3-propanediol in *S. cerevisiae* is one of the approaches applied for producing it. However, it was found that the yeast did not produce measurable amount of 1,3-PD in the fermentation broth when subjected to co-expression of 1,3-PD genes (*dhaB3*, *dhaB3a*, *dhaB4*, and *dhaB4a*) of *Klebsiella pneumoniae* encoding glycerol dehydratase, along with *dhaT* gene of *K. pneumoniae* encoding 1,3-PD oxidoreductase (Cameron et al. 1998). Coenzyme vitamin B12 is required for the activity of glycerol dehydratase. The strategy failed as yeast cells are incapable of synthesis of coenzyme vitamin B12, neither can it be transported across the cells. On the other hand, recombinant *S. cerevisiae* strain expressing the *dhaB* gene of *Klebsiella* and *yqhD* gene of *E. coli* has been reported to produced 1,3-PD (Ma et al. 2007).

9.5.3 Organic Acids

Organic acids like lactic acid, pyruvic acid, and succinic acid have been produced by yeast. The most significant advantage of using yeast cells for producing organic acids is its tolerance to low pH and reduced risk of contamination.

9.5.3.1 Lactic Acid

Pure lactic acid production is an interesting area of industrial biotechnology as it is required for the synthesis of biodegradable polymer materials (Porro et al. 1999). Polymerization of lactic acid to form polylactic acid is the most interesting use of lactic acid in industries (Ghaffar et al. 2014). Synthesis of lactic acid can be done industrially by two ways:

- Chemically
- By microbial fermentation

Higher productivity of lactic acid is achieved by continuous cultures of microbes (Ghaffar et al. 2014). Diverse group of microbes have been exploited for lactic acid production. Lactic acid bacteria require a large number of complex nutrients for their growth, and further their growth is inhibited by the accumulated end product, particularly at an acidic pH (Dato et al. 2014). Low pH is developed by the presence of lactic acid that inhibits microbial activities. To neutralize the effect of lactic acid, $Ca(OH)_2$, $CaCO_3$, NaOH, and NH_4OH are added during industrial processes. Thus, the requirement for neutralization of lactic acid is the major drawback during such fermentation process (Porro et al. 1999). In addition, recovery procedure for the separation of the product and its purification from the fermentation broth poses serious problems (Abdel-Rahman et al. 2013). Thus, instead of using lactic acid bacteria, yeast can be a real alternative for the synthesis of lactic acid, as yeast can tolerate an acidic environment. Lactic acid has various applications like in food industry as an acidulant and a flavoring agent, and it is widely used as pH buffering agent and as food preservative. Industrial applications of lactic acid also include chemical and biological conversion of useful chemicals like acrylic acid, 1,2-propanediol, pyruvic acid, and lactate ester because of the presence of hydroxyl and carboxylic groups in lactic acid (Dato et al. 2014). Further, cultivation of yeast using mineral media facilitates in recovery of lactic acid from the fermentation broth (Abdel-Rahman et al. 2013).

Engineering of Yeast for Lactic Acid Production

High production of lactic acid can be achieved by the substitution of ethanol production pathway. Activity of pyruvate decarboxylase and/or pyruvate dehydrogenase is deleted (Abdel-Rahman et al. 2013) as pyruvate decarboxylase gene is required for the ethanol production and lactate dehydrogenase is necessary for the conversion of pyruvate to lactic acid (Ilmén et al. 2013).

During stress conditions, cells change their metabolic pathway. Most of the metabolic engineering focuses on enzymes—like their addition, deletion, or amplification. However, the lack of full knowledge of cellular mechanism and metabolic pathways poses major obstacle in engineering of the genes coding for enzymes in these pathways. Cofactors are one of the important factors for any biochemical reactions, so changes in cofactors have effects on metabolic networks and also induce changes at the transcriptional levels. Nowadays, engineered yeasts are available for lactic acid production. However, engineered yeast does not produce net ATP because of the energy utilized to export lactate (Dato et al. 2014).

Different Strategies for Lactic Acid Overproduction in Yeast

- Using metabolically engineered strains of S. *cerevisiae*
- Selection of strains under stress conditions (higher intracellular pH)

- Overexpression of the hexose transporters (e.g., Hxt1p and Hxt7p) for increased glucose uptake hence leading to enhanced lactic acid productivity (Dato et al. 2014).
- Deletion of pyruvate decarboxylase gene(s). Targeting lactate dehydrogenase gene is one of the strategies to modify glycolytic flux toward the synthesis of lactic acid. Inactivating or suppressing the enzymes (e.g., pyruvate decarboxylase activity) responsible for the manufacture of ethanol can be done to yield higher amount of lactic acid. This method stops the production of ethanol (Porro et al. 1999). *S. cerevisiae* produces ethanol in the presence of excess glucose, a major concern in commercial lactic acid production. Pyruvate decarboxylase gene is required for the ethanol production so its deletion gives rise to high lactic acid production on glucose-based media. Removal of PDC1 gene decreases the PDC activity.
- Introduction of L-lactate dehydrogenase (LDH) gene and the deletion of genes encoding for pyruvate decarboxylase1 (PDC1), glycerol-3-phosphate dehydrogenase (GPD1), and L-lactate cytochrome-c oxidoreductase (CYB2) increase the production of lactic acid. Moreover, further removal of NADH dehydrogenase genes (NDE1 and NDE2) results in the lactic acid production more efficiently (Lee et al. 2015).
- *S*-Adenosylmethionine (SAM or AdoMet) plays a central role as a coenzyme in most of the metabolism processes. AdoMet is a methyl donor causing DNA methylation; such methylated DNA is unable to function properly hence switching off of the respective gene. There are two SAM synthetase genes in *S. cerevisiae*, named SAM1 and SAM2. Deletion of SAM2 gene has been shown to result in higher lactic acid production (Dato et al. 2014).

Various studies have shown that *S. cerevisiae* having heterologous gene encoding lactate dehydrogenase promotes lactic acid productivity. Some nonconventional yeast species have also been used like *Kluyveromyces lactis*, *Pichia stipitis*, *Candida boidinii*, and *Candida utilis* which have good acid tolerance and metabolize carbohydrate. Introduction of LDH gene in these strains increases the production of lactic acid (Ilmén et al. 2013). Introduction of bovine L-lactate dehydrogenase gene (*LDH*) into a wild-type *K. lactis* yeast strain has been shown to increase the production of lactic acid; in this case ethanol is also produced. Engineered *K. lactis* strain was also developed having LDH gene along with deleted pyruvate decarboxylase gene. It results in the high productivity of lactic acid. *K. lactis* strain having deleted PDC gene produces no ethanol, but this strain grows on glucose-based media (Porro et al. 1999).

Nonconventional yeast like *C. sonorensis* has also been genetically modified using vectors. *C. sonorensis* is a methylotrophic, acid-tolerant yeast that produces ethanol by the fermentation of glucose. It requires simple nutrients and exploits different carbon sources together with xylose and arabinose, sugars having five carbon atoms. Lactic acid production in the organism can be enhanced by expressing L-lactate dehydrogenase (LDH) from *Lactobacillus helveticus* (Ilmén et al. 2013).

9.5.3.2 Pyruvic Acid

To produce higher amount of acetyl-CoA carboxylic acid, higher amount of pyruvate titer is required (Pronk et al. 1996). In yeast, pyruvate is present at the point where it can be converted into ethanol. When sugars are present in excess, then the pathway generally produces ethanol under aerobic conditions. Several strategies have been employed to enhance the production of pyruvate and minimize by-products, as the pyruvate is present at the branch point between respiratory metabolism of sugars and alcoholic fermentation (Pronk et al. 1996; Xu et al. 2012). By using genetic engineering, carbon can be diverted from ethanol to pyruvate in *S. cerevisiae* (Abbott et al. 2009). Pyruvate or pyruvic acid would ultimately lead to acetyl coenzyme A (acetyl-CoA) biosynthesis which functions as a central precursor molecule in yeast and has myriad industrial applications (Fig. 9.2).

Commercial products derived from the acetyl-CoA:

- Polyhydroxybutyrates—a polymer having various applications
- 1-Butanol—used as biofuel and chemical building block
- Waxes—as lubricants and additives for cosmetics
- Isoprenoids—used as biofuels, pharmaceuticals, nutraceuticals, perfumes, etc.
- Polyketides—as chemotherapeutic drugs for cancer

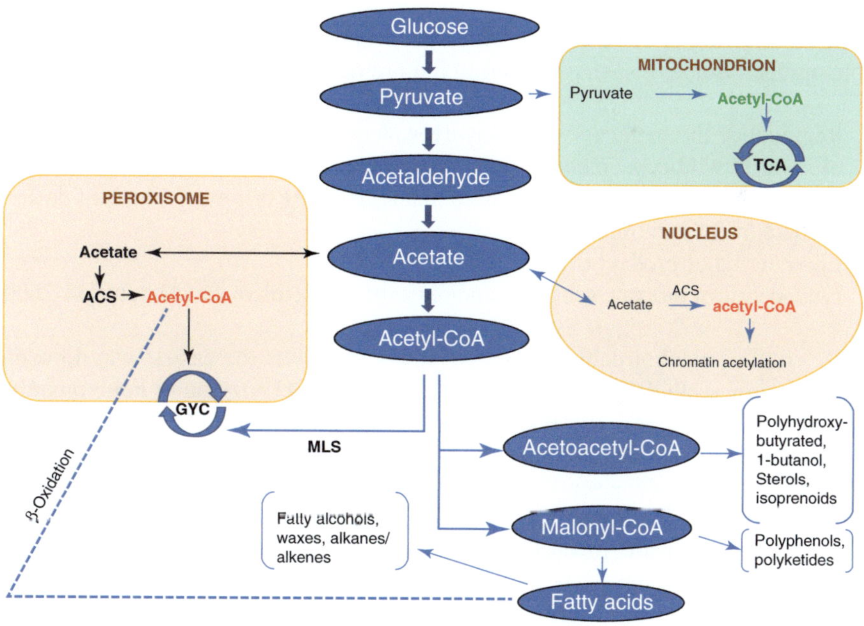

Fig. 9.2 Overview of the pathway showing metabolism of acetyl-CoA in yeast. *ACS* acetyl-CoA synthetase, *GYC* glyoxylate cycle, MiLS malate synthase, *TCA* tricarboxylic acid cycle (Krebs cycle)

- Polyphenols—used as nutraceuticals and antioxidants
- Alkanes/alkenes and fatty alcohols—can be used as advanced biofuels (e.g., diesel for trucks and airplanes)

During glycolysis the glucose is converted into pyruvate, which goes into the mitochondria for respiration. In mitochondria, the pyruvate dehydrogenase complex (PDH) converts pyruvate to acetyl-CoA, followed by the oxidation of acetyl-CoA by TCA cycle. Pyruvate decarboxylase (PDC) then converts pyruvate to acetaldehyde in the cytosol, which is further converted to acetate or ethanol by the action of aldehyde dehydrogenase (ALD) or alcohol dehydrogenase (ADH), respectively. Acetate can enter inside the cell nucleus or peroxisomes. The acetate is converted to acetyl-CoA by the action of acetyl-CoA synthetase (ACS) inside these organelles. Inside nucleus, the acetyl-CoA functions for acetylation of the histones, while inside peroxisomes, it enters the glyoxylate cycle (GYC). Cytosol acetyl-CoA also enters the glyoxylate cycle through malate synthase (MLS) action. During the glyoxylate cycle, 1 mol of malate from 2 mol acetyl-CoA is formed, and then malate enters inside mitochondria where oxidation of malate takes place. This is the important route for growth on ethanol or acetate. Acetyl-CoA present in the cell cytosol can also lead to the formation of valuable intermediates like acetoacetyl-CoA or malonyl-CoA which can be used for various valuable biotechnological products (Nielsen et al. 2014) including biosynthesis of lipids such as ergosterol, fatty acids, etc.

Metabolic Engineering of Yeast for Pyruvic Acid Overproduction

Production of compounds like pyruvate and lactate meets major challenges in metabolic engineering of *S. cerevisiae* which need to be addressed for obtaining desirable product. Some of such engineering requirements are:

- Removal of the pathway which involves alcoholic fermentation (dissemination of sugar in wild-type *S. cerevisiae*) (Abbott et al. 2009)
- Engineering of metabolic pathway so that it gives rise to elevated level of desired product in *S. cerevisiae*
- Engineering of product export
- Engineering of product, substrate, and/or environment tolerance (Abbott et al. 2009)

Deletion of alcohol dehydrogenase (ADH) activity is the easiest way to avoid conversion of acetaldehyde to ethanol. Yeast have ADH enzyme in large quantity, and these enzymes are used in specific pathways (Chen et al. 2012). Some of the specific product pathway like butanol biosynthesis depends on the activity of alcohol dehydrogenase (Nielsen et al. 2014). Regulation of acetyl-CoA synthetase activity is equally important for overproduction of pyruvate as the enzyme influences redirection of flux toward cytosolic acetyl-CoA. Two acetyl-CoA synthetase (ACS) genes are present in yeast, ACS-1 and the second one is ACS2. The location of protein formed by these genes relies on the type of carbon source. These genes are active only in the cytosol. Presence of glucose represses the activity of ACS1, but its presence is required to express ACS2. Posttranscriptional regulation of ACS by acetylation is found in bacteria but not in the yeast.

Overexpression of aldehyde dehydrogenase (ALD, mainly ALD6) and ADH2 is also a successful strategy to increase the production of acetyl-CoA and acetaldehyde, respectively (Shiba et al. 2007). Moreover, the deletion of genes for citrate synthase (CIT2) and malate synthase (MLS1) avoids the entry of acetyl-CoA into glyoxylate cycle, hence facilitating the formation of products derived from acetyl-CoA (Chen et al. 2012).

9.5.3.3 Malate and Succinate

Malate and succinate are useful in the sustainable production of various chemicals. They have versatile structures (Werpy et al. 2004). Carboxylating anaplerotic pathways are required to increase the production of these compounds, which in turn convert three carbon compounds formed during glycolysis into malate and succinate (four-carbon backbones) (Abbott et al. 2009). Initially the focus was on filamentous fungus *Aspergillus flavus* for malate production via biotechnological routes because *A. flavus* naturally produces malate. However, for industrial purposes, use of *A. flavus* is discouraged because of the risk of aflatoxin production (Hesseltine et al. 1966). Aflatoxin production is incompatible with process safety. While *S. cerevisiae* wild-type strain is the poor malate producer (Fatichenti et al. 1984), overexpression of fumarase enzyme can increase the malate production in yeast cells, by enhancing the activity of malate dehydrogenase (Abbott et al. 2009).

9.5.4 Sugar Alcohol

Xylitol is a five-carbon sugar alcohol like sucrose; however, xylitol is the preferred sugar under various situations as it has low calorific value in comparison to other sugars (hence useful for calorie-conscious people), is useful for diabetics and hyperglycemic patients as it has insulin-independent route of metabolism, and may also be useful in maintenance of oral health (Miyasawa et al. 2003) and ear infections (Uhari et al. 1998).

On industrial scale, xylitol is produced by hydrogenation of pure xylose, which can be obtained from hydrolysate of hardwood. Metabolic engineering attempts have also been made for industrial production of xylitol using yeast cells. Simultaneously deletion of transketolase isogenes and xylulokinase gene was done in yeast, along with overexpression of genes for *Pichia stipitis* xylitol dehydrogenase and endogenous sugar phosphate dephosphorylase. This engineered yeast cell produced xylitol along with another sugar alcohol ribitol (Toivari et al. 2007). Although metabolic engineering approaches are possible for sugar alcohol production, exemplary improvements are required for utilization of the strategies at commercial scale.

9.5.5 L-G3P

Microbial production of L-glycerol 3-phosphate (L-G3P) holds great promise since it can replace the unstable dihydroxyacetone phosphate (DHAP) in a single-pot enzymatic carbohydrate synthesis. In enzymatic and chemical synthesis, phosphorylated intermediates have an important role as they act as a starting material. Microbes synthesize phosphorylated intermediates, but there are various challenges associated with the microbial production. Some of them are cell metabolism, homeostasis effects, and efficient retention of the product inside the host cell.

Genetically modified *S. cerevisiae* is used for the accumulation of L-G3P. This is achieved by engineering metabolic pathway (Nguyen et al. 2004). For example, Popp et al. engineered a strain by overexpressing *GPD1* and simultaneously deleting both isogenes encoding glycerol 3-phosphatase. They observed that the level of intracellular L-G3P increased by 200 mM (Popp et al. 2008).

S. cerevisiae strain-lacking genes GPP1 and GPP2 along with overexpression of L-glycerol 3-phosphate dehydrogenase (GPD1) allowed accumulation of L-G3P up to concentrations of about 200 mM for a short period. Extracellular L-G3P was also found during this fermentation. Although specific L-G3Pase activity was completely inhibited, the strain could produce glycerol indicating some dephosphorylation process as a mechanism to relieve cells of L-G3P. Upscaling could be achieved through fed-batch fermentation, where glucose was fed repeatedly and aerobic phase was followed by anaerobic phase. This enable the production of about 325 mg total L-G3P per liter of broth.

9.5.6 Terpenoids

Terpenoids constitute a large family of natural products (Ignea et al. 2011) with more than 55,000 terpenoids identified (Zhuang 2013). Terpenoids have various attracting applications like in flavors, fragrances, drugs, and alternative fuels. Taxol, a diterpene, is widely recognized for its chemotherapeutic properties (Zhuang and Chappell 2015). Terpenoids are synthesized chemically, but it is often costly and inefficient. Hence, various efforts have been done to produce this interesting class of molecules by biological means. *S. cerevisiae* can be used as an original factory for terpenoid biosynthesis (Ignea et al. 2011). Two C5 precursors are utilized for the biosynthesis of terpenoids: isopentenyl diphosphate (IPP) and dimethylallyl diphosphate (DMAPP). In yeast and mammals, acetyl-CoA converts into mevalonic acid and then finally gives rise to IPP (Ignea et al. 2011). Prenyltransferase enzymes act on IPP and synthesize higher-order building blocks: geranyl pyrophosphate (GPP), a 10-carbon compound; farnesyl pyrophosphate (FPP), a 15-carbon compound; and geranyl pyrophosphate (GGPP), a 20-carbon compound. In yeast, FPP is the most desired product of the pathway that is utilized for the biosynthesis of sterols. GPP, FPP, and GGPP are the substrate of terpene synthases (Kampranis and Makris 2012). Ergosterol is the chief product of the yeast mevalonate (MVA) pathway. Sterols, mainly ergosterols, are important for the growth of yeast as it is an important component of its cell membrane, leading to yeast cell death in the absence of ergosterol (Fig. 9.3).

HMGR

MVA pathway is governed by the concentration of HMG-CoA reductase, i.e., 3-hydroxy-3-methyl-glutaryl-CoA reductase (HMGR—indicated as HMG1 in the

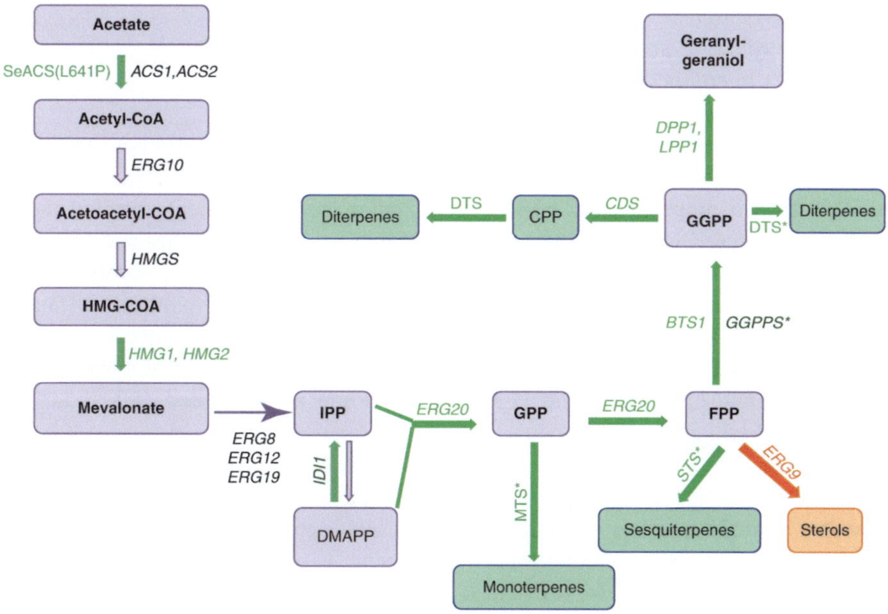

Fig. 9.3 Terpene biosynthesis in yeast. Figure indicating genes involved and metabolic engineering interventions employed. Upregulated yeast genes indicated in *green* and downregulated yeast genes in *orange*. Enzymes with yield or specificity enhancement or altered by protein engineering are indicated by superscripted *asterisks*. *CPP* copalyl diphosphate, *CDS* copalyl diphosphate synthase, *DTS* diterpene synthase, *MTS* monoterpene synthase, *SeACS(L641P) Salmonella enterica* acetyl-CoA synthase mutant L641P

figure). HMG-CoA produces mevalonic acid with the reduction of NADPH by the action of HMGR (Kampranis and Makris 2012). Yeast has two enzymes of HMGR: Hmg1p and Hmg2p. Studies showed that the elevated levels of pathway products led to degradation of HMG proteins. Thus, incorporation of point mutation (substituting lysine 6 by an arginine (K6R)) in Hmg2p was done to make the enzyme resistant to ubiquitination. HMG2 (K6R) showed enhanced level of monoterpene and sesquiterpene production (Kampranis and Makris 2012).

ERG9

The ERG9 gene codes for squalene synthase. Synthesis of squalene in the presence of squalene synthase is the major step because of which isoprenoid substrates cannot be utilized for production of commercial molecules, and hence its suppression is desirable. As a complete knockout of ERG9 gene is lethal for the organism itself, strategies for suppression of ERG9 gene have been devised using MET3 promoter (P_{MET3}), expression of which is suppressed by more than 0.5 concentration of methionine leading to reduced transcription of ERG9. This mutation increases the production of the sesquiterpenes cubebol, valencene, and patchoulol (Kampranis and Makris 2012).

Monoterpenoids

Monoterpenoids are produced by expressing linalool synthase and geraniol synthase genes in engineered yeasts (Oswald et al. 2007).

Sesquiterpene

Sesquiterpenes are produced in the MVA pathway upon cyclization or ionization of an intermediate, farnesyl diphosphate (FPP). Sesquiterpenes have multiple chemotherapeutic properties for wide application in several clinical manifestations including cancer (Ro et al. 2006). Sesquiterpenes are not biosynthesized by wild-type yeast as all the FPP produced are channeled to the production of sterols, mainly ergosterol. Thus, engineering of genes responsible for sterol synthesis (Zhuang 2013) is used for the desired sesquiterpene production. Overexpression of truncated HMGR (tHMGR), downregulation of ERG9, and overexpression of the yeast farnesyl diphosphate synthase gene (ERG20) are some of the successful strategies employed (Farhi et al. 2011).

For the biosynthesis of terpenes, a SUE (sterol uptake enhancement) mutant was constructed which could uptake sterol from external source (Bourot and Karst 1995; Shianna et al. 2001), thus making the endogenous ergosterol biosynthetic pathway nonessential. A knockout mutation of ERG9 gene was created in the SUE mutant. As a result of this, FPP, DMAPP, and IPP pools get diverted toward the production of monoterpenes, sesquiterpenes, diterpenes, and triterpenes which are 10-, 15-, 20-, and 30-carbon compounds, respectively (Zhuang 2013).

9.5.7 Ergosterol and Other Steroids

Ergosterol is found in the membrane of yeast and steryl esters are present as cytosolic lipid particles. Steryl esters are the storage form of sterols. Ergosterol and sterol synthesis pathways are very complex, and a large number of other products are also formed during this pathway. Production of sterols can be enhanced through the application of metabolic engineering. Recombinant *S. cerevisiae* was utilized for synthesis of hydrocortisone, a major adrenal glucocorticoid of mammals. Thirteen genes were engineered and assembled for developing a self-sufficient biosynthetic pathway which was expressed in a single yeast strain (Szczebara et al. 2003).

9.5.8 Fatty Acids and Their Derivatives

Fatty acids are present in all living organisms in the form of fatty acyl-acyl carrier protein (acyl-ACP), fatty acyl-coenzyme A ester (acyl-CoA), storage lipids, eicosanoids, and unesterified or free fatty acids (Peralta-Yahya et al. 2012). Long-chain carboxylic acids constitute the fatty acids, which are used by the cells for chemical as well as energy storage functions. Natural wax esters contain long-chain fatty acids (Stöveken et al. 2005). Wax esters are used as lubricants, diesel, linoleum, printing inks, cosmetics, candles, and polishes (Jetter and Kunst 2008). Fossil sources are used for the chemical synthesis of wax esters from plant and animal tissue, which is the major limitation of chemical synthesis of wax esters (Kalscheuer et al. 2006); moreover, due to the increasing need, cost, and harmful effect of wax esters on the environment, it becomes mandatory to have a renewable, economically beneficial, and eco-friendly resource of fatty acids. "The natural petroleum" can be obtained from microbes, which can act as an alternative for the production of cost-effective and eco-friendly fatty acids and their derivatives. As compared to the fatty acid-derived products, fatty acids have limited use in industry due to the ionic nature of carboxylic acid. Microbes naturally cannot produce fatty acid-derived products with higher yield. Application of metabolic engineering of microbes plays an important role in synthesis of chemicals which are derived from fatty acid-derived chemicals such as fatty alkanes, lactones, fatty acid methyl esters, fatty acid ethyl esters, fatty alcohols, and hydroxy fatty acids (Lennen and Pfleger 2013; Peralta-Yahya et al. 2012; Runguphan and Keasling 2014).

Cyanobacteria and microalgae are attractive hosts for the production of various biofuel-derived compounds including 1-butanol (Lan and Liao 2012), isobutanol, isobutyraldehyde (Atsumi et al. 2009), and 2-methyl-1-butanol (Shen and Liao 2012), due to their unique property of converting solar energy and carbon dioxide into fuels. However, due to their complicated growth and difficulty in genetic manipulation, the use of these organisms is limited. These shortcomings can be overcome by using microbes.

Oleaginous yeast has high oil and lipid content comprising about 20% of the dry weight of oleaginous yeast (Ratledge and Wynn 2002). Oleaginous yeast is fast growing and can utilize diverse carbon sources for its growth. Fatty acid and lipid content of the oleaginous yeast varies from species to species. It can be 40% or 70% of the total dry weight of species, under nutrition-limiting condition (Beopoulos et al. 2009). Yeasts, namely, *Rhodotorula glutinis*, *Cryptococcus albidus*, *Lipomyces starkeyi*, and *Candida curvata*, have high lipid content (Meng et al. 2009).

Fatty acid biosynthesis primarily produces fatty acids having 16- and 18-carbon atoms, and introduction of double bond in long-chain fatty acids gives rise to unsaturated fatty acids. The relative ratio of chain lengths and the degree of unsaturation of these fatty acids vary widely depending on the microorganisms. *S. cerevisiae* primarily synthesizes 16- and 18-carbon-containing saturated and monounsaturated fatty acids. *S. cerevisiae* is comprised of one fatty acid desaturase and a Δ9-desaturase (OLE1) due to which it is responsible for the synthesis of monounsaturated palmitoleic (C16:1) and oleic (C18:1n-9) acids (Stukey et al. 1989).

9.5.8.1 Approach for the Synthesis of Fatty Acids and Their Derivatives

The main product of glycolytic pathway is pyruvate. Pyruvate dehydrogenase is responsible for the conversion of pyruvate to acetyl-coA. Acetyl-coA then enters into citrate-pyruvate shuttle pathway in which acetyl-coA together with oxaloacetate of TCA cycle condenses to citrate. When level of citrate increases in the mitochondria, citrate enters into the cytoplasm and gets converted into acetyl-coA and oxaloacetate. Acetyl-CoA converts into malonyl-CoA in the presence of acetyl-CoA carboxylase (ACC). One equivalent of acetyl-CoA and 7–8 equivalents of malonyl-CoA are condensed into C16–C18 fatty acyl-CoAs by the action of fatty acid synthase (FAS) complex. Regulation of fatty acid synthesis inside the cell is done at multiple levels (Tehlivets et al. 2007). High levels of long-chain acyl-CoA inhibit the synthesis of fatty acid, whereas low concentration of long-chain acyl-coA inhibits ACC which prevents the accumulation of large amount of fatty acids. Overproduction of fatty acids can be achieved by the overexpression of endogenous or heterologous acyl carrier protein (ACP) and acyl-CoA thioesterase (Fig. 9.4).

The main objective of metabolic engineering is to improve the production of different metabolites from microbes. Metabolic engineering can be achieved by alternation of metabolic pathway for desired metabolite. Productivity of desired metabolite can be enhanced by altering the native pathway or by introducing completely new pathways. Metabolic engineering can also be done for better host strains by altering the endogenous as well as heterogeneous expression of genes and enzymes, or by genetic perturbations for modification of the promoter activity of a given gene (Ostergaard et al. 2000).

The important fatty acids in yeast include stearic acid, myristic acid, palmitic acid, oleic acid, linoleic acid, and linolenic acid (Li et al. 2007). It has been found that C/N ratio, pH, temperature, and other environmental factors affect the lipid production by a yeast cell. At C/N ratio of 150, the oleaginous yeast *L. starkeyi*

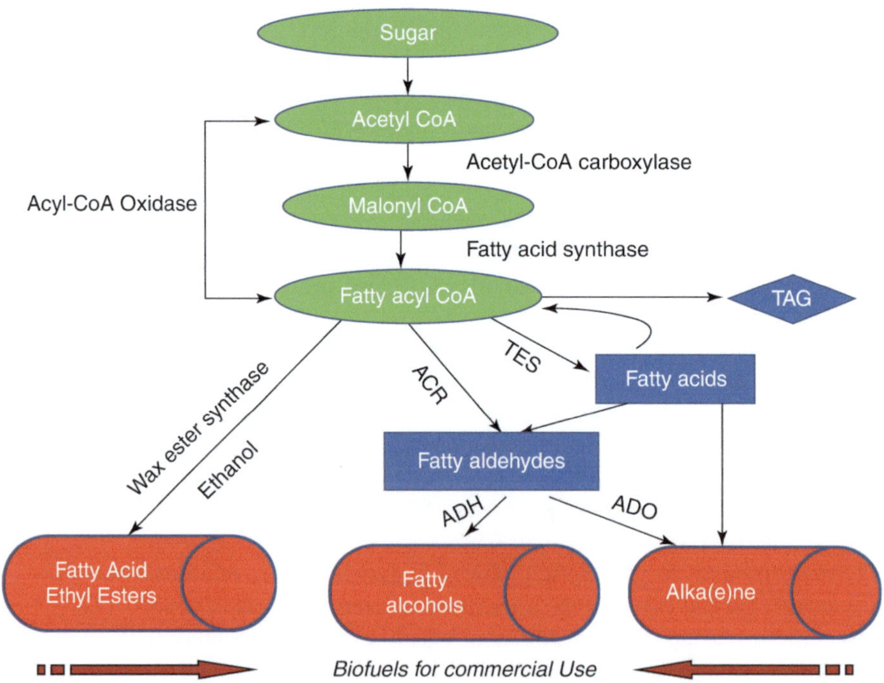

Fig. 9.4 Metabolic pathways for the synthesis of fatty acid-derived biofuel and chemical biosynthesis. *TES* thioesterase, *ACR* fatty acyl-CoA reductase, *ADO* aldehyde-deformylating oxygenase, *ADH* alcohol dehydrogenase

delivered 68% lipid content, whereas at C/N ratio of 60, it delivered 40% lipid content when it has been grown on digested sewage sludge. At low temperature, *S. cerevisiae* cells have increased level of degree of unsaturation but have reduced length of fatty acids (Beltran et al. 2008). It was observed that when *L. starkeyi* was cultivated at higher temperatures, there was increase in the cellular lipid content, but with low degree of unsaturation. *Rhodotorula glutinis* accumulates lipids up to 49% of cell dry weight when subjected to optimal batch fermentation conditions (Dai et al. 2007), while under continuous culture, its growth rate decreases as the lipid content and lipid yield increase (Alvarez et al. 1992).

9.5.8.2 Fatty Acid Ester

Biodiesel is comprised of different fatty acid short-chain alkyl esters having diverse fatty acid carbon chain lengths. Fatty acid methyl esters (FAMEs) and fatty acid ethyl esters (FAEEs) can be utilized as commercial "biodiesel" fuel.

These esters are synthesized by transesterification of vegetable oils and an alcohol (methanol or ethanol) with the aid of a catalyst (Lang et al. 2001; Röttig et al. 2010). They are attracting significant attention due to their renewability and environmental friendliness (Lam et al. 2010). The most important enzyme in fatty acid ester is wax ester synthase which catalyzes the esterification reaction of acetyl-coAs and alcohol.

Acetyl-CoA is the main component of fatty acid biosynthetic pathway in yeast, which is also a central component for several cellular biosynthetic pathways (Chen et al. 2012). Hence, overproduction of fatty acids is required which can be achieved by engineering of yeast in such a way that the production of acetyl-coA is increased, with a simultaneous decrease in its use in other cellular pathways. Another important component for bioester synthesis is wax ester synthase (WS) enzyme, catalyzing the transesterification reaction between saturated or unsaturated acyl-coA and alcohols. As *S. cerevisiae* is the ideal organism for metabolic engineering, three different types of wax synthases from plants, bacteria, and mammals were heterologously expressed in *S. cerevisiae.* WS from *Marinobacter hydrocarbonoclasticus* showed the highest ethanol production and tolerance compared to the other WSs and could permit the engineered *S. cerevisiae* to produce biodiesel to a concentration of 6.3 mg/L (Barney et al. 2012; Shi et al. 2012).

During the synthesis of FAEEs, ethanol concentration is one of the most critical parameter. On one hand, the rate of reaction and the yield can be enhanced by high ethanol concentration, while it may also inhibit activity of the enzyme (Lim et al. 2013).

Metabolic Engineering of Fatty Acid Esters

1. Overexpression of *Marinobacter hydrocarbonoclasticus* DSM 8798 wax ester synthase in *S. cerevisiae* and upregulation of acetyl-coA carboxylase yielded 8.2 mg/L FAEE (Shi et al. 2012).
2. Eliminating pathways that compete with the FAEE formation such as triacylglycerol pathway, sterile ester formation, and beta-oxidation pathway leads to increase in the FAEE yield up to 17.2 mg/L (Valle-Rodríguez et al. 2014). Yield of FAEE further increases with the expression of wax ester synthase into the above strain.

3. Integration of yeast acyl-CoA-binding protein and glyceraldehyde-3-phosphate dehydrogenase ($NADP^+$ dependent) that can route glyceraldehyde-3-phosphate produced by glycolytic pathway and xylulose-5-phosphate from pentose phosphate pathway into acyl-coA pathway from *Streptococcus mutans* increases FAEE production up to 47.6 mg/L (Shi et al. 2012).
4. FAEE production requires acyl-coA and cofactor NADPH. Acyl-coA is produced by precursors of acetyl-coA. One approach used for overproduction of FAEE was utilization of oxidative part of pentose phosphate pathway for the production of acetyl-coA that produces two molecules of NADPH per molecule of glucose. *A. nidulans* ack (acetate kinase) or *B. subtilis* pta (phosphotransacetylase) expression was combined with expression of *A. nidulans* xpkA (xylulose-5-phosphate phosphoketolase), and *M. hydrocarbonoclasticus* ws2 was expressed in *S. cerevisiae* which leads to increase in the production of FAEE. Xylulose-5-phosphate is converted to acetyl phosphate and glyceraldehyde-3-phosphate in the presence of xpkA; *ack* catalyzes the conversion of acetyl phosphate to acetate; in the presence of *pta*, acetyl phosphate gets converted to acetyl-coA (de Jong et al. 2014).

9.5.8.3 Fatty Alkanes

Fatty acid alkanes act as biofuels and these can be straight and branched chain. Alkane biosynthesis includes two pathways:

(1) Acyl-ACP reduced to fatty aldehyde with the help of enzyme fatty acyl-ACP reductase. Fatty aldehyde is converted to fatty alkanes by the action of aldehyde-deformylating oxygenase.

(2) Reduction and decarboxylation of fatty acids to produce alkanes. Heterologous expression of genes involved in alkane synthesis such as *Clostridium acetobutylicum* fatty acyl-ACP reductase and *Arabidopsis thaliana* fatty aldehyde decarbonylase in *E. coli* gives rise to high alkane production; however, there is only one report of *S. cerevisiae* engineering for the synthesis of alkanes that involves the integration of aldehyde decarboxylase CER-1 and acyl-coA reductase CER-3 genes in *S. cerevisiae* INVSur4# which synthesizes long-chain fatty acid alkanes (Bernard et al. 2012).

9.5.8.4 Fatty Alcohols

The use of fatty alcohols is widespread in various areas like in detergents, skin care products, cosmetics, medicine, and potential biofuel (Liu et al. 2014). Fatty aldehydes are utilized for the production of fatty alcohols. Fatty aldehydes are formed either by fatty acyl-ACP (Schirmer et al. 2010), acyl-CoA (Reiser and Somerville 1997), or fatty acid (Akhtar et al. 2013) by the action of fatty acyl-ACP reductase, acyl-CoA reductase, or carboxylic acid reductase, respectively. Metabolic engineering of *S. cerevisiae* by integration of *Mus musculus* NADPH-dependent bifunctional fatty acyl-CoA reductase produces fatty alcohols up to 47.4 mg/L. The production of fatty alcohols further increased up to 98 mg/L by enhancement of fatty acid synthesis and NADPH supply by introduction of malic enzyme from *Mucor circinelloides* in *S. cerevisiae* (Runguphan and Keasling 2014); however, the production of fatty alcohols from *S. cerevisiae* is comparatively less than *E. coli*; therefore, more efforts are required for overproduction of fatty alcohols in *S. cerevisiae*.

9.5.9 Yeast as Source of Proteins

Proteins are building blocks of organism acting as biocatalysts and involved in cellular functions like immune response, cell cycle, cell signaling, etc. Traditionally, bacterial and other eukaryotic hosts have been used based on their inherent ability to synthesize a desired product (compound, enzymes, biopharmaceuticals, etc.). With development of biotechnological tools, desired compounds can be synthesized from nonnatural sources tapping recombinant DNA technology. Insulin, a human protein, was the first of the biomolecules to be synthesized using bacterial host cells of *E. coli* by Herbert Boyer in 1977. There are several advantages of native and recombinant proteins in biopharmaceutical industry, enzyme industry, and agricultural industry as the products in diverse areas like health, nutrition, textiles, leather, paper, polymers, plastics, etc.

For the production of recombinant proteins, yeast species have been popularly used as an industrial host. Using yeast for engineering has been proved advantageous as it is a unicellular organism facilitating easy genetic manipulations. The growth of yeast is rapid and also facilitates eukaryotic posttranslational modifications. As compared to other more complex eukaryotic organisms, it is more economical to use yeast as an expression system. In addition, yeast expression system produces high cell densities and high protein titers and does not contain pyrogens, pathogens, or viral inclusions. However, yeast has a strong fermentative metabolism, and proteins produced by *S. cerevisiae* are often hyperglycosylated.

Higher eukaryotic proteins carry sialylated O-linked chains; hence, modification of mammalian protein, i.e., glycosylation by *S. cerevisiae*, is not suitable due to the presence of only mannose residues at O-linked oligosaccharides. Moreover, the glycosylated protein by yeast may be responsible for immunological problems due to reduced receptor binding as the proteins get over-glycosylated at N-linked sites. Commercially available products which are synthesized using *S. cerevisiae* are insulin, hirudin, granulocyte-macrophage colony-stimulating factor (GM-CSF), hepatitis B surface antigen, urate oxidase, glucagons, and platelet-derived growth factor. These hyperglycosylated products are maintained within the periplasmic space and undergo partial degradation. It is difficult to remove desired product from the degradation products. Due to these disadvantages, alternative hosts for the development of expression system come into existence since 1980. These alternative hosts are nonconventional yeast such as *Hansenula polymorpha*, *P. pastoris*, *Schizosaccharomyces pombe*, *Kluyveromyces lactis*, *K. marxianus*, *Arxula adeninivorans*, *Yarrowia lipolytica*, *Pichia methanolica*, and *Zygosaccharomyces bailii*.

For protein secretion, strain engineering of strains can also be done, which is focused primarily on:

1. Posttranslational glycosylation
2. Protein folding
3. Intracellular protein trafficking pathway
4. Quality control in endoplasmic reticulum

9.5.9.1 Engineering of Protein Folding and Quality Control in ER

Various events inside the cell take place before the translocation of secretory proteins outside the cell:

1. Formation of intracellular protein and translocation of this protein to endoplasmic reticulum lumen through sec6 translocon (Anelli and Sitia 2008; Ellgaard and Helenius 2003)
2. Two categories of proteins get translocated into ER lumen: polypeptide and glycopeptides. With the aim of folding to native structure, nascent polypeptides go into ER-resident chaperone binding protein (BiP; encoded by Kar2), whereas ER chaperone calnexin (encoded by CNE1) is responsible to do correct folding and N-glycan processing of nascent glycoproteins.
3. Inside ER, various covalent modifications take place, including signal sequence processing and N-glycosylation.
4. After folding of protein in ER, the proteins get translocated to Golgi apparatus for further modification. Proteins which are improperly folded or aggregated are degraded by the mechanism called as ER-associated protein degradation (ERAD). QC system recognizes these proteins and enables them to bind the BiP complex for degradation in the cytosol. Unfolded protein response (UPR) gets induced on binding of misfolded proteins to BiP. This binding inhibits transcription and translocation of the target protein due to the proteolytic action of ERAD (Fig. 9.5).

Bovine prochymosin has been produced using engineered yeast having overexpression of BiP and disruption of the Golgi-resident calcium ATPase-encoding gene, PMR1 (Harmsen et al. 1996). Engineering of protease that causes the degradation of heterologous proteins after secretion can have a vital role in the overproduction of proteins. Environmental stress mainly high-density fermentation processes is responsible for the induction of yeast's proteases. The expression of genes required for the maturation and activation of proteases can be downregulated for the efficient product recovery. Extracellular and intracellular protease activity of fission yeast has been reduced by co-deletion of psp3 and isp6, vacuole protease genes (Idiris et al. 2006). However, this approach was based only on deletion of some proteases and cannot control the protein degradation completely; hence, another approach employing multiple modifications of protease has been applied.

Glycosylation is necessary for the synthesis of heterologous protein. Yeasts have the ability to perform posttranslational modification process; however, yeast cannot be used for the production of glycoproteins of humans and animals because it is not able to modify the protein according to human and animal glycosylation structure. To overcome this disadvantage, glycosylation engineering is the efficient method.

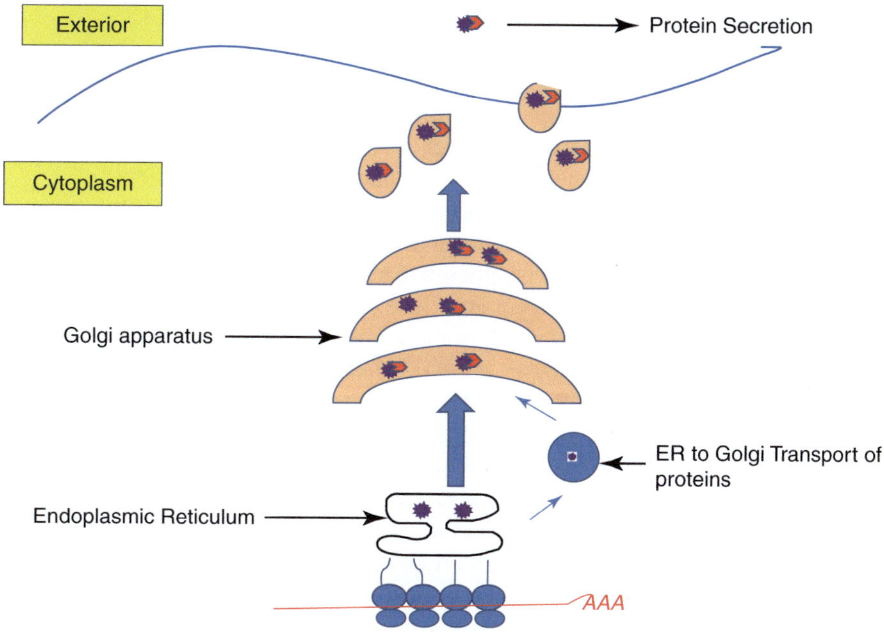

Fig. 9.5 Pathway showing secretion of proteins

Glycosylation engineering includes the deletion of yeast-specific glycosyltransferase and introduction of genes which are accountable for the synthesis of human-like sugar nucleotide and their transport. Engineering of *P. pastoris* by deletion of glycosyltransferase genes alg3 and och1 and introduction of glycosylation-related genes such as MnsI, GnTI, GnTII, GalT, and SiaT lead to the production of humanized sugar chain (Hamilton et al. 2003, 2006; Hamilton and Gerngross 2007). It has been observed that the introduction of enzyme responsible for the synthesis of CMP-sialic acid and sialyltransferase resulted in complex glycoproteins with sialic acid onto terminal β-1,4-galactose sugars (Hamilton et al. 2006).

9.5.10 Yeast as a Biocatalyst

Yeast as a whole-cell biocatalyst can catalyze a variety of chemical reaction including:

- Reduction of C=O bond
- Reduction of C=C bond
- Oxidation and racemization reaction
- Hydrolase enzymes
- Formation of C=C bond

Applications of using wild yeast as whole-cell biocatalyst:

1. Acytoin-type condensation of benzaldehyde which gives (1*R*)-phenylacetylcarbinol which is then further converted to ephedrine and pseudo-ephedrine. These products have been used as bronchodilating agent and decongestant for asthma and treatment for hay fever (Cheetham 1994).
2. *Rhodotorula rubra* L-phenylalanine ammonia lyase is accountable for the synthesis of L-phenylalanine from trans-cinnamic acid and ammonia. L-Phenylalanine has several applications like in the synthesis of artificial sweeteners (aspartame) and macrolide antibiotics (rutamycin B) (Liese et al. 2006).

Disadvantages of using wild-type yeast as whole-cell catalyst:

1. Substrates used by yeast are nonnatural, posing toxicity.
2. Only a small fraction of auxiliary substrates are being used for cofactor recycle, i.e., most of the substrates are being metabolized that makes the product recovery very complicated if the product is not secreted.
3. A large number of dehydrogenases are present that have overlapping substrate specificity with different stereoselectivities.
4. Low yield and undesirable hyperglycosylation.

To compensate these limitations, recombinant yeast has been used as a biocatalyst. Genetic engineering helped in overcoming the limitation of using wild yeast as the biocatalyst. With the aim to improve the selectivity of whole-cell biocatalyst, some alternative approaches to classical techniques have been employed like gene knockout technique and overexpression of heterologous enzymes, i.e., removing the host enzyme's catalytic activities or adding the desired catalytic property, respectively. Primarily, mutants either having one gene deleted or overexpressed were studied thoroughly, and then combinatorial approaches were applied which expand the spectrum of valuable whole-cell biocatalysts. Oxidoreductases were the first of the enzymes in *S. cerevisiae* to be studied. Thus, "first-generation" yeast strains having deletion or overexpression of an enzyme were mutants involving fatty acid synthase (Fasp), aldo-keto reductase (Ypr1p), or acetoxy-ketone reductase (Gre2p), while the "second-generation" yeast strains have a combination of gene deletion and overexpression (Rodríguez et al. 2001). Some examples of using recombinant yeast are:

1. *S. cerevisiae* ATCC 26403 gene knockout mutant lacking β-keto reductase produces ethyl (*R*)-4-chloro-3-hydroxybutanoate with improved optical purity.
2. Resveratrol acts as an antioxidant and has cancer chemoprotective properties (Jang et al. 1997, Subbaramaiah et al. 1998). To improve the synthesis of resveratrol, transgenic yeast strain was created, and for this *Candida molischiana* bglN gene was integrated in *S. cerevisiae* genome which codes for β-glucosidase.
3. *Pichia pastoris* was engineered by overexpressing the *Rhodotorula glutinis* epoxide hydrolase resulting in tenfold increased activity toward (*R*)-styrene oxide conversion in comparison to native *R. glutinis* cells (Lee et al. 2004).

9.6 Perspective

Tools and techniques to manipulate yeast for development of a robust cell factory have been discussed in the chapter. *S. cerevisiae* serves as a striking cell factory, and the market value of its products is constantly increasing. Be it human insulin or organic acids like lactic acid or pyruvate, its commercial applications have gotten the attention of every researcher. Diverse bottlenecks while engineering such strains hamper the development, and to deal with such problems, a more advanced and modern approach is needed. Systems biology, metabolic engineering, and synthetic biology are together being employed for the production of cell factories and to tackle such issues. This integrated approach is based on DBTA cycles. Recent years have seen much advancement in these disciplines and, thus, have led to the development of technologies that can design, construct, and analyze such cell factories more efficiently. For example, MGPS enables network fine-tuning and can precisely control the expression levels of multiple genes. On the other hand, omics tools are very important for the analysis of engineered microorganisms. They help in studying the behavior of the system and exhaustive analysis of the cell at numerous levels of biological organization. These strategies promise development of more robust and efficient cell factories which have a much faster generation time. In the coming future, more yeast-derived bioproducts are expected to reach the market.

Acknowledgments The authors wish to thank the Administration of Jaypee University of Information Technology Waknaghat, Solan (H.P.), for providing necessary infrastructure and facilities and research fellowship to Poonam. Financial assistance provided by DST-SERB project (Sanction Order No. SR/FT/LS-117/2012) is also gratefully acknowledged.

References

Abbott DA, Zelle RM, Pronk JT, Van Maris AJ (2009) Metabolic engineering *of Saccharomyces cerevisiae* for production of carboxylic acids: current status and challenges. FEMS Yeast Res 9:1123–1136. doi:10.1111/j.1567-1364.2009.00537.x

Abdel-Rahman MA, Tashiro Y, Sonomoto K (2013) Recent advances in lactic acid production by microbial fermentation processes. Biotechnol Adv 31:877–902. doi:10.1016/j.biotechadv.2013.04.002

Akhtar MK, Turner NJ, Jones PR (2013) Carboxylic acid reductase is a versatile enzyme for the conversion of fatty acids into fuels and chemical commodities. Proc Natl Acad Sci U S A 110:87–92. doi:10.1073/pnas.1216516110

Albers E, Larsson C, Lidén G, Niklasson C, Gustafsson L (1996) Influence of the nitrogen source on *Saccharomyces cerevisiae* anaerobic growth and product formation. Appl Environ Microbiol 62:3187–3195

Albertsen L, Chen Y, Bach LS, Rattleff S, Maury J, Brix S, Nielsen J, Mortensen UH (2011) Diversion of flux toward sesquiterpene production in *Saccharomyces cerevisiae* by fusion of host and heterologous enzymes. Appl Environ Microbiol 77:1033–1040. doi:10.1128/AEM.01361-10

Albertyn J, van Tonder A, Prior BA (1992) Purification and characterization of glycerol-3-phosphate dehydrogenase of *Saccharomyces cerevisiae*. FEBS Lett 308:130–132. doi:10.1016/0014-5793(92)81259-O

Alvarez R, Rodriguez B, Romano J, Diaz A, Gomez E, Miro D, Navarro L, Saura G, Garcia J (1992) Lipid accumulation in *Rhodotorula glutinis* on sugar cane molasses in single-stage continuous culture. World J Microbiol Biotechnol 8:214–215. doi:10.1007/BF01195853

Anelli T, Sitia R (2008) Protein quality control in the early secretory pathway. EMBO J 27:315–327. doi:10.1038/sj.emboj.7601974

Atsumi S, Higashide W, Liao JC (2009) Direct photosynthetic recycling of carbon dioxide to isobutyraldehyde. Nat Biotechnol 27:1177–1180. doi:10.1038/nbt.1586

Bailey JE (1991) Toward a science of metabolic engineering. Science 252:1668–1675. doi:10.1126/science.2047876

Barney BM, Wahlen BD, Garner E, Wei J, Seefeldt LC (2012) Differences in substrate specificities of five bacterial wax ester synthases. Appl Environ Microbiol 78:5734–5745. doi:10.1128/AEM.00534-12

Barrett CL, Kim TY, Kim HU, Palsson BO, Lee SY (2006) Systems biology as a foundation for genome-scale synthetic biology. Curr Opin Biotechnol 17:488–492. doi: http://dx.doi.org/10.1016/j.copbio.2006.08.001

Beltran G, Novo M, Guillamón JM, Mas A, Rozès N (2008) Effect of fermentation temperature and culture media on the yeast lipid composition and wine volatile compounds. Int J Food Microbiol 121:169–177. doi:10.1016/j.ijfoodmicro.2007.11.030

Beopoulos A, Cescut J, Haddouche R, Uribelarrea J-L, Molina-Jouve C, Nicaud J-M (2009) *Yarrowia lipolytica* as a model for bio-oil production. Prog Lipid Res 48:375–387. doi:10.1016/j.plipres.2009.08.005

Bernard A, Domergue F, Pascal S, Jetter R, Renne C, Faure J-D, Haslam RP, Napier JA, Lessire R, Joubès J (2012) Reconstitution of plant alkane biosynthesis in yeast demonstrates that Arabidopsis ECERIFERUM1 and ECERIFERUM3 are core components of a very-long-chain alkane synthesis complex. Plant Cell 24:3106–3118. doi:10.1105/tpc.112.099796

Bourot S, Karst F (1995) Isolation and characterization of the *Saccharomyces cerevisiae* SUT1 gene involved in sterol uptake. Gene 165:97–102

Bro C, Regenberg B, Förster J, Nielsen J (2006) In silico aided metabolic engineering of *Saccharomyces cerevisiae* for improved bioethanol production. Metab Eng 8:102–111. doi:10.1016/j.ymben.2005.09.007

Cameron D, Altaras N, Hoffman M, Shaw A (1998) Metabolic engineering of propanediol pathways. Biotechnol Prog 14:116–125. doi:10.1021/bp9701325

Cheetham P (1994) Case studies in applied biocatalysis—from ideas to products. Applied biocatalysis. Harwood Academic Publishers, Chur, pp 47–108

Chen Y, Siewers V, Nielsen J (2012) Profiling of cytosolic and peroxisomal acetyl-CoA metabolism in *Saccharomyces cerevisiae*. PLoS One 7:e42475. doi:10.1371/journal.pone.0042475

Chigira Y, Oka T, Okajima T, Jigami Y (2008) Engineering of a mammalian O-glycosylation pathway in the yeast *Saccharomyces cerevisiae*: production of O-fucosylated epidermal growth factor domains. Glycobiology 18:303–314. doi:10.1093/glycob/cwn008

Christen S, Sauer U (2011) Intracellular characterization of aerobic glucose metabolism in seven yeast species by 13C flux analysis and metabolomics. FEMS Yeast Res 11:263–272. doi:10.1111/j.1567-1364.2010.00713.x

Dai CC, Tao J, Xie F, Dai YJ, Zhao M (2007) Biodiesel generation from oleaginous yeast Rhodotorula glutinis with xylose assimilating capacity. Afr J Biotechnol 6(18):2130–2134

Dato L, Berterame N, Ricci M, Paganoni P, Palmieri L, Porro D, Branduardi P (2014) Changes in SAM2 expression affect lactic acid tolerance and lactic acid production in *Saccharomyces cerevisiae*. Microb Cell Factories 13:147. doi:10.1186/s12934-014-0147-7

David F, Siewers V (2014) Advances in yeast genome engineering. FEMS Yeast Res 15:1–14. doi:10.1111/1567-1364.12200

de Jong BW, Shi S, Siewers V, Nielsen J (2014) Improved production of fatty acid ethyl esters in *Saccharomyces cerevisiae* through up-regulation of the ethanol degradation pathway and expression of the heterologous phosphoketolase pathway. Microb Cell Factories 13:39. doi:10.1186/1475-2859-13-39

Ellgaard L, Helenius A (2003) Quality control in the endoplasmic reticulum. Nat Rev Mol Cell Biol 4:181–191. doi:10.1038/nrm1052

Faber K (2011) Biotransformations in organic chemistry: a textbook. Springer Science & Business Media, Berlin

Farhi M, Marhevka E, Masci T, Marcos E, Eyal Y, Ovadis M, Abeliovich H, Vainstein A (2011) Harnessing yeast subcellular compartments for the production of plant terpenoids. Metab Eng 13:474–481. doi:10.1016/j.ymben.2011.05.001

Fatichenti F, Farris GA, Deiana P, Ceccarelli S (1984) Malic acid production and consumption by selected of *Saccharomyces cerevisiae* under anaerobic and aerobic conditions. Appl Microbiol Biotechnol 19:427–429

Foss EJ, Radulovic D, Shaffer SA, Goodlett DR, Kruglyak L, Bedalov A (2011) Genetic variation shapes protein networks mainly through non-transcriptional mechanisms. PLoS Biol 9:e1001144. doi:10.1371/journal.pbio.1001144

Geertman J-MA, van Maris AJ, van Dijken JP, Pronk JT (2006) Physiological and genetic engineering of cytosolic redox metabolism in *Saccharomyces cerevisiae* for improved glycerol production. Metab Eng 8:532–542

Ghaffar T, Irshad M, Anwar Z, Aqil T, Zulifqar Z, Tariq A, Kamran M, Ehsan N, Mehmood S (2014) Recent trends in lactic acid biotechnology: a brief review on production to purification. J Radiat Res Appl Sci 7:222–229. doi:10.1016/j.jrras.2014.03.002

Gombert AK, dos Santos MM, Christensen B, Nielsen J (2001) Network identification and flux quantification in the central metabolism of *Saccharomyces cerevisiae* under different conditions of glucose repression. J Bacteriol 183:1441–1451. doi:10.1128/JB.183.4.1441-1451.2001

Graf A, Dragosits M, Gasser B, Mattanovich D (2009) Yeast systems biotechnology for the production of heterologous proteins. FEMS Yeast Res 9:335–348. doi:10.1111/j.1567-1364.2009.00507.x

Grünberg R, Serrano L (2010) Strategies for protein synthetic biology. Nucleic Acids Res 38:2663–2675. doi:10.1093/nar/gkq139

Gustafsson C, Govindarajan S, Minshull J (2004) Codon bias and heterologous protein expression. Trends Biotechnol 22:346–353. doi: http://dx.doi.org/10.1016/j.tibtech.2004.04.006

Hamilton SR, Gerngross TU (2007) Glycosylation engineering in yeast: the advent of fully humanized yeast. Curr Opin Biotechnol 18:387–392

Hamilton SR, Bobrowicz P, Bobrowicz B, Davidson RC, Li H, Mitchell T, Nett JH, Rausch S, Stadheim TA, Wischnewski H (2003) Production of complex human glycoproteins in yeast. Science 301:1244–1246. doi:10.1126/science.1088166

Hamilton SR, Davidson RC, Sethuraman N, Nett JH, Jiang Y, Rios S, Bobrowicz P, Stadheim TA, Li H, Choi BK (2006) Humanization of yeast to produce complex terminally sialylated glycoproteins. Science 313:1441–1443

Harmsen M, Bruyne M, Raue H, Maat J (1996) Overexpression of binding protein and disruption of the PMR1 gene synergistically stimulate secretion of bovine prochymosin but not plant thaumatin in yeast. Appl Microbiol Biotechnol 46:365–370. doi:10.1007/BF00166231

Hartner FS, Ruth C, Langenegger D, Johnson SN, Hyka P, Lin-Cereghino GP, Lin-Cereghino J, Kovar K, Cregg JM, Glieder A (2008) Promoter library designed for fine-tuned gene expression in *Pichia pastoris*. Nucleic Acids Res 36:e76. doi:10.1093/nar/gkn369

Hesseltine C, Shotwell OL, Ellis J, Stubblefield R (1966) Aflatoxin formation by *Aspergillus flavus*. Bacteriol Rev 30:795

Idiris A, Tohda H, Bi KW, Isoai A, Kumagai H, Giga-Hama Y (2006) Enhanced productivity of protease-sensitive heterologous proteins by disruption of multiple protease genes in the fission yeast *Schizosaccharomyces pombe*. Appl Microbiol Biotechnol 73:404–420

Ignea C, Cvetkovic I, Loupassaki S, Kefalas P, Johnson CB, Kampranis SC, Makris AM (2011) Improving yeast strains using recyclable integration cassettes, for the production of plant terpenoids. Microb Cell Factories 10:1–18. doi:10.1186/1475-2859-10-4

Ilmén M, Koivuranta K, Ruohonen L, Rajgarhia V, Suominen P, Penttilä M (2013) Production of L-lactic acid by the yeast *Candida sonorensis* expressing heterologous bacterial and fungal lactate dehydrogenases. Microb Cell Factories 12:53. doi:10.1186/1475-2859-12-53

Isaacs FJ, Dwyer DJ, Collins JJ (2006) RNA synthetic biology. Nat Biotechnol 24:545–554

Jang M, Cai L, Udeani GO, Slowing KV, Thomas CF, Beecher CW, Fong HH, Farnsworth NR, Kinghorn AD, Mehta RG (1997) Cancer chemopreventive activity of resveratrol, a natural product derived from grapes. Science 275:218–220. doi:10.1126/science.275.5297.218

Jeppsson M, Johansson B, Jensen PR, Hahn-Hägerdal B, Gorwa-Grauslund MF (2003) The level of glucose-6-phosphate dehydrogenase activity strongly influences xylose fermentation and inhibitor sensitivity in recombinant *Saccharomyces cerevisiae* strains. Yeast 20:1263–1272. doi:10.1002/yea.1043

Jetter R, Kunst L (2008) Plant surface lipid biosynthetic pathways and their utility for metabolic engineering of waxes and hydrocarbon biofuels. Plant J 54:670–683. doi:10.1111/j.1365-313X.2008.03467.x

Jewett MC, Hofmann G, Nielsen J (2006) Fungal metabolite analysis in genomics and phenomics. Curr Opin Biotechnol 17:191–197. doi:10.1016/j.copbio.2006.02.001

Kalscheuer R, Stöveken T, Luftmann H, Malkus U, Reichelt R, Steinbüchel A (2006) Neutral lipid biosynthesis in engineered *Escherichia coli*: jojoba oil-like wax esters and fatty acid butyl esters. Appl Environ Microbiol 72:1373–1379. doi:10.1128/AEM.72.2.1373-1379.2006

Kampranis SC, Makris AM (2012) Developing a yeast cell factory for the production of terpenoids. Comput Struct Biotechnol J 3:1–7. doi:10.5936/csbj.201210006

Kavšček M, Stražar M, Curk T, Natter K, Petrovič U (2015) Yeast as a cell factory: current state and perspectives. Microb Cell Factories 14:94. doi:10.1186/s12934-015-0281-x

Keasling JD (2010) Manufacturing molecules through metabolic engineering. Science 330:1355–1358. doi:10.1126/science.1193990

Kim IK, Roldão A, Siewers V, Nielsen J (2012) A systems-level approach for metabolic engineering of yeast cell factories. FEMS Yeast Res 12:228–248. doi:10.1111/j.1567-1364.2011.00779

Klein-Marcuschamer D, Yadav VG, Ghaderi A, Stephanopoulos GN (2010) De novo metabolic engineering and the promise of synthetic DNA. Biosyst Eng 120:101–131. doi:10.1007/10_2009_52

Krivoruchko A, Siewers V, Nielsen J (2011) Opportunities for yeast metabolic engineering: Lessons from synthetic biology. Biotechnol J 6:262–276. doi:10.1002/biot.201000308

Kvitek DJ, Sherlock G (2011) Reciprocal sign epistasis between frequently experimentally evolved adaptive mutations causes a rugged fitness landscape. PLoS Genet 7:e1002056. doi:10.1371/journal.pgen.1002056

Labrou NE (2010) Random mutagenesis methods for in vitro directed enzyme evolution. Curr Protein Pept Sci 11:91–100

Lam MK, Lee KT, Mohamed AR (2010) Homogeneous, heterogeneous and enzymatic catalysis for transesterification of high free fatty acid oil (waste cooking oil) to biodiesel: a review. Biotechnol Adv 28:500–518. doi:10.1016/j.biotechadv.2010.03.002

Lan EI, Liao JC (2012) ATP drives direct photosynthetic production of 1-butanol in cyanobacteria. Proc Natl Acad Sci U S A 109:6018–6023. doi:10.1073/pnas.1200074109

Lang X, Dalai AK, Bakhshi NN, Reaney MJ, Hertz P (2001) Preparation and characterization of bio-diesels from various bio-oils. Bioresour Technol 80:53–62. doi:10.1016/S0960-8524(01)00051-7

Lashkari DA, DeRisi JL, McCusker JH, Namath AF, Gentile C, Hwang SY, Brown PO, Davis RW (1997) Yeast microarrays for genome wide parallel genetic and gene expression analysis. Proc Natl Acad Sci U S A 94:13057–13062

Lee EY, Yoo S-S, Kim HS, Lee SJ, YK O, Park S (2004) Production of (S)-styrene oxide by recombinant *Pichia pastoris* containing epoxide hydrolase from *Rhodotorula glutinis*. Enzym Microb Technol 35:624–631. doi:10.1016/j.enzmictec.2004.08.016

Lee JY, Kang CD, Lee SH, Park YK, Cho KM (2015) Engineering cellular redox balance in *Saccharomyces cerevisiae* for improved production of L-lactic acid. Biotechnol Bioeng 112:751–758. doi:10.1002/bit.25488

Lee KS, Hong ME, Jung SC, Ha SJ, BJ Y, Koo HM, Park SM, Seo JH, Kweon DH, Park JC (2011) Improved galactose fermentation of *Saccharomyces cerevisiae* through inverse metabolic engineering. Biotechnol Bioeng 108:621–631. doi:10.1002/bit.22988

Lee TI, Rinaldi NJ, Robert F, Odom DT, Bar-Joseph Z, Gerber GK, Hannett NM, Harbison CT, Thompson CM, Simon I (2002) Transcriptional regulatory networks in *Saccharomyces cerevisiae*. Science 298:799–804. doi:10.1126/science.1075090

Lee W, DaSilva NA (2006) Application of sequential integration for metabolic engineering of 1,2-propanediol production in yeast. Metab Eng 8:58–65. doi:10.1016/j.ymben.2005.09.001

Lennen RM, Pfleger BF (2013) Microbial production of fatty acid-derived fuels and chemicals. Curr Opin Biotechnol 24:1044–1053. doi:10.1016/j.copbio.2013.02.028

Li M, Borodina I (2014) Application of synthetic biology for production of chemicals in yeast *Saccharomyces cerevisiae*. FEMS Yeast Res 15:1–12. doi:10.1111/1567-1364.12213

Li W, Du W, Y-h L, D-h L, Z-b Z (2007) Enzymatic transesterification of yeast oil for biodiesel fuel production. Chin J Process Eng 7:133

Liese A, Seelbach K, Wandrey C (2006) Industrial biotransformations. Wiley, New York

Lim Y, Jang Y, Kim K (2013) Production of a high concentration of ethanol from potato tuber by high gravity fermentation. Food Sci Biotechnol 22:441–448. doi:10.1007/s10068-013-0099-4

Liu J-J, Ding W-T, Zhang GC, Wang JY (2011) Improving ethanol fermentation performance of *Saccharomyces cerevisiae* in very high-gravity fermentation through chemical mutagenesis and meiotic recombination. Appl Microbiol Biotechnol 91:1239–1246. doi:10.1007/s00253-011-3404-2

Liu R, Zhu F, Lu L, Fu A, Lu J, Deng Z, Liu T (2014) Metabolic engineering of fatty acyl-ACP reductase-dependent pathway to improve fatty alcohol production in *Escherichia coli*. Metab Eng 22:10–21. doi:10.1016/j.ymben.2013.12.004

Lu C, Jeffries T (2007) Shuffling of promoters for multiple genes to optimize xylose fermentation in an engineered *Saccharomyces cerevisiae* strain. Appl Environ Microbiol 73:6072–6077

Luetz S, Giver L, Lalonde J (2008) Engineered enzymes for chemical production. Biotechnol Bioeng 101:647–653. doi:10.1002/bit.22077

Ma Z, Rao Z, Shen W, Fang H, Zhuge J (2007) Construction of recombinant *Saccharomyces cerevisiae* producing 1, 3-propanediol by one step method. Wei Sheng Wu Xue Bao=Acta Microbiol Sin 47:598–603

Martin CH, Nielsen DR, Solomon KV, Prather KLJ (2009) Synthetic metabolism: engineering biology at the protein and pathway scales. Chem Biol 16:277–286. doi:10.1016/j.chembiol.2009.01.010

Meng X, Yang J, Xu X, Zhang L, Nie Q, Xian M (2009) Biodiesel production from oleaginous microorganisms. Renew Energy 34:1–5

Miyasawa H, Iwami Y, Mayanagi H, Takahashi N (2003) Xylitol inhibition of anaerobic acid production by *Streptococcus mutans* at various pH levels. Oral Microbiol Immunol 18:215–219. doi:10.1034/j.1399-302X.2003.00068.x

Mustacchi R, Hohmann S, Nielsen J (2006) Yeast systems biology to unravel the network of life. Yeast 23:227–238. doi:10.1002/yea.1357

Nakayama K-i, Nagasu T, Shimma Y, Kuromitsu J, Jigami Y (1992) OCH1 encodes a novel membrane bound mannosyltransferase: outer chain elongation of asparagine-linked oligosaccharides. EMBO J 11:2511

Nasr F, Bécam A-M, Slonimski P, Herbert C, Brown S, De Nay D (1995) Artificial antisense RNA regulation of YBR1012 (YBR136w), an essential gene from *Saccharomyces cerevisiae* which is important for progression through G1/S. Mol Gen Genet 249:51–57

Nevoigt E, Kohnke J, Fischer CR, Alper H, Stahl U, Stephanopoulos G (2006) Engineering of promoter replacement cassettes for fine-tuning of gene expression in *Saccharomyces cerevisiae*. Appl Environ Microbiol 72:5266–5273. doi:10.1128/AEM.00530-06

Nguyen HTT, Dieterich A, Athenstaedt K, Truong N, Stahl U, Nevoigt E (2004) Engineering of *Saccharomyces cerevisiae* for the production of L-glycerol 3-phosphate. Metab Eng 6:155–163. doi:10.1016/j.ymben.2004.02.005

Nielsen J (2001) Metabolic engineering. Appl Microbiol Biotechnol 55:263–283

Nielsen J (2014) Synthetic biology for engineering acetyl coenzyme a metabolism in yeast. mBio 5:e02153-02114. doi:10.1128/mBio.02153-14

Nielsen J, Jewett MC (2008) Impact of systems biology on metabolic engineering of *Saccharomyces cerevisiae*. FEMS Yeast Res 8:122–131

Nielsen J, Fussenegger M, Keasling J, Lee SY, Liao JC, Prather K, Palsson B (2014) Engineering synergy in biotechnology. Nat Chem Biol 10:319–322. doi:10.1038/nchembio.1519

Norbeck J, Påhlman AK, Akhtar N, Blomberg A, Adler L (1996) Purification and characterization of two isoenzymes of dl-glycerol-3-phosphatase from *Saccharomyces cerevisiae* identification of the corresponding GPP1 and GPP2 genes and evidence for osmotic regulation of Gpp2p expression by the osmosensing mitogen-activated protein kinase signal transduction pathway. J Biol Chem 271:13875–13881

Olivares-Hernández R, Bordel S, Nielsen J (2011) Codon usage variability determines the correlation between proteome and transcriptome fold changes. BMC Syst Biol 5:33. doi:10.1186/1752-0509-5-33

Ostergaard S, Olsson L, Nielsen J (2000) Metabolic engineering of *Saccharomyces cerevisiae*. Microbiol Mol Biol Rev 64:34–50. doi:10.1128/MMBR.64.1.34-50.2000

Oswald M, Fischer M, Dirninger N, Karst F (2007) Monoterpenoid biosynthesis in *Saccharomyces cerevisiae*. FEMS Yeast Res 7:413–421. doi:10.1111/j.1567-1364.2006.00172.x

Partow S, Siewers V, Bjørn S, Nielsen J, Maury J (2010) Characterization of different promoters for designing a new expression vector in *Saccharomyces cerevisiae*. Yeast 27:955–964. doi:10.1002/yea.1806

Peralta-Yahya PP, Zhang F, Del Cardayre SB, Keasling JD (2012) Microbial engineering for the production of advanced biofuels. Nature 488:320–328. doi:10.1038/nature11478

Popp A, Nguyen H, Boulahya K, Bideaux C, Alfenore S, Guillouet S, Nevoigt E (2008) Fermentative production of l-glycerol 3-phosphate utilizing a *Saccharomyces cerevisiae* strain with an engineered glycerol biosynthetic pathway. Biotechnol Bioeng 100:497–505. doi:10.1002/bit.21777

Porro D, Bianchi MM, Brambilla L, Menghini R, Bolzani D, Carrera V, Lievense J, Liu C-L, Ranzi BM, Frontali L (1999) Replacement of a metabolic pathway for large-scale production of lactic acid from engineered yeasts. Appl Environ Microbiol 65:4211–4215

Prather KLJ, Martin CH (2008) De novo biosynthetic pathways: rational design of microbial chemical factories. Curr Opin Biotechnol 19:468–474. doi:10.1016/j.copbio.2008.07.009

Pronk JT, Steensma HY, van Dijken JP (1996) Pyruvate metabolism in *Saccharomyces cerevisiae*. Yeast 12(16):1607–1633

Ratledge C, Wynn JP (2002) The biochemistry and molecular biology of lipid accumulation in oleaginous microorganisms. Adv Appl Microbiol 51:1–52

Redding-Johanson AM, Batth TS, Chan R, Krupa R, Szmidt HL, Adams PD, Keasling JD, Lee TS, Mukhopadhyay A, Petzold CJ (2011) Targeted proteomics for metabolic pathway optimization: application to terpene production. Metab Eng 13:194–203. doi:10.1016/j.ymben.2010.12.005

Reiser S, Somerville C (1997) Isolation of mutants of *Acinetobacter calcoaceticus* deficient in wax ester synthesis and complementation of one mutation with a gene encoding a fatty acyl coenzyme A reductase. J Bacteriol 179:2969–2975

Ro D-K, Paradise EM, Ouellet M, Fisher KJ, Newman KL, Ndungu JM, Ho KA, Eachus RA, Ham TS, Kirby J (2006) Production of the antimalarial drug precursor artemisinic acid in engineered yeast. Nature 440:940–943. doi:10.1038/nature04640

Rodríguez S, Kayser MM, Stewart JD (2001) Highly stereoselective reagents for β-keto ester reductions by genetic engineering of baker's yeast. J Am Chem Soc 123:1547–1555. doi:10.1021/ja0027968

Röttig A, Wenning L, Bröker D, Steinbüchel A (2010) Fatty acid alkyl esters: perspectives for production of alternative biofuels. Appl Microbiol Biotechnol 85:1713–1733. doi:10.1007/s00253-009-2383-z

Runguphan W, Keasling JD (2014) Metabolic engineering of *Saccharomyces cerevisiae* for production of fatty acid-derived biofuels and chemicals. Metab Eng 21:103–113. doi:10.1016/j.ymben.2013.07.003

Sauer U (2001) Evolutionary engineering of industrially important microbial phenotypes. Metab Eng. Springer, Berlin, pp 129–169

Sauer U (2006) Metabolic networks in motion: 13C-based flux analysis. Mol Syst Biol 2:62. doi:10.1038/msb4100109

Saven JG (2011) Computational protein design: engineering molecular diversity, nonnatural enzymes, nonbiological cofactor complexes, and membrane proteins. Curr Opin Chem Biol 15:452–457. doi:10.1016/j.cbpa.2011.03.014

Schirmer A, Rude MA, Li X, Popova E, Del Cardayre SB (2010) Microbial biosynthesis of alkanes. Science 329:559–562. doi:10.1126/science.1187936

Seino A, Yanagida Y, Aizawa M, Kobatake E (2005) Translational control by internal ribosome entry site in *Saccharomyces cerevisiae*. Biochim Biophys Acta 1681:166–174

Shen CR, Liao JC (2012) Photosynthetic production of 2-methyl-1-butanol from CO_2 in cyanobacterium *Synechococcus elongatus* PCC7942 and characterization of the native acetohydroxyacid synthase. Energy Environ Sci 5:9574–9583. doi:10.1039/C2EE23148D

Shi S, Valle-Rodríguez JO, Khoomrung S, Siewers V, Nielsen J (2012) Functional expression and characterization of five wax ester synthases in *Saccharomyces cerevisiae* and their utility for biodiesel production. Biotechnol Biofuels 5:1. doi:10.1186/1754-6834-5-7

Shianna KV, Dotson WD, Tove S, Parks LW (2001) Identification of a UPC2 homolog in *Saccharomyces cerevisiae* and its involvement in aerobic sterol uptake. J Bacteriol 183:830–834. doi:10.1128/JB.183.3.830-834.2001

Shiba Y, Paradise EM, Kirby J, Ro DK, Keasling JD (2007) Engineering of the pyruvate dehydrogenase bypass in *Saccharomyces cerevisiae* for high-level production of isoprenoids. Metab Eng 9:160–168. doi:10.1016/j.ymben.2006.10.005

Stephanopoulos G, Vallino JJ (1991) Network rigidity and metabolic engineering in metabolite overproduction. Science 252:1675–1681. doi:10.1126/science.1904627

Stöveken T, Kalscheuer R, Malkus U, Reichelt R, Steinbüchel A (2005) The wax ester synthase/acyl coenzyme A: diacylglycerol acyltransferase from *Acinetobacter sp.* strain ADP1: characterization of a novel type of acyltransferase. J Bacteriol 187:1369–1376. doi:10.1128/JB.187.4.1369-1376.2005

Straub L (2011) Beyond the transcripts: what controls protein variation? PLoS Biol 9:e1001146. doi:10.1371/journal.pbio.1001146

Stukey JE, McDonough VM, Martin CE (1989) Isolation and characterization of OLE1, a gene affecting fatty acid desaturation from *Saccharomyces cerevisiae*. J Biol Chem 264:16537–16544

Subbaramaiah K, Chung WJ, Michaluart P, Telang N, Tanabe T, Inoue H, Jang M, Pezzuto JM, Dannenberg AJ (1998) Resveratrol inhibits cyclooxygenase-2 transcription and activity in phorbol ester-treated human mammary epithelial cells. J Biol Chem 273:21875–21882. doi:10.1074/jbc.273.34.21875

Szczebara FM, Chandelier C, Villeret C, Masurel A, Bourot S, Duport C, Blanchard S, Groisillier A, Testet E, Costaglioli P (2003) Total biosynthesis of hydrocortisone from a simple carbon source in yeast. Nat Biotechnol 21:143–149. doi:10.1038/nbt775

Tang WL, Zhao H (2009) Industrial biotechnology: tools and applications. Biotechnol J 4:1725–1739. doi:10.1002/biot.200900127

Tavares S, Grotkjær T, Obsen T, Haslam RP, Napier JA, Gunnarsson N (2011) Metabolic engineering of *Saccharomyces cerevisiae* for production of eicosapentaenoic acid, using a novel Δ5-desaturase from Paramecium tetraurelia. Appl Environ Microbiol 77:1854–1861. doi:10.1128/AEM.01935-10

Tehlivets O, Scheuringer K, Kohlwein SD (2007) Fatty acid synthesis and elongation in yeast. Biochim Biophys Acta 1771:255–270. doi:10.1016/j.bbalip.2006.07.004

Tian J, Ma K, Saaem I (2009) Advancing high-throughput gene synthesis technology. Mol BioSyst 5:714–722. doi:10.1039/b822268c

Tochigi Y, Sato N, Sahara T, Wu C, Saito S, Irie T, Fujibuchi W, Goda T, Yamaji R, Ogawa M (2010) Sensitive and convenient yeast reporter assay for high-throughput analysis by using a secretory luciferase from *Cypridina noctiluca*. Anal Chem 82:5768–5776. doi:10.1021/ac100832b

Toivari MH, Ruohonen L, Miasnikov AN, Richard P, Penttilä M (2007) Metabolic engineering of *Saccharomyces cerevisiae* for conversion of D-glucose to xylitol and other five-carbon sugars and sugar alcohols. Appl Environ Microbiol 73:5471–5476. doi:10.1128/AEM.02707-06

Tyo KE, Alper HS, Stephanopoulos GN (2007) Expanding the metabolic engineering toolbox: more options to engineer cells. Trends Biotechnol 25:132–137

Tyo KE, Kocharin K, Nielsen J (2010) Toward design-based engineering of industrial microbes. Curr Opin Microbiol 13:255–262. doi:10.1016/j.mib.2010.02.001

Tyo KE, Nevoigt E, Stephanopoulos G (2011) 6 directed evolution of promoters and tandem gene arrays for customizing RNA synthesis rates and regulation. Methods Enzymol 497:135. doi:10.1016/B978-0-12-385075-1.00006-8

Uhari M, Kontiokari T, Niemelä M (1998) A novel use of xylitol sugar in preventing acute otitis media. Pediatrics 102:879–884

Usaite R, Wohlschlegel J, Venable JD, Park SK, Nielsen J, Olsson L, Yates Iii JR (2008) Characterization of global yeast quantitative proteome data generated from the wild-type and glucose repression *Saccharomyces cerevisiae* strains: the comparison of two quantitative methods. J Proteome Res 7:266–275. doi:10.1021/pr700580m

Valle-Rodríguez JO, Shi S, Siewers V, Nielsen J (2014) Metabolic engineering of *Saccharomyces cerevisiae* for production of fatty acid ethyl esters, an advanced biofuel, by eliminating non-essential fatty acid utilization pathways. Appl Energy 115:226–232. doi:10.1016/j.apenergy.2013.10.003

van Maris AJ, Luttik MA, Winkler AA, van Dijken JP, Pronk JT (2003) Overproduction of threonine aldolase circumvents the biosynthetic role of pyruvate decarboxylase in glucose-limited chemostat cultures of *Saccharomyces cerevisiae*. Appl Environ Microbiol 69:2094–2099. doi:10.1128/AEM.69.4.2094-2099.2003

Velculescu VE, Zhang L, Vogelstein B, Kinzler KW (1995) Serial analysis of gene expression. Science 270:484–487. doi:10.1126/science.270.5235.484

Villas-Bôas S, Akesson M, Moxley J, Stephanopoulos G, Nielsen J (2005) High-throughput metabolic state analysis: the missing link in integrated functional genomics of yeasts. Biochem J 388:669–677. doi:10.1042/BJ20041162

Voloshchuk N, Montclare JK (2010) Incorporation of unnatural amino acids for synthetic biology. Mol BioSyst 6:65–80. doi:10.1039/b909200p

Wang Z, Zhuge J, Fang H, Prior BA (2001) Glycerol production by microbial fermentation: a review. Biotechnol Adv 19:201–223. doi:10.1016/S0734-9750(01)00060-X

Werpy T, Petersen G, Aden A, Bozell J, Holladay J, White J, Manheim A, Eliot D, Lasure L, Jones S (2004) Top value added chemicals from biomass. Volume 1-Results of screening for potential candidates from sugars and synthesis gas. DTIC Document

Wolf JJ, Dowell RD, Mahony S, Rabani M, Gifford DK, Fink GR (2010) Feed-forward regulation of a cell fate determinant by an RNA-binding protein generates asymmetry in yeast. Genetics 185:513–522. doi:10.1534/genetics.110.113944

Xia X, Holcik M (2009) Strong eukaryotic IRESs have weak secondary structure. PLoS One 4:e4136. doi:10.1371/journal.pone.0004136

Xu G, Hua Q, Duan N, Liu L, Chen J (2012) Regulation of thiamine synthesis in *Saccharomyces cerevisiae* for improved pyruvate production. Yeast 29:209–217. doi:10.1002/yea.2902

Zhu H, Bilgin M, Bangham R, Hall D, Casamayor A, Bertone P, Lan N, Jansen R, Bidlingmaier S, Houfek T (2001) Global analysis of protein activities using proteome chips. Science 293:2101–2105. doi:10.1126/science.1062191

Zhuang X (2013) Engineering novel terpene production platforms in the yeast *Saccharomyces cerevisiae*. University of Kentucky, Lexington

Zhuang X, Chappell J (2015) Building terpene production platforms in yeast. Biotechnol Bioeng 112:1854–1864. doi:10.1002/bit.25588

Yeast as a Model System to Study Human Diseases

10

Ashu M. Poswal and Adesh K. Saini

Abstract

Saccharomyces cerevisiae or yeast has served as a toolbox to understand human physiology and has gathered significant attention over the last three decades. Given that scientists have extensively studied it and there is significant homology of its genes to humans and conservation of important cellular pathways in both, yeast has become one of the most popular model systems to study human diseases. Here in the current chapter, we are dealing with the yeast as a model organism for three important human health aspects, viz., neurodegenerative diseases, aging, and cancer. These human health conditions follow very complex pathways, and it will be worth observing how a simple unicellular organism is used to study these health conditions. It has been found that many of the genes in yeast can be very easily deleted and complemented with human orthologues. Now if the human genes thus complemented contain a set of mutations responsible for a health condition, then the molecular details for the cause of disease can be worked out using yeast system, which then can be precisely extrapolated in human systems that are otherwise difficult to do. Not only this, yeast model expressing defective human proteins can also be used in the screening of metabolites or novel compounds that could suppress the phenotypic defects owing to mutations.

Keywords

Yeast • Neurodegenerative disease • Aging • Cancer • Human orthologues

A.M. Poswal • A.K. Saini, Ph.D. (✉)
Faculty of Applied Sciences and Biotechnology, Shoolini University,
Solan 173 229, Himachal Pradesh, India
e-mail: sainiade@gmail.com

V.C. Kalia, A.K. Saini (eds.), *Metabolic Engineering for Bioactive Compounds*,
DOI 10.1007/978-981-10-5511-9_10

10.1 Introduction

Due to less practicability of steering direct study on human for their different health conditions or diseases, researchers were always looking for alternatives that are not only cheap but also scientifically accepted. After three decades of intensive research, a wide array of model organisms are now their first choices. A number of model organisms are currently used as a disease model system, for example, *Drosophila melanogaster*, transgenic mice, *Caenorhabditis elegans*, *Saccharomyces cerevisiae*, etc.; however, over the recent past years, *S. cerevisiae* "the budding yeast" has emerged as one of the most frequently used disease model organisms (Franssens et al. 2013; Laurent et al. 2015).

Yeast is a unicellular organism and the basic features of its genome are:

- Genome size: 12.5 Mb + rDNA (rDNA usually 1–2 Mb)
- Chromosome no.: 16
- Mitochondrial DNA: 75 kb
- Extrachromosomal DNA: 2 μm (Olson 1991)

Yeast cells exist as both haploid and diploid forms which are morphologically similar but differ in many other important ways. First haploid cells contain 16 chromosomes and the diploid cells contain 32 chromosomes. Second, round-shaped haploid cells are smaller as compared to the ovoid-shaped elongated diploid cells. Third, haploids and diploids have budding patterns different from each other. Buds in haploid cells emerge from the site adjacent to the previous bud, whereas in diploid cells they emerge from either end of the mother cell.

Transformation of yeast can be done with quite ease, and the plasmids that are most commonly used for yeast transformation are:

- ARS (autonomously replicating sequence) plasmid with 1) chromosomal DNA origin of replication and 2) 2 μm origin of replication.
- CEN (centromere) plasmid: It is ARS with centromere which ensures the segregation to both daughter and mother cells (Amberg et al. 2005).

Though yeast is a unicellular organism, due to the common ancestry with humans, its genes still show approximately 60% sequence homology with human orthologue. If we talk about the human genes that are associated with different diseases, there is over 25% of close homologue in yeast. Although higher model organisms have more homology to humans, yeast models have other advantages over them like:

1. Swift growth having a doubling time around 90–180 min
2. Easy genetic modification and DNA transformation
3. Very efficient homologous recombination
4. Genome-wide study of specific screenable phenotypes or cellular processes
5. Easily accessible and no ethical issues
6. Economically beneficial (Franssens et al. 2013)

All these advantages encouraged researchers to choose yeast as a mimic of human cell line model system over higher organism. So in the recent past, budding yeast has been used frequently as a disease model in relevance to neurological disorders, cancers, aging-related diseases, etc. (Simon and Bedalov 2004; Piper 2006; Franssens et al. 2013).

10.2 Yeast Model for Human Neurodegenerative Disease

Till date, many neurodegenerative complications have been copied in yeast depending on the presence or absence of yeast orthologue of human disease-associated gene (Table 10.1). When a yeast orthologue of human gene is present, then the yeast gene can be knocked out or overexpressed to ascertain the loss or gain of functional phenotypes like Friedreich's ataxia, a disease which is caused by frataxin reduction. In humans frataxin is a product of *FRDA* gene (or *FXN*) (Khurana and Lindqist 2010). Yeast contains human *FRDA* homologue *Yfh1p* whose decreased expression or function is related to Friedreich's ataxia. Studies with *Yfh1p* have been used to determine the function of frataxin. Deletion of *Yfh1p* in yeast cells results in iron accumulation in mitochondria, mitochondrial dysfunction, and oxidative stress (Orr and Zoghbi 2007; Pereira et al. 2012). Interestingly, it was found that introduction of human *FRDA* rescued the yeast cell from these phenotypes in *Yfh1p* null background (Sherman 2002; Khurana and Lindqist 2010).

In case, when disease-associated genes are not represented as yeast orthologues, the functional characterization is performed via heterologous expression in wild-type strains (Menezes et al. 2015). The most common human neurological diseases that do not have a yeast homologue are polyglutamine or polyQ diseases. These include Huntington's disease and synucleinopathies (like Parkinson's disease, Alzheimer's disease, amyotrophic lateral sclerosis, prion diseases, and tauopathies) (Khurana and Lindqist 2010). Proteins associated with these diseases are α-synuclein and Lrrk2 in Parkinson's disease, tau and β-amyloid (Aβ) peptide in Alzheimer's disease, and huntingtin protein with expanded polyglutamine/polyQ tracts in Huntington's disease. These proteins were widely analyzed using yeast cells for their phenotypes such as aggregation or toxicity (Pereira et al. 2012). Many of the abovementioned polyQ neurodegenerative disorders, like Huntington's disease, are caused by unstable extension of CAG repeats. Huntington's disease is caused by the expansion of CAG repeats in the first exon (eX1) in *IT-15* gene (Ehrnhoefer et al. 2006; Mason and Giorgini 2011) that forms huntingtin protein carrying expanded polyQ tract that must carry a minimum of ~35 glutamines to set off the disease (Khurana and Lindqist 2010).

It was found that overexpression of HTTeX1 in neuronal aggregates leads to Huntington's disease in mouse. Using a yeast model where fragments of HTTeX1 were overexpressed and followed by polyQ tracts of variable lengths resulted in polyQ length-dependent aggregation of HTTeX1 and its higher-order complex formation leading to transcriptional dysregulation (Meriin et al. 2002; Khurana and Lindqist 2010), dysfunction of mitochondria and redox stress, and fragmentation of

Table 10.1 Human neurodegenerative disorders and their associated proteins studied in yeast model system

Disease	Inheritance	Human gene	Protein(s) involved	Yeast orthologue	References
Batten's disease	Autosomal recessive	*CLN3*	CLN3 (ceroid lipofuscinosis, neuronal 3)	YHC3/BTN1	Pereira et al. (2012); Kinarivala and Trippier (2015)
Friedreich's ataxia	Autosomal recessive	*FXN*	Frataxin	YFH1	Pereira et al. (2012); Khurana and Lindqist (2010)
Ataxia-telangiectasia	Autosomal recessive	*ATM*	ATM (ataxia-telangiectasia mutated)	TEL1, MEC1	Pereira et al. (2012); Li and Jiang (2015)
Niemann-Pick disease	Autosomal recessive	*NPC1*	NPC1 (Niemann-Pick disease, type C1)	NCR1	Pereira et al. (2012); Vilaça et al. (2014)
Hereditary spastic paraplegia	Autosomal dominant, or autosomal recessive, or cross-linked recessive	*SPG7*	mAAA proteases (Afg3l2 and paraplegin)	mAAA proteases (Yta10 and Yta12)	Pereira et al. (2012); Khurana and Lindqist (2010)
Creutzfeldt-Jakob disease	Autosomal dominant	*PRNP*	PrP (prion *protein*)	No orthologue SOD1	Pereira et al. (2012); Khurana and Lindqist (2010)
		SOD1	SOD1 (superoxide dismutase 1)		
Amyotrophic lateral sclerosis	Autosomal dominant	*TARDBP*	TDP-43 (Tar DNA-binding protein-43)	No orthologue	Johnson et al. (2008); Pereira et al. (2012); Fushimi et al. (2011); Ju et al. (2011)
		FUS	FUS (fused in sarcoma)/TLS (translocated in sarcoma)	No orthologue	
Parkinson's disease	Unknown	*SNCA*	α-Synuclein	No orthologue	Pereira et al. (2012)
		PARK8	Lrrk2 (leucine-rich repeat kinase 2)	No orthologue	Xiong et al. (2010)
Alzheimer's disease	Autosomal dominant	*MAPT*	Tau	No orthologue	Pereira et al. (2012); Porzoor and Macreadie (2013)
		APP	β-amyloid	No orthologue	
Huntington's disease	Autosomal dominant	*HTT*	Huntington	No orthologue	Pereira et al. (2012); Mason and Giorgini (2011)

DNA (Giorgini et al. 2005; Sokolov et al. 2006). There are various applications of this kind of yeast model. Firstly is to gain the mechanistic insight of Huntington's disease at cellular level. Secondly, this model can be used to screen suppressors of related phenotypes like to design a yeast cell-based high-throughput screen for pharmacological polyQ aggregation inhibitors. In this regard a 16,000-compound library screen resulted in the identification of C2–8, a highly potent analog that inhibited polyQ-related phenotypes. This study was further taken to neurons and has shown promising results by rescuing polyQ-dependent neurodegeneration in vivo in higher organism including worms and *Drosophila* suggesting that C2–8 is a strong candidate for developing drug against neurodegenerative diseases (Zhang et al. 2005). Furthermore, ~5000 natural products were screened and confirmed in yeast and fly model showing that epigallocatechin 3-gallate strongly rescued the aggregation of polyQ-expanded HTTeX1 in a dose-dependent manner and improved photoreceptor degeneration and motor function and reduced polyQ-mediated toxicity in vivo (Ehrnhoefer et al. 2006).

Synucleinopathies are neurodegenerative diseases distinguished by the irregular accumulation of α-synuclein aggregates in neurons, nerve fibers, or glial cells. Yeast models of synucleinopathies rely upon the heterologous or homologous expression of the genes encoding α-syn, LRRK2, Parkin, EIF4G1, DJ1 VPS35, and ATP13A (Nieoullon 2011; Menezes et al. 2015). A variety of synucleinopathies including Pakinson's disease are characterized by the appearance of Lewy bodies which are cytoplasmic inclusions of proteinaceous nature in neurons (Schaeffer 2010). Lewy bodies are insoluble aggregates and are mostly composed of α-synuclein which is of 14 kDa MW protein (Menezes et al. 2015). Based on these studies, different forms of α-synuclein were expressed in yeast model. In one of the models, α-synuclein of wild-type nature or mutant forms were linked to green fluorescent protein. The expression of these variants was controlled from a regulatable (galactose (Gal)-inducible) promoter to study α-syn toxicity in yeast cell. Based on the dosage, the expression resulted into no toxicity, intermediate toxicity, and high toxicity which is directly proportional to the expression of α-synuclein (Khurana and Lindqist 2010). This system was used to identify small molecules to suppress the phenotypes of α-synuclein expression. It resulted into discovery of a class of structurally related 1,2,3,4-tetrahydroquinolinones, (–)-epigallocatechin 3-gallate (EGCG), and quercetin that are able to reduce α-synuclein toxicity (Griffioen et al. 2006; Su et al. 2010). These studies in yeast also discovered a natural antioxidant, ascorbic acid, which was found to reduce the number of yeast cells bearing α-synuclein inclusions (Fernandes et al. 2014). In a remarkable molecular study using yeast as a toolbox, it has been shown that mechanistically the toxicity of α-synuclein and beta-synuclein are similar. They found that similar to α-synuclein, beta-synuclein is toxic due to its ability to form cytosolic inclusions and has similar impairment in vesicular trafficking and increase in oxidative stress (Tenreiro et al. 2016).

Another most prevalent human health tauopathy is Alzheimer's disease (Delacourte and Buee 2000), and it is characterized by the deposition of β-amyloid peptides also called as amyloid or senile plaques. In this condition the amyloid precursor protein is cleaved by secretases which are specific proteases that results in the formation of β-amyloid peptide (Pereira et al. 2012). In this situation also, yeast model has been used for high-throughput screening to screen for new proteases. This had resulted into the identification of caspase-3 and caspase-8 using yeast galactose reporter system (Gunyuzlu et al. 2000). GFP-tagged β-amyloid in yeast model was also used in which β-amyloid misfolding contributes to GFP misfolding and consequent loss of fluorescence, and this yielded folate as an inhibitor of β-amyloid misfolding (Macreadie et al. 2008). Bagriantsev and Liebman used high-throughput screening of small-molecule inhibitors and identified two antioligomeric compounds: AO-11 (2-(4-methoxyphenyl)-1,3-benzoxazol-5-amine) and AO-15 (1-(2,3-dimethylphenyl)-4-(2-furoyl) piperazine) (Bagriantsev and Liebman 2006).

10.3 Yeast as a Model of Aging

The budding yeast is particularly amenable to study aging because of its relatively short life span, availability of genetic manipulation techniques, and its use in high-throughput screening technology specifically designed for yeast. The *S. cerevisiae* is considered to have two phases in life span: firstly, a replicative stage where dividing mother cell produces daughter cells and, secondly, the chronological life span defined as the competence of stationary cultures to preserve viability (Kaeberlein 2006).

The replicative phase of yeast is analog to the replicative life span of mammalian fibroblast and lymphocyte. The yeast replicative model of aging showed that the proliferation and accumulation of an extrachromosomal rDNA circle (ERC) is a major key player in replicative aging of yeast cells (Sinclair and Guarente 1997). This rDNA locus consists of 100–200 tandem copies of a 9 kilobase pairs of *rDNA* locus on XII chromosome of yeast. Extrachromosomal rDNA circle is generated via an incident of homologous recombination within the rDNA repeat (Steinkraus et al. 2008). At the time of cell division, the extrachromosomal rDNA circles are mostly retained within the mother cell, and consequently their number increases in the nucleus with each event of cell division, leading to their accumulation which eventually causes aging or senescence of the mother cell (Kaeberlein 2010).

Several mutations in yeast have been identified that affect the replicative life span of yeast, e.g., overexpression of *SIR2P*. Sir2p belongs to a family called eukaryotic protein deacetylases collectively known as the sirtuins (Kaeberlein and Powers 2007). In yeast, overproduction of Sir2p sirtuin reduces ERC accumulation by inhibiting rDNA stability that slows replicative aging in yeast (Kaeberlein 2010). Using yeast cells, recently it has been shown that deletion of TOR1 lead to the extension of longevity via decreasing autophagy which is against the loss of Sir2p which has opposite effects (Guedes et al. 2016). Another candidate protein in yeast is Fob1 (replication fork block protein) that has also been found to affect longevity of yeast replicative life span by controlling ERC level (Kaeberlein 2010).

Kaeberlein and colleagues measured the effects of 50 mutations in the long-lived strain known as BY4742 (Kaeberlein et al. 2004). It was found that null strains for either *FOB1*, *GPA2*, *SCH9*, *HXK2*, or *GPR1* genes or the overexpression of SIR2 leads to any significant increase in replicative longevity in the BY4742 background. Out of five gene deletions, *fob1*Δ acts by repressing ERC formation (Defossez et al. 1999). Against this, calorie restriction was found to be affected by *gpa2*Δ, *gpr1*Δ or *hxk2*Δ. Calorie restriction, defined by a diminution in nutrient availability, has been demonstrated to extend life span in yeast, nematodes, fruit flies, and rodents (Masoro 2005). Replicative phase of yeast cells with low ERC levels is extended by reducing glucose in the culture medium leading to calorie restriction or by diminished TOR signaling (Steffen et al. 2008; Kaeberlein 2010). The fifth protein involved in aging belongs to protein kinase, Sch9p, and its deletion has showed increased replicative longevity in yeast model (Fabrizio et al. 2004; Kaeberlein et al. 2005). Beside these factors, other genes involved in replicative aging are RAS1/RAS2 and LAG1/LAG2. RAS proteins play opposite functions in the regulation of both replicative and chronological longevity. It has been shown that the deletion of *ras1*Δ strain has increased replicative life span against *ras2*Δ strain which has shorter replicative life span and extended chronological life span (Kaeberlein and Powers 2007). Interestingly, it has been shown that *lag1*Δ, which affects ceramide synthesis in yeast or overexpression of LAG2 in yeast (unknown function), promotes replicative life span (D'Mello et al. 1994; Steinkraus et al. 2008).

Other studies conducted on yeast have identified proteins involved in the maintaining the chronological phase of life span. It has been found that the nutrient-responsive target of rapamycin (TOR) signaling pathway of yeast is a vital component to mediate replicative and chronological life span by using genome-wide screenings of single-gene deletion in ~4900 isogenic yeast strains (Kaeberlein et al. 2005; Powers et al. 2006). TOR activity was inversely related to replicative and chronological life span since mutations that decrease TOR activity were found to increase both (Kaeberlein et al. 2005; Powers et al. 2006). In *S. cerevisiae*, TOR acts in harmony with Sch9 and protein kinase A, to synchronize the cellular response to nutrient availability. In agreement to this deletion of nutrient-responsive kinase, Sch9 or adenylate cyclase (Cyr1) enhances both replicative and chronological life span (Kaeberlein and Powers 2007). Chronological life span is also increased with increased expression of mitochondrial SOD2 (superoxide dismutase) that is requisite for complete life span lengthening via Cyr1, Sch9, or Ras2 mutations (Fabrizio et al. 2003; Longo and Fabrizio 2012) and sufficient to boost chronological phase of life span when overexpressed simultaneously with SOD1 protein (Burtner et al. 2009; Longo and Fabrizio 2012).

10.4 Yeast as a Model of Cancer

Since most cancer-related pathways and corresponding genes are homologous between yeast and humans, yeast could be a good model to understand the cancer physiology (Table 10.2) (Simon and Bedalov 2004; Hekmat-Scafe et al. 2016) and to screen novel anticancer compounds. It has been found that hyperactivation of cyclin-dependent kinase (CDK) is related to several human cancers. Its hyperactivation leads to the inactivation of retinoblastoma tumor suppressor pathway (Ortega et al.

Table 10.2 Functional homologues between yeast and human and their cancer-related pathways

Cancer-related pathway	Human genes	Yeast genes	Reference
DNA-damage checkpoint	ATM (ataxia-telangiectasia mutated), ATR	RAD53 (radiation sensitive 53)	Simon and Bedalov (2004)
Replication checkpoint	BLM (Bloom syndrome), WRN1 (Werner syndrome)	SGS1 (slow growth suppressor 1)	
Mitotic spindle assembly checkpoint	BUB1(budding uninhibited by benzimidazoles 1), BUBR1 (Budding uninhibited by benzimidazole-related 1)	BUB1 (budding uninhibited by benzimidazoles 1)	
Mismatch repair	MLH1 (mutL homolog 1)	MLH1 (mutL homolog 1)	
Repair of DNA double-strand breaks	BRCA1 (breast cancer 1)	RAD50 (radiation sensitive 50), RAD52 (radiation sensitive 52)	
G1- to S-phase transition	Cyclin D1, cyclin E	CLN2(Cyclin 2)	
Response to mitogenic stimuli	TOR (*target of rapamycin*)	TOR1 (*target of rapamycin 1*)	

2002). Higher expression of CDK4 has been related to the cancer, and in this regard flavopiridol which is an inhibitor of this CDK4 kinase was identified (Liu et al. 2004). Interestingly, flavopiridol or co-expression of INK4A (human CDK4–cyclin-D1 inhibitor or p16) inhibitor could restore the growth arrest of yeast cells overexpressing CDK4 (Moorthamer et al. 1998). Poly (ADP-ribose) polymerases (PARP) are considered to be the promising targets for cancer therapy. Yeast cell-based screens has been designed, especially against BRCA-associated cancers, to identify inhibitors against protein tankyrase 1 (a member of PARP family) that is a telomere-associated factor using its heterologous expression (Yashiroda et al. 2010). Expression of tankyrase 1 in yeast cells leads to growth inhibition which is the basis of the screening by identifying molecules that could lead to reversal of growth inhibition. Using this technique, a flavone was identified as an inhibitor that can rescue the effects of tankyrase 1 overexpression in yeast cells by restoring the growth. Subsequent assays indicated that the flavone not only inhibited the effects of tankyrase 1 in yeast cells but also in mammalian cells (Yashiroda et al. 2010; Gao et al. 2014).

TOR proteins are highly conserved in both humans and yeast (Gao et al. 2014). The TOR-dependent network controls cell growth in response to nutrient signals and metabolism, and its deregulation may lead to corresponding diseased states (Jacinto and Hall 2003). With yeast-based search, Jing and colleagues identified SMERs which stands for small-molecule enhancers of rapamycin that target cell growth control. More than 400 compounds were identified that showed "no growth" phenotype in the presence of a suboptimal rapamycin concentration, whereas subsequent unrelated screens identified 86 potential SMERs considered to be synthetic lethal with rapamycin but with less toxicity alone at concentrations used (Aghajan et al. 2010). Consequently, five structurally distinct molecules were selected for further genomic, genetic, and biochemical analysis, and it has been found that one of the SMERs (SMER3) inhibited Skp1-Cullin-F-box (SCF, Met30, a member of the SCF E3-ligase family) (Aghajan et al. 2010; Gao et al. 2014).

10.5 Perspective

Yeast has the potential to be one of the finest models to study different human diseases apart from the diseases discussed in this chapter. It would be desirable to design a single yeast model system to carry out the analysis of different human diseases or to screen the compounds which can be used as drugs against them. In the abovementioned success stories, it is quite clear that yeast cells can indeed be used to study the effects of several mutation.

References

Aghajan M, Jonai N, Flick K, Fu F, Luo ML, Cai XL, Ouni I, Pierce N (2010) Chemical genetics screen for enhancers of rapamycin identifies a specific inhibitor of an SCF family E3 ubiquitin ligase. Nat Biotechnol 28:738–742. doi:10.1038/nbt.1645

Amberg DC, Burke DJ, Strathern JN (eds) (2005) Methods in yeast genetics. Cold Spring Harbor Laboratory Press, Woodbury. 2p, 44p

Bagriantsev S, Liebman S (2006) Modulation of Aβ42 low-n oligomerization using a novel yeast reporter system. BMC Biol 4:32. doi:10.1186/1741-7007-4-32

Burtner CR, Murakami CJ, Kennedy BK, Kaeberlein M (2009) A molecular mechanism of chronological aging in yeast. Cell Cycle 8:1256–1270. doi:10.4161/cc.8.8.8287

Defossez PA, Prusty R, Kaeberlein M, Lin SJ, Ferringo P, Silver PA (1999) Elimination of replication block protein Fob1 extends the lifespan of yeast mother cells. Mol Cell 3:447–455. doi:10.1016/s1097-2765(00)80472-4

Delacourte A, Buee L (2000) Tau pathology: a marker of neurodegenerative disorders. Curr Opin Neurol 13:371–376. doi:10.1097/00019052-200008000-00002

D'Mello NP, Childress AM, Franklin DS, Kale SP, Pinswasdi C, Jazwinski SM (1994) Cloning and characterization of LAG1, a longevity-assurance gene in yeast. J Biol Chem 269:15451–15459

Ehrnhoefer DE, Duennwald M, Markovic P, Wacker JL, Engemann S, Roark M, Legleiter J, Marsh JL, Thompson LM, Lindquist S, Muchowski PJ, Wanker EE (2006) Green tea(−)-epigallocatechin-gallate modulates early events in Huntington misfolding and reduces toxicity in Huntington's disease models. Hum Mol Genet 15:2743–2751. doi:10.1093/hmg/ddl210

Fabrizio P, Liou LL, Moy VN, Diaspro A, Selverstone Valentine J, Gralla EB, Longo VD (2003) SOD2 function downstream of Sch9 to extend longevity in yeast. Genetics 163:35–46

Fabrizio P, Pletcher SD, Minois N, Vaupel JW, Longo VD (2004) Chronological aging-independent replicative lifespan regulation by Msn2/Msn4 and Sod2 in *Saccharomyces cerevisiae*. FEBS Lett 557:136–142. doi:10.1016/s0014-5793(03)01462-5

Fernandes JT, Tenreiro S, Gameiro A, Chu V, Outeiro TF, Conde JP (2014) Modulation of alpha synuclein toxicity in yeast using a novel microfluidic-based gradient generator. Lab Chip 14:3949–3957. doi:10.1039/c4lc00756e

Franssens V, Bynens T, Van den Brande J, Vandermeeren K, Verduyckt M, Winderickx J (2013) The benefits of humanized yeast models to study Parkinson's disease. Oxidative Med Cell Longev 2013:1–9. doi:10.1155/2013/760629

Fushimi K, Long C, Jayaram N, Chen X, Li L, JY W (2011) Expression of human FUS/TLS in yeast leads to protein aggregation and cytotoxicity, recapitulating key features of FUS proteinopathy. Protein Cell 2:141–149. doi:10.1007/s13238-011-1014-5

Gao G, Chen L, Huang C (2014) Anti-cancer drug discovery: update and comparisons in *Yeast, Drosophila,* and *Zebrafish*. Curr Mol Pharmacol 7:44–51. doi:10.2174/18744672076661407 02113629

Giorgini F, Guidetti P, Nguyen Q, Bennett SC, Muchowski PJ (2005) A genomic screen in yeast implicates kynurenine 3-monooxygenase as a therapeutic target for Huntington disease. Nat Genet 37:526–531. doi:10.1038/ng1542

Griffioen G, Duhamel H, Damme VN, Pellens K, Zabrocki P, Pannecouque C, van Leuven F, Winderickx J, Wera S (2006) A yeast based model of alpha-synucleinopathy identifies

compounds with therapeutic potential. Biochim Biophys Acta 1762:312–318. doi:10.1016/j.bbadis.2005.11.009
Guedes A, Ludovico P, Sampaio-Marques B (2016) Caloric restriction alleviates alpha-synuclein toxicity in aged yeast cells by controlling the opposite roles of Tor1 and Sir2 on autophagy. Mech Ageing Dev 61:270–276
Gunyuzlu PL, White WH, Davis GL, Hollis GF, Toyn JH (2000) A yeast genetic assay for caspase cleavage of the amyloid-β precursor protein. Mol Biotechnol 15:29–38. doi:10.1385/mb:15:1:29
Hekmat-Scafe DS, Brownell SE, Seawell PC, Malladi S, Imam JF, Singla V, Bradon N, Cyert MS, Stearns T (2016) Using yeast to determine the functional consequences of mutations in the human p53 tumor suppressor gene: an introductory course-based undergraduate research experience in molecular and cell biology. Biochem Mol Biol Educ 45(2):161–178. doi:10.1002/bmb.21024
Jacinto E, Hall MN (2003) Tor signalling in bugs, brain and brawn. Nat Rev Mol Cell Biol 4:117–126. doi:10.1038/nrm1018
Johnson BS, McCaffery JM, Lindquist S, Gitler AD (2008) A yeast TDP-43 proteinopathy model: exploring the molecular determinants of TDP-43 aggregation and cellular toxicity. Proc Natl Acad Sci U S A 105:6439–6444. doi:10.1073/pnas.0802082105
Ju S, Tardiff DF, Han H, Divya K, Zhong Q, Maquat LE, Bosco DA, Hayward LJ, Brown RH Jr, Lindquist S, Ringe D, Petsko GA (2011) A yeast model of FUS/TLS-dependent cytotoxicity. PLoS Biol 9: e1001052. doi: 10.1371/journal.pbio.1001052
Kaeberlein M (2010) Lessons on longevity from budding yeast. Nature 464:513–519. doi:10.1038/nature08981
Kaeberlein M, Kirkland KT, Fields S, Kennedy BK (2004) Sir2-independent life span extension by calorie restriction in yeast. PLoS Biol 2:E296. doi:10.1371/journal.pbio.0020296
Kaeberlein M, Powers RW (2007) Sir2 and calorie restriction in yeast: a skeptical perspective. Ageing Res Rev 6:128–140. doi:10.1016/j.arr.2007.04.001
Kaeberlein M (2006) Longevity and aging in the budding yeast. In: Michael C (ed) Handbook of models for human aging. Elsevier, Amsterdam, pp 207–217. doi:10.1016/B978-012369391-4/50019-9
Kaeberlein M, Powers RW, Steffen KK, Westman EA, Hu D, Dang N, Kerr EO, Kirkland KT, Fields S, Kennedy BK (2005) Regulation of yeast replicative life span by TOR and Sch9 in response to nutrients. Science 310:1193–1196. doi:10.1126/science.1115535
Khurana V, Lindqist S (2010) Modelling neurodegeneration in *Saccharomyces cerevisiae*: why cook with baker's yeast? Nat Rev Neurosci 11:436–449. doi:10.1038/nrn2809
Kinarivala N, Trippier PC (2015) Progress in the development of small molecule therapeutics for the treatment of neuronal ceroid Lipofuscinoses (NCLs). J Med Chem 59(10):4415–4427. doi:10.1021/acs.jmedchem.5b01020
Laurent JM, Young JH, Kachroo AH, Marcotte EM (2015) Efforts to make and apply humanized yeast. Brief Funct Genomics 15:155–163. doi:10.1093/bfgp/elv041
Li J, Jiang D (2015) The role of epigenomics in the neurodegeneration of ataxia-telangiectasia. Epigenomics 7:137–141. doi:10.2217/epi.14.81
Liu G, Gandara DR, Lara PN, Raghavan D, Doroshow JH, Twardowski P, Kantoff P, Oh W, Kim K, Wilding G (2004) A phase II trial of flavopiridol (NSC #649890) in patients with previously untreated metastatic androgen-independent prostate cancer. Clin Cancer Res 10:924–928. doi:10.1158/1078-0432.ccr-03-0050
Longo VD, Fabrizio P (2012) Chronological aging in *Saccharomyces cerevisiae*. Subcell Biochem 57:101–121. doi: 10.1007/978-94-007-2561-4_5
Macreadie L-MM, Mohotti S, Shapira D, Bennett L, Varghese J (2008) Validation of folate in a convenient yeast assay suited for identification of inhibitors of Alzheimer's amyloid-β aggregation. J Alzheimers Dis 15:391–396
Mason RP, Giorgini F (2011) Modeling Huntington disease in yeast: perspectives and future directions. Prion 5:269–276. doi:10.4161/pri.18005
Masoro EJ (2005) Overview of caloric restriction and ageing. Mech Ageing Dev 126:913–922. doi:10.1016/j.mad.2005.03.012

Menezes R, Tenreiro S, Macedo D, Claudia N, Tiago S, Outeiro F (2015) From the baker to the bedside: yeast models of Parkinson's disease. Microb Cell 2:262–279. doi:10.15698/mic2015.08.219

Meriin AB, Zhang X, He X, Newnam GP, Chernoff YO, Sherman MY (2002) Huntington toxicity in yeast model depends on polyglutamine aggregation mediated by a prion-like protein Rnq1. J Cell Biol 157:997–1004. doi:10.1083/jcb.200112104

Moorthamer M, Panchal M, Greenhalf W, Chaudhuri B (1998) The p16INK4A protein and flavopiridol restore yeast cell growth inhibited by Cdk4. Mol Cell Biol Res Commun 250:791–797. doi:10.1006/bbrc.1998.9348

Nieoullon A (2011) Neurodegenerative diseases and neuroprotection: current views and prospects. J Appl Biomed 9:173–183. doi:10.2478/v10136-011-0013-4

Olson MV (1991) Genome structure and organization in *Saccharomyces cerevisiae*. In: Broach JR, Pringle JR, Jones EW (eds) The molecular and cellular biology of the yeast *Saccharomyces*, Genome dynamics, protein synthesis and energetics, vol vol. 1. Cold Spring Harbor Laboratory Press, Cold Spring Harbor. 2p

Orr HT, Zoghbi HY (2007) Trinucleotide repeat disorders. Annu Rev Neurosci 30:575–621. doi:10.1146/annurev.neuro.29.051605.113042

Ortega S, Malumbres M, Barbacid M (2002) Cyclin dependent kinases, INK4 inhibitors and cancer. Biochim Biophys Acta 1602:73–87. doi:10.1016/s0304-419x(02)00037-9

Pereira C, Bessa C, Soares J, Leao M, Saraiva L (2012) Contribution of yeast models to neurodegeneration research. J Biomed Biotechnol 2012:1–2. doi:10.1155/2012/941232

Piper PW (2006) Long-lived yeast as a model for ageing research. Yeast 23:215–226. doi:10.1002/yea.1354

Porzoor A, Macreadie IG (2013) Application of yeast to study the tau and amyloid-β abnormalities of Alzheimer's disease. J Alzheimers Dis 35:217–225. doi:10.3233/JAD-122035.

Powers RW, Kaeberlein M, Seth DC, Kennedy BK, Fields S (2006) Extension of chronological life span in yeast by decreased TOR pathway signaling. Genes Dev 20:174–184. doi:10.1101/gad.1381406

Schaeffer SWJ (2010) The synaptic pathology of α- synuclein aggregation in dementia with Lewy bodies, Parkinson's disease and Parkinson's disease dementia. Acta Neuropathol 120:131–143. doi:10.1007/s00401-010-0711-0

Simon JA, Bedalov A (2004) Opinion-yeast as a model system for anticancer drug discovery. Nat Rev Cancer 4:481–488. doi:10.1038/nrc1372

Sinclair DA, Guarente L (1997) Extrachromosomal rDNA circles—a cause of aging in yeast. Cell 91:1033–1042. doi:10.1016/s0092-8674(00)80493-6

Sherman F (2002) Getting started with yeast. Methods Enzymol 350:3–41. doi:10.1016/s0076-6879(02)50954-x

Sokolov S, Pozniakovsky A, Bocharova N, Knorre D, Severin F (2006) Expression of an expanded polyglutamine domain in yeast causes death with apoptotic markers. Biochim Biophys Acta 1757:660–666. doi:10.1016/j.bbabio.2006.05.004

Steffen KK, MacKay VL, Kerr EO, Tsuchiya M, Hu D (2008) Yeast life span extension by depletion of 60S ribosomal subunits is mediated by Gcn4. Cell 133:292–302. doi:10.1016/j.cell.2008.02.037

Steinkraus KA, Kaeberlein M, Kennedy BK (2008) Replicative aging in yeast. Annu Rev Cell Dev Biol 24:29–54. doi:10.1146/annurev.cellbio.23.090506.123509

Su LJ, Auluck PK, Outeiro TF, Yeger-Lotem E, Kritzer JA, Tardiff DF, Strathearn KE, Liu F, Cao S, Hamamichi S, Hill KJ, Caldwell KA, Bell GW, Fraenkel E, Cooper AA, Caldwell GA, McCaffery JM, Rochet JC, Lindquist S (2010) Compounds from an unbiased chemical screen reverse both ER-to-Golgi trafficking defects and mitochondrial dysfunction in Parkinson's disease models. Dis Model Mech 3:194–208. doi:10.1242/dmm.004267

Tenreiro S, Rosado-Ramos R, Gerhardt E, Favretto F, Magalhães F, Popova B, Becker S, Zweckstetter M, Braus GH, Outeiro TF (2016) Yeast reveals similar molecular mechanisms underlying alpha- and beta-synuclein toxicity. Hum Mol Genet 25:275–290. doi:10.1093/hmg/ddv470

Vilaça R, Silva E, Nadais A, Teixeira V, Matmati N, Gaifem J, Hannun YA, Sá Miranda MC, Costa V (2014) Sphingolipid signalling mediates mitochondrial dysfunctions and reduced chronological lifespan in the yeast model of Niemann-Pick type C1. Mol Microbiol 91:438–451. doi:10.1111/mmi.12470

Xiong Y, Coombes CE, Kilaru A, Li X, Gitler AD, Bowers WJ, Dawson VL, Dawson TM, Moore DJ (2010) GTPase activity plays a key role in the pathobiology of LRRK2. PLoS Genet 6:e1000902. doi:10.1371/journal.pgen.1000902

Yashiroda Y, Okamoto R, Hatsugai K, Takemoto Y, Goshima N, Saito T, Hamamoto M, Sugimoto Y, Osada H, Seimiya H, Yoshida M (2010) A novel yeast cell-based screen identifies flavone as a tankyrase inhibitor. Biochem Biophys Res Commun 394:569–573. doi:10.1016/j.bbrc.2010.03.021

Zhang X, Smith DL, Meriin AB, Engemann S, Russel DE, Roark M, Washington SL, Maxwell MM, Marsh JL, Thompson LM, Wanker EE, Young AB, Housman DE, Bates GP, Sherman MY, Kazantsev AG (2005) A potent small molecule inhibits polyglutamine aggregation in Huntington's disease neurons and suppresses neurodegeneration in vivo. Proc Natl Acad Sci U S A 102:892–897. doi:10.1073/pnas.0408936102

Part V

Metabolic Engineering Strategies for Bioactive Compounds

Strategies for Gene Expression in Prokaryotic and Eukaryotic System

11

Raman Thakur and Jata Shankar

Abstract

In the recent years, a large number of recombinant or heterologous proteins of human interest have been commercially produced using different prokaryotic and eukaryotic host cells. This is possible due to the rapid development of genetic engineering technologies. Among prokaryotic expression system, *Escherichia coli* is the most suitable expression host for foreign gene expression and protein production. *E. coli* is preferred over other bacterial host for gene expression because it has short life, is easier to grow in inexpensive medium, has high cell density, has well-known genetic makeup, and could be genetically manipulated. The major disadvantage of *E. coli* as a host is that it lacks posttranslational modifications that are required for human proteins. On the other hand, yeasts are excellent host for expression of foreign gene and heterologous protein expression. Yeasts utilize the advantage of unicellular organism because they are easy to genetically manipulate and have the capacity of mRNA splicing and posttranslational modifications for eukaryotic organisms. Among higher eukaryotic organisms, plant cell expression systems are now used as an expression system for vaccines, antigens, and antibodies due to their low production cost. On the other hand, animal or mammalian expression systems are utilized for the synthesis of therapeutic heterologous proteins. The production of proteins utilizing mammalian or animal cell expression system are able to do posttranslational modification, proper protein folding, and product assembly, which are required for biologically functional proteins. The present chapter discusses the strategies of gene expression in prokaryotic and eukaryotic host cell for the recombinant protein expression.

R. Thakur • J. Shankar (✉)
Genomics Laboratory, Department of Biotechnology and Bioinformatics, Jaypee University of Information Technology, Solan 173 234, Himachal Pradesh, India
e-mail: jata.shankar@juit.ac.in; jata_s@yahoo.com

V.C. Kalia, A.K. Saini (eds.), *Metabolic Engineering for Bioactive Compounds*, DOI 10.1007/978-981-10-5511-9_11

Keywords
Prokaryotic system • Eukaryotic system • Gene • Protein • Expression • Escherichia coli • Yeasts • Cell lines

11.1 Introduction

Gene is the fundamental unit of genetic information that is an essential component for the synthesis of functional parts, i.e., proteins. A gene consists of coding (exon) and noncoding (intron) sequences along with promoter and terminator sequences. The expression of a gene can be described in the form of mRNA production and proteins (Khan 2013). Nowadays, there are different ranges of expression vectors available for the expression of heterologous or recombinant proteins. Such expression systems/vectors include prokaryotic (*Escherichia coli*), lower eukaryotic (yeasts), and higher eukaryotic (plant cells and animal or mammalian cells) expression systems. Each of these expression systems have there advantages, e.g., user friendly, able to do posttranslational modifications, and low production cost (Yin et al. 2007). Among different bacterial (prokaryotic) expression systems, *E. coli* expression system is mostly used for the expression of different heterologous or recombinant proteins. *E. coli* is the most common single cell or bacterial cell host for recombinant gene expression since it is probably the most studied prokaryotic cell system using different functional genomics technologies. Further, knowledge about its biochemical properties and available genetic information has facilitated gene cloning in *E. coli*. The advantage of having *E. coli* as expression system is due to its rapid growth rate in simple and inexpensive medium. Further, purification of recombinant or heterologous proteins have been greatly simplified by fusion proteins (affinity tags). However, the expression in *E. coli* have serious problems, such that it does not glycosylate (N- or O-glycosylation) the proteins (Reiser et al. 1990; Rai and Padh 2001; Rosano and Ceccarelli 2014). On the other hand, yeasts also significantly gained attention for the production of recombinant or heterologous proteins especially for human proteins (Wuest et al. 2012). Among yeasts, different methylotrophic (*Pichia pastoris*, *Candida boidinii*, *P. methanolica*, etc.) and non-methylotrophic (*Saccharomyces cerevisiae*, *Yarrowia lipolytica*, *Pichia stipitis*, etc.) yeasts are used for protein production (Cos et al. 2006; Kroll et al. 2010; Zhang et al. 2010). Yeasts have gained attention for the production of recombinant proteins because they share common features such as biochemical and genetic feature with higher eukaryotic cells. Further, they are simple to grow at a large scale in inexpensive medium and achieve high cell densities because they are single-celled microorganisms. Some of the yeasts are genetically well characterized, and their genetic maps are available such as for *Saccharomyces cerevisiae* and *Schizosaccharomyces pombe* (Fooks et al. 1995). Yeasts can be easily manipulated, and recombinant yeasts can be selected by complementation method. Further, yeasts are safe for fermentation processes because they are free of endotoxins as compared to other fungal species which are known to produce mycotoxins (Shankar 2013), thus requiring minimal toxicological studies. Furthermore, stronger promoters are available which help to increase the heterologous gene expression more inexpensively than bacterial expression system. But the main disadvantage of low eukaryotic (yeast) expression system is that they perform N- and O-linked glycosylation on target proteins in a different way than higher eukaryotes like animal cells. But advances in glycoengineering of yeasts leading to humanized N-glycosylation showed great potential (Yin et al. 2007). Thus yeast can be considered as powerful recombinant host for heterologous gene expression system.

In the past decade, plants have gained importance for the synthesis of recombinant or heterologous proteins. Recently, antibodies, human blood products, vaccines, etc. are derived from plant system reaching to commercial level. For biomedical or therapeutic relevant proteins, plant-derived expression systems offer safety as well as low production cost advantages over animal cell or mammalian cell expression systems (Fischer et al. 2004; Shanmugaraj and Ramalingam 2014).

As the demand of biomedical relevant proteins have been increased, to meet the demand of increasing therapeutic agents for human uses, the mammalian or animal cells have gained importance for the production of biomedical relevant therapeutic proteins such as interferon, hormones, etc. (Li et al. 2010). The mammalian or animal cell expression system has several advantages over prokaryotic or low eukaryotic expression system. They promote protein N- and O-linked glycosylation, signal synthesis, and even secrete proteins particularly eukaryotic proteins. Various expression systems in cell lines are developed for the expression of heterologous gene or protein production such as CHO (Chinese hamster ovary). Other animal cell line systems for recombinant or heterologous protein production are human embryonic kidney (HEK) 293, hamster kidney cells, etc. The main disadvantages of animal or mammalian cell line expression systems are potential contamination with animal or mammalian viruses and high-cost and complicated technologies (Yin et al. 2007; Dalton and Barton 2014). Further, transgenic animal expression system has been also developed. The recombinant DNA can be transferred to fertilize animal ovum; after that this DNA molecule is integrated into animal genome expressing the foreign gene as protein in specific tissue or organ in the host depending upon promoters. The main drawback of transgenic animal expression system is that it requires longer production period and higher cost. So this system is only used for medically important purposes but not for large-scale industrial purposes for the reason of having low yield and high cost value (Reichenstein et al. 2001; Soler et al. 2005).

In this chapter, the heterologous (foreign) gene expression or heterologous protein production strategies in prokaryotic cell (*E. coli*), lower eukaryotic cells (yeasts), and higher eukaryotic cells (plant cells and animal or mammalian cells) have been discussed.

11.2 Prokaryotic Expression System

Escherichia coli expression system is one of the important bacterial expression system in laboratory investigation and in the development of marketable products and also acts as the standard system among different expression systems. *E. coli* expression system also provides platform for engineering of proteins as well as structural analyses of proteins. Among different bacteria, *E. coli* expression system is the most accepted bacterial system for the manufacturing of homologous, heterologous, and recombinant proteins, which does not require posttranslational modifications of proteins (Rai and Padh 2001). In glucose-salts medium, the doubling time of the *E. coli* cell is approximately 20 min. Within few hours of culture in optimum conditions, *E. coli* culture reaches to stationary condition, though expression of heterologous or recombinant proteins may alter the doubling time or generation time by imposing a metabolic load on the microorganism (Bentley et al. 1990; Sezonov et al. 2007). *E. coli* cells easily achieve high cell density. It has been estimated that *E. coli* in liquid media has theoretical density limit of approximately 1×10^{13} bacteria (viable)/mL. However, in complex medium, exponential growth leads to decrease in cell densities from the actual required number of cells. In the simple laboratory setup, i.e., in batch culture at 37 °C (in Luria Broth medium), the upper limit is $<1 \times 10^{10}$ cells/mL, which is under the theoretical limit (<0.1%). Thus, to obtain high density of cell for the expression of heterologous or recombinant protein, there is the need of high-cell-density culture methods (Pope and Kent 1996; Shiloach and Fass 2005; Sezonov et al. 2007).

Escherichia coli is now considered as an ideal organism for expression system. Several strategies have been developed due to the extensive knowledge on genetics and physiology of *E. coli*. Complex media with rich source of nutrients are now available from inexpensive sources. Transformation of foreign DNA in *E. coli* is easy and effective. Transformation of plasmid with foreign DNA in *E. coli* strains can be achieved in a short period of time (approximately 5 min) (Pope and Kent 1996). Different strains of *E. coli* are now available for recombinant protein production; these strains are genetically manipulated which make them user friendly to scale up the fermentation. Purification of the recombinant proteins is now eased by manufacturing fusion proteins along with the desired protein and can be purified through affinity column.

However, expression of protein in *E. coli* does has shortcomings. It does not allow posttranslational modifications of the recombinant proteins. *E. coli* cell is not capable of glycosylating proteins (e.g., N- or O-linked glycosylation), though a bacterial strain such as *Neisseria meningitidis* has been reported to provide O-linked glycosylation to the endogenous proteins (Rai and Padh 2001). According to the PubMed and Protein Data Bank (PDB) among all prokaryotic and eukaryotic host cells, *E. coli* is the most important host cell for the expression of recombinant proteins. Most of the crystal structures of proteins that are deposited in PDB have been obtained from *E. coli* expression system (Fernández and Vega 2016). Crystal structure of characterized protein may serve as a template for homology modeling for the mutated protein or for the protein where three-dimensional structure is not available (Thakur and Shankar 2016).

11.2.1 *E. coli* Strains Used as Host

E. coli is a gram-negative bacterium and commonly used for production of heterologous proteins (Terpe 2006). Since, the organism is well known that helped to its establishment in most of the laboratory environments. Thus, it makes no surprise that *E. coli* expression systems are the choice of heterologous or recombinant protein production that does not require posttranslational modifications (Baeshen et al. 2015).

A wide variety of *E. coli* strains or *E. coli* mutants have been identified as well as characterized. Mostly the strains that are currently being used in rDNA experiments or for the heterologous protein production are obtained from a single *E. coli* strain: *E. coli* K-12, which was identified from the diphtheria patient feces in the year of 1922 (Casali 2003). Currently, the most common *E. coli* strain for the heterologous protein production is BL21 (DE3) which is from the K-12 lineage (Daegelen et al. 2009). Commonly used *E. coli* K-12 mutants for heterologous protein production are AD494, a trxB mutant, which provides the formation of cytoplasmic disulfide bond; JM 83, commonly used for recombinant proteins that need to be secreted to periplasm; and HMS174, with recA mutant that helps to provide the stability of plasmid and Rif resistance (Terpe 2006). In comparison with K-12 strains, derivatives of BL21 are *ompT* and *Lon* protease deficient. Lon protease showed degradation of many recombinant proteins. Another mutant of BL21 strain of *E. coli* contains a gene that encodes for outer membrane protease (*OmpT*). This protease generally degrades extracellular proteins, and the *E. coli* cells in the culture access the available amino acids. Degradation of proteins by these proteases possess problem to the host cell in expression of recombinant proteins. Post-cell lysis, *OmpT* proteases may also degrade recombinant or heterologous proteins. In addition, loss of plasmid can be avoided due to mutation of hsdSB available in the B834 strain. As B834 is also a source of BL21, disruption of DNA methylation followed by degradation could be achieved. Nowadays, the BL21 (DE3) and their derivatives are commonly used for recombinant expression of the desired proteins (Gottesman 1996; Rosano and Ceccarelli 2014).

11.2.2 Plasmids for Expression System in *E. coli*

Plasmids used for the expression of heterologous proteins in different bacterial host, e.g., *E. coli*, are the result of combination of different promoters and selection markers. These plasmids also contain origin of replications, different cloning sites for foreign genes, and a tagged protein to remove the recombinant protein. Today there is large number of expression vectors available, and it depends upon individual need which one of them to be selected. For the selection of expression vector, the following features have to be carefully evaluated:

1. Promoters
2. Selection markers
3. Affinity tags and tag removals for recombinant proteins
4. Replicons

11.2.2.1 Promoters

There are many promoter systems of *E .coli* for protein expression, but only some of them are used for the expression of proteins in *E. coli*. A useful promoter should be strong and must be tightly regulated (low basal expression level system); it must be easily transferable among different strains of *E. coli* that are used further for the expression of proteins in the host. Further, it should not cost much as well as should not be dependent on ingredients for culturing that are not easily available (Casali 2003; Terpe 2006):

Lac promoter systems: The utilization of lactose in *E. coli* is key component of the *lac* operon system that regulates the expression. The mechanism of *lac* operon system makes such expression vectors user friendly. Lactose sugar allows the induction of transcription of a foreign gene in the *E. coli* expression system followed by protein synthesis (Casali 2003; Rosano and Ceccarelli 2014). There are many commercially available promoters that utilize *lac*-derived regulatory element and are used for the expression system. For example, the commercial vectors that contain lac or tac promoters for recombinant gene expression are the pUC series (lacUV5 promoter) and vectors from pMAL series (tac promoter) (Casali 2003). But the main disadvantage of lac promoter is that it is a weak promoter and is not considered where higher recombinant protein expression is required. Also, induction of lac operon finds difficulty where glucose as carbon source is used, and if both lactose and glucose as carbon are used, lac promoter is not enough to be considered inducible until the glucose in the medium is fully exhausted. At low concentration of glucose, cyclic AMP is synthesized in the cell, which is one of the important component for complete lac operon activation. This whole process is known as catabolic repression (Epstein et al. 1975; Ullmann et al. 1976). To achieve the expression of proteins in glucose-containing media, a mutant has been constructed that reduces sensitivity to catabolite repression, i.e., lacUV5 promoter. But lacUV5 promoter and its derivatives are weak promoters and are not much useful for the synthesis of heterologous and foreign proteins. Then, hybrids of other promoters and lac promoter are constructed, e.g., tac promoter consisting of trp promoter (−35 region) and lac promoter (−10 region). This tac promoter has ten times more performance capacity compared to lacUV5 promoter (de Boer et al. 1983). To reduce all these problems with lac promoter, other system was constructed, e.g., T7 RNA polymerase system (de Boer et al. 1983).

T7 RNA polymerase system: T7 RNA polymerase system is the most widely used expression system in which the polymerase elongates chains more efficiently (five times) in comparison with *E. coli* polymerase. T7 RNA and *E. coli* RNA polymerases recognize entirely different promoters (Terpe 2006). Gene encoding T7 RNA polymerase in *E. coli* strain is regulated by a derivative of L8-UV5 lac promoter. Thus, this polymerase induces high level of recombinant proteins in both homologous and heterologous BL21 (DE3) strain of *E. coli* (Grossman et al. 1998; Pan and Malcolm 2000). Two point mutations in L8-UV5 lac promoter allow L8-UV5 lac promoter to be different from wild-type promoter. To increase promoter strength, point mutations at 10 regions were made, and the mutation decreases the dependency of lac promoter on cyclic AMP. There is also a third point mutation that makes the promoter stronger which decreases its response to glucose (Grossman et al. 1998). Such mutation helps the stimulation of polymerase with IPTG (isopropyl β-D-1-thiogalactopyranoside) and is not influenced by glucose as carbon source present in the medium. The T7 RNA polymerase system present in the pET vector is the most popular for recombinant protein expression. The target protein through this system can represent up to 50% out of the total cell protein (Dubendorff and Studier 1991; Studier 1991; Pan and Malcolm 2000). In this system, a gene is cloned under a promoter that is identified by T7 RNA polymerase (T7 RNAP). T7 RNAP is an active polymerase and should be present in another plasmid. Generally, it is inserted in the genome of bacteria in a prophage (λDE3) encoding for the T7 RNAP under lacUV5 promoter. Thus, analog of lactose IPTG or lactose is used to induce the desired gene in this system (Pan and Malcolm 2000; Daegelen et al. 2009).There are other promoter systems that are widely used for the expression of proteins, for example, phage promoter PL, tetA promoter/operator system, PBAD promoter induced by L-arabinose, and L-rhamnose-inducible rha PBAD promoter. So these different promoter systems can be used in different expression vectors in *E. coli* for heterologous or recombinant protein expression (Dubendorff and Studier 1991).

11.2.2.2 Selection Markers

The introduction or transformation of vectors in *E. coli* cells is a difficult process. Thus, it is important to select those cells that have the capacity to receive a vector for protein expression (Korpimaki et al. 2003). Further, cells grow well that do not have a plasmid over those cells which have the desired plasmid because cells with plasmid have to replicate the plasmid DNA as well as the chromosomal DNA. It is particularly important when cells contain plasmid with high copy number or large-sized plasmids. Thus, a selective pressure is required in order to maintain the plasmid in the cells. In most cases, plasmids contain a gene that is resistant to antibiotic and considered as a selectable marker. Thus, bacterial cells without plasmids with selection marker will not survive upon addition of the appropriate antibiotic and a desired cell culture with a plasmid can be produced. In the *E. coli* expression system, there are many plasmids with antibiotic resistance genes. A *bla* gene provides resistance against ampicillin and produces a periplasmic β-lactamase enzyme involved in inactivation β-lactam ring present in β-lactam derivative ampicillin antibiotic. Other selective agents are chloramphenicol and kanamycin. Another antibiotic tetracycline showed more stability in the culture in comparison with other antibiotics (Shaw 1983; Roberts 1996; Korpimaki et al. 2003). The spreading of antibiotic resistance cells and to some extent the cost of antibiotics are major concerns for the use of these antibiotics in large-scale cultures. So many efforts have been going on for the development of antibiotic-free plasmid systems (Fair and Tor 2014). For example, an essential gene for the growth of cells can be removed from genome of bacteria and then cloned into a plasmid instead of an antibiotic resistance gene. Thus, after cell division, the bacteria that do not contain plasmids generally die. There are many plasmid-addiction systems based on mode of functions such as toxin/antitoxin, metabolism, and operator-repressor titration systems. These plasmid-addiction-based systems showed success in large-scale fermentation; however, plasmid-addiction systems for expression studies are still in fancy (Kroll et al. 2010).

11.2.2.3 Affinity Tags

For the purification of recombinant proteins, affinity tags are considered most suitable tools where prior knowledge of its biochemical function of protein is not available. In other words, recombinant protein purification via affinity tags does not require the knowledge of function of the desired protein. Affinity tags are mainly exogenous amino acid sequences of short peptides that have high affinity for special biological or chemical ligands linked. These affinity tags bind to small ligands that are either linked to solid support or immobilized protein partner (Lichty et al. 2005). However, there are reports that biological activity or structure of a chimeric protein or heterologously expressed protein may be affected by these tags (Khan et al. 2012). There are now different constructs on which these peptides can be positioned either on N-terminal or C-terminal. There are different tags that are frequently used for protein expression and purification system like His tag, FLAG, Strep tagII, HA-tag, c-myc, T7-tag, NusA, etc. (Arnau et al. 2006; Rosano and Ceccarelli 2014). Also commercially monoclonal antibodies are available for all these tags, and they can be detected by Western blot. Further affinity tags allow one-step purification of the desired expressed proteins because resin that binds specifically with particular tag proteins is available (Rosano and Ceccarelli 2014).

Further, it is also important to remove these tags because they can interfere with the structure stability or biological function of the desired protein. These tags can be removed either chemically or by enzymatic cleavage. Different enzymes like enterokinase and the tobacco etch virus (TEV) protease are used to remove affinity tags (Blommel and Fox 2007; Waugh 2011). After removal of tags, purified protein could be subjected to functional characterization such as biological functions, cellular components, and molecular functions (Tiwari et al. 2016)

11.2.2.4 Replicons

Replicons are an important parameter when heterologous gene expression or its product is required. Replicons control the copy number of vectors that are chosen for gene expression. Plasmid vectors undergo autonomous replication and contain replicons and origin of replication together with other control elements. The yield of the heterologous proteins depends upon the copy number of a plasmid which acts as vector. Sometimes high copy number of plasmid increases the metabolic burden that leads to decrease in the growth rate of bacteria and also causes instability of a plasmid (del Solar and Espinosa 2000). Different vectors use different origin of replication such as pMNB1 (15–60 copies per cell) used by PET series of vectors. Another origin of replication CO1E1 (15–20 copies per cell) is used by pQE vector. Both of these are incompatible to each other and cannot be used in the same host because they competed for replication machinery (Camps 2010). Expression of two proteins in a single host can be achieved by two plasmid systems such as pACYC and pBAD series that contain p15A origin of replication and have 10–12 copies per cell (Sathiamoorthy and Shin 2012; Rosano and Ceccarelli 2014; Chakravartty and Cronan 2015). Further, with the use of pSC101 plasmid, triple protein expression can be achieved. As low as <5 copies (per cell) of pSC101 plasmid can be present in a host cell (Nordstrom 2006).

11.3 Yeast Expression System

Yeast is a unicellular lower eukaryotic microorganism, which possesses similar biochemical and genetic makeup compared to those of higher eukaryotic cell system and is beneficial eukaryotic system for the expression of heterologous proteins. Yeast cells can rapidly grow with higher cell densities, are easy to grow, and are not expensive. They are *S. cerevisiae*, *Pichia stipitis*, and *Pichia pastoris* (*P. pastoris*). *S. cerevisiae* (baker's yeast) is probably the most studied eukaryotic cell system using different functional genomics technologies. The genetics of yeasts and physiological characteristics are well reported, and proteins that are posttranslationally modified show similar mechanism to that of plants. Being a good host for heterologous protein expression, it also has some limitations like less production of recombinant proteins; also gene from foreign organism brings stress to the yeast cells which causes hyperglycosylation of the foreign proteins. Further, it has been observed that there are limitations of strong promoter that can be overcome by using *P. pastoris* and *S. cerevisiae.* Yeast system for expression of a eukaryotic gene has emerged as a leading eukaryotic expression system for heterologous protein expression and for function characterization. Yeast expression system provides various genomic and proteomic tools to investigate heterologous expression of proteins, even at subcellular level, such as expression of proteins at the cell organelle level (Popa et al. 2016).

The heterologous expression of protein in *S. cerevisiae* began in the 1980s. *S. cerevisiae* expression system has more advantages than *E. coli* expression system due to posttranslational modification, or exogenous proteins may be targeted to membrane-bound compartments of cell and easily isolated for biochemical experiments (Ton and Rao 2004). Further, heat-shock proteins are categorized as a class of proteins that assist in posttranslational modification of other proteins that are necessary to perform routine biological function or during stress conditions (Tiwari et al. 2015). Furthermore, the yeasts used for the expression of heterologous protein production are divided into two categories. One is nonmethylotrophic yeast strains and the other is methylotrophic yeasts. The methylotrophic yeast species use common carbon and energy source, i.e., methanol.

Methylotrophic yeasts are initially used for the synthesis of single-cell proteins, but after that yeast cells have emerged as production house to express foreign proteins due to two main reasons. Firstly, yeast cells can grow rapidly in mass of cells even if the condition of fermentation processes is not that much favorable, and secondly, methylotrophic yeasts produce methanol-oxidizing enzymes (i.e., alcohol oxidase) under control of very strong and strictly regulated promoters that are in high demand (Mattanovich et al. 2012). The list of methylotrophic or nonmethylotrophic yeasts which are used as host cells for production of heterologous proteins is given in Table 11.1.

Table 11.1 Methylotrophic and nonmethylotrophic yeasts

Methylotrophic	Nonmethylotrophic	References
Pichia pastoris	*Saccharomyces cerevisiae*	Cos et al. (2006), Mokdad-Gargouri et al. (2012)
Pichia methanolica	*Zygosaccharomyces rouxii*	Nakagawa et al. (2006), Ogawa et al. (1990)
Candida boidinii	*Kluyveromyces lactis*	Yurimoto and Sakai (2009), Rocha et al. (2011)
Hansenula polymorpha	*Yarrowia lipolytica*	Guengerich et al. (2004), Chuang et al. (2010)

The expression of foreign protein in eukaryotic cells requires cloning and transformation of heterologous gene in a yeast vector that encodes for the desired protein and expression cassette with promoter and other transcriptional elements. Transformation of plasmid containing the desired gene is needed to express the protein in the suitable host cells (Reiser et al. 1990). The heterologous genes can be inserted in host cell by means of episomal or autonomous plasmid vectors and by means of integrative approaches. In the case of integrative approach, the heterologous gene can be inserted into host cell chromosomal DNA by recombination, but it requires the DNA homologue. On the other side, heterologous genes can be replicated autonomously along with the plasmids. A number of expression vectors have been developed for the heterologous protein expression in *S. cerevisiae* or other yeast host cells. Episomal vectors are one of them. Episomal vectors have origin of replication, and they replicate autonomously in the host cell. These vectors contain different components.

Origin of replication: Origin of replication is essential for the replication, and episomal vectors replicate autonomously outside the chromosome. Episomal vectors used two μ-ori or ars1-ori from *S. cerevisiae* and *S. pombe*, respectively. *Ars1* has been identified as an autonomous replicating sequence (ARS) for *S. pombe* that helps the plasmid to replicate extrachromosomally and maintain its stability (Wright et al. 1986).

One of the critical factors to express recombinant protein in host cells depends upon transcription rate of the foreign gene. To enhance the transcription efficiency, different varieties of homologous or heterologous promoters are available. Among these promoters only homologous promoters are preferable because heterologous promoters lead to poor yield of recombinant proteins. The availability of stronger promoters such as GAL1/10, which commonly use to study the virulence proteins from pathogen in *S. cerevisiae*, for example, T3E, a type III effector virulence protein (Popa et al. 2016). Different promoters are given in Table 11.2 used for gene expression in yeast cells such as constitutive promoter glyceraldehyde-3-phosphate dehydrogenase (PGAP) and inducible alcohol oxidase 1 promoter (PAOX) used in *Pichia pastoris* (Yu et al. 2016; Mattanovich et al. 2012).

Another important factor is the selection markers. The transformation of heterologous genes in yeast cell can be achieved by lithium acetate method, spheroplast preparation, and the use of electroporation for transformation of yeast cell with vectors carrying heterologous gene for expression. The efficiency of transformation depends upon the host cell (Kawai et al. 2010). After the transformation of host cell with plasmid vector containing the desired gene for expression, it is important to select the transformed host cell with that plasmid. So selection markers are required for this purpose. The most commonly used markers are *LEU2* and *URA3* gene of *S. cerevisiae*

Table 11.2 Yeast species and their constitutive and induced promoter

Yeast species	Constitutive promoter	Induced promoter
Saccharomyces cerevisiae	GAPDH, PGK, TP1, ENO	GAL1-10, GAL7, ADH2
Pichia pastoris	GAP	AOX1

which show complementation with *S. pombe* mutants' *leu1* and *Ura4*. Other commonly used markers are *ade6*, *his31*, and *his71*, and some other antibiotic agents such as phleomycin and bleomycin are also used as selection markers (Nasser et al. 2003).

Another approach used for heterologous gene expression in yeast is protein fusion technology. Protein fusion technology has emerged as a new tool for heterologous protein expression in yeast such as heterologous expression of granulocyte colony-stimulating factor (G-CSF) in *P. pastoris* with the help of protein fusion technology. In this technology a well-known extremely expressed protein of host (yeast) fused with heterologous protein to be expressed in a vector. The heterologous proteins are separated by linkers that allow their proper folding during synthesis in yeast cells (Sigar et al. 2016).

11.4 Strategies for Gene Expression in Plant Cells

Plant cells provide a cost-effective and convenient expression system for the expression of heterologous or we can say that for the production of valuable recombinant protein products. In last few years, several recombinant proteins have been produced through plant expression system for human welfare. The therapeutic or biomedical important protein production in plant cells can be achieved by the use of stable transgenic lines of plants or by the use of plant suspension cultures or by use of transient expression systems (Fischer et al. 1999). For the development of stable transgenic plants, heterologous gene expression needs transgenic seeds, and the main disadvantage of this system is that it requires long time and needs sterile environment to generate stable transgenic seeds. These days transient expression system (agroinfiltration and viral vector expression system) is used for the expression of heterologous gene or heterologous protein production in plant cells. The main advantages of transient gene expression system in plant cells are less time for transient expression of heterologous gene in plant cells and more protein expression and accumulation, and also there are no biosafety concerns (Kapila et al. 1997; Daniell et al. 2009), whereas the plant cell suspension culture requires downstream processing of secreted protein in culture medium, which ultimately increases the product cost (Twyman et al. 2003). Here we only discuss the transient gene expression strategies in plant cells. Heterologous gene expression in plants needs a suitable host system and an efficient gene transformation system.

11.4.1 Transient Gene Expression in Plant Cells

Transient gene expression in plants needs genetic transformation of plants with the desired gene whose product is required. Plant genetic transformation is achieved by the use of two different approaches:

1 Vector-mediated gene transfer in plants (indirect method)
2 Direct gene transfer in plants

11.4.1.1 Vector-Mediated Gene Transfer in Plants

Different vectors are used for the transformation of heterologous gene in plants; Ti-plasmid is present in *Agrobacterium tumefaciens*, which is most commonly used for transfer of recombinant DNA into plant cells which is known as agroinfiltration (Gelvin 2003; Sainsbury and Lomonossoff 2014). *A. tumefaciens* is a ubiquitous soil organism that infects plant that belongs to *Rhizobiaceae* family (Hao et al. 2012). It causes plant crown gall disease in this family of plant. This crown gall disease results from the transfer, integration, and expression of some set of genes for Ti-plasmid. So Ti-plasmid of *A. tumefaciens* is mainly utilized as a vehicle for gene transfer for plant cells. The transformation resulted from transmit of T-DNA from Ti-plasmid into the host plant cell and other several virulence gene both coded by chromosomal and Ti-plasmid (vir) also transfer to plant cells (Gohlke and Deeken 2014).

T-DNA is present in Ti-plasmid and bordered by 25-bp direct repeats known as right and left T-DNA border. Two vir proteins, vir D2 and virD1, nick both borders of strands of T-DNA present in Ti-plasmid. It results in a single-stranded T-DNA that is further transferred along with various Vir genes into the cytoplasm of host cell. VirD4 and VirB proteins help in the formation of a channel for transformation of this single-stranded T-DNA. Further, VirD2 and VirE2 proteins import T-DNA into the nucleus that is subsequently inserted into the genome of host. The removal of some gene from T-DNA does not impair its T-DNA transfer ability but affects tumor formation in plants. The *Agrobacterium* strains that are not oncogenic are generally called "disarmed." Two important advancements have been made in *Agrobacterium*-mediated transformation of T-DNA: first, the choice of transformation of foreign gene in plant cells, and construction of binary Ti-vectors and different disarmed *Agrobacterium* strains. The strains of *Agrobacterium* with Ti-plasmid used for transformation in plant cells are classified by resident Ti-plasmid and chromosomal background. Significant modifications have been done in virulence of *Agrobacterium* strains, which further expand to different plant species, where T-DNA-mediated transformation can be achieved (Eckardt 2004; Lee and Gelvin 2008). *Agrobacterium*-mediated gene transfer in plants can be performed by the use of vacuum infiltration or syringe infiltration. In vacuum infiltration, the plant tissue is submerged into *A. tumefaciens* culture under decreased pressure and is further rapidly depressurized (Tague and Mantis 2006). This is the most preferable method for gene transfer in lettuce, *Arabidopsis*, etc. (Leuzinger et al. 2013). These days this technique is used for the production of vaccine antigens and antibodies in various plant species (Both et al. 2013).

Virus-Based Expression Vectors for Plants

Various plant viruses are used as vectors to transfer heterologous gene in plants for the desired protein production. The desired or heterologous gene is cloned in plant virus genetic material mostly preferably in plant RNA viruses, and then these viruses allowed infecting plants to produce heterologous proteins (Shanmugaraj and Ramalingam 2014). There are two viral vector systems for gene expression in plants: first is independent virus and second is minimal virus vector system. Independent virus vectors are replication-competent vectors that can be inoculated to plants as viral RNA or virus particles and infect the phloem tissues of a host plant, whereas minimal virus vector systems are also replication competent but modified by inserting the desired gene in place of virus-encoded open reading frame. The independent virus vector systems for heterologous gene expression in plant cells are derived from the genome of tobacco mosaic virus (TMV), comoviruses (e.g., cowpea mosaic virus), potexviruses (e.g., potato virus X), and other RNA viruses of plants. Most of these virus vector systems are functional with *Nicotiana* or other herbaceous species of plants (Pogue et al. 2002; Lico et al. 2008). A minimal virus vector system has been developed by the use of *tobamovirus*, *bromovirus*, *comovirus*, and *geminivirus* genomes (Pogue et al. 2002). The major advantage of using virus-based vectors over *Agrobacterium* (non-replicating T-DNA) gene expression system is that after infecting plant host, they act as replicon and increase their copy number by multiplying in plant host cells. Thus, they increase the level of heterologous or foreign genes in plants (Canto 2016).

11.4.1.2 Direct Gene Transfer in Plants

There are different direct gene transfer methods for plant cells. Chemical methods like polyethylene glycol treatment of plant protoplast can easily take up the desired DNA molecules from medium which can be finally integrated into host genome. Further electroporation can be utilized for the transformation of plant cells because thick cell wall of plant cells restricts the movement of DNA molecules. Electrical pulses are applied to protoplast of pant cells that are placed in a solution containing the desired DNA which is to be transferred. These high-voltage pulses induce the creation of micropores in cell membrane of plant cells that allow transfer of the desired DNA molecule into cell which is further integrated into the host genome. Microprojectile or particle bombardment is another technique for the transfer of the desired gene into plant cells. This technique uses 1–2-μm-sized gold or tungsten metal particles coated with the desired DNA molecules. These particles are accelerated with high speed after loading into a particle gun and targeted to plant cell suspension, plant tissues, or callus cultures. These particles enter the plant cell along with the desired DNA molecules which are further integrated into host genomes (Kikkert et al. 2005; Potter and Heller 2010; Liu and Vidali 2011). So these strategies can be employed for transient gene expression in plant cells.

Multiple-Gene and Multiple-Protein Expression System in Plants

The multiple-gene transformation in plant cells can be achieved by the use of different strategies like linked co-transformation, unlinked co-transformation, artificial chromosomes, and multiple-gene expression system. Multiple-gene expression system uses the site-specific integration of transgene in plant hosts at specific loci using site-specific nuclease system. Such systems as clustered regulatory interspaced short palindromic repeats (CRISPR-Cas9), zinc finger nucleases (ZFNs), and transcription activator-like effector nucleases (TALENs) help in the proper integration of foreign DNA in plant host cells leading to multiple-protein expression of integrated multiple genes in plants (Ferrer et al. 2016).

Another technology which is emerged in the last few years for multiple-gene delivery in eukaryotic and prokaryotic host cells is automated unrestricted DNA recombineering for multigene delivery (ACEMBL) system. This system uses the multiple-fragment cloning technology independent to sequences and ligation and is further combined with site-specific recombination of multigene at defined loci in eukaryotic and prokaryotic host cells that leads to multiple-protein expression in respective host cells (Nie et al. 2016).

11.5 Mammalian or Animal Cell Expression System

Recently, the need of highly purified recombinant proteins are increasing. To meet the demand of increasing recombinant proteins, the *E. coli* has been considered as a cheap method for required recombinant proteins. However, the number of therapeutic or biomedical important proteins failed to express as well as failed to fold into functional proteins in prokaryotic expression system because most of these proteins require posttranslational modifications for their proper function and biological activities, which are not available in prokaryotic expression system. However, there are different expression systems available from eukaryotes for recombinant or heterologous protein expression that require posttranslational modifications. Among them, mammalian or animal cell expression system is required for therapeutic or biomedical relevant proteins.

11.5.1 Requirement for Gene Expression in Mammalian System

Gene expression depends on its transcription efficiency. Transcription of genes depends upon the RNA polymerase interaction with promoters that finally leads to production of the desired protein product. The gene expression system in mammalian or animal has different requirements. They are as follows:

1. Cell line for mammalian or animal expression system
2. Transfer system of foreign gene into mammalian or animal cell
   (a) Vector-mediated gene transfer
   (b) Chemical-mediated gene transfer
3. Medial selection
4. Selection marker for mammalian or animal cells

11.5.1.1 Cell Lines for Mammalian or Animal Expression System

There are different mammalian cell lines that are used for the expression of the desired proteins. Among all cell lines, few of the popular cell lines are CHO (Chinese hamster ovary) cell line, HEK293 (adenovirus-transformed human kidney cell line), and PER.C6 cell line. Other popular cell lines are NS0 and Sp2/0 mouse myeloma cell lines (Griffin et al. 2007; Khan 2013; Bandaranayake and Almo 2014).

11.5.1.2 Chinese Hamster Ovary (CHO) Cell Line

Among all cell lines, the most popular cell line for industrial point of view is CHO. CHO cell lines were isolated by Theodore Puck in the late 1950s (Bandaranayake and Almo 2014). Initially these cell lines were used for cytogenetic studies due to their low chromosome numbers (2n = 22). Plasminogen activator (tPA) was the first recombinant therapeutic protein produced from these cell lines (Deschenes et al. 1998). The wide use of the CHO cell line expression platform is due to its incomparable adaptability allowing cell line growth in suspension cultures at higher cell densities and ease of adaptation to serum-free medium. In addition, CHO cell lines have been reported free from replication of virus such as human immunodeficiency virus (HIV), influenza, polio, herpes, and measles, thus making CHO cell lines ideal for regulatory clearance from different regulatory authorities (Bandaranayake and Almo 2014).

11.5.1.3 HEK293

HEK293 is the second leading cell lines for commercial use. It was developed by the transformation of human embryonic kidney (HEK) cells by adenovirus DNA (Ad5 serotype of adenovirus) with the help of calcium transfection technique developed by Graham and van der Eb. This transformation leads to integration of E1A and E1B genes of adenovirus into human embryonic kidney cells. There are many variants of original HEK293 cell line, such as 293N3S developed for suspension growth in fermenters for large-scale production of adenovirus and 293S that was adapted to grow under serum-free conditions. Further two additional variants of HEK293, 293-T cell line and 293-E cell line, have been developed to improve transient gene expression. These cell lines express SV40 large T antigen and Epstein-Barr virus EBNA1 protein, respectively. Further, these two cell lines support the replication of plasmids which contain SV40 or EBV oriP origin of replication and prolong the expression of target gene (Graham and van der Eb 1973; Khan 2013; Bandaranayake and Almo 2014; Dyson 2016).

Other cell lines used for gene expression for the synthesis of the desired proteins are PER.C6, DuKX-XII, CAD, P19, HT1080, etc. (Khandelia et al. 2011; Wuest et al. 2012). The mammalian cell lines could be transfected with the use of either polyethyleneimine (PEI) or calcium phosphate. Among these, a different mammalian cell line, human embryonic kidney 293 (HEK293), can exhibit the highest level of PEI-mediated transfection. Thus, it becomes a common choice for the recombinant protein production (Khan 2013).

11.5.1.4 Transfer System of Foreign Gene into Mammalian or Animal Cell

Expression Vectors for Mammalian or Animal Cells

Another essential component for mammalian or animal gene expression system is vector system. Vectors are self-replicating DNA construct to transport the desired DNA fragment for the production of required proteins. Vectors are the vehicles in which the desired gene is cloned and further transformed for gene expression. For the translation of heterologous gene in mammalian cells or cell lines, the vectors derived from mammalian or animal virus are used0. Such virus as Simian virus 40 (SV40), adenovirus, baculovirus, vaccinia virus, polyomavirus, herpesvirus, and papovavirus are used for vector construction. In order to construct an expression system, the essential need is to choose an efficient promoter and markers for the selection. The following vector systems are used for gene expression in mammalian or animal cell system (Khan 2013; Bandaranayake and Almo 2014).

Adenovirus-Based Vectors

Adenoviruses are non-enveloped, medium-sized virus, composed of nucleocapsid, and consist of double-stranded linear DNA. The genome of adenovirus can be used as a cloning vector. The E1 as well as E3 region of genomic DNA of adenovirus can be deleted that makes virus incompetent for replication, and thus for replication of viral genome, it depends on the host cells (Berkner 1992). An expression cassette system has been placed in the place of deleted E1 region. In this cassette, a recombinant or heterologous gene is placed along with the additional late promoter or with an exogenous promoter, e.g., cytomegalovirus (Fooks et al. 1995). Lately, pAdBM5 series vectors were constructed having enhancers upstream to late promoter region. It has been an advantage that adenovirus can grow well in suspension cell cultures that helps large-scale production. Reports are available for the adeno-associated viral vectors for recombinant use (Fooks et al. 1995).

Vaccinia Virus-Based Vectors

To eradicate smallpox virus, vaccinia virus has been utilized as a vector for development of various immunotherapeutic agents as well as vaccine production (Yuan et al. 2015). Vaccinia virus is a large orthopoxvirus having double-stranded DNA genome and a complex virion structure. Vaccinia virus multiplies in the cytoplasm of infected host cells. The host cells having vaccinia virus generated approximately 5000 virus particles in each infected cell that causes high-level expression of heterologous or recombinant proteins (Khan 2013). Thus, this vector system has been used by different industries such as Pasteur-Merieux and in human prothrombin by Immuno AG.3 for large-scale production of medically important proteins, e.g., HIV-1 rgp160 protein (Falkner et al. 1992).

Plasmid-Based Gene Expression Vectors for Mammalian Cells

Plasmids can be used for foreign gene expression in mammalian or animal cells. Plasmids are extra chromosomal DNA elements. They can be converted to vectors that carry foreign genes for its expression in mammalian host. They contain promoters from cytomegalovirus (CMV) and SV40 which are considered strong promoters for gene expression (Khan 2013).

Other vectors used for gene delivery in mammalian or animal expression system are lentiviral vectors (LVs) derived form from HIV type 1 virus and baculovirus vectors because they efficiently infect arthropods particularly insects. Two baculovirus Bombyx morinuclear polyhedrosis virus (BmNPV) and *Autographa californica* multiple nuclear polyhedrosis virus (AcMNPV) are developed as vectors and are used for expression of proteins in insect cell lines particularly from *Spodoptera frugiperda* (e.g., Sf9, Sf21) and silkworm (Khan 2013; Bandaranayake and Almo 2014; Dalton and Barton 2014).

Chemical-Mediated Gene Transfer

Another approach to deliver the foreign gene to mammalian or animal cells is chemical-mediated gene transfer. This system uses in expensive transfection agents such as PEI and calcium phosphate. For large-scale transient gene expression, calcium phosphate and PEI-mediated gene transfer have been shown to be successful. Under suitable conditions, DNA transfer occurs by the formation of calcium phosphate and PEI complexes, and such complexes with foreign DNA are endocytosed by the cells. Further the transfection efficiency of PEI can be increased with the help of suitable cultivation medium having dextran, iron (III) citrate, and heparin sulfate. PEI is one of the most suitable and common chemical agents used as transfecting agent and has a long successful record (Derouazi et al. 2004; Jin et al. 2014; Dyson 2016).

11.5.1.5 Media Formulation for Gene Expression in Mammalian or Animal Cells

A number of media formulations are available for expression cell lines. For example, Dulbecco's Modified Eagle's Medium contains different ingredients such as sodium pyruvate, glucose, and 5–10% fetal bovine serum (FBS) for the growth of HEK293 cell lines. Another medium used for the suspension cell culture growth is Joklik's modified minimal essential medium that contains 5–10% either horse serum or fetal bovine serum. The selection of media depends upon the type of cell culture, method of cell culture either suspension or adherent, and the absence or presence of affinity tags for purification of produced proteins (Graham 1987; Racher et al. 1995). These days, serum-free media are used for suspension cell culture. Cell clumping can be minimized with the use of serum-free media in comparison with the use of classical media with serum. There is a problem in the use of serum-free media because defined chemicals for serum-free media contain hydrolysates of plant or animal proteins apart from purified transferrin or insulin (Derouazi et al. 2004; van der Valk et al. 2010). In past, most of the mammalian or animal cell lines use either fetal bovine serum or bovine serum (5–10%). However, only 10% bovine serum provides best cell growth rates and viability of cultured cells, and fetal bovine serum (5%) is usually added for regular use. By using HEK293, horse serum (10%) can provide good cell growth and viability in liquid culture (Graham 1987). Despite all these, only fetal bovine serum is extensively used for overexpression and for high growth and for better cell viability rate to obtain high yield of the desired protein. In addition, to increase protein expression, culture medium can be supplemented with other chemical components like sodium butyrate or valproic acid. Addition of sodium butyrate provides around 50% increase in protein yield by blocking histone deacetylation (Allen et al. 2008).

11.5.1.6 Selection Marker for Transformed Mammalian or Animal Cells

A suitable marker is also necessary for the selection of mammalian cells or animals cells that contain the desired gene for the desired protein production. The most common selection reagents are mainly antibiotics, for example, neomycin, puromycin, hygromycin, blasticidin S, and zeocin. Neomycin and hygromycin are the most frequently used drugs for the selection of cloned mammalian cells that contain the desired gene for expression, and these drug makers are used in a number of commercially available vectors as selection markers (Becker and Cooper 2013). Hygromycin is the most efficient drug that kills undesired cells within few days which results in high number of clones representing transgene or gene of interest; however, puromycin is also frequently used as a selection marker. Puromycin resistance is provided by enzyme puromycin N-acetyltransferase, and in the case of hygromycin, it is provided by hygromycin B phosphotransferase (Dalton and Barton 2014). It has been recently identified that zeocin is one of the best agents for the selection during development of human cells/cell lines in comparison with neomycin, hygromycin, and puromycin (Lanza et al. 2013).

Reporter Markers for Gene Expression

Reporter makers can help to monitor the expression of gene in mammalian or animal cells or in vivo. Green fluorescent protein (GFP) acts as a reporter marker for monitoring of gene or expressed protein in vivo. GFP when used as a C-terminal fusion protein, it is utilized for the extension of expressed protein as well as for accumulation. GFP is not good for the proteins that are secreted, but several laboratories co-expressed this fluorescent protein, and intracellular fluorescence of this protein can also help to identify expression of gene (Dalton and Barton 2014).

Fusion Tags

Fusion tags are used for purification of expressed proteins and to maintain proper folding and stability of the proteins. Cytoplasmic proteins are expressed with fusion proteins, but only few fusion tags are available for secreted proteins; commonly used fusion tags are constant domain or crystallized fragment of IgG, histidine tag, etc. (Carter et al. 2010; Zhang et al. 2010). Fc tag protein is fused with carboxyl-terminus of the expressed protein which increases expression of protein. Histidine is also commonly used for purification of protein from media and also increased protein expression (Zhang et al. 2010).

Advances in Gene Expression in Animal or Mammalian Cells

The selection of transformed cells using drugs after transient transfection has been the traditional method for foreign DNA delivery and for creation of producer cell lines and further production of recombinant proteins in mammalian or animal cells. These approaches are successful but often take more time 4–6 months for obtaining producer cell lines. In the last decade different approaches have been developed to enhance the efficiency of delivery and reduce time for obtaining producer cell lines such as transient gene expression, recombinase-mediated cassette exchange, and viral delivery. Another strategy for foreign gene expression is transgenic animals system. The foreign gene can be transferred to fertilized animal ovum for expression. But the main disadvantage of use of transgenic animals for gene expression is that it requires longer period and has low yield and high cost (Soler et al. 2005; Maksimenko et al. 2013).

Application of Mammalian or Animal Cell Expression System

The mammalian or animal cell expression systems are used for the production of various therapeutic agents like monoclonal antibodies, various hormones, tissue plasminogen activator, follicle-stimulating hormone, interferon, viral vaccine production, etc.

Conclusion

E. coli expression system is the most suitable expression system among prokaryotic gene expression systems for the production of heterologous proteins. Different strategies make *E. coli* a good expression system such as *E. coli* expression system having strong promoter system like lac promoter and T7 RNA polymerase system. Further, it also has good replicon and selection system to control plasmid copy number and to select transformed cells such as pMNB1 and CO1E1 replicons (Rosano and Ceccarelli 2014). In yeasts, the foreign gene expression can be achieved by the use of episomal vector system having good origin of replication (Ars1) and strong promoters (GAPDH, PGK, TP1, GAL1, etc.) (Mattanovich et al. 2012). Plant cells provided cost-effective expression system to produce recombinant proteins. Foreign gene expression in plant can be achieved by the use of plant suspension culture, stable transgenic plants, and transient expression system. Among them transient expression system is the most suitable and can be achieved by using strategies like agroinfiltration and viral vector system (Twyman et al. 2003; Daniell et al. 2009). Animal or mammalian expression system introduces proper folding of proteins, posttranslational modifications, and final protein product which is required for biological function of the recombinant proteins. Different expression cell lines have been developed for the production of heterologous proteins such as HEK293 and CHO, and also there are good expression vector systems for animal or mammalian cells such as pAdBM5 vector (Khan 2013). The transgenic animal strategies can be employed for the expression of heterologous proteins, but it requires longer production period and has low yield and high cost.

Opinion

The information compiled in this chapter encompasses a number of different strategies of gene expression or homologous and heterologous protein production in prokaryotic, low eukaryotic, and higher eukaryotic cells. Among prokaryotic cell expression system, *E. coli* expression system is the most convenient due to its rapid growth in inexpensive medium, genetic manipulation, and low downstream protein purification system. But it still has a major disadvantage; it does not glycosylate heterologous proteins. But this problem can be solved by the use of another bacterial expression system that can perform protein glycosylation or eukaryotic protein glycosylation pathway and can be engineered in *E. coli*. Further, to obtain glycosylate proteins, yeast cells can be used as expression system especially *S. cerevisiae* because it can be easily genetically manipulated and have complete genetic map. The combination of recent development in synthetic biology can be utilized to engineer *S. cerevisiae* strain that efficiently increase the expression of foreign genes, whereas plant transient expression system is preferred for the production of low-cost edible vaccines and antibodies. In the last 20 years, protein expression system from animal or mammalian cell has been used for the production of biomedical therapeutic proteins due to their exact posttranslational modification and proper folding of proteins to maintain their biological activities. Still, there is need of newer cell lines and efficient vectors and technologies to express the foreign genes to attain good biological properties or activities of the expressed proteins. Protein engineering or expression technologies are still in their infancy to get the desired set of proteins. So there is need to achieve such technologies to construct such strains of bacteria, yeast, and animal cell lines to get efficiently active proteins to meet the demand in future in a cost-effective manner.

Acknowledgment We are thankful to the Department of Biotechnology and Bioinformatics, Jaypee University of Information Technology, for providing facilities.

References

Allen MJ, Boyce JP, Trentalange MT, Treiber DL, Rasmussen B, Tillotson B, Davis R, Reddy P (2008) Identification of novel small molecule enhancers of protein production by cultured mammalian cells. Biotechnol Bioeng 100:1193–1204. doi:10.1002/bit.21839

Arnau J, Lauritzen C, Petersen GE, Pedersen J (2006) Current strategies for the use of affinity tags and tag removal for the purification of recombinant proteins. Protein Expr Purif 48:1–13. doi:10.1016/j.pep.2005.12.002

Baeshen MN, Al-Hejin AM, Bora RS, Ahmed MM, Ramadan HA, Saini KS, Baeshen NA, Redwan EM (2015) Production of biopharmaceuticals in E. coli: Current scenario and future perspectives. J Microbiol Biotechnol 25:953–962. doi:10.4014/jmb.1412.12079

Bandaranayake AD, Almo SC (2014) Recent advances in mammalian protein production. FEBS Lett 588:253–260. doi:10.1016/j.febslet.2013.11.035

Becker B, Cooper MA (2013) Aminoglycoside antibiotics in the 21st century. ACS Chem Biol 8:105–115. doi:10.1021/cb3005116

Bentley WE, Mirjalili N, Andersen DC, Davis RH, Kompala DS (1990) Plasmid-encoded protein: the principal factor in the "metabolic burden" associated with recombinant bacteria. Biotechnol Bioeng 35:668–681. doi:10.1002/bit.260350704
Berkner KL (1992) Expression of heterologous sequences in adenoviral vectors. In: Muzyczka N (ed) Viral expression vectors. Springer, Berlin, pp 39–66. 978-3-642-75608-5
Blommel PG, Fox BG (2007) A combined approach to improving large-scale production of tobacco etch virus protease. Protein Expr Purif 55:53–68. doi:10.1016/j.pep.2007.04.013
Both L, van Dolleweerd C, Wright E, Banyard AC, Bulmer-Thomas B, Selden D, Altmann F, Fooks AR, Ma JK (2013) Production, characterization, and antigen specificity of recombinant 62-71-3, a candidate monoclonal antibody for rabies prophylaxis in humans. FASEB J 27:2055–2065. doi:10.1096/fj.12-219964
Camps M (2010) Modulation of ColE1-like plasmid replication for recombinant gene expression. Recent Pat DNA Gene Seq 4:58–73
Canto T (2016) Transient expression systems in plants: potentialities and constraints. In: Vega MC (ed) Advanced technologies for protein complex production and characterization. Springer, Cham, pp 287–301. doi:10.1007/978-3-319-27216-0_18
Carter J, Zhang J, Dang TL, Hasegawa H, Cheng JD, Gianan I, O'Neill JW, Wolfson M, Siu S, Qu S, Meininger D, Kim H, Delaney J, Mehlin C (2010) Fusion partners can increase the expression of recombinant interleukins via transient transfection in 2936E cells. Protein Sci 19:357–362. doi:10.1002/pro.307
Casali N (2003) Escherichia coli host strains. In: Casali N, Preston A (eds) E coli plasmid vectors, Methods in molecular biology. Humana, Clifton, pp 27–48. doi:10.1385/1-59259-409-3:27
Chakravartty V, Cronan JE (2015) A series of medium and high copy number arabinose-inducible *Escherichia coli* expression vectors compatible with pBR322 and pACYC184. Plasmid 81:21–26. doi:10.1016/j.plasmid.2015.03.001
Chuang LT, Chen DC, Nicaud JM, Madzak C, Chen YH, Huang YS (2010) Co-expression of heterologous desaturase genes in *Yarrowia lipolytica*. N Biotechnol 27:277–282. doi:10.1016/j.nbt.2010.02.006
Cos O, Ramon R, Montesinos JL, Valero F (2006) Operational strategies, monitoring and control of heterologous protein production in the methylotrophic yeast *Pichia pastoris* under different promoters: a review. Microb Cell Fact 5:17. doi:10.1186/1475-2859-5-17
Daegelen P, Studier FW, Lenski RE, Cure S, Kim JF (2009) Tracing ancestors and relatives of Escherichia coli B, and the derivation of B strains REL606 and BL21(DE3). J Mol Biol 394:634–643. doi:10.1016/j.jmb.2009.09.022
Dalton AC, Barton WA (2014) Over-expression of secreted proteins from mammalian cell-lines. Protein Sci 23:517–525. doi:10.1002/pro.2439
Daniell H, Singh ND, Mason H, Streatfield SJ (2009) Plant-made vaccine antigens and biopharmaceuticals. Trends Plant Sci 14:669–679. doi:10.1016/j.tplants.2009.09.009
de Boer HA, Comstock LJ, Vasser M (1983) The tac promoter: a functional hybrid derived from the trp and lac promoters. Proc Natl Acad Sci U S A 80:21–25
del Solar G, Espinosa M (2000) Plasmid copy number control: an ever-growing story. Mol Microbiol 37:492–500
Derouazi M, Girard P, Van Tilborgh F, Iglesias K, Muller N, Bertschinger M, Wurm FM (2004) Serum-free large-scale transient transfection of CHO cells. Biotechnol Bioeng 87:537–545. doi:10.1002/bit.20161
Deschenes I, Finkle CD, Winocour PD (1998) Effective use of BCH-2763, a new potent injectable direct thrombin inhibitor, in combination with tissue plasminogen activator (tPA) in a rat arterial thrombolysis model. Thromb Haemost 80:186–191
Dubendorff JW, Studier FW (1991) Controlling basal expression in an inducible T7 expression system by blocking the target T7 promoter with lac repressor. J Mol Biol 219:45–59
Dyson MR (2016) Fundamentals of expression in mammalian cells. In: Vega MC (ed) Advanced technologies for protein complex production and characterization. Springer, Cham, pp 217–224. doi:10.1007/978-3-319-27216-0_14

Eckardt NA (2004) Host proteins guide *Agrobacterium*-mediated plant transformation. Plant Cell 16:2837–2839

Epstein W, Rothman-Denes LB, Hesse J (1975) Adenosine 3′:5′-cyclic monophosphate as mediator of catabolite repression in *Escherichia coli*. Proc Natl Acad Sci U S A 72:2300–2304

Fair RJ, Tor Y (2014) Antibiotics and bacterial resistance in the 21st century. Perspect Medicin Chem 6:25–64. doi:10.4137/pmc.s14459

Falkner FG, Turecek PL, MacGillivray RT, Bodemer W, Scheiflinger F, Kandels S, Mitterer A, Kistner O, Barrett N, Eibl J (1992) High level expression of active human prothrombin in a vaccinia virus expression system. Thromb Haemost 68:119–124

Ferrer A, Arró M, Manzano D, Altabella T (2016) Strategies and methodologies for the co-expression of multiple proteins in plants. In: Vega MC (ed) Advanced technologies for protein complex production and characterization. Springer, Cham, pp 263–285. doi:10.1007/978-3-319-27216-0_17

Fischer R, Drossard J, Commandeur U, Schillberg S, Emans N (1999) Towards molecular farming in the future: moving from diagnostic protein and antibody production in microbes to plants. Biotechnol Appl Biochem 30:101–108

Fischer R, Stoger E, Schillberg S, Christou P, Twyman RM (2004) Plant-based production of biopharmaceuticals. Curr Opin Plant Biol 7:152–158. doi:10.1016/j.pbi.2004.01.007

Fooks AR, Schadeck E, Liebert UG, Dowsett AB, Rima BK, Steward M, Stephenson JR, Wilkinson GW (1995) High-level expression of the measles virus nucleocapsid protein by using a replication-deficient adenovirus vector: induction of an MHC-1-restricted CTL response and protection in a murine model. Virology 210:456–465. doi:10.1006/viro.1995.1362

Fernández FJ, Vega MC (2016) Choose a suitable expression host: a survey of available protein production platforms. In: Vega MC (ed) Advanced technologies for protein complex production and characterization. Springer, Cham, pp 15–24. doi:10.1007/978-3-319-27216-0_2

Gelvin SB (2003) Agrobacterium-mediated plant transformation: the biology behind the "gene-jockeying" tool. Microbiol Mol Biol Rev 67:16–37

Gohlke J, Deeken R (2014) Plant responses to *Agrobacterium tumefaciens* and crown gall development. Front Plant Sci 5:155. doi:10.3389/fpls.2014.00155

Gottesman S (1996) Proteases and their targets in *Escherichia coli*. Annu Rev Genet 30:465–506. doi:10.1146/annurev.genet.30.1.465

Graham FL (1987) Growth of 293 cells in suspension culture. J Gen Virol 68(Pt 3):937–940. doi:10.1099/0022-1317-68-3-937

Graham FL, van der Eb AJ (1973) A new technique for the assay of infectivity of human adenovirus 5 DNA. Virology 52:456–467. doi:10.1016/0042-6822(73)90341-3

Griffin TJ, Seth G, Xie H, Bandhakavi S, WS H (2007) Advancing mammalian cell culture engineering using genome-scale technologies. Trends Biotechnol 25:401–408. doi:10.1016/j.tibtech.2007.07.004

Grossman TH, Kawasaki ES, Punreddy SR, Osburne MS (1998) Spontaneous cAMP-dependent derepression of gene expression in stationary phase plays a role in recombinant expression instability. Gene 209:95–103

Guengerich L, Kang HA, Behle B, Gellissen G, Suckow M (2004) A platform for heterologous gene expression based on the methylotrophic yeast *Hansenula polymorpha*. In: Kück U (ed) Genetics and biotechnology, The Mycota, vol 2. Springer, Berlin, pp 273–287. doi:10.1007/978-3-662-07426-8_14

Hao X, Xie P, Johnstone L, Miller SJ, Rensing C, Wei G (2012) Genome sequence and mutational analysis of plant-growth-promoting bacterium *Agrobacterium tumefaciens* CCNWGS0286 isolated from a zinc-lead mine tailing. Appl Environ Microbiol 78:5384–5394. doi:10.1128/aem.01200-12

Jin L, Zeng X, Liu M, Deng Y, He N (2014) Current progress in gene delivery technology based on chemical methods and nano-carriers. Theranostics 4:240–255. doi:10.7150/thno.6914

Kapila J, De Rycke R, Van Montagu M, Angenon G (1997) An *Agrobacterium*-mediated transient gene expression system for intact leaves. Plant Sci 122:101–108. doi:10.1016/S0168-9452(96)04541-4

Kawai S, Hashimoto W, Murata K (2010) Transformation of *Saccharomyces cerevisiae* and other fungi: methods and possible underlying mechanism. Bioeng Bugs 1:395–403. doi:10.4161/bbug.1.6.13257

Khan F, Legler PM, Mease RM, Duncan EH, Bergmann-Leitner ES, Angov E (2012) Histidine affinity tags affect MSP1(42) structural stability and immunodominance in mice. Biotechnol J 7:133–147. doi:10.1002/biot.201100331

Khan KH (2013) Gene expression in mammalian cells and its applications. Adv Pharm Bull 3:257–263. doi:10.5681/apb.2013.042

Khandelia P, Yap K, Makeyev EV (2011) Streamlined platform for short hairpin RNA interference and transgenesis in cultured mammalian cells. Proc Natl Acad Sci U S A 108:12799–12804. doi:10.1073/pnas.1103532108

Kikkert JR, Vidal JR, Reisch BI (2005) Stable transformation of plant cells by particle bombardment/biolistics. Methods Mol Biol 286:61–78

Korpimaki T, Kurittu J, Karp M (2003) Surprisingly fast disappearance of beta-lactam selection pressure in cultivation as detected with novel biosensing approaches. J Microbiol Methods 53:37–42

Kroll J, Klinter S, Schneider C, Voss I, Steinbuchel A (2010) Plasmid addiction systems: perspectives and applications in biotechnology. J Microbial Biotechnol 3:634–657. doi:10.1111/j.1751-7915.2010.00170.x

Lanza AM, Kim do S, Alper HS (2013) Evaluating the influence of selection markers on obtaining selected pools and stable cell-lines in human cells. Biotechnol J 8:811–821. doi:10.1002/biot.201200364

Lee LY, Gelvin SB (2008) T-DNA binary vectors and systems. Plant Physiol 146:325–332. doi:10.1104/pp.107.113001

Leuzinger K, Dent M, Hurtado J, Stahnke J, Lai H, Zhou X, Chen Q (2013) Efficient agroinfiltration of plants for high-level transient expression of recombinant proteins. J Vis Exp. doi:10.3791/50521

Li F, Vijayasankaran N, Shen AY, Kiss R, Amanullah A (2010) Cell culture processes for monoclonal antibody production. MAbs 2:466–479

Lichty JJ, Malecki JL, Agnew HD, Michelson-Horowitz DJ, Tan S (2005) Comparison of affinity tags for protein purification. Protein Expr Purif 41:98–105. doi:10.1016/j.pep.2005.01.019

Lico C, Chen Q, Santi L (2008) Viral vectors for production of recombinant proteins in plants. J Cell Physiol 216:366–377. doi:10.1002/jcp.21423

Liu YC, Vidali L (2011) Efficient polyethylene glycol (PEG) mediated transformation of the moss Physcomitrella patens. J Vis Exp. doi:10.3791/2560

Maksimenko OG, Deykin AV, Khodarovich YM, Georgiev PG (2013) Use of transgenic animals in biotechnology: prospects and problems. Acta Naturae 5:33–46

Mattanovich D, Branduardi P, Dato L, Gasser B, Sauer M, Porro D (2012) Recombinant protein production in yeasts. Methods Mol Biol 824:329–358. doi:10.1007/978-1-61779-433-9_17

Mokdad-Gargouri R, Abdelmoula-Soussi S, Hadiji-Abbes N, Amor IY, Borchani-Chabchoub I, Gargouri A (2012) Yeasts as a tool for heterologous gene expression. Methods Mol Biol 824:359–370. doi:10.1007/978-1-61779-433-9_18

Nakagawa T, Inagaki A, Ito T, Fujimura S, Miyaji T, Yurimoto H, Kato N, Sakai Y, Tomizuka N (2006) Regulation of two distinct alcohol oxidase promoters in the methylotrophic yeast *Pichia methanolica*. Yeast 23:15–22. doi:10.1002/yea.1334

Nasser MW, Pogoja V, Abdin MZ, Jain SK (2003) Evaluation of yeast as an expression system. Ind J Biotechnol 2:477–493

Nie Y, Chaillet M, Becke C, Haffke M, Pelosse M, Fitzgerald D, Collinson I, Schaffitzel C, Berger I (2016) ACEMBL tool-kits for high-throughput multigene delivery and expression in prokaryotic and eukaryotic hosts. In: Vega MC (ed) Advanced technologies for protein complex production and characterization. Springer, Cham, pp 27–42. doi:10.1007/978-3-319-27216-0_3

Nordstrom K (2006) Plasmid R1—replication and its control. Plasmid 55:1–26. doi:10.1016/j.plasmid.2005.07.002

Ogawa Y, Tatsumi H, Murakami S, Ishida Y, Murakami K, Masaki A, Kawabe H, Arimura H, Nakano E, Motai H (1990) Secretion of *Aspergillus oryzae* alkaline protease in an osmophilic yeast, *Zygosaccharomyces rouxii*. Agric Biol Chem 54:2521–2529

Pan SH, Malcolm BA (2000) Reduced background expression and improved plasmid stability with pET vectors in BL21 (DE3). Biotechniques 29:1234–1238

Pogue GP, Lindbo JA, Garger SJ, Fitzmaurice WP (2002) Making an ally from an enemy: plant virology and the new agriculture. Annu Rev Phytopathol 40:45–74. doi:10.1146/annurev.phyto.40.021102.150133

Popa C, Coll NS, Valls M, Sessa G (2016) Yeast as a heterologous model system to uncover type III effector function. PLoS Pathog 12(2):e1005360. doi:10.1371/journal.ppat.1005360

Pope B, Kent HM (1996) High efficiency 5 min transformation of *Escherichia coli*. Nucleic Acids Res 24:536–537

Potter H, Heller R (2010) Transfection by electroporation. In: Current protocols in molecular biology. Wiley, Hoboken, pp 9.3.1–9.3.10. doi:10.1002/0471142727.mb0903s92

Racher A, Fooks A, Griffiths JB (1995) Culture of 293 cells in different culture systems: cell growth and recombinant adenovirus production. Biotechnol Tech 9:169–174. doi:10.1007/bf00157073

Rai M, Padh H (2001) Expression systems for production of heterologous proteins. Curr Sci 80:1121–1128

Reichenstein M, Gottlieb H, Damari GM, Iavnilovitch E, Barash I (2001) A new beta-lactoglobulin-based vector targets luciferase cDNA expression to the mammary gland of transgenic mice. Transgenic Res 10:445–456

Reiser J, Glumoff V, Kälin M, Ochsner U (1990) Transfer and expression of heterologous genes in yeasts other than Saccharomyces cerevisiae. In: Applied molecular genetics, Advances in biochemical engineering/biotechnology, vol 43. Springer, Berlin, pp 75–102. doi:10.1007/BFb0009080

Roberts MC (1996) Tetracycline resistance determinants: mechanisms of action, regulation of expression, genetic mobility, and distribution. FEMS Microbiol Rev 19:1–24

Rocha SN, Abrahao-Neto J, Cerdan ME, Gombert AK, Gonzalez-Siso MI (2011) Heterologous expression of a thermophilic esterase in Kluyveromyces yeasts. Appl Microbiol Biotechnol 89:375–385. doi:10.1007/s00253-010-2869-8

Rosano GL, Ceccarelli EA (2014) Recombinant protein expression in Escherichia coli: advances and challenges. Front Microbiol 5:172. doi: 10.3389/fmicb.2014.00172

Sainsbury F, Lomonossoff GP (2014) Transient expressions of synthetic biology in plants. Curr Opin Plant Biol 19:1–7. doi: http://dx.doi.org/10.1016/j.pbi.2014.02.003

Sathiamoorthy S, Shin JA (2012) Boundaries of the origin of replication: creation of a pET-28a-derived vector with p15A copy control allowing compatible coexistence with pET vectors. PLoS One 7:e47259. doi:10.1371/journal.pone.0047259

Sezonov G, Joseleau-Petit D, D'Ari R (2007) *Escherichia coli* physiology in Luria-Bertani broth. J Bacteriol 189:8746–8749. doi:10.1128/jb.01368-07

Shaw WV (1983) Chloramphenicol acetyltransferase: enzymology and molecular biology. CRC Crit Rev Biochem 14:1–46

Shankar J (2013) An overview of toxins in Aspergillus associated with pathogenesis. Int J Life Sci Biotechnol Pharm Res 2:16–31

Shanmugaraj BM, Ramalingam S (2014) Plant expression platform for the production of recombinant pharmaceutical proteins. Austin J. Biotechnol Bioeng 1:4

Sigar M, Maity N, Mishra S (2016) Enhancing granulocyte colony stimulating factor expression in Pichia pastoris via fusion with human serum albumin. Prep Biochem Biotechnol 47(4):364–370. doi:10.1080/10826068.2016.1252922

Shiloach J, Fass R (2005) Growing E. coli to high cell density—a historical perspective on method development. Biotechnol Adv 23:345–357. doi:10.1016/j.biotechadv.2005.04.004

Soler E, Le Saux A, Guinut F, Passet B, Cohen R, Merle C, Charpilienne A, Fourgeux C, Sorel V, Piriou A, Schwartz-Cornil I, Cohen J, Houdebine LM (2005) Production of two vaccinating recombinant rotavirus proteins in the milk of transgenic rabbits. Transgenic Res 14:833–844. doi:10.1007/s11248-005-1771-0

Studier FW (1991) Use of bacteriophage T7 lysozyme to improve an inducible T7 expression system. J Mol Biol 219:37–44

Tague BW, Mantis J (2006) In planta Agrobacterium-mediated transformation by vacuum infiltration. Methods Mol Biol 323:215–223. doi:10.1385/1-59745-003-0:215

Terpe K (2006) Overview of bacterial expression systems for heterologous protein production: from molecular and biochemical fundamentals to commercial systems. Appl Microbiol Biotechnol 72:211–222. doi:10.1007/s00253-006-0465-8

Thakur R, Shankar J (2016) In silico analysis revealed high-risk single nucleotide polymorphisms in human Pentraxin-3 Gene and their impact on innate immune response against microbial pathogens. Front Microbiol 7:192. doi:10.3389/fmicb.2016.00192

Tiwari S, Thakur R, Shankar J (2015) Role of heat-shock proteins in cellular function and in the biology of fungi. Biotechnol Res Int 2015:132635., 11 p. doi:10.1155/2015/132635

Tiwari S, Thakur R, Goel G, Shankar J (2016) Nano-LC-Q-TOF analysis of proteome revealed germination of Aspergillus flavus conidia is accompanied by MAPK signalling and cell wall modulation. Mycopathologia:1–18. doi:10.1007/s11046-016-0056-x

Ton VK, Rao R (2004) Functional expression of heterologous proteins in yeast: insights into Ca^{2+} signaling and Ca^{2+}-transporting ATPases. Am J Physiol Cell Physiol 287:C580–C589. doi:10.1152/ajpcell.00135.2004

Twyman RM, Stoger E, Schillberg S, Christou P, Fischer R (2003) Molecular farming in plants: host systems and expression technology. Trends Biotechnol 21:570–578. doi:10.1016/j.tibtech.2003.10.002

Ullmann A, Tillier F, Monod J (1976) Catabolite modulator factor: a possible mediator of catabolite repression in bacteria. Proc Natl Acad Sci U S A 73:3476–3479

van der Valk J, Brunner D, De Smet K, Fex Svenningsen A, Honegger P, Knudsen LE, Lindl T, Noraberg J, Price A, Scarino ML, Gstraunthaler G (2010) Optimization of chemically defined cell culture media—replacing fetal bovine serum in mammalian in vitro methods. Toxicol In Vitro 24:1053–1063. doi:10.1016/j.tiv.2010.03.016

Waugh DS (2011) An overview of enzymatic reagents for the removal of affinity tags. Protein Expr Purif 80:283–293. doi:10.1016/j.pep.2011.08.005

Wright A, Maundrell K, Heyer WD, Beach D, Nurse P (1986) Vectors for the construction of gene banks and the integration of cloned genes in *Schizosaccharomyces pombe* and *Saccharomyces cerevisiae*. Plasmid 15:156–158. doi:10.1016/0147-619X(86)90051-X

Wuest DM, Harcum SW, Lee KH (2012) Genomics in mammalian cell culture bioprocessing. Biotechnol Adv 30:629–638. doi:10.1016/j.biotechadv.2011.10.010

Yin J, Li G, Ren X, Herrler G (2007) Select what you need: a comparative evaluation of the advantages and limitations of frequently used expression systems for foreign genes. J Biotechnol 127:335–347. doi:10.1016/j.jbiotec.2006.07.012

Yuan M, Gao X, Chard LS, Ali Z, Ahmed J, Li Y, Liu P, Lemoine NR, Wang Y (2015) A marker-free system for highly efficient construction of vaccinia virus vectors using CRISPR Cas9. Mol Ther Methods Clin Dev 2:15035. doi:10.1038/mtm.2015.35

Yurimoto H, Sakai Y (2009) Methanol-inducible gene expression and heterologous protein production in the methylotrophic yeast *Candida boidinii*. Biotechnol Appl Biochem 53:85–92. doi:10.1042/ba20090030

Yu XW, Xu Y, Xiao R (2016) Lipases from the genus Rhizopus: characteristics, expression, protein engineering and application. Prog Lipid Res 64:57–68. doi:10.1016/j.plipres.2016.08.001

Zhang J, Carter J, Siu S, O'Neill JW, Gates AH, Delaney J, Mehlin C (2010) Fusion partners as a tool for the expression of difficult proteins in mammalian cells. Curr Pharm Biotechnol 11:241–245

Green Chemistry Approach Towards Nanoparticle Synthesis

12

Neetu Phougat, Manish Kumar, Reena V. Saini, and Anil Kumar Chhillar

Abstract

Nanoparticle synthesis using plant extracts have numerous applications in different fields such as therapeutics (antimicrobials, antiviral, anticancerous), industries (catalyst, dye degradation), biosensors, agriculture, nanocomposites, etc. Plants contain primary (carbohydrates, proteins) as well as secondary metabolites (alkaloids, tannins, saponins, flavonoids, phenols, etc.) which not only have medicinal values but also possess potential for nanoparticle synthesis. Plant metabolites act as reducing and capping agents for nanoparticles. The plant extract-based synthesis is economical and controlled in terms of shape and size of nanoparticles. The plant-based nanoparticles are free from toxic contaminants and, thus, more compatible for therapeutic applications. Plant-mediated nanoparticle synthesis is one of the best alternatives, economically as well as environmentally.

Keywords

Antimicrobial • Anticancerous • Green chemistry • Nanoparticles • Plant metabolites

N. Phougat • M. Kumar • A.K. Chhillar (✉)
Centre for Biotechnology, MD University, Rohtak, Haryana 124 001, India
e-mail: anil.chhillar@gmail.com

R.V. Saini
Animal Biotechnology Laboratory, Faculty of Applied Sciences and Biotechnology, Shoolini University, Solan, Himachal Pradesh, India

V.C. Kalia, A.K. Saini (eds.), *Metabolic Engineering for Bioactive Compounds*, DOI 10.1007/978-981-10-5511-9_12

12.1 Introduction

The story began in 1959 at the meeting of American Physical Society, when a Nobel Prize winner, Richard A. Frey, gave a lecture "There is plenty of room at the bottom". He is considered as the father of the nanotechnology idea (Feynman 1992). The word "nano" originates from Greek and means "dwarf". Reducing the overall dimensions of a single particle to nanoscale changes its properties and gives it unique physical, chemical, and biological features. 100 nm is a threshold dimension. Beneath this value, the relation between the surfaces of a particle to its mass is big enough to alter the properties of such particle. Nanoparticles have increased surface area as the particles size decreased considerably. Most of the atoms are open to their environment and freely accessible for reactions. At nanoscale, particles exhibit different physical, optical and chemical properties (Likus et al. 2013).

The most commercialized metal for nanoparticle synthesis is silver (Sintubin et al. 2012). Antibacterial properties of silver have been known since ancient times. In ancient Egypt, silver bars were put into water, which was drunk as a medicine for ulcers. Food, wine and other beverages had been stored in silver vessels in order to prevent them from getting spoilt. Soldiers in Roman legions used to put silver coins on their wounds to accelerate their healing. The use of silver decreased with the increasing discoveries in the field of antibiotics. However, soon the antibiotic abuse led to increasing cases of resistance among all kinds of pathogenic and non-pathogenic micro-organisms. As microbes developed resistance strategies to cope with the effect of antimicrobials, mankind has been forced again to look at therapies of ancient era, viz. therapies based on phytochemicals and noble metals like silver, gold, copper, etc. Recent advancements involved conversion of metal salts into their corresponding nanoparticles by various means, viz. chemical methods (reduction using harsh chemicals), physical methods (synthesis using physical means, e.g. laser ablation) and biological methods (using oxidizing biochemicals present in the living organisms or phytochemicals) (Borase et al. 2014). Nanoparticle synthesis via biological agents is advantageous over physical and chemical methods in terms of their eco-friendly and more economical synthesis (Shankar et al. 2004; Dahl et al. 2007). As the field of medicine is one of the main target areas of the nanoparticles, so safe and non-toxic synthesis is highly recommended to avoid any cross-reactivity.

The metal-reducing properties of plant extracts have been known since the early 1900s although the exact mechanism involved was not clearly known. During the last decades, nanoparticle synthesis via plant extracts has drawn much attention due to their less expensive and scalable nature as compared to other physical, chemical or even biological methods which involved micro-organisms. The plant extract concoction has capability to reduce the metal salts and led to formation of nanoparticles. Phytochemicals also stabilize the newly synthesized nanoparticles by acting as capping agents (Kumar and Yadav 2009; Iravani 2011; Dhillon et al. 2012; Gan and

Li 2012). The properties of the synthesized nanoparticles depend upon the plant extract used, as the phytochemical composition of different extracts is entirely different and also different concentrations are used during synthesis. Typically, the green synthesis of nanoparticles has been reported to involve mixing of preferably aqueous extracts with aqueous solution of metal salt. The reduction reaction completes in few minutes to few hours at room temperature.

This chapter is focused on the various methods of synthesis and applications of nanoparticles, mainly silver nanoparticles.

12.2 Methods of Nanoparticle Synthesis

The nanoparticles can be synthesized by either a top-down approach (size reduction by chemical or physical treatments) or a bottom-up approach (assembly of smaller particles) (Fig. 12.1) (Meyers et al. 2006; Borase et al. 2014). One of the major limitations of the top-down approach is the defects in morphology of nanoparticles which determines the surface chemistry and the other important physical properties (Thakkar et al. 2010). Ag-NPs are generally produced by chemical, physical and biological methods (Table 12.1). All the methods have their own pros and cons such as cost, time, quality, stability, flaw, etc.

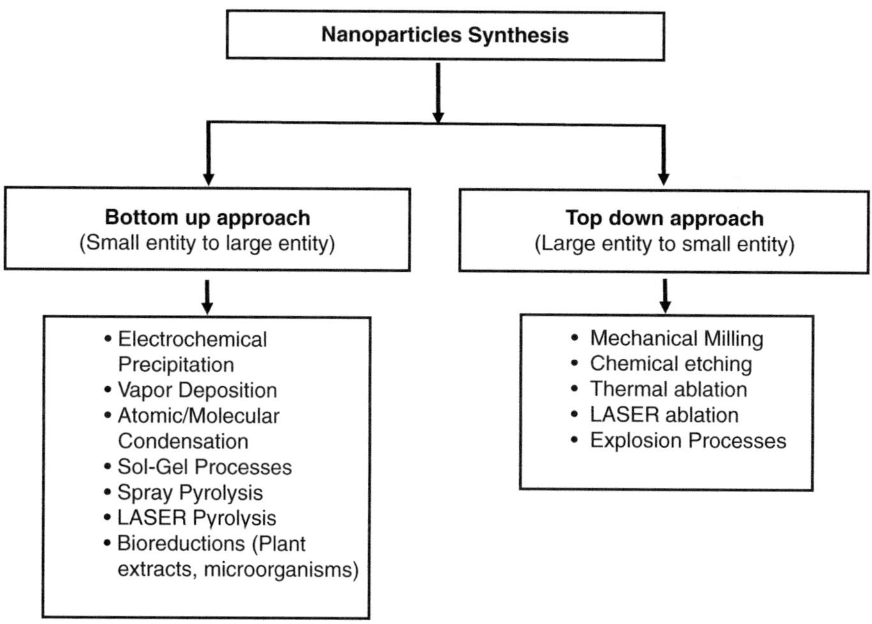

Fig. 12.1 Different approaches for synthesis of nanoparticles

Table 12.1 Comparative account of different parameters used for Ag-NP synthesis by different methods (phytochemical, microbial, chemical and physical)

Sr. no.	Parameter of synthesis	Method of synthesis			
		Phytochemical	Microbial	Chemical	Physical
1.	Reducing Agent	Proteins, phenolic compounds, alkaloids, carbohydrates, terpenoids	Proteins, enzyme (nitrate reductase), carbohydrates	Sodium borohydrate, hydrazine, trisodium citrate, polyols	Physical vapours Arc discharge, UV irradiation, gamma radiation, X-ray
2.	Stabilizing Agent	Proteins, phenolic compounds, alkaloids, saponins, terpenoids	Proteins, enzyme (nitrate reductase), starch	Polyvinyl aniline, Polyvinyl pyrrolidone (PVP), starch, sodium dodecyl sulphate (SDS), etc.	Physical vapours Arc discharge, UV irradiation, gamma radiation
3.	Minimum requirement for nanosynthcsis	Rotary shaker, flasks	Rotary shaker, flasks, media for microbial growth, autoclave, laminar airflow	Rotary shaker, flasks, precursor chemicals, temperature, pH maintenance	Sophisticated instrument for generating radiation and high temperature maintenance
4.	Time required for nanosynthesis	Depend on plant extract (from a few minutes to hours)	Depend on microbial species (more 24–120 h)	Depend on chemical route (within a few minutes)	Depend on method of synthesis (within a few minutes)
5.	Stability in aqueous media	Stable	Relatively stable	Less stable	Less stable
6.	Toxicity of reducing agents, capping agents and synthesized Ag-NPs	Less toxic (mostly non-toxic)	Toxic	Toxic	Toxic

12.2.1 Chemical Methods

The principal prerequisites for chemical synthesis of Ag-NPs are precursor (silver nitrate), reducing agent (sodium borohydrate) and capping agent (polyvinylpyrrolidone) (Ledwith et al. 2007; Xiuyan et al. 2012).

The major advantage of chemically synthesized nanoparticles is defined by shape and size. However, the use of hazardous chemicals, harsh reaction parameters (high temperature and pressure) and toxic by-products create serious environmental issues (Zhu et al. 2000; Wang et al. 2002; Brichkin et al. 2008). Further, their use is also restricted in the field of medicines as there are chances of contamination by toxic chemicals used in synthesis which may adhere on the surface of newly synthesized nanoparticles. Additionally, the chemically synthesized Ag-NPs tend to become hydrophobic. Therefore, there are stability and solubility issues in their application in living system (Mafune et al. 2000).

12.2.2 Physical Methods

Similar to chemical methods, the physical methods (plasma catalysis, laser ablation, etc.) of synthesis also employ harsh and hazardous conditions (temperature, pressure and light) and critical instruments which not only are costly but also require trained manpower for handling. The instability issue still remains the same as with chemical methods (Bae et al. 2002; Malynych and Chumanov 2003; Shahverdi et al. 2007).

12.2.3 Biological Methods

The shortcomings of chemical and physical methods of Ag-NP synthesis lead to exploration for safer, cheaper and easier alternative methods. So, we have to turn towards biological methods of nanoparticle synthesis which involves the use of micro-organisms and plants.

12.2.3.1 Ag-NP Synthesis Using Micro-organisms

Various micro-organisms and algae have been reported for their employment in Ag-NP synthesis such as *Pseudomonas stutzeri* (Klaus et al. 1999), *Lactobacillus* spp. (Nair and Pradeep 2002), *Cochliobolus lunatus* (Salunkhe et al. 2011), *Fusarium oxysporum* (Duran et al. 2007), *Trichoderma* spp. (Thakkar et al. 2010), *Aspergillus* spp. (Binupriya et al. 2010) and *Verticillium* spp. (Mukherjee et al. 2001), *Sargassum wightii* Greville, *Chaetomorpha linum* (Kannan et al. 2013) and *Padina gymnospora*. Main drawbacks associated with the employment of micro-organisms are costly media for the maintenance of microbial culture, pH and temperature sensitivity, lengthy and time-consuming purification process of Ag-NPs (through cell lysis) and contamination of Ag-NPs with micro-organisms adherence on the surface. So, their uses are also restricted in medicinal formulations due to risks of infection.

12.2.3.2 Ag-NP Synthesis Using Plant Extracts: Green Chemistry Approach

Ag-NP synthesis via green chemistry approach has many promising applications in various fields. The sizes of these Ag-NPs may vary from 2 to 4000 nm, whereas the shapes may vary from round to rod and rectangular to polygonal. Plant families

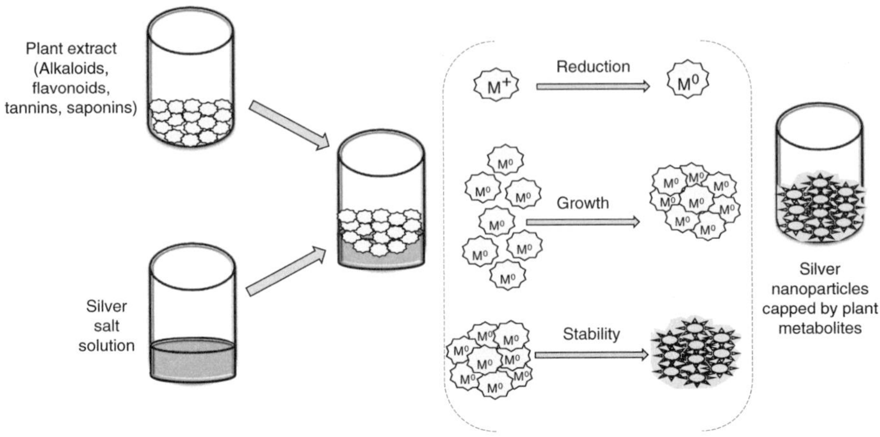

Fig. 12.2 Schematic representation of green synthesis of silver nanoparticles

like Euphorbiaceae, Rutaceae, Poaceae, Myrtaceae, Solanaceae and Asteraceae seem to have great potential for synthesizing Ag-NPs. Euphorbiaceae family-mediated Ag-NPs have good antibacterial and insecticidal activities (Agrawal et al. 2008; Krishnaraj et al. 2010). Further, the Apocynaceae family-mediated Ag-NPs have been found to possess anti-*plasmodial* and mosquitocidal activity. A large number of studies conducted have a vast amount of information which is necessary to synthesize Ag-NPs of definite shape, size and activity (Mittal et al. 2013). Figure 12.2 shows an illustrative outline of synthesis of Ag-NPs using plant extracts.

Large-scale nanoparticle synthesis via green chemistry approach has been considered to be comparatively easy and less time-consuming as plant extract isolation and purification processes are efficient and feasible. Various studies on the extract such as *Cibotium barometz* root and *Musa paradisiaca* peel have shown antioxidant, antimicrobial activity against *Escherichia*, *Staphylococcus aureus*, *Salmonella enterica*, *Pseudomonas aeruginosa* and multiple antibiotic-resistant (MAR) *Enterococcus faecalis* (Vijayakumar et al. 2016; Wang et al. 2016). Studies on *Azadirachta indica* concluded that the rate of metal ion reduction has been found to be faster as compared to reduction by micro-organisms (Mohanpuria et al. 2008; Iravani 2011). Modifications in the genetic makeup of plants using genetic engineering approach for increased production of the plant metabolites which are responsible for nanoparticle synthesis can further make plant-based synthesis more effective. The plant-based reducing and capping agents generate uniform, stable and well-dispersed nanoparticles as compared to other microbial or physicochemical methods. These Ag-NPs have been reported to be more suitable for therapeutic purposes (Mohanty et al. 2013; Borase et al. 2013).

Ag-NPs of required morphology can be easily acquired by using different concentrations and proportions of plant extract and precursor with variation in other parameters, viz. reaction time, temperature, etc. Hence, plant-mediated synthesis is more flexible to achieve desired nanoparticle profile (Tripathi et al. 2009; Bar et al. 2009; Iravani 2011). Plant-mediated nanoparticle synthesis is cost-effective due to easy availability of plants, and also extract preparation is simple and economical as compared to maintenance of microbial culture. The capping and reducing agents of medicinal plants not only diminished the toxicity issues but may also increase the medicinal properties of the nanoparticles (Li et al. 2008). Along with these, the problem of aggregation of nanoparticles can be easily addressed by large amount of plant metabolites.

The use of Ag-NPs is increasing day by day. Even with high cost and environmental issues, physicochemical routes are widely adapted for synthesis of nanoparticles. So, in this chapter, emphasis is given to the plant-mediated Ag-NP synthesis as they provide a safer, cheaper, more effective and compatible route for nanoparticle synthesis.

12.3 Selection Criteria for Suitable Phytofactory for Nanoparticle Synthesis

12.3.1 Availability

It has been crucial to study the availability and topographical distribution of plant, i.e. when and where the particular plant grows. Also, it has to be understood which plant part (leaves, flowers, fruits, stems, barks, roots) is having more amount of active metabolites useful for nanoparticle synthesis (Borase et al. 2014).

12.3.2 Phytochemical Profile

Plants have abundant amount of active phytochemicals which differs from plant to plant and species to species according to their environmental conditions. For bulk production of phyto-nanoparticles, priority should be given to find out the exact mechanism of plant metabolite action, i.e. how and which metabolite/metabolites is/are responsible for nanoparticle synthesis. Generally, plant metabolites such as alkaloids, flavonoids, terpenoids, saponins, tannins, phenolics, quinines, polyphenols, aldehydes, ketones, starch, proteins, etc. are mainly responsible for metal salt reduction and nanoparticle stabilization. Thus, metabolite profile should be considered as a key feature for determining the potential of a plant for nanoparticle synthesis (Lukman et al. 2011).

12.3.3 Synthesis Duration and Stability

It is important to determine how much time is required for nanoparticle synthesis. The time taken could be decreased either by addition of extract or increasing temperature and shaking speed. For utilization of nanoparticles, their stability should be checked periodically (Satyavani et al. 2011).

12.3.4 Purification

It is important that the active phyto-metabolite can be easily and economically isolated and purified. For increased production of active metabolite, genetic engineering could be used as a tool. Genetically engineered plant could produce large amount of particular metabolite which is necessary for large-scale and economical production of nanoparticles (Gan and Li 2012; Iravani 2011).

12.3.5 Seasonal Variation

Due to inconsistent seasonal variations, there are chances of variation in phytochemical profile which could affect phyto-nanosynthesis (Singh et al. 2013). So, plant should be robust enough to tackle these variations. To determine the effect of seasonal variations, metabolite profile according to different seasons should be analysed.

12.4 Effects of Physicochemical Parameters on Green Synthesis of Ag-NPs

12.4.1 pH

Many studies have showed that pH can alter significantly the shape, size, stability and rate of reaction in phyto-nanoparticles (Sathishkumar et al. 2009; Velmurugan et al. 2013). In every biochemical reaction including nanoparticle synthesis, pH plays a significant role. Anionic groups interact with cationic silver ions. So, pH of the medium significantly alters their interaction. The reduction of silver ions to nanosilver is affected by binding of silver ions to phytochemicals which in turn is directly affected by the pH. It was observed that at alkaline pH, small and stable silver nanoparticles were obtained with *Cinnamon zeylanicum*, while at acidic pH, there was increase in the size of nanoparticles due to aggregation. In case of alkaline pH, more functional groups on phytochemicals were available which leads to faster reduction and stabilization of newly synthesized nanoparticles. Another study provides more insight on the role of pH. During tyrosine-mediated synthesis, at pH 10, the colourless silver salt solution turns yellow indicating synthesis of nanoparticles, whereas at neutral and acidic pH, no such reaction takes place (Selvakannan et al. 2004). It was also found in the case of *A. indica*-mediated synthesis that at alkaline pH, more stable nanoparticles were formed (Tripathi et al. 2009). It occurs due to complete charging of nanoparticles at alkaline condition as a large number of OH groups present can change the charge on nanoparticles and thus maximize repulsion and stability.

12.4.2 Temperature

Increase in temperature increases the activation energy of the any kind of reaction. Various reports have concluded that nanoparticle synthesis has increased considerably with the rise in temperature (Ghosh et al. 2012). It was also observed that change in temperature also changed the shape of the nanoparticles synthesized (Lukman et al. 2011). It was presumed that temperature may augment nucleation of Ag-NPs and also leads to more stable nanoparticles by regulating aggregation. The thermostable phyto-metabolites could be the key factors for enhanced yield at higher temperature.

12.4.3 Time of Reaction and Ratio of Precursors

Apart from pH and temperature, the time of reaction and ratio of precursors also play a significant role. Increased reaction time has been reported to increase nanoparticle synthesis (Krishnaraj et al. 2010). The proportion of plant extract to metal precursor is also crucial for Ag-NP synthesis and also determines shape, size and stability of Ag-NPs (Gan and Li 2012). It was observed that Ag-NP synthesis increased with increasing garlic extract while keeping the quantity of metal precursor (silver nitrate) constant. Christensen et al. (2011) kept *Murraya koenigii* leaf extract amount constant and added silver nitrate in increasing ratios and found that combination having least silver nitrate (1:20) yield large number of Ag-NPs and with the increased amount of precursor, quantity of Ag-NPs got decreased. As quantity of precursor is increased, the availability of active molecules decreased, thereby resulting in lesser Ag-NP synthesis. On the other hand, Krishnaraj et al. (2010) explained that less amount of plant extract synthesized more Ag-NPs in case of *A. indica*. This may be due to the interference of large amount of extract during stabilization which leads to agglomeration, resulting in lesser Ag-NP synthesis.

12.4.3.1 Mechanism of Ag-NP Synthesis

Though the exact mechanism of reduction and stabilization is not clearly understood so far, various studies suggested that different functional groups (such as hydroxyl, carbonyl and amine) present in the plant metabolites react and reduce the metal ions to form nanoparticles (Vigneshwaran et al. 2006; Sathishkumar et al. 2009; Kora et al. 2010; Marimuthu et al. 2011).

In the case of *Jatropha curcas*, the latex having peptides curcacyclin A and curcacyclin B and enzyme curcain is responsible for nanoparticle synthesis (Bar et al. 2009). Further, the carbonyl groups of proteins have strong binding affinity towards metals. The role of amines is confirmed by FTIR studies (decreased band intensity after metal reduction and shift of band after binding of nanoparticles with (NH) C=O group).

In the case of fungus *F. oxysporum*, the role of enzyme α-NADPH-dependent nitrate reductase in development of nanoparticles is established by SPR (peak at 413 nm) of the mixture of enzyme and $AgNO_3$ (Kumar et al. 2007). Further, no SPR peak was evident in the absence of α-NADPH, 4-hydroxyquinoline and phytochelatin. So, it was concluded that formation of NADP by reduction of α-NADPH is essential. Also, 4-hydroxyquinoline acts as electron shuttle and transfers the electrons generated during the conversion of Ag^{2+} to Ag^0.

12.5 Characterization of Nanoparticles

Nanoparticles are characterized on the basis of their size, shape and surface area. The uniformity of the nanoparticles is one of the important parameter. Different techniques are used for characterization of nanoparticles such as UV–visible spectrophotometry, dynamic light scattering (DLS), electron microscopy (EM), Fourier transform infrared spectroscopy (FTIR), powder X-ray diffraction (XRD) and energy-dispersive spectroscopy (EDS) (Shahverdi et al. 2011).

The most commonly used technique for characterization is UV–visible spectroscopy. Metal nanoparticles in the size range of 2–100 nm can be characterized using light wavelengths in the range of 300–800 nm (Huang and Yang 2004). For silver and gold nanoparticle characterization, 400–450 and 500–550 nm wavelengths are used, respectively (Shankar et al. 2004). The surface charge and size of nanoparticles is determined by DLS. For morphological characterization, electron microscopy is widely used (Schaffer et al. 2009). FTIR is used to determine the surface chemistry, i.e. functional groups or other residues attached to the nanoparticles (Chithrani et al. 2006). XRD is used for the crystal structure and phase identification (Sun et al. 2000). EDS is used to analyse the elemental composition of metallic nanoparticles (Strasser et al. 2010).

12.6 Applications of Ag-NPs

12.6.1 Antimicrobial Potential

Plant extract-based Ag-NPs are well established as antimicrobials. Ag-NPs acts as antimicrobial by either causing structural damage (Kim et al. 2007) or generation of reactive oxygen species or interfering with DNA replication or reacting with the thiol groups of enzymes (Feng et al. 2000; Li et al. 2008) (Fig. 12.3). Bacterial cell

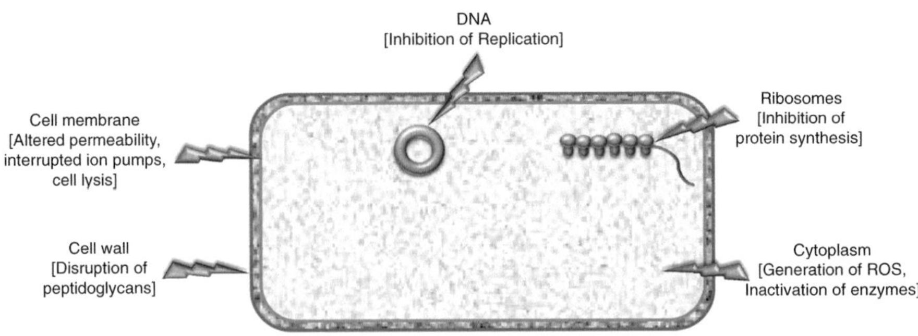

Fig. 12.3 Mechanisms of the antibacterial activity of silver nanoparticles

wall contains cysteine possessing reactive thiol groups (−SH). Silver ions interact with cysteine residues leading to protein or enzyme inactivation. Apart from sulphur, silver has strong affinity towards phosphorus as well. As cell wall contains sulphur and phosphorus, so silver can interact, form complexes and alter their properties (Gordon et al. 2010). Deposition of Ag-NPs on the bacterial cell can affect permeability of cell membrane resulted in disturbances in transport of ions and other substances through sodium/potassium pumps. The diminished activity of membrane results in disruption of the cell. Also Ag-NPs can penetrate into a bacterial cell causing damage to its intracellular structures. Denaturation of ribosomes leads to the inhibition of protein synthesis (Jung et al. 2008). Silver ions can bind to the bases constructing DNA. This condensation with DNA leads to its inability to replicate preventing the bacterial reproduction. Ag-NPs binding to bacterial wall create a coat disrupting moves of bacterial flagella. This multidimensional antibacterial activity of silver and Ag-NPs is the key to low bacterial resistance rates. The antibacterial action of silver ions and Ag-NPs starts with the binding to peptidoglycans of bacterial cell wall. Mammalian cells do not possess cell wall covering cell membrane. However, other mechanisms of Ag-NPs action can affect both bacterial and human cells. The next mode of Ag-NPs action uses its catalytic properties of generating free protons. They interact with disulphide bonds breaking them and leading to the dysfunction of integral proteins of outer cell membrane which are responsible for interactions between bacteria and their environment, stability of bacterial cells or binding of various substrates. Silver blocks some metabolic reactions taking place in cells. Silver combines with thiol residues of enzymes inactivating them. Ag-NPs induce production of reactive oxygen species, which also take part in a destruction of bacterial cell (Lok et al. 2007). Ag-NPs also influence the cell wall of fungi and interact with proteins of proteinolipid core of viruses, e.g. HIV-1 (Lara et al. 2010).

Silver exhibits higher toxicity to micro-organisms as compared to mammalian tissues. The increased cases of antimicrobial drug resistance are a serious problem and cause attention towards the proper use of Ag-NPs in antimicrobial therapy. Numerous studies have confirmed the role of plant-synthesized Ag-NPs as antimicrobials. Ag-NPs obtained using *Acalypha indica* plant extract act against *Escherichia coli* and *Staphylococcus aureus* (Krishnaraj et al. 2010). Similarly, the Ag-NPs synthesized from several plants of Euphorbiaceae family (*J. curcas*, *J. gossypifolia*, *Pedilanthus tithymaloides*, *Euphorbia milii*, *Euphorbia hirta*, etc.) were reported to have potent activity against *E. coli*, *S. aureus*, *S. epidermidis* and *Micrococcus luteus*. Some other studies showed the synergistic effects of plant-synthesized Ag-NPs in combination with standard antibiotics against multidrug-resistant bacteria (Birla et al. 2009, Singh et al. 2013). Ag-NPs also act as bio-insecticides. Ag-NPs synthesized from *Pityriasis rubra* and *Pergularia daemia* latex were effective against resistant larvae of *Anopheles aegypti* and *A. stephensi*, vectors of dengue and malaria.

Mosquito larvicidal properties were also exhibited by *E. hirta* synthesized Ag-NPs against *A. stephensi* (Priyadarshini et al. 2012), *Nelumbo nucifera*-mediated Ag-NPs against malarial and filarial vectors (Santoshkumar et al. 2011). *Ocimum canum*-synthesized Ag-NPs showed acaricidal activity against *Hyalomma anatolicum*, and *Hyalomma marginatum isaaci*. *Tinospora cordifolia* leaf extract-synthesized Ag-NPs displayed inhibitory action against head louse, *Pediculus humanus* (Jayaseelan and Rahuman 2011). Theoretically, there are negligible chances of bacterial resistance unlike antibiotic therapy. However, Radzig and co-workers point out that consistent and long-term usage of Ag-NPs may generate Ag-NP-resistant species of bacteria (Radzig et al. 2013).

12.6.2 Antiviral and Anticancerous Potential

Plant-mediated Ag-NPs are more promising for biomedical field as they are free from toxic contaminants. Anticancerous potential of *Citrullus colocynthis* Ag-NPs was tested by Satyavani et al. (2011). The reason lying behind this type of activity is that inside the cells, Ag-NPs enhanced the caspase-3 activity which leads to apoptosis. Antiviral potential of *Zingiber officinale* Ag-NPs is also evaluated. Ag-NPs bind to Gp-120 glycoprotein of HIV-1 and thus inhibit virus interaction with the host cells (Elechiguerra et al. 2005). Ag-NPs also exhibit a unique property of high blood compatibility which makes them a suitable candidate for drug delivery and biosensors (Kumar et al. 2012). Ag-NPs are also used in implants and wound dressings (Furno et al. 2004; Rodriguez-Carmona and Villaverde 2010).

12.6.3 Tissue Culture Technique

For tissue culture techniques, a biocompatible scaffold having mechanical strength is required for cellular growth. It was found that the protein adsorption and osteoblast adhesion was more in nano-ceramics than in micro-ceramics. Besides that, Ag-NPs also have antibacterial and biocidal potential. So, a nanocomposite of biocompatible polymer with Ag-NPs may protect the scaffold from contamination (Borase et al. 2014).

12.6.4 Water Purification

Ag-NPs can be used for water purification as they have antimicrobial properties (Jain and Pradeep 2005). Heidarpour et al. (2011) have found complete removal of *E. coli* (10^3 cfu/mm) from water after filtration through silver-coated polypropylene. To avoid flushing away of nanoparticles, immobilized polymer nanoconjugates should be used.

12.6.5 Agriculture

Nanoencapsulated Ag-NPs can constantly release the water and nutrients which increase their bioavailability. In addition, nanoparticles have potential to be used in plant genetic engineering for transfer of genes. It was also reported that Ag-NPs can improve growth and antioxidant potential of *Brassica juncea* seedlings (Sharma et al. 2012).

12.6.6 Biosensor

Management of toxic metals, pollutants, pesticides, etc. and their detection at lower concentrations are crucial to circumvent the effect on nontarget organisms. For their quantitative detection, various techniques such as high-performance liquid chromatography and gas chromatography are used. The main disadvantage of these techniques is high cost and tedious field analysis. Ag-NPs can act as ideal biosensors for environmental pollutants as well as for biomolecules (glucose, uric acid, etc.) as they possess distinct optical characteristics, localized SPR and high molar extinction coefficient. Fan et al. (2009) reported that starch-stabilized Ag-NPs can selectively recognize Hg^{2+} (detection limit ~5 ppb). Similarly, Wang et al. (2010) also proved the potential of Ag-NPs in Hg^{2+} detection (detection limit 17 nM). This highly specific detection mechanism involves binding of Hg^{2+} over Ag-NPs surface leading to blue shift of characteristic SPR peak of Ag-NPs. *A. indica* Ag-NPs were employed in detection of very low levels of copper with high sensitivity. Apart from Ag-NPs, gold nanoparticles were also studied for lead, (Chai et al. 2010), polynucleotide and copper (II) ions (Storhoff et al. 1998) detection. So, potential of plant extract-synthesized nanoparticles could be used for pollutant detection.

12.6.7 Catalyst

Extremely small sizes and high surface to volume ratios are ideal properties for a catalyst, and Ag-NPs possess both of them. Several catalysts cause pollution. Nitroaromatic compounds used in dyes, pesticides and explosive industries are one of the main priority pollutants as listed by the US Environmental Protection Agency. These pollutants are highly soluble and stable in water, resist degradation and get accumulated in the soil. Ag-NPs were found very effective in the reduction of 4-nitrophenol and p-nitroaniline which have many applications in dyes and pharmaceutical industries (Borase et al. 2013). The reduction mechanism of nitroaromatic compounds by Ag-NPs involves transfer of electrons from borohydride to nitroaromatic compounds. The catalytic activity of plant extract-mediated Ag-NPs has found to be more prominent due to capping of Ag-NPs by plant metabolites which further promote catalysis (Gangula et al. 2011).

12.6.8 Nanocomposite

Nanocomposite is the matrix (organic and inorganic compounds) in which nanoparticles are dispersed. The polymer–metal nanocomposites drew extensive attention due to their unique properties (electronic properties, barrier properties, flammability, resistance) and potential applications such as in biosensor. The electron-rich polymer (organic/inorganic) acts as a scaffold for the nanoparticles or in other terms provides stability to the electron-deficient metal ions. The polymer provides high surface, protection and scaffold for attachment of the metal ions (Hatchett and Josowicz 2008). Integration of plant-based nanosilver in different polymers (polyvinylpyrrolidone, polytetrafluoroethylene) yields Ag-NP–polymer nanocomposites having improved antimicrobial, electrical and mechanical properties (Egger et al. 2009). Further advantage was provided by the more eco-friendly and biocompatible nature of the plant-based Ag-NPs.

12.6.9 Dye Degradation

The release of dyes in waste water from textile industries creates serious environmental issues. Ag-NPs arise as an alternative to traditional dye degradation techniques using carbon sorption and flocculation. Methyl orange degradation by *Ulva lactuca* Ag-NPs was also reported (Kumar et al. 2013). Similarly, Ag-NPs from *Terminalia chebula* fruit extract converted methylene blue to leucomethylene blue (Edison and Sethuraman 2012).

The dye degradation occurs due to electron relay effect created by Ag-NPs. The Ag-NPs facilitate electron transfer from the plant extract to dye. The potential of Ag-NPs for large-scale dye degradation is yet to be explored (Savage and Diallo 2005).

12.6.10 Miscellaneous Applications

Besides these applications, Ag-NPs also play important roles in various other fields such as in textile industry as antifungal agent (Gittard et al. 2010) and nanofilters for purification of various compounds at large scale such as lactic acid (Li et al. 2007) (Fig. 12.4). Other important field of applications include electronics, aerospace, solar panels, cloud seeding, etc.

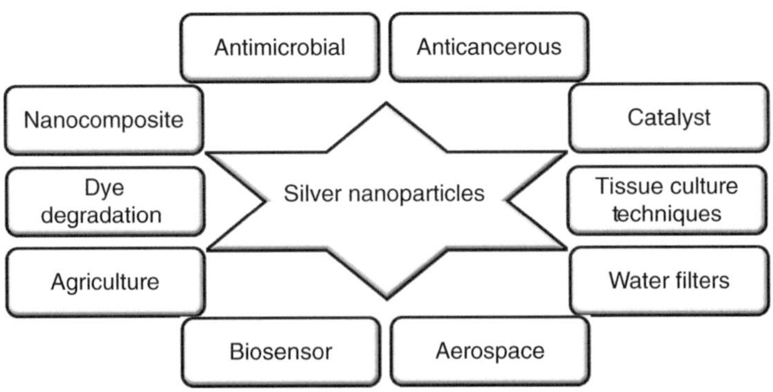

Fig. 12.4 Applications of silver nanoparticles

12.7 Perspectives

The exploitation of plant extracts for the synthesis of Ag-NPs requires thorough investigation of geographical distribution and phytochemical profile of plants under all environmental conditions. Earlier data collected identified promising plant families like Euphorbiaceae, Moraceae, Apocynaceae, etc. for nanoparticle production. Plant metabolites such as alkaloids, flavonoids, terpenoids, saponins, tannins, quinines, polyphenols, aldehydes, ketones, starch, proteins, etc. have been reported to be mainly responsible for metal salt reduction and stabilization of nanoparticles. Phytosynthesized Ag-NPs have many applications in various fields such as antimicrobial, antiviral, anticancerous, catalyst, biosensor, dye degradation, etc. The Ag-NPs also provide an eco-friendly alternative to traditional hazardous compounds. Discovering the exact mechanism of nanoparticle synthesis using plant extracts and further genetic engineering of plant for production of most active metabolite responsible for Ag-NP synthesis are the promising areas of research which require extensive focus.

Acknowledgement The authors wish to thank the University Grant Commission, New Delhi for UGC-BSR Senior Research Fellowship.

References

Agrawal AA, Lajeunesse MJ, Fishbein M (2008) Evolution of latex and its constituent defensive chemistry in milkweeds (Asclepias): a phylogenetic test of plant defense escalation. Entomol Exp Appl 128:126–138. doi:10.1111/j.1570-7458.2008.00690.x

Bae CH, Nam SH, Park SM (2002) Formation of silver nanoparticles by laser ablation of a silver target in NaCl solution. Appl Surf Sci 197:628–634. doi:10.1016/S0169-4332(02)00430-0

Bar H, Bhui DK, Sahoo GP, Sarkar P, De SP, Misra A (2009) Green synthesis of silver nanoparticles using latex of *Jatropha curcas*. Colloids Surf A Physicochem Eng Asp 339:134–139. doi:10.1016/j.colsurfa.2009.02.008

Binupriya AR, Sathishkumar M, Yun SI (2010) Myco-crystallization of silver ions to nanosized particles by live and dead cell filtrates of *Aspergillus oryzae* var. *viridis* and its bactericidal activity towards *Staphylococcus aureus* KCCM 12256. Ind Eng Chem Res 49:852–858. doi:10.1021/ie9014183

Birla SS, Tiwari VV, Gade AK, Ingle AP, Yadav AP, Rai MK (2009) Fabrication of silver nanoparticles by Phoma glomerata and its combined effect against *Escherichia coli*, *Pseudomonas aeruginosa* and *Staphylococcus aureus*. Lett Appl Microbiol 48:173–179. doi:10.1111/j.1472-765X.2008.02510.x

Borase HP, Patil CD, Sauter IP, Rott MB, Patil SV (2013) Amoebicidal activity of phytosynthesized silver nanoparticles and their *in vitro* cytotoxicity to human cells. FEMS Microbiol Lett 345:127–231. doi:10.1111/1574-6968.12195

Borase HP, Salunke BK, Salunkhe RB, Chandrashekhar DP, Hallsworth JE, Kim BS, Patil SV (2014) Plant extract: a promising biomatrix for ecofriendly, controlled synthesis of silver nanoparticles. Appl Biochem Biotechnol 173:1–29. doi:10.1007/s12010-014-0831-4

Brichkin SB, Spirin MG, Nikolenko LM, Nikolenko DY, Gak VY, Ivanchikhina AV, Razumov VF (2008) The use of reversed micelles for the synthesis of nanoparticles. High Energ Chem 42:516–521. doi:10.1134/S0018143908070035

Chai F, Wang C, Wang T, Li L, Su Z (2010) Colorimetric detection of Pb^{2+} using glutathione functionalized gold nanoparticles. Appl Mater Inter 2:1466–1470. doi:10.1021/am100107k

Chithrani BD, Ghazani AA, Chan WC (2006) Determining the size and shape dependence of gold nanoparticle uptake into mammalian cells. Nano Lett 6:662–668. doi:10.1021/nl052396o

Christensen L, Vivekanandhan S, Misra M, Mohanty AM (2011) Biosynthesis of silver nanoparticles using *Murraya koenigii* (curry leaf): an investigation on the effect of broth concentration in reduction mechanism and particle size. Adv Mater Let 2:429–434. doi:10.5185/amlett.2011.4256

Dahl JA, Maddux BL, Hutchison JE (2007) Toward greener nanosynthesis. Chem Rev 107:2228–2269. doi:10.1021/cr050943k

Dhillon GS, Brar SK, Kaur S, Verma M (2012) Green approach for nanoparticle biosynthesis by fungi: current trends and applications. Crit Rev Biotechnol 32:49–73. doi:10.3109/07388551.2010.550568

Duran N, Marcato PD, De'Souza GIH, Alves OL, Esposito E (2007) Antibacterial effect of silver nanoparticles produced by fungal process on textile fabrics and their effluent treatment. J Biomed Nanotechnol 3:203–208. doi:10.1166/jbn.2007.022

Edison TJI, Sethuraman MG (2012) Instant green synthesis of silver nanoparticles using *Terminalia chebula* fruit extract and evaluation of their catalytic activity on reduction of methylene blue. Process Biochem 47:1351–1357. doi:10.1016/j.procbio.2012.04.025

Egger S, Lehmann RP, Height MJ, Loessner MJ, Schuppler M (2009) Antimicrobial properties of a novel silver-silica nanocomposite material. Appl Environ Microbiol 75:2973. doi:10.1128/AEM.01658-08

Elechiguerra JL, Burt JL, Morones JR, Camacho-Bragado A, Gao X, Lara HH, Yacaman MJ (2005) Interaction of silver nanoparticles with HIV-1. J Nanobiotechnol 3:6. doi:10.1186/1477-3155-3-6

Fan Y, Liu Z, Wang L, Zhan J (2009) Synthesis of starch-stabilized Ag nanoparticles and Hg^{2+} recognition in aqueous media. Nanoscale Res Lett 4:1230–1235. doi:10.1007/s11671-009-9387-6

Feng QL, Wu J, Chen GQ, Cui FZ, Kim TN, Kim JO (2000) A mechanistic study of the antibacterial effect of silver ions on *Escherichia coli* and *Staphylococcus aureus*. J Biomed Mater Res 52:662–668. doi: 10.1002/1097-4636(20001215)52:4<662::AID-JBM10>

Feynman RP (1992) There's plenty of room at the bottom. J Microelectromech Syst 1:60–66. doi:10.1109/84.128057

Furno F, Morley KS, Sharp BL, Arnild PL, Howdle SM (2004) Silver nanoparticles and polymeric medical devices: a new approach to prevention of infection? J Antimicrob Chemother 54:1019–1024. doi:10.1093/jac/dkh478

Gan PP, Li SFY (2012) Potential of plant as a biological factory to synthesize gold and silver nanoparticles and their applications. Rev Environ Sci Biotechnol 11:169–206. doi:10.1007/s11157-012-9278-7

Gangula A, Podila R, Karanam L, Janardhana C, Rao AM (2011) Catalytic reduction of 4-nitrophenol using biogenic gold and silver nanoparticles derived from *Breynia rhamnoides*. Langmuir 27:15268–15274. doi:10.1021/la2034559

Ghosh S, Patil S, Ahire M, Kitture R, Kale S, Pardesi K, Cameotra SS, Bellare J, Dhavale DD, Jabgunde A, Chopade BA (2012) Synthesis of silver nanoparticles using *Dioscorea bulbifera* Tuber extract and evaluation of its synergistic potential in combination with antimicrobial agents. Int J Nanomedicine 7:483–496. doi:10.2147/IJN.S24793

Gittard SD, Hojo D, Hyde GK, Scarel G, Narayan RJ, Parsons GN (2010) Antifungal textiles formed using silver deposition in supercritical carbon dioxide. J Mater Eng Perform 19:368–373. doi:10.1007/s11665-009-9514-7

Gordon O, Vig Slenters T, Brunetto PS, Villaruz AE, Sturdevant DE, Ott M (2010) Silver coordination polymers for prevention of implant infection: thiol interaction, impact on respiratory chain enzymes, and hydroxyl radical induction. Antimicrob Agents Chemother 54:4208–4218. doi:10.1128/AAC.01830-09

Hatchett DW, Josowicz M (2008) Composites of intrinsically conducting polymers as sensing nanomaterials. Chem Rev 108:746–769. doi:10.1021/cr068112h

Heidarpour F, Wan Ab Karim Ghani WA, Fakhru'l-Razi A, Sobri S, Heydarpour V, Zargar M, Mozafari MR (2011) Complete removal of pathogenic bacteria from drinking water using nano silver-coated cylindrical polypropylene filters. J Clean Technol Environ 13:499–507. doi:10.1007/s10098-010-0332-2

Huang HZ, Yang XR (2004) Synthesis of polysaccharide-stabilized gold and silver nanoparticles: a green method. Carbohydr Res 339:2627–2631. doi:10.1016/j.carres.2004.08.005

Iravani S (2011) Green synthesis of metal nanoparticles using plants. Green Chem 13:2638–2650. doi:10.1039/C1GC15386B

Jain P, Pradeep T (2005) Potential of silver nanoparticle-coated polyurethane foam as an antibacterial water filter. Biotechnol Bioeng 90:59–63. doi:10.1002/bit.20368

Jayaseelan C, Rahuman AA (2011) Acaricidal efficacy of synthesized silver nanoparticles using aqueous leaf extract of *Ocimum canum* against *Hyalomma anatolicum anatolicum* and *Hyalomma marginatum isaaci*. Ixodidae, Acari. doi:10.1007/s00436-011-2559-1

Jung WK, Koo HC, Kim KW, Shin S, Kim SH, Park YH (2008) Antibacterial activity and mechanism of action of the silver ion in *Staphylococcus aureus* and *Escherichia coli*. Appl Environ Microbiol 74:2171–2178. doi:10.1128/AEM.02001-07

Kannan RR, Arumugam R, Ramya D, Manivannan K, Anantharaman P (2013) Green synthesis of silver nanoparticles using marine macroalgae *Chaetomorpha linum*. Appl Nanosci 3:229–233. doi:10.1007/s13204-012-0125-5

Kim JS, Kuk E, Yu KN, Kim JH, Park SJ, Lee HJ, Kim SH, Park YK, Park YH, Huwang CY, Kim YK, Lee YS, Jeong DH, Cho MH (2007) Antimicrobial effects of silver nanoparticles. Nanomedicine 3:95–101. doi:10.1016/j.nano.2006.12.001

Klaus T, Joerger R, Olsson E, Granqvist CG (1999) Silver-based crystalline nanoparticles, microbially fabricated. Proc Natl Acad Sci U S A 96:13611–13614. doi:10.1073/pnas.96.24.13611

Kora AJ, Sashidhar RB, Arunachalam J (2010) Gum kondagogu (*Cochlospermum gossypium*): a template for the green synthesis and stabilization of silver nanoparticles with antibacterial application. Carbohydr Polym 82:670–679. doi:10.1016/j.carbpol.2010.05.034

Krishnaraj C, Jagan EG, Rajasekar S, Selvakumar P, Kalaichelvan PT, Mohan N (2010) Synthesis of silver nanoparticles using *Acalypha indica* leaf extracts and its antibacterial activity against water borne pathogens. Colloids Surf B Biointerfaces 76:50–56. doi:10.1016/j.colsurfb.2009.10.008

Kumar KP, Paul W, Sharma CP (2012) Green synthesis of silver nanoparticles with *Zingiber officinale* extract and study of its blood compatibility. BioNanoScience 2:144–152. doi:10.1007/s12668-012-0044-7

Kumar P, Govindraju M, Senthaamilselvi S, Premkumar K (2013) Photocatalytic degradation of methyl orange dye using silver (Ag) nanoparticles synthesized from *Ulva lactuca*. Colloids Surf B Biointerfaces 103:658–661. doi:10.1016/j.colsurfb.2012.11.022

Kumar SA, Abyaneh MK, Gosavi SW, Kulkarni SK, Pasricha R, Ahmad A, Khan MI (2007) Nitrate reductase-mediated synthesis of silver nanoparticles from AgNO3. Biotechnol Lett 29:439–445. doi:10.1007/s10529-006-9256-7

Kumar V, Yadav SK (2009) Plant-mediated synthesis of silver and gold nanoparticles and their applications. J Chem Technol Biotechnol 84:151–157. doi:10.1002/jctb.2023

Lara HH, Ayala-Nuñez NV, Ixtepan-Turrent L, Rodriguez-Padilla C (2010) Mode of antiviral action of silver nanoparticles against HIV. J Nanobiotechnol 8:1. doi:10.1186/1477-3155-8-1

Ledwith DM, Whelan AM, Kelly JM (2007) A rapid, straight-forward method for controlling the morphology of stable silver nanoparticles. J Mater Chem 17:2459–2464. doi:10.1021/J100214A025

Li Q, Mahendra S, Lyon DY, Brunet L, Liga MV, Li D, Alvarez PJJ (2008) Antimicrobial nanomaterials for water disinfection and microbial control: potential applications and implications. Water Res 42:4591–4602. doi:10.1016/j.watres.2008.08.015

Li S, Shen Y, Xie A, Yu X, Qiu L, Zhang L, Zhang Q (2007) Green synthesis of silver nanoparticles using *Capsicum annuum* L. extracts. Green Chem 9:852–858. doi:10.1039/b615357g

Likus W, Bajor G, Siemianowicz K (2013) Nanosilver—does it have only one face? Acta Biochim Pol 60:495–501. doi:10.5772/60956

Lok CN, Ho CM, Chen R, He QY, WY Y, Sun H (2007) Silver nanoparticles: partial oxidation and antibacterial activities. J Biol Inorg Chem 12:527–534. 10.1007/s00775-007-0208-z
Lukman AI, Gong B, Marjo CE, Roessner U, Harris AT (2011) Facile synthesis, stabilization, and anti-bacterial performance of discrete Ag nanoparticles using *Medicago sativa* seed exudates. J Colloid Interface Sci 353:433–444. doi:10.1016/j.jcis.2010.09.088
Mafune F, Kohno J, Takeda Y, Kondow T, Sawabe H (2000) Structure and stability of silver nanoparticles in aqueous solution produced by laser ablation. J Phys Chem B 104:8333–8337. doi:10.1021/jp001803b
Malynych SZ, Chumanov G (2003) Vacuum deposition of silver island films on chemically modified surfaces. J Vac Sci Technol A 21:723–727. doi:10.1116/1.1570837
Marimuthu S, Rahuman AA, Rajakumar G, Santhoshkumar T, Kirthi AV, Jayaseelan C, Bagavan A, Zahir AA, Elango G, Kamaraj C (2011) Evaluation of green synthesized green silver nanoparticles against parasites. Parasitol Res 108:1541–1549. doi:10.1007/s00436-010-2212-4
Meyers MA, Mishra A, Benson DJ (2006) Mechanical properties of nanocrystalline materials. Prog Mater Sci 51:427–556. doi:10.1016/j.pmatsci.2005.08.003
Mittal AK, Chisti Y, Banerjee UC (2013) Synthesis of metallic nanoparticles using plant extracts. Biotechnol Adv 31:346–356. doi:10.1016/j.biotechadv.2013.01.003
Mohanpuria P, Rana NK, Yadav SK (2008) Biosynthesis of nanoparticles: technological concepts and future applications. J Nanopart Res 10:507–517. doi:10.1007/s11051-007-9275-x
Mohanty S, Jena P, Mehta R, Pati R, Banerjee B, Patil S, Sonawane A (2013) Cationic antimicrobial peptides and biogenic silver nanoparticles kill mycobacteria without eliciting DNA damage and cytotoxicity in mouse macrophages. Antimicrob Agents Chemother 57:3688–3698. doi:10.1128/AAC.02475-12
Mukherjee P, Ahmad A, Mandal D, Senapati S, Sainkar SR, Khan MI (2001) Fungus mediated synthesis of silver nanoparticles and their immobilization in the mycelial matrix: a novel biological approach to nanoparticle synthesis. Nano Lett 1:515–519. doi:10.1021/nl0155274
Nair B, Pradeep T (2002) Coalescence of nanoclusters and formation of submicron crystallites assisted by *Lactobacillus strains*. Cryst Growth Des 2:293–298. doi:10.1021/cg0255164
Priyadarshini KA, Murugan K, Panneerselvam C, Ponarulselvam S, Hwang J, Nicoletti M (2012) Biolarvicidal and pupicidal potential of silver nanoparticles synthesized using Euphorbia hirta against *Anopheles stephensi* Liston (Diptera: Culicidae). Parasitol Res 111:997–1006. doi:10.1007/s00436-012-2924-8
Radzig MA, Nadtochenko VA, Koksharova OA, Kiwi J, Lipasova VA, Khmel IA (2013) Antibacterial effects of silver nanoparticles on gram-negative bacteria: influence on the growth and biofilms formation, mechanism of action. Colloids Surf B Biointerfaces 102:300–306. doi:10.1016/j.colsurfb.2012.07.039
Rodriguez-Carmona E, Villaverde A (2010) Nanostructured bacterial materials for innovative medicines. Trends Microbiol 18:423–430. doi:10.1016/j.tim.2010.06.007
Salunkhe RB, Patil SV, Patil CD, Salunke BK (2011) Larvicidal potential of silver nanoparticles synthesized using fungus *Cochliobolus lunatus* against *Aedes aegypti* (Linnaeus, 1762) and *Anopheles stephensi* Liston (Diptera; Culicidae). Parasitol Res 109:823–831. doi:10.1007/s00436-011-2328-1
Sathishkumar M, Sneha K, Won SW, Cho CW, Kim S, Yun YS (2009) *Cinnamon zeylanicum* bark extract and powder mediated green synthesis of nano-crystalline silver particles and its bactericidal activity. Colloids Surf B Biointerfaces 73:332–338. doi:10.1016/j.colsurfb.2009.06.005
Satyavani K, Gurudeeban S, Ramanathan T, Balasubramanian T (2011) Biomedical potential of silver nanoparticles synthesized from calli cells of *Citrullus colocynthis* (L.) Schrad. J Nanobiotechnol 9:43. doi:10.1186/1477-3155-9-43
Savage N, Diallo MS (2005) Nanomaterials and water purification: opportunities and challenges. J Nanopart Res 7:331–342. doi:10.1007/s11051-005-7523-5
Schaffer B, Hohenester U, Trugler A, Hofer F (2009) High-resolution surface plasmon imaging of gold nanoparticles by energy-filtered transmission electron microscopy. Phys Rev B 79:article 041401. doi:10.1103/PhysRevB.79.041401
Selvakannan PR, Swami A, Srisathiyanarayanan D, Shirude PS, Pasricha R, Mandale AB, Sastry M (2004) Synthesis of aqueous au core-Ag shell nanoparticles using tyrosine as a pH-dependent

reducing agent and assembling phase-transferred silver nanoparticles at the air-water interface. Langmuir 20:7825–7836. doi:10.1021/la049258j

Shahverdi AR, Minaeian S, Shahverdi HR, Jamalifar H, Nohi AA (2007) Rapid synthesis of silver nanoparticles using culture supernatants of *Enterobacteria*: a novel biological approach. Process Biochem 42:919–923. doi:10.1016/j.procbio.2007.02.005

Shahverdi AR, Shakibaie M, Nazari P (2011) Basic and practical procedures for microbial synthesis of nanoparticles. In: Rai M, Duran N (eds) Metal nanoparticles in microbiology. Springer, Berlin, pp 177–196. doi: 10.1007/978-3-642-18312-6_8

Shankar SS, Rai A, Ahmad A, Sastry M (2004) Rapid synthesis of au, Ag, and bimetallic au core–Ag shell nanoparticles using Neem (*Azadirachta indica*) leaf broth. J Colloid Interface Sci 275:496–502. doi:10.1016/j.jcis.2004.03.003

Sharma P, Bhatt D, Zaidi MGH, Saradhi PP, Khanna PK, Arora S (2012) Silver nanoparticle-mediated enhancement in growth and antioxidant status of *Brassica juncea*. Appl Biochem Biotechnol 167:2225–2233. doi:10.1007/s12010-012-9759-8

Singh R, Wagh P, Wadhwani S, Gaidhani S, Kumbhar A, Bellare J, Chopade BA (2013) Synthesis, optimization, and characterization of silver nanoparticles from *Acinetobacter calcoaceticus* and their enhanced antibacterial activity when combined with antibiotics. Int J Nanomedicine 8:4277–4290. doi:10.2147/IJN.S48913

Sintubin L, Verstraete W, Boon N (2012) Biologically produced nanosilver: current state and future perspectives. Biotechnol Bioeng 109:2422–2436. doi:10.1002/bit.24570

Storhoff JJ, Elghanian R, Music RC, Mirkin CA, Robert L, Letsinger RL (1998) Enzymatic manipulation of DNA-gold nanoparticles. J Am Chem Soc 120:1959–1964. doi:10.1021/ja972332i

Strasser P, Koh S, Anniyev T, Greeley J, More K, Yu C (2010) Lattice-strain control of the activity in dealloyed core–shell fuel cell catalysts. Nat Chem 2:454–460. doi:10.1038/nchem.623

Sun S, Murray C, Weller D, Folks L, Moser A (2000) Monodisperse FePt nanoparticles and ferromagnetic FePt nanocrystal superlattices. Science 287:1989–1992. doi:10.1126/science.287.5460.1989

Thakkar KN, Mhatre SS, Parikh RY (2010) Biological synthesis of metallic nanoparticles. Nanomed Nanotechnol Biol Med 6:257–262. doi:10.1016/j.nano.2009.07.002

Tripathi A, Chandrasekaran N, Raichur AM, Mukherjee A (2009) Antibacterial applications of silver nanoparticles synthesized by aqueous extract of *Azadirachta indica* (Neem) leaves. J Biomed Nanotechnol 5:93–98. doi:10.1166/jbn.2009.038

Velmurugan P, Lee SM, Iydroose M, Lee KJ, BT O (2013) Pine cone-mediated green synthesis of silver nanoparticles and their antibacterial activity against agricultural pathogens. Appl Microbiol Biotechnol 97:361–368. doi:10.1007/s00253-012-3892-8

Vigneshwaran N, Nachane RP, Balasubramanya RH, Varadarajan PV (2006) A novel one-pot 'green' synthesis of stable silver nanoparticles using soluble starch. Carbohydr Res 341:2012–2018. doi:10.1016/j.carres.2006.04.042

Vijayakumar S, Vaseeharan B, Malaikozhundan B, Gopi N, Ekambaram P, Pachaiappan R, Velusamy P, Murugan K, Benelli G, Suresh Kumar R, Suriyanarayanamoorthy M (2016) Therapeutic effects of gold nanoparticles synthesized using Musa Paradisiaca peel extract against multiple antibiotic resistant enterococcus faecalis biofilms and human lung cancer cells (A549). Microb Pathog S0882-4010(16):30714–30718. doi:10.1016/j.micpath.2016.11.029

Wang D, Markus J, Wang C, Kim YJ, Mathiyalagan R, Aceituno VC, Ahn S, Yang DC (2016) Green synthesis of gold and silver nanoparticles using aqueous extract of Cibotium barometz root. Artif Cells Nanomed Biotechnol 4:1–8. Epub ahead of print

Wang TC, Rubner MF, Cohen RE (2002) Polyelectrolyte multilayer nanoreactors for preparing silver nanoparticle composites: controlling metal concentration and nanoparticle size. Langmuir 18:3370–3375. doi:10.1021/la015725a

Wang Y, Yang F, Yang X (2010) Colorimetric detection of mercury (II) ion using unmodified silver nanoparticles and mercury-specific oligonucleotides. Appl Mater Interfac 2:339–342. doi:10.1021/am9007243

Xiuyan L, Jun S, Ai D, Zhihua Z, Guohua G, Huiyu Y, Jiandong W (2012) Facile synthesis of silver nanoparticles with high concentration via a CTAB-induced silver mirror reaction. Colloids Surf A Physicochem Eng Asp 400:73–79. doi:10.1016/j.colsurfa.2012.03.002

Zhu JJ, Liu SW, Palchik O, Gedanken A (2000) Shape-controlled synthesis of silver nanoparticles by pulse sonoelectrochemical methods. Lengmuir 16:6396–6399. doi:10.1021/la991507u

The Antiproliferative and Antibacterial Effect of *Moringa oleifera*-Mediated Gold Nanoparticles: A Review

13

Charlette Tiloke and Anil A. Chuturgoon

Abstract

South Africa's (SA's) mortality rates are increasing due to increased cancer diagnosis. In addition, bacterial infections are on the rise with increasing antibacterial resistance. Current chemotherapies and antibacterial agents are expensive with many side-effects; therefore alternative therapies are actively being investigated. A medicinal tree, *Moringa oleifera* (MO), is found throughout SA and used in traditional treatment of a variety of ailments including cancer and microbial infections. All parts of the tree are used for medicinal purposes including leaves, flowers, bark, seeds and seedpods (drumstick). The use of MO extracts has been beneficial for people across many communities as a food source in addition to its medicinal properties. Nanoparticles have a wide range of biomedical applications and are showing potential as anticancer and antibacterial agents. Their relative smaller size (1–100 nm) facilitates their mode of action against diseases. Nanoparticles are synthesised chemically or via plant extracts which are environmentally friendly and less toxic. MO can be used for nanoparticle synthesis such as gold nanoparticles (AuNPs) using a one-pot green synthesis technique. Therapeutic applications of AuNPs have been assessed both in vitro and in vivo

C. Tiloke
Discipline of Medical Biochemistry and Chemical Pathology, School of Laboratory Medicine and Medical Sciences, College of Health Sciences, University of KwaZulu-Natal, Private Bag 7, Congella, Durban 4013, South Africa

Department of Chemistry, Faculty of Applied Sciences, Durban University of Technology, Durban 4001, South Africa

A.A. Chuturgoon (✉)
Discipline of Medical Biochemistry and Chemical Pathology, School of Laboratory Medicine and Medical Sciences, College of Health Sciences, University of KwaZulu-Natal, Private Bag 7, Congella, Durban 4013, South Africa
e-mail: chutur@ukzn.ac.za

V.C. Kalia, A.K. Saini (eds.), *Metabolic Engineering for Bioactive Compounds*,
DOI 10.1007/978-981-10-5511-9_13

with positive outcomes. The review aims to discuss the therapeutic potential of AuNPs as anticancer and antibacterial agents.

Keywords
Moringa oleifera • Gold nanoparticles • Antiproliferation • Antibacterial

13.1 Introduction

In South Africa (SA), the healthcare burden is compounded by the presence of infectious diseases such as human immunodeficiency virus (HIV) infection/acquired immune deficiency syndrome (AIDS) and *Mycobacterium tuberculosis* (MtB) infections in addition to cancer and other noncommunicable diseases (Mayosi et al. 2009). It was estimated that in 2010, 34 million people were HIV positive and 67% were from sub-Saharan Africa (Granich et al. 2012). SA has the largest population of HIV-positive individuals globally (Hontelez et al. 2013). The introduction of antiretroviral treatment (ART) increased life expectancy in these patients. However several challenges do exist such as accessibility to treatment, cost and toxicity associated with adverse drug reactions that have been a concern which affects patient compliance to therapy (Mudzviti et al. 2012). Tuberculosis (TB) is the second leading cause of death worldwide following HIV infection (Bailo et al. 2015). Current therapies have increasing drug resistance; therefore investigations into bacteriostatic and bactericidal drugs are actively sought. In addition to multidrug-resistant (MDR) and extremely drug-resistant (XDR) strains, total drug-resistant strains are also emerging, which is of growing concern. According to the World Health Organization, approximately 9 million people were diagnosed with TB, of which 1.5 million died in 2013 (Baddeley et al. 2014). The TB mortality includes 360,000 patients who were also HIV positive.

The risk of cancer in South Africans is increasing which can be attributed to their current lifestyle and burden of these infectious diseases (Bello et al. 2011). Cancer is also ranked as one of the main causes of death accounting for 21.7% of all noncommunicable diseases. In SA, cancer incidences are on the rise (Sheridan and Collins 2013), and this is due to smoking, occupational exposure, diet, air pollution and infectious diseases such as HIV infection/AIDS. Conventional cancer treatment is surgical removal of tumour, radiation and chemotherapy; however there are many side-effects in addition to the non-specificity (Boisselier and Astruc 2009). Cancer such as lung, liver and oesophageal and leukaemia is commonly diagnosed worldwide. Cancer cell evades apoptosis and is a possible target for chemotherapeutic agents. Apoptosis or programmed cell death is a vital process in embryonic development and tissue homoeostasis and is regulated by aspartate-specific cysteine proteases (caspases) (Fan et al. 2005; Sharma et al. 2011; Circu and Aw 2012). The progression of apoptosis is due to signal cascades. Apoptosis occurs through two pathways, namely, death receptor-mediated procaspase activation pathway (extrinsic pathway) and mitochondrion-mediated procaspase activation pathway (intrinsic pathway).

Homoeostasis in the body is maintained by following a process of cell growth, division and death. This ensures a healthy state by regulating cell number and tissue size (Hengartner 2000). Cells that threaten homoeostasis are signalled for cell death. Aberration to this process results in the initiation of cancer as abnormal cells are not removed and continue to proliferate. Cancer cells evade apoptosis and proliferate. One of their mechanisms by which they proliferate is via oncogene activation, e.g. c-myc, and tumour suppressor gene inactivation, e.g. p53. In addition, regulation of genes and proteins determines cellular fate which can also aid in the development of chemotherapeutic agents. This is evident by the alternate splicing mechanism as the variant of a gene being translated results in differing protein expression. Caspase-9, an initiator caspase, undergoes alternate splicing via SRp30a, causing two splice variants—*caspase-9a* (pro-apoptotic) and *caspase-9b* (anti-apoptotic) (Massiello and Chalfant 2006; Shultz et al. 2010, 2011). The variant expression is a determinant of the activation of the intrinsic pathway to ensure cellular death. Targeting c-myc, p53 and splice variants which ultimately influence the apoptotic cascade can be effective for cancer therapeutics.

Antibacterial resistance is also a growing concern in many countries (Vinoth et al. 2012). Antibacterial resistance increases the severity of infections with poor patient outcomes. In developing countries such as SA, MDR bacterial infections are increasing (Mendelson 2014). This can be attributed to the increased use of antibiotics leading to resistance. The costs associated with antibacterial agents are high which led to investigations into cost-effective and alternate medicines (Vinoth et al. 2012). The plant kingdom has tremendous potential as it is often utilised by many people for traditional medicines. Nowadays, pharmacological drugs that are developed have been obtained through natural crude and bioactive compounds. It was determined that majority of the population heavily rely on traditional plant-based therapies for their primary care as rural settings lack basic healthcare and financial resources to maintain a healthy lifestyle.

SA has a diverse range of plants and trees which are readily available for traditional medicinal use (Rybicki et al. 2012). More than 3000 species are used for herbal remedies which have shown potential in treatment of various ailments including cancer. A growing interest into the local biodiversity of our plant kingdom and their medicinal potential has led to the expansion of indigenous knowledge and medical use. Although modern medicine has advanced tremendously, the rate of disease progression especially in developing countries is also advancing with uncontrollable side-effects. Traditional medicinal plants and trees may provide positive outcomes in chronic diseases with minimal side-effects. Of the many traditional medicinal trees available, *Moringa oleifera* (MO) is used by many South Africans (Goyal et al. 2007; Djakalia et al. 2011). It is often known as the drumstick tree as it is readily available and grows well in both rural and urban areas. Almost all parts of the tree, viz. leaves (MOE), flowers (MF), seeds, seedpods (drumstick) and bark, are used in traditional treatments as there are many bioactive compounds such as niazimicin, gallic acid, glucosinolates and isothiocyanates present (Fahey 2005; Goyal et al. 2007; Mishra et al. 2011). In addition, MO also contains vitamins, amino acids, protein, potassium and calcium (Anwar et al. 2007). MO is used traditionally because it displays antioxidant, anti-inflammatory, anticancer, anti-atherosclerotic and antiulcer activity (Anwar and Bhanger 2003; Fahey 2005). It also has hypotensive, hypocholesterolaemic, hypoglycaemic and immune-boosting effects. Traditional healers in SA utilise MO as part of their treatment of various ailments; however their exact mode of action has not been fully elucidated. Awodele et al. (2012) investigated the toxicity of MOE in male Wistar albino mice (Awodele et al. 2012). Haematological and biochemical analysis showed that MOE was relatively safe when administered orally. The seedpod (drumstick) extract exerted a chemopreventative effects against colon carcinogenesis in mice (Budda et al. 2011). Singh et al. (2009) showed that MOE, seeds and seedpod (drumstick) possess antioxidant potential and anti-quorum-sensing activity against *Chromobacterium violaceum* (*C. violaceum*) (Singh et al. 2009).

Nanoparticles (NPs) are showing huge promise as anticancer and antibacterial agents (Prasad and Elumalai 2011) because of their characteristic small properties (1–100 nm) (Rai et al. 2009). There are many different types of nanoparticles being developed. Metals such as gold (Au) are readily available for use in the synthesis of NPs. These NPs can be synthesised using plant extracts such as MO—in a cost-effective and environmentally friendly synthesis. The use of plant extracts in the synthesis of nanoparticles is favourable for large-scale production. In addition, it is less harmful as toxic chemicals are eliminated from the synthesis steps. NPs have a wide range of biomedical applications from cancer imaging, diagnosis and drug delivery for therapy (Cai et al. 2008). Gold nanoparticles (AuNPs) induced a concentration-dependent decrease in cell viability and induced apoptosis in Vero cells (Chueh et al. 2014). In MRC-5 cells, AuNPs influenced oxidative stress, DNA damage and repair and cell cycle progression. However in NIH3T3 cells, the synthesised AuNPs induced autophagy. Azam et al. (2009) synthesised AuNPs using citric acid and CTAB in a one-step microwave irradiation method (Azam et al. 2009). The synthesised AuNPs were 1–22 nm and displayed antibacterial activity against gram-negative *Escherichia coli* (*E. coli*). The AuNPs had a larger surface-to-volume ratio which aided in its antibacterial effect. *Mentha piperita* (Lamiaceae) was used for the bio-reduction of silver nitrate and chloroauric acid for the synthesis of silver nanoparticles (AgNPs) and AuNPs (MubarakAli et al. 2011). AgNPs and AuNPs showed antibacterial activity against *Staphylococcus aureus* (*S. aureus*) and *E. coli*. Anand et al. (2014) successfully synthesised AuNPs using MF extracts which showed anticancer and catalytic activity (Anand et al. 2014). The synthesised stable nanoparticles caused the catalytic reduction of nitroaniline and nitrophenol. It also showed a dose-dependent decline in A549 lung cancer cell viability with no cytotoxic effect on normal healthy peripheral blood mononuclear cells (PBMCs).

Current cancer and antibacterial therapies are expensive and poorly attainable. This has led to patients resorting to the use of traditional medicine as an alternate source for treatment especially in the rural areas of SA. There is no cure for cancer, and with existing therapies, patients experience a wide range of side-effects. Traditional medicine such as the use of MO can reduce infections and deaths by providing an alternative to existing therapies. Medicinal plants possess anticancer and antimicrobial potential and can be used for the synthesis of NPs. The review aims to present and discuss the potential of MO and synthesised NPs as anticancer and antibacterial agents. Furthermore, their biochemical mechanism underlying their use will be assessed for chemotherapeutic and antibacterial drug development.

13.2 Nanoparticles: Diverse Biological and Pharmacological Applications

Nanotechnology is a rapidly evolving field (Faraji and Wipf 2009). Through the development of NPs, it has impacted and improved a vast range of sectors. There are many NPs that are synthesised including metals, liposomes, micelles, polymers and antibodies (Fig. 13.1) (Kanapathipillai et al. 2014). Due to their physiochemical properties, they have a wide range of applications in medical diagnosis, pharmaceutical drug development, medical treatment, electronic devices and commercial products (Kuppusamy et al. 2015). The development of Ag and Au NPs have led to their use in electronics, as catalysts for chemical reactions, in chemotherapeutic drugs, and as anti-inflammatory, antioxidant, antibacterial and antiviral agents. In addition, they can be used in the treatment of acute and chronic diseases.

Existing chemotherapeutic agent 5-fluorouracil (5-FU) (Gopinath et al. 2008) is used for the treatment of colon, head, neck and breast cancer. Human solid tumours are resistant to 5-FU and therefore requires gene therapy to convert 5-FU to 5-flurouridinemonophosphate (5-FUMP) which sensitises the cancer cells to treatment. The incorporation of NPs with existing drugs can potentiate the anticancer effects with minimal harmful side-effects, thereby increasing positive outcomes.

There are various routes of administration of NPs which include through intravenous, inhalation and ingestion (AshaRani et al. 2009). Through these routes, the NPs are transported to various parts of the body via the lymphatic system. There is no safety regulation for NP's use due to limited scientific studies (AshaRani et al. 2009). NPs' interaction with cells, absorption, distribution and excretion as well as their toxicity-associated outcomes require further investigations.

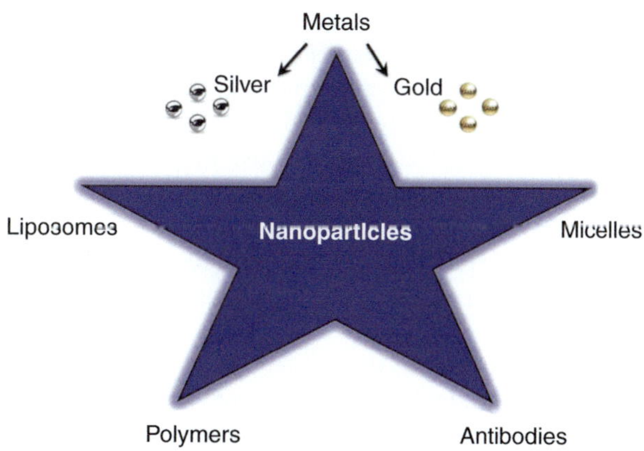

Fig. 13.1 The various types of nanoparticles with multiple functionality

Existing cancer therapies are non-specific with many side-effects (Markman et al. 2013). Chemotherapy resistance is another problem we are currently facing as the cancer microenvironment aids in the resistance. Cancer cells increase their expression of the ATP-binding cassette (ABC) superfamily and P-glycoproteins which remove anticancer drugs such as paclitaxel and doxorubicin, thereby decreasing their concentrations in the tumour. The microenvironment further assists cancer cell proliferation via the synthesis of growth factors and chemokines. It also facilitates cancer cell metastasis and the evasion of apoptosis. Therefore developing NPs that specifically target cancer cells will be beneficial.

13.3 Therapeutic Interventions: Possible Targets in Cancer and Bacterial Infections

p53, a tumour suppressor gene and transcription factor, is a regulator of the apoptotic cascade (Radhakrishna Pillai et al. 2004; Nithipongvanitch et al. 2007). Through the activation of a pro-apoptotic molecule Bax, it signals the mitochondria to depolarise releasing cytochrome c which forms an apoptosome with APAF-1, ATP and procaspase-9. This leads to the activation of initiator caspase-9 which thereby activates executioner caspase-3/7 for cell death (Bertram 2001; Radhakrishna Pillai et al. 2004). A defective p53 gene is generally seen in more than 70% of cancers, thereby affecting the cell cycle checkpoint resulting in cell proliferation (Bertram 2001). Reactive oxygen species, DNA damage and inappropriate oncogene activation cause the activation of p53 leading to cell cycle arrest and induction of the apoptotic cascade. Oncogenes, e.g. c-myc, cause cancer formation due to increased cell proliferation (Bertram 2001). The cell continues to divide and proliferate through the increased expression of c-myc evading the apoptotic cascade (Vicente-Duenas et al. 2013). Lung, breast and colon cancers express high levels of c-myc and could possibly be a target for chemotherapeutic agents. Alternate splicing of genes is a potential target for chemotherapeutic agents as approximately 90% of genes have alternate splice isoforms (Pajares et al. 2007; Shankarling and Lynch 2010). This results in various proteins with different functionality. However, this process is under homoeostasis, and the nonsense-mediated mRNA decay (NMD) pathway is responsible for removal of dysregulated isoforms. Cancer cells increase splice forms which enable them to continually grow and proliferate. An example of alternate splice isoforms is pro-apoptotic *caspase-9a* (exon 3, 4, 5, 6 cassette inclusion) and anti-apoptotic *caspase-9b* (exon 3, 4, 5, 6 cassette exclusion) generated through the alternate splicing of *caspase-9* (Massiello and Chalfant 2006). This is regulated via SRp30a which thereby determines whether a cell undergoes apoptosis. Cancer cells increase the expression of epidermal growth factor receptor (EGFR) which increases signal transduction and cell proliferation (Kanapathipillai et al. 2014). In addition human epidermal growth factor receptor 2 (HER2) is overexpressed in breast, ovarian and uterine cancers. By targeting these overexpressed receptors, NPs will be able to be efficiently taken up by the cells to execute its therapeutic effect.

Pathogenic bacteria are responsible for the increased rate of bacterial infection with antibiotic resistance (Yacoby and Benhar 2007). The bacterium replicates at a fast rate with mutations assisting their survival. Current drug therapies are non-specific as it targets the microorganism in addition to host normal healthy cells. Therefore the development if antibacterial agents requires the identification of target molecules, the quantity and concentration required and specificity of the target. In addition the bacterial structure and replication are potential targets for therapy. This will enable antibacterial agents to be more efficient and effective in their mode of action preventing severe infections and potential deaths.

13.4 Nanoparticle Synthesis: Green vs. Chemical

Nanoparticles have unique properties which can be attributed to the physical, chemical or biological processes that are used during the NP's synthesis (Makarov et al. 2014; Krithiga et al. 2015). This enables the development of NPs of different size, shape and charge aiding in their therapeutic effect (Kuppusamy et al. 2015). Nanoparticles are extremely small and their size ranges between 1 and 100 nm (Rai et al. 2009). Chemical synthesis of NPs is expensive, labour-intensive, not environmentally friendly and toxic to living organisms. The toxic effect is due to the strong chemicals used at the reduction and stabilisation steps during the synthesis of the NPs (Kuppusamy et al. 2015). Therefore alternative methods for synthesis were sought which are cost-effective and environmentally friendly (Makarov et al. 2014). Nanoparticles require stabilisation which can be conducted chemically or via the use of plant extracts. The synthesis of NPs involves a redox reaction. This can be facilitated by plant extracts, yeast, fungi, bacteria and algae (green synthesis of nanoparticles) which have the ability to reduce metal ions into NPs. The use of plant extracts such as leaves and flowers is favourable as cost for cultivation is relatively low and advantageous for large-scale production (Prasad and Elumalai 2011). It also does not require any special preparation or isolation (Shankar et al. 2004). The reduction of metal ions by plant extracts for the synthesis of NPs can be attributed to plant components such as polyphenols, phenolic acids, alkaloids, sugars and terpenoids (Fig. 13.2) (Makarov et al. 2014).

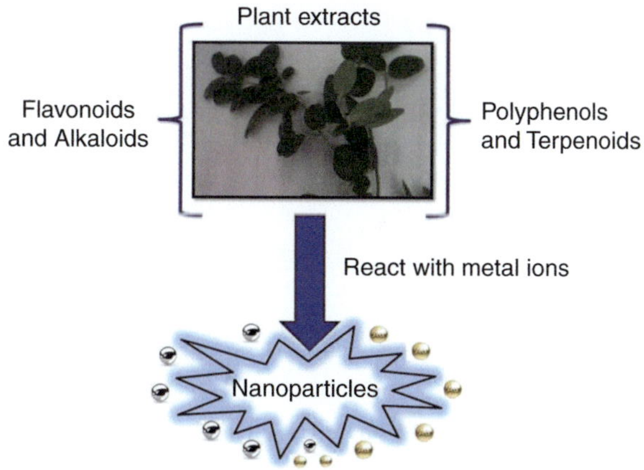

Fig. 13.2 The potential of plant extracts in nanoparticle synthesis

A one-pot green synthesis technique can be used for the synthesis of NPs (Gengan et al. 2013; Govender et al. 2013; Anand et al. 2014). Gengan et al. (2013) used a one-pot green synthesis technique for the biosynthesis of AgNPs by *Albizia adianthifolia* leaf extract which caused the reduction of Ag^{+} ions to Ag^{0} ions (Gengan et al. 2013). The phytochemical composition of the leaf extract such as saponins, glycosides, flavonoids, amines and reducing sugars have the potential as reducing agents for the synthesis of AgNPs. In addition, biochemical analysis of the synthesised AgNPs showed that it is able to induce apoptosis in A549 cells (human lung carcinoma cell line) (Govender et al. 2013). Anand et al. (2014) synthesised AuNPs using MF extract which caused the reduction of Au^{+} ions to Au^{0} ions (Anand et al. 2014). This reduction of ions for the synthesis of the AuNPs was attributed to the flower phytochemical composition. MO is a medicinal tree which is currently showing huge potential as a chemotherapeutic, antibacterial agent and in the synthesis of NPs.

13.5 *Moringa oleifera:* Antibacterial and Chemotherapeutic Potential

For centuries, trees and plants have been utilised for dietary supplementation and medicinal properties. These medicinal trees and plants play a pivotal role in the traditional treatment of a variety of diseases. One such tree is MO, indigenous to India, and is found throughout SA (Fig. 13.3) (Goyal et al. 2007; Djakalia et al. 2011).

MO belongs to the Plantae kingdom and is commonly known as the drumstick tree (Table 13.1) (Kasolo et al. 2011; Vinoth et al. 2012; Patel et al. 2014). Almost all parts of the tree, viz. MOE, MF, seeds, seedpods (drumstick) and bark, are used

Fig. 13.3 *Moringa oleifera* tree found locally

Table 13.1 Scientific classification of *Moringa oleifera*

Scientific classification		References
Kingdom	Plantae	Sanchez-Machado et al. (2010)
Division	Magnoliophyta	Kasolo et al. (2011)
Class	Magnoliopsida	Mishra et al. (2011)
Order	Brassicales	Sreelatha et al. (2011)
Family	Moringaceae	Sharma et al. (2011)
Genus	*Moringa*	Anand et al. (2014)
Species	*oleifera*	Belliraj et al. (2015)

either for dietary supplementation or medicinal properties. However MOE and MF possess the highest nutritional value with a wide range of bioactive compounds that display anticancer, anti-inflammatory and antimicrobial activity, among several others (Fahey 2005; Goyal et al. 2007; Sanchez-Machado et al. 2010; Sreelatha et al. 2011).

Vinoth et al. (2012) showed that the phytochemical analysis of MOE revealed it contains flavonoids, alkaloids, terpenoids, saponins, glycosides and tannins (Vinoth et al. 2012). Aqueous, chloroform and ethanol extracts of MOE were assessed for antibacterial activity and showed that the aqueous extract had a maximum zone of inhibition against *S. aureus* with no effect on *E. coli*, *Pseudomonas aeruginosa* (*P. aeruginosa*) and *Salmonella typhi* (*S. typhi*). The chloroform extract had a maximum zone of inhibition against *S. typhi* and *E. coli* with no effect on *S. aureus* and *P. aeruginosa*. The ethanol extract had a maximum zone of inhibition against *S. typhi* and a minimum zone of inhibition against *E. coli*. Nepolean et al. (2009) showed that ethanolic MF extract can be used as an antibacterial agent against *E. coli*, *Klebsiella pneumoniae* (*K. pneumoniae*) and *S. aureus* (Nepolean et al. 2009).

Listeria monocytogenes (*L. monocytogenes*) is a food-borne pathogen which causes listeriosis (Eruteya and Badon 2014). The ethanolic and aqueous extracts of MOE and seeds were assessed for in vitro antilisterial activity. The aqueous extract of both MOE and seeds inhibited *L. monocytogenes* growth. However this was not seen in the ethanol extract demonstrating the potential in treatment of listeriosis.

Cancer is considered one of the main causes of mortality across the world (Globocan 2012; Mendis et al. 2014). Existing chemotherapeutic agents are not effective in treating cancers; therefore complementary and alternative therapies are actively being sought. The aqueous extracts of MOE showed antiproliferative effects in cancerous human alveolar epithelial cells in vitro (Tiloke et al. 2013). MOE significantly increased oxidative stress by increasing reactive oxygen species and decreasing Nrf2 and glutathione (GSH) levels. This led to significant DNA damage and p53 activation of the apoptotic pathway in the A549 lung cancer cells. It showed a possible use as a complementary and alternate medicine in the treatment of lung cancer. In addition, the aqueous MOE had antiproliferative effects in human SNO oesophageal cancer cells (Tiloke et al. 2016a). At an IC_{50} 389.2 μg/mL, MOE for 24 h significantly increased lipid peroxidation leading to oxidative stress and DNA damage. Apoptotic markers such as phosphatidylserine externalisation, caspase-9, caspase-3/7, Smac/DIABLO and cleavage of PARP-1 were significantly elevated.

Cellular ATP levels were significantly reduced validating the execution of apoptosis in SNO cancer cells. Aqueous MOE also showed potential against human pancreatic cancer in vitro (Berkovich et al. 2013). MOE inhibited Panc-1 cell growth. MOE influenced the cell cycle as it caused the accumulation of cells in the G1 phase. It also affected the NFkB pathway by reducing expression of p65, p-IkBα and IkBα. Interestingly it worked synergistically with cisplatin to cause cytotoxicity. Methanol extract of MF showed a dose-dependent decline in PC3 (pancreatic cancer) cell viability with no effect on normal Vero cells (Inbathamizh and Padmini 2013). This growth inhibition effect was attributed to the constituents of the flower extract.

The ethanolic seed extract displayed anticancer potential in Raji cells by the inhibition of Epstein-Barr virus early antigen activation (Guevara et al. 1999). In addition the bioactive compound, niazimicin, showed anticancer activity against two-stage carcinogenesis in mice. An ethanol-water extract of MOE was assessed for cytotoxicity against cancerous HeLa cells (Jafarain et al. 2014). A dose-dependent decline in cell viability was seen. The cytotoxic effect of MOE was greater than 50%. Nair and Varalakshmi (2011) showed that the aqueous MOE displayed anticancer activity in HeLa cells (Nair and Varalakshmi 2011). MOE showed a dose-dependent decline in cell viability with no cytotoxicity on normal healthy PBMCs.

13.6 *Moringa oleifera:* Potential in Nanoparticle Synthesis

MO has shown potential as an antibacterial and anticancer agent. These properties can be attributed to the phytochemical composition of the extracts (Table 13.2).

The phytochemical analysis of MOE showed that it contains a variety of compounds such as phenolic acids and flavonoids including catechol, itaconic acid and gallic acid (Luqman et al. 2012; Belliraj et al. 2015; El Sohaimy et al. 2015). Flavonoids are composed of flavones, flavanones, flavonols, isoflavonoids and anthocyanins (Makarov et al. 2014). These compounds play a role in chelation and

Table 13.2 Phytochemical constituents of *Moringa oleifera*

	Moringa oleifera tree				
	MOE	MF	Seeds	Seedpod (drumstick)	References
Phytochemical constituents	4-(4′-*O*-acetyl-α-L-rhamnosyloxy) benzyl isothiocyanate	Protein	Alkaloids	Alkaloids	Guevara et al. (1999)
	Niaziminin A and B	Ascorbic acid	Terpenoids	Terpenoids	Luqman et al. 2012
	Niazimicin	Poly-saccharide	Sugar	Quinines	Mishra et al. (2011)
	Niazirin and niazirinin	D-mannose and D-glucose	Saponin and sterols	Saponin and sterols	Belliraj et al. (2015)

MOE leaves, *MF* flowers

are able to reduce metal ions into nanoparticles. The flavonoid conversion from enol-form to keto-form releases a hydrogen atom which reduces the metal ions to form the nanoparticles. The *Ocimum basilicum* extracts contained flavonoids such as luteolin and rosmarinic acid. The conversion of these flavonoids from their enol-form to keto-form resulted in the reduction of the Ag^+ ions for the formation of AgNPs.

MF aqueous extracts contain sugars such as D-mannose and D-glucose (Goyal et al. 2007). Polysaccharides are also present and become hydrolysed into D-glucose, D-glucuronic acid and G-galactose. The sugars present in the plant extracts aid in the synthesis of nanoparticles (Makarov et al. 2014). Monosaccharides such as glucose have the ability to cause reduction and can be utilised in the synthesis of nanoparticles. In addition, fructose (monosaccharide with a keto-group) can be used for the synthesis of Au and Ag nanoparticles. The sugar aldehyde group can be oxidised to a carboxyl group. This occurs via OH^- nucleophilic addition resulting in the reduction of metal ions. This was seen for the synthesis of AuNPs using a magnolia vine extract.

MO can be used for the synthesis of AgNPs and AuNPs. The MOE extracts were used for the synthesis of AgNPs and displayed antimicrobial activity (Prasad and Elumalai 2011). Mubayi et al. (2012) used MOE and silver nitrate for the synthesis of AgNPs (Mubayi et al. 2012). The synthesised AgNPs had antimicrobial activity against bacteria (gram-positive and gram-negative) such as the clinical strain of *E. coli*, *S. aureus*, *K. pneumoniae* and *Enterococcus faecalis* (*E. faecalis*). The seedpod (drumstick) extract was used in the synthesis of AgNPs (Sangeetha and Ponnuswami 2014). The AgNPs were characterised using ultraviolet-visible (UV-vis) spectroscopy, transmission electron microscopy (TEM) and Fourier transform infrared spectroscopy (FTIR) which showed the successful synthesis of AgNPs, and it was environmentally friendly. Chakraborty et al. (2013) used MOE for the synthesis of stabilised AuNPs (Chakraborty et al. 2013). MOE caused the reduction of chloroauric acid which produced stable nanoparticles (20–60 nm), therefore showing potential in the synthesis of nanoparticles. AuNPs synthesised from MOE showed significant antiproliferative effects in A549 lung cancer cells (Tiloke et al. 2016b). The novel AuNPs activated the alternate splicing of *caspase-9* in A549 lung cancer cells leading to the induction and execution of the apoptotic cascade. In addition, the AuNPs significantly reduced oncogenes and increased tumour suppressor genes further ensuring the cancer cell death. Bionanofibres were isolated from MO gum having size of 20 μm (Mehetre et al. 2015). The bionanofibres can be utilised as a drug delivery system by which drugs can be loaded on it for therapeutic target therapies. However limited scientific studies are available for the use of MO in the nanoparticle synthesis and their biomedical applications.

13.7 Silver Nanoparticles

Silver nanoparticles are most effective against microbial infections (Rai et al. 2009). It displays antibacterial and antiviral activity. It is used for the treatment of burns, wounds and various cosmetics. However silver can be cytotoxic to mammalian cells and can cause argyrosis. Due to antibiotic-resistant bacteria, healthcare professionals are reducing the prescription of antibiotics. Therefore the use of AgNPs and Ag-associated products has increased.

When silver reacts with moisture which is found in wounds and on the skin, it becomes ionised (Rai et al. 2009). The ionised silver targets and binds to proteins. The bacterial nuclear membrane and cell wall undergo changes caused by silver. It also binds to RNA and DNA from the bacteria preventing bacterial replication. Silver nanoparticles have a large surface area which enables greater binding capacity. They are also able to attach and penetrate cellular membranes. These nanoparticles target sulphur-containing proteins, DNA, respiratory chain, cell division and replication. In addition, the silver nanoparticles release silver ions further ensuring bactericidal effects.

Bacterial infection of wounds is a major problem experienced throughout the world (Velazquez-Velazquez et al. 2014). Due to the antibacterial property of AgNPs, commercial wound dressings have been impregnated with silver nanoparticles. It showed antibacterial and anti-biofilm activity against *P. aeruginosa* increasing wound healing outcomes.

Silver nanoparticles can also be used as anticancer agents. IMR-90 cell line (normal human lung fibroblasts) and U251 cell line (human glioblastoma cells) were used to assess the effect of AgNPs (AshaRani et al. 2009). The uptake of AgNPs was via endocytosis, and there was a uniform distribution in the cytoplasm and nucleus. AgNPs inhibited cellular proliferation of human glioblastoma cells, whereas in normal human lung fibroblasts, cellular recovery occurred. The AgNPs targeted intracellular calcium levels and cellular morphology preventing cancer cell proliferation.

Gopinath et al. (2008) showed that AgNPs induced apoptosis in cancerous HT29 cells and non-cancerous BHK21 cells (Gopinath et al. 2008). It caused mitochondrial membrane damage and endonuclease cleavage of DNA. The biological synthesised silver nanoparticles displayed anticancer effect in Dalton's lymphoma ascites (DLA) cells as there was a dose-dependent cytotoxicity seen (Sriram et al. 2010). It also activated caspase-3 activity for the execution of apoptosis. In addition AgNPs caused nuclear fragmentation. Acute toxicity testing in vivo showed that no effect on body weight and haematological parameters was seen.

Prasad and Elumalai (2011) used aqueous MOE for the synthesis of AgNPs (ML_{AgNP}) (Prasad and Elumalai 2011). MOE caused the reduction of silver ions—Ag^+ to Ag^0. It resulted in the formation of AgNPs of size 57 nm. The ML_{AgNP} caused maximum zone of inhibition against *Candida tropicalis* (*C. tropicalis*) and *S. aureus*. In addition, the minimum zone of inhibition was observed against *K. pneumoniae* and *Candida krusei* (*C. krusei*). AgNPs from MO stem bark was successfully synthesised and showed anticancer activity in human cervical cancer cells (HeLa cells) (Vasantha et al. 2014). The synthesised AgNPs caused HeLa cell apoptosis visible via morphological changes. The HeLa cells underwent membrane blebbing and shrinkage which resulted in significant apoptotic bodies. In addition, an increase in early apoptotic cells (Annexin V/PI stain) and intracellular reactive oxygen species levels was seen. MOE was used for the reduction and stabilisation of

AgNPs from aqueous silver nitrate (Das et al. 2013). The synthesised AgNPs were 40–50 nm and had an ultraviolet-visible spectrum absorption peak of 415 nm. The biomedical assessment of the synthesised AgNPs showed that it possessed antibacterial activity. Particularly AgNPs had potential against bacteria (gram-positive and gram-negative) such as *E. faecalis*, *Enterobacter cloacae* (*E. cloacae*), *E. coli*, *K. pneumoniae*, *Proteus vulgaris* (*P. vulgaris*) and *S. aureus*.

Silver nanoparticles have shown to have potential as antibacterial and anticancer agents. However exposure to large amounts of silver can result in argyrosis (Rai et al. 2009). In addition, the synthesis of AgNPs is not environmentally friendly. A study by Burd et al. (2007) showed that wound dressings impregnated with AgNPs have caused cytotoxicity to keratinocytes and fibroblasts (Burd et al. 2007). In vivo assessment of AgNPs in zebrafish embryos showed that the AgNPs were deposited in various organs which led to deformity (AshaRani et al. 2009). The different sizes of AgNPs were assessed in a rat liver cell line (BRL 3A) for 24 h (Hussain et al. 2005). The AgNPs caused mitochondrial dysfunction with increase in LDH levels in BRL 3A cells. Silver nanoparticles were cytotoxic due to oxidative stress. A significant elevation of reactive oxygen species and a significant reduction in GSH were noted. Although silver nanoparticles have shown to be beneficial in treatment, it is also toxic with many side-effects. Toxicity, efficacy and safety associated with silver have been a major concern; therefore alternate metals such as gold for synthesis of nanoparticles are now being investigated.

13.8 Gold Nanoparticles

The exploration into biocompatible and stable nanoparticles has led to the development of AuNPs (Sardar et al. 2009). AuNPs are inert and stable which makes them more biocompatible, can act as probes for the detection and diagnosis of cancer (Boisselier and Astruc 2009), and can be conjugated to antibodies which further can facilitate cancer cell targeting as certain cancers overexpress receptor proteins such as EGFR. AuNPs were synthesised for the diagnosis and treatment of Alzheimer's, HIV infection, hepatitis B, TB, rheumatoid arthritis and diabetes mellitus type 2. AuNPs play an important role in photothermal cancer therapy as AuNPs are readily absorbed, are non-toxic and require less irradiation energy. The smaller AuNPs (10–30 nm) are easily distributed and absorbed by cancer cells. AuNPs (10–40 nm) also display antibacterial activity against gram-positive *S. aureus*.

AuNPs are favoured over AgNPs as they are less toxic and more biocompatible (Schlinkert et al. 2015). The conventional method of administration of chemotherapeutic agents is intravenous, and surface charge of the nanoparticle is important as the NPs with a positive charge remain longer in the bloodstream. It has been shown that the mechanism by which nanoparticles exert the anticancer effect is via oxidative stress. The induction of oxidative stress has resulted in apoptosis. This is important as cancer cells evade apoptosis and continue to proliferate. Therefore the development of NPs to specifically target and induce apoptosis is imperative.

AuNPs have potential in nasopharyngeal carcinoma therapy (Lan et al. 2013). The AuNPs (20.5 nm) caused a dose-dependent decline in the cancer cell viability. The TUNEL assay showed that the AuNPs induced nasopharyngeal cancer cell apoptosis. Belliraj et al. (2015) used MO seedpod (drumstick) in the synthesis of AuNPs (Belliraj et al. 2015). The seedpod extract caused the reduction of chloroauric acid which resulted in the AuNP's formation. The synthesised AuNPs protected the liver cells ($HepG_2$ cells) against acetaminophen- and carbon tetrachloride-induced toxicity. Using an aqueous extract of *Garcinia mangostana*, AuNPs were synthesised as the components such as flavonoids, anthocyanins and benzophenones reduced the Au metal ions (Lee et al. 2016). The synthesised AuNPs had an absorption peak of 540–550 nm and were spherical in shape (size = 32.96–5.25 nm). The newly synthesised AuNPs were environmentally friendly, thereby showing potential as therapeutic agents. *Peltophorum pterocarpum* flower was used in synthesis of AuNPs (Balamurugan et al. 2016). The flower extract was used as an agent for the reduction and capping of the AuNPs. The synthesised AuNPs had an absorption of 560 nm and size of 5–50 nm.

13.9 Therapeutic Potential of Gold Nanoparticles

13.9.1 Antibacterial

AuNPs conjugated to antibodies and antibiotics are effective against protozoa and bacteria (Burygin et al. 2009). AuNPs conjugated to vancomycin inhibited bacterial growth. MtB has a major impact on many people worldwide (Zhou et al. 2012). The TB vaccine bacillus Calmette-Guérin (BCG) which was prepared from attenuated *Mycobacterium bovis* (*M. bovis*) has lost its virulence against bovine TB. Through the emergence of drug resistance, NPs have been developed and showing potential as antibacterial agents. The interaction between NPs and bacterium has led to absorption, membrane damage and toxicity. AuNPs and AgNPs on *E. coli* were investigated and demonstrated antibacterial activity. The study showed that PAH-capped AuNPs caused bacterial cell lysis, and AgNPs were toxic due to their chemical composition.

Due to the NP's physical property, it facilitated efficient binding and interactions with bacteria (Li et al. 2014). NPs conjugated to existing antibiotics via covalent or non-covalent bonds will enable to combat MDR bacteria. Both methods have improved the efficacy of the NPs as it enhanced internalisation and concentration at site of bacterial infection. AuNPs (2 nm) showed antimicrobial activity against multiple strains of uropathogens and methicillin-resistant *S. aureus* (Li et al. 2014). Interestingly after 20 generations, there were no signs of AuNP's bacterial resistance. It was compatible with mammalian cells with greater than 80% cell viability.

Malathi et al. (2013) used chitosan for the synthesis of AuNPs. The synthesised AuNPs served as a carrier for hydrophobic drugs such as rifampicin, first-line drug for TB. The size of the AuNPs was 50 nm with a surface plasmon resonance of

526 nm. The loading efficiency of rifampicin was 71%, and assessment of the drug release showed that it followed a Higuchi model. It also was effective against gram-positive and gram-negative bacteria such as *Bacillus subtilis* (*B. subtilis*) and *P. aeruginosa*, respectively.

13.9.2 Anticancer

AuNPs were synthesised using bioactive compounds from aqueous *Gymnema sylvestre* extracts (Arunachalam et al. 2014). The synthesised AuNP's size and shape were characterised using UV-vis spectroscopy, FTIR, scanning electron microscopy (SEM), X-ray diffraction (XRD) and energy dispersive X-ray analysis (EDAX). The 3-(4,5-Dimethyl-2-thiazolyl)-2,5-diphenyl-2H-tetrazolium bromide (MTT) assay was used to access the effect of the synthesised AuNPs on cell viability in human colon adenocarcinoma cells (HT-29 cells). The study showed that the bioactive compound's anticancer activity may be increased through conjugation with AuNPs facilitating enhanced uptake in cells.

Kang et al. (2010) assessed the effects of AuNPs on the cancer cell's nucleus (Kang et al. 2010). The nucleus is the main hub of the cell controlling all cellular processes; therefore by targeting the nucleus using AuNPs, the cell's DNA becomes compromised, negatively affecting downstream pathways. The synthesised AuNPs caused cytokinesis arrest in human oral squamous cell carcinoma (HSC). This was evident by the formation of binucleate cells at late mitosis stages which resulted in inhibition of cell division. In addition, DNA damage also occurred which culminated in apoptosis. The effect of AuNPs on human dermal fibroblasts showed that AuNPs penetrate plasma membranes (Mironava et al. 2010). It accumulated in vacuoles. The study showed that size of AuNPs will determine the mode of cellular uptake. AuNPs of size 13 nm enter the cell via phagocytosis, whereas larger AuNPs (45 nm) enter via clathrin-mediated endocytosis. The AuNPs caused disruption to the cytoskeleton filament, collagen and fibronectin. In addition, the cells were able to recover once the AuNPs were removed.

13.10 Gold Nanoparticles: In Vitro Evaluation

Mohan et al. (2013) synthesised citrate and polyethyleneimine AuNPs (Mohan et al. 2013). The citrate AuNPs caused apoptosis in A549 lung cancer cells as changes in nuclear morphology were seen. Also, there was significant fragmentation of the DNA. The polyethyleneimine AuNPs did not induce apoptosis in human breast cancer cells (MCF-7 cells) and A549 cells. AuNP's mode of action is via oxidative stress (Lim et al. 2011). The AuNPs induced HeLa cell death via reactive oxygen species leading to lipid and protein damage. It also targeted the mitochondria and reduced mitochondrial function. AuNPs also influenced inflammatory and cell cycle genes. AuNPs of size 20 nm caused oxidative stress in MRC-5 cells. It also decreased cell cycle genes such as cyclin B1 and B2.

Song et al. (2013) investigated the uptake and localisation of AuNPs and glucose-capped AuNPs in HeLa and MCF-7 cancer cells (Song et al. 2013). The study showed that the glucose-capped AuNPs had a greater uptake in cells in comparison to plain AuNPs. Endocytosis occurred and the AuNPs localised in the cytoplasm. They concluded that glucose-capped AuNPs showed potential in diagnosing and treating cancer. AuNPs were also assessed in human breast epithelial cancer cells (MCF-7 cells) (Selim and Hendi 2012). After exposure to AuNPs for 24 h, key apoptotic markers were elevated for the induction of apoptosis in the MCF-7 cells. p53, Bax, caspase-9 and caspase-3 were increased. In addition anti-apoptotic bcl-2 was decreased to effectively execute apoptosis. *Carica papaya* and *Catharanthus roseus* were used singly and in combination for the synthesis of AuNPs (Muthukumar et al. 2016). The proteins, flavonoids and alkaloids from the plant extracts were responsible for the synthesis of the AuNPs. In addition, their shape was mainly spherical and displayed anticancer and antibacterial activity. The combination of the extracts worked synergistically thereby enhancing their therapeutic effect.

The effect of AuNPs conjugated to anti-epidermal growth factor receptor was investigated in non-cancerous epithelial cell line (HaCaT) and two cancerous oral epithelial cell lines (HOC313 clone 8 and HSC3) (El-Sayed et al. 2005). The conjugated AuNPs had a 600% greater affinity to the cancerous cells than non-cancerous cells. This resulted in better surface plasmon resonance absorption which could be used for diagnosis. AuNPs with and without sodium citrate residues on their surface were assessed in A549 and NCIH441 cells (Uboldi et al. 2009). AuNPs with sodium citrate residues decreased cell viability. However, with an increase in sodium citrate residues on AuNPs, it caused lactate dehydrogenase (LDH) release. Using TEM, AuNPs entered the cells via endocytosis within 3 h of exposure. In addition, the sodium citrate residues did not influence the uptake of the AuNPs. Green synthesis of AuNPs using *Cassia tora* was assessed in colon cancer cells (Col320 cells) (Abel et al. 2015). Aqueous *Cassia tora* leaf extract was used in the reduction of tetrachloroauric acid which resulted in AuNPs. The AuNPs were characterised and assessed for antibacterial, antioxidant and anticancer activity. The synthesised AuNPs displayed greater bioavailability with a size of 41 nm. AuNPs decreased Col320 cell viability and decreased reactive oxygen species—decreasing lipid peroxidation and nitric oxide production. Antioxidant, catalase activity was significantly elevated. The study showed the possible use of *Cassia tora*-mediated AuNPs against colon cancer.

13.11 Gold Nanoparticles: In Vivo Evaluation

AuNPs exert their effect in vivo via active and passive targeting (Tomar and Garg 2013). Active targeting recognises ligands and receptors on target cells, whereas passive targeting depends on the diffusion through blood vessels. AuNP-exposed *Mytilus edulis* (blue mussel) showed elevated levels of reactive oxygen species after 24 h (Tedesco et al. 2010). Biodistribution of AuNPs is important for its mode of action as size, shape and charge determine which area of the body it accumulates (De Jong et al. 2008). Smaller AuNPs (10 nm) are distributed throughout the body via blood circulation and are able to interact with many organs including the heart, kidney, liver, immune system, brain and reproductive system. In addition, larger AuNPs (50–250) are distributed to the liver and spleen.

Effective therapy for the brain is challenging due to the blood-brain barrier (Siddiqi et al. 2012). It is composed of endothelial cells with tight junctions that facilitate and restrict the transport of molecules across. The effect of AuNPs on the brain is important to be determined. The biochemical effects of AuNPs were assessed in rat brain (Siddiqi et al. 2012). Male Wistar rats were injected with 20 μg/kg body weight of 20 nm AuNPs. There was a significant increase in oxidative stress in the rat brain—increasing lipid peroxidation and decreased glutathione peroxidase levels. There was also an increase in 8-hydroxydeoxyguanosine, heat shock protein 70 and caspase-3 activity. Serotonin, dopamine and interferon-gamma (IFN-γ) were decreased. The study showed that the AuNPs influenced inflammation, DNA damage and apoptosis in rats.

Visaria et al. (2006) synthesised 33 nm polyethylene glycol-coated colloidal AuNPs conjugated to tumour necrosis factor-α (TNF-α) (Visaria et al. 2006). TNF-α is a cytokine which has anticancer activity. The concentration by which TNF-α was conjugated was to ensure anticancer activity with minimal systemic toxicity. SCK murine mammary carcinoma model (mammary carcinoma in A/J mice) was treated with the synthesised AuNPs and subjected to heating (42.5 °C; water bath). There was a significant reduction in tumour growth in the mice. It also reduced blood flow to the tumour significantly reducing their nutrient supply for cell proliferation.

Conclusion

Cancer mortality and bacterial infections are increasing at a rapid rate. The high cancer mortality rates are due to late-stage detection which increases the aggressive nature of cancer cells. Therefore, early detection and diagnosis will allow for treatment with positive outcomes. Unfortunately, the cost associated with early detection and subsequent treatments is relatively high, in addition to the many side-effects, which necessitates for complementary and alternative medicines (CAM). CAM can be beneficial as it is not only cost-effective but also induces cancerous cell death. In addition, antibacterial agents are also expensive and resistance to these agents is increasing. The use of CAM for antibacterial activity can also be beneficial. Furthermore, plant extracts can be used for the synthesis of AuNPs in an environmentally friendly synthesis. AuNPs are being favoured over AgNPs as they are biocompatible and less cytotoxic. Their biomedical application is endless and is showing positive results as anticancer and antibacterial agents. MO, a medicinal tree, is found in SA with a wide range of traditional uses. MO extracts can be used for the synthesis of NPs and have shown medical usage. In conclusion, SA's medicinal tree MO is beneficial displaying a wide range of therapeutic effects.

13.12 Perspectives

In SA, majority of the population are relatively poor and rely on traditional medicines for many ailments such as cancer and bacterial infections. The medicinal tree, MO, forms part of the remedies by many traditional healers. MO has been utilised in many communities throughout the years, and their traditional uses have been passed down through generations. Although the extracts have seen to be beneficial with positive outcomes, their exact mechanism of action was largely to be understood. Nature has provided us with the unique multipurpose tree, and both the crude and isolated bioactive compounds have potential for therapies. Nanoparticles are also showing potential and can be easily synthesised using plant extracts such as MO. Current therapies have many side-effects, and therefore the use of natural resources in conjunction with their synthesised NPs will be beneficial for treatment. In addition, by using natural alternatives for synthesis, it eliminates chemical usage which is beneficial for the environment. Their therapeutic mode of action ranges from inhibition of bacterial and cell growth and the induction of cancer cell apoptosis. This is important especially for patient target therapy as existing drugs are nonspecific. It is therefore recommended that MO and their synthesised NPs show potential for therapies.

Acknowledgement Dr. C. Tiloke acknowledges the prestigious postdoctoral scholarship from the National Research Foundation, SA.

References

Abel EE, Poonga PRJ, Panicker SG (2015) Characterization and *in vitro* studies on anticancer, antioxidant activity against colon cancer cell line of gold nanoparticles capped with *Cassia tora* SM leaf extract. Appl Nanosci 6:121–129. doi:10.1007/s13204-015-0422-x

Anand K, Gengan RM, Phulukdaree A, Chuturgoon A (2014) Agroforestry waste *Moringa oleifera* petals mediated green synthesis of gold nanoparticles and their anti-cancer and catalytic activity. J Ind Eng Chem 21:1105–1111. doi:10.1016/j.jiec.2014.05.021

Anwar F, Bhanger MI (2003) Analytical characterization of *Moringa oleifera* seed oil grown in temperate regions of Pakistan. J Agric Food Chem 51:6558–6563. doi:10.1021/jf0209894

Anwar F, Latif S, Ashraf M, Gilani AH (2007) *Moringa oleifera*: A food plant with multiple medicinal uses. Phytother Res 21:17–25. doi:10.1002/ptr.2023

Arunachalam KD, Arun LB, Annamala SK, Arunachalam AM (2014) Biofunctionalized gold nanoparticles synthesis from *Gymnema sylvestre* and its preliminary anticancer activity. Int J Pharm Pharm Sci 6:423–430

AshaRani PV, Hande MP, Valiyaveettil S (2009) Anti-proliferative activity of silver nanoparticles. BMC Cell Biol 10:65. doi:10.1186/1471-2121-10-65

Awodele O, Oreagba IA, Odoma S, da Silva JAT, Osunkalu VO (2012) Toxicological evaluation of the aqueous leaf extract of *Moringa oleifera* Lam. (Moringaceae). J Ethnopharmacol 139:330–336. doi:10.1016/j.jep.2011.10.008

Azam A, Ahmed F, Arshi N, Chaman M, Naqvi AH (2009) One step synthesis and characterization of gold nanoparticles and their antibacterial activities against *E. coli* (ATCC 25922 strain). Int J Theor Appl Sci 1:1–4

Baddeley A, Dean A, Dias HM, Falzon D, Floyd K, Baena IG, Gilpin C, Glaziou P, Hiatt T, Law I, Lienhardt C, Likhite N, Nguyen L, Siroka A, Sismanidis C, Timmi H, van Gemert W, Zignol M, Raviglione M (2014) Global tuberculosis report 2014. WHO Press, Switzerland. 171p

Bailo R, Bhatt A, Ainsa JA (2015) Lipid transport in *Mycobacterium tuberculosis* and its implications in virulence and drug development. Biochem Pharmacol 96:159–167. doi:10.1016/j.bcp.2015.05.001

Balamurugan M, Kaushik S, Saravanan S (2016) Green synthesis of gold nanoparticles by using *Peltophorum pterocarpum* flower extracts. Nano Biomed Eng 8:213–218. doi:10.5101/nbe.v8i4.p213-218

Belliraj TS, Nanda A, Ragunathan R (2015) *In-vitro* hepatoprotective activity of *Moringa oleifera* mediated synthesis of gold nanoparticles. J Chem Pharm Res 7:781–788

Bello B, Fadahun O, Kielkowski D, Nelson G (2011) Trends in lung cancer mortality in South Africa: 1995-2006. BMC Public Health 11:1–5. doi:10.1186/1471-2458-11-209

Berkovich L, Earon G, Ron I, Rimmon A, Vexler A, Lev-Ari S (2013) *Moringa oleifera* aqueous leaf extract down-regulates nuclear factor-kappaB and increases cytotoxic effect of chemotherapy in pancreatic cancer cells. BMC Complement Altern Med 13:1–7. doi:10.1186/1472-6882-13-212

Bertram JS (2001) The molecular biology of cancer. Mol Aspects Med 21:167–223. doi:10.1016/S0098-2997(00)00007-8

Boisselier E, Astruc D (2009) Gold nanoparticles in nanomedicine: preparations, imaging, diagnostics, therapies and toxicity. Chem Soc Rev 38:1759–1782. doi:10.1039/b806051g

Budda S, Butryee C, Tuntipopipat S, Rungsipipat A, Wangnaithum S, Lee J, Kupradinun P (2011) Suppressive effects of *Moringa oleifera* Lam pod against mouse colon carcinogenesis induced by azoxymethane and dextran sodium sulfate. Asian Pac J Cancer Prev 12:3221–3228

Burd A, Kwok CH, Hung SC, Chan HS, Gu H, Lam WK, Huang L (2007) A comparative study of the cytotoxicity of silver-based dressings in monolayer cell, tissue explant, and animal models. Wound Repair Regen 15:94–104. doi:10.1111/j.1524-475X.2006.00190.x

Burygin GL, Khlebtsov BN, Shantrokha AN, Dykman LA, Bogatyrev VA, Khlebtsov NG (2009) On the enhanced antibacterial activity of antibiotics mixed with gold nanoparticles. Nanoscale Res Lett 4:794–801. doi:10.1007/s11671-009-9316-8

Cai W, Gao T, Hong H, Sun J (2008) Applications of gold nanoparticles in cancer nanotechnology. Nanotechnol Sci Appl 1:17–32. doi:10.2147/NSA.S3788

Chakraborty A, Das K, Sinha M, Dey S, Bhattacharjee S (2013) *Moringa oleifera* leaf extract mediated green synthesis of stabilized gold nanoparticles. J Bionanosci 7:1–5. doi:10.1166/jbns.2013.1149

Chueh PJ, Liang R, Lee Y, Zeng Z, Chuang S (2014) Differential cytotoxic effects of gold nanoparticles in different mammalian cell lines. J Hazard Mater 264:303–312. doi:10.1016/j.jhazmat.2013.11.031

Circu ML, Aw TY (2012) Glutathione and modulation of cell apoptosis. Biochim Biophys Acta Mol Cell Res 1823:1767–1777. doi:10.1016/j.bbamcr.2012.06.019

Das S, Parida UK, Bindhani BK (2013) Green biosynthesis of silver nanoparticles using *Moringa oleifera* L. leaf. Int J Nanotechnol Appl 3:51–62

De Jong WH, Hagens WI, Krystek P, Burger MC, Sips AJAM, Geertsma RE (2008) Particle size-dependent organ distribution of gold nanoparticles after intravenous administration. Biomaterials 29:1912–1919. doi:10.1016/j.biomaterials.2007.12.037

Djakalia B, Guichard BL, Soumaila D (2011) Effect of *Moringa oleifera* on growth performance and health status of young post-weaning rabbits. Res J Poultry Sci 4:7–13. doi:10.3923/rjpscience.2011.7.13

El Sohaimy SA, Hamad GM, Mohamed SE, Amar MH, Al-Hindi RR (2015) Biochemical and functional properties of *Moringa oleifera* leaves and their potential as a functional food. GARJAS 4:188–199

El-Sayed IH, Huang X, El-Sayed MA (2005) Surface plasmon resonance scattering and absorption of anti-EGFR antibody conjugated gold nanoparticles in cancer diagnostics: applications in oral cancer. Nano Lett 5:829–834. doi:10.1021/nl050074e

Eruteya OC, Badon B (2014) Antilisterial activity of ethanolic and aqueous extracts of the leaves and seeds of *Moringa oleifera*. Asia J Appl Microbiol 1:60–65

Fahey JW (2005) *Moringa oleifera*: A review of the medical evidence for its nutritional, therapeutic, and prophylactic properties. Part 1. Trees for Life J 1:5

Fan T, Han L, Cong R, Liang J (2005) Caspase family proteases and apoptosis. Acta Biochim Biophys Sin 37:719–727. doi:10.1111/j.1745-7270.2005.00108x

Faraji AH, Wipf P (2009) Nanoparticles in cellular drug delivery. Bioorg Med Chem 17:2950–2962. doi:10.1016/j.bmc.2009.02.043

Gengan RM, Anand K, Phulukdaree A, Chuturgoon A (2013) A549 lung cell line activity of biosynthesized silver nanoparticles using *Albizia adianthifolia* leaf. Colloids Surf B Biointerfaces 105:87–91. doi:10.1016/j.colsurfb.2012.12.044

Globocan (2012) Estimated cancer incidence, mortality, prevalence and disability-adjusted life years (DALYs) worldwide in 2012 (http://globocan.iarc.fr/)

Gopinath P, Gogoi SK, Chattopadhyay A, Ghosh SS (2008) Implications of silver nanoparticle induced cell apoptosis for *in vitro* gene therapy. Nanotechnology 19:1–10. doi:10.1088/0957-4484/19/7/075104

Govender R, Phulukdaree A, Gengan RM, Anand K, Chuturgoon AA (2013) Silver nanoparticles of *Albizia adianthifolia*: the induction of apoptosis in human lung carcinoma cell line. BMC. J Nanobiotechnol 11:1–9. doi:10.1186/1477-3155-11-5

Goyal BR, Agrawal BB, Goyal RK, Mahta AA (2007) Phyto-pharmacology of *Moringa oleifera* Lam an overview. Nat Prod Radiance 6:347–353

Granich R, Kahn JG, Bennett R, Holmes CB, Garg N, Serenata C, Sabin ML, Makhlouf-Obermeyer C, Mack CDF, Williams P, Jones L, Smyth C, Kutch KA, Ying-Ru L, Vitoria M, Souteyrand Y, Crowley S, Korenromp EL, Williams BG (2012) Expanding ART for treatment and prevention of HIV in South Africa: Estimated cost and cost-effectiveness 2011-2050. PLoS One 7:1–15. doi:10.1371/journal.pone.0030216

Guevara AP, Vargas C, Sakurai H, Fujiwara Y, Hashimoto K, Maoka T, Kozuka M, Ito Y, Tokuda H, Nishino H (1999) An antitumor promoter from *Moringa oleifera* Lam. Mutat Res 440:181–188. doi:10.1016/S1383-5718(99)00025-X

Hengartner MO (2000) The biochemistry of apoptosis. Nature 407:770–776. doi:10.1038/35037710

Hontelez JAC, Lurie MN, Barnighausen T, Bakker R, Baltussen R, Tanser F, Hallett TB, Newell M, de Vlas SJ (2013) Elimination of HIV in South Africa through expanded access to antiretroviral therapy: a model comparison study. PLoS Med 10:1–14. doi:10.1371/journal.pmed.1001534

Hussain SM, Hess KL, Gearhart JM, Geiss KT, Schlager JJ (2005) *In vitro* toxicity of nanoparticles in BRL 3A rat liver cells. Toxicol In Vitro 19:975–983. doi:10.1016/j.tiv.2005.06.034

Inbathamizh L, Padmini E (2013) Evaluation of growth inhibitory potential of *Moringa oleifera* flowers on PC3 cell lines. Asian J Pharm Clin Res 6:60–64

Jafarain A, Asghari G, Ghassami E (2014) Evaluation of cytotoxicity of *Moringa oleifera* Lam. callus and leaf extracts on Hela cells. Adv Biomed Res 3:1–5. doi:10.4103/2277-9175.140668

Kanapathipillai M, Brock A, Ingber DE (2014) Nanoparticle targeting of anti-cancer drugs that alter intracellular signaling or influence the tumor microenvironment. Adv Drug Deliv Rev 79-80:107–118. doi:10.1016/j.addr.2014.05.005

Kang B, Mackey MA, El-Sayed MA (2010) Nuclear targeting of gold nanoparticles in cancer cells induces DNA damage, causing cytokinesis arrest and apoptosis. J Am Chem Soc 132:1517–1519. doi:10.1021/ja9102698

Kasolo JN, Bimenya GS, Ojok L, Ogwal-okeng JW (2011) Phytochemicals and acute toxicity of *Moringa oleifera* roots in mice. J Pharmacogn Phytother 3:38–42

Krithiga N, Rajalakshmi A, Jayachitra A (2015) Green synthesis of silver nanoparticles using leaf extracts of *Clitoria ternatea* and *Solanum nigrum* and study of its antibacterial effect against common nosocomial pathogens. J Nanosci 2015:1–8. doi:10.1155/2015/928204

Kuppusamy P, Ichwan SJA, Parine NR, Yusoff MM, Maniam GP, Govindan N (2015) Intracellular biosynthesis of Au and Ag nanoparticles using ethanolic extract of *Brassica oleracea* L. and studies on their physicochemical and biological properties. J Environ Sci 29:151–157. doi:10.1016/j.jes.2014.06.050

Lan M, Hsu Y, Hsu C, Ho C, Lin J, Lee S (2013) Induction of apoptosis by high-dose gold nanoparticles in nasopharyngeal carcinoma cells. Auris Nasus Larynx 40:563–568. doi:10.1016/j.anl.2013.04.011

Lee KX, Shameli K, Miyake M, Kuwano N, Khairudin NBBA, Mohamad SEB, Yew YP (2016) Green synthesis of gold nanoparticles using aqueous extract of *Garcinia mangostana* fruit peels. J Nanomater 2016:1–7. doi:10.1155/2016/8489094

Li X, Robinson SM, Gupta A, Saha K, Jiang Z, Moyano DF, Sahar A, Riley MA, Rotello VM (2014) Functional gold nanoparticles as potent antimicrobial agents against multi-drug-resistant bacteria. ACS Nano 8:10682–10686. doi:10.1021/nn5042625

Lim ZJ, Li JJ, Ng C, Yung L, Bay B (2011) Gold nanoparticles in cancer therapy. Acta Pharmacol Sin 32:983–990. doi:10.1038/aps.2011.82

Luqman S, Srivastava S, Kumar R, Maurya AK, Chanda D (2012) Experimental assessment of *Moringa oleifera* leaf and fruit for its antistress, antioxidant, and scavenging potential using *in vitro* and *in vivo* assays. Evid Based Complement Alternat Med 2012:1–12. doi:10.1155/2012/519084

Makarov VV, Love AJ, Sinitsyna OV, Makarova SS, Yaminsky IV, Taliansky ME, Kalinina NO (2014) "Green" nanotechnologies: synthesis of metal nanoparticles using plants. Acta Naturae 6:35–44

Malathi S, Balakumaran MD, Kalaichelvan PT, Balasubramanian S (2013) Green synthesis of gold nanoparticles for controlled delivery. Adv Mat Lett 4:933–940. doi:10.5185/amlett.2013.5477

Markman JL, Rekechenetskiy A, Holler E, Ljubimova JY (2013) Nanomedicine therapeutic approaches to overcome cancer drug resistance. Adv Drug Deliv Rev 65:1866–1879. doi:10.1016/j.addr.2013.09.019

Massiello A, Chalfant CE (2006) SRp30a (ASF/SF2) regulates the alternative splicing of caspase-9 pre-mRNA and is required for ceramide-responsiveness. J Lipid Res 47:892–897. doi:10.1194/jlr.C600003-JLR200

Mayosi BM, Flisher AJ, Lalloo UG, Sitas F, Tollman SM, Bradshaw D (2009) The burden of non-communicable diseases in South Africa. Lancet 374:934–947. doi:10.1016/S0140-6736(09)61087-4

Mehetre G, Pande V, Kendre P (2015) Isolation and characterization of bionanofibers from *Moringa oleifera* gum as a platform for drug delivery. Nanosci Nanotechnol Res 3:1–5. doi:10.12691/nnr-3-1-1

Mendelson M (2014) Patient empowerment as a driver of hand hygiene practice: time for patients in South Africa to have their say. South Afr J Infect Dis 29:1–4

Mendis S, Armstrong T, Bettcher D, Branca F, Lauer J, Mace C, Poznyak V, Riley L, Da Costa E, Silva V, Stevens G (2014) Global status report on noncommunicable diseases. WHO, Geneva. 280p

Mironava T, Hadjiargyrou M, Simon M, Jurukovski V, Rafailovich MH (2010) Gold nanoparticles cellular toxicity and recovery: Effect of size, concentration and exposure time. Nanotoxicology 4:120–137. doi:10.3109/17435390903471463

Mishra G, Singh P, Verma R, Kumar S, Srivastav S, Jha KK, Khosa RL (2011) Traditional uses, phytochemistry and pharmacological properties of *Moringa oleifera* plant: An overview. Pharm Lett 3:141–164

Mohan JC, Praveen G, Chennazhi KP, Jayakumar R, Nair SV (2013) Functionalised gold nanoparticles for selective induction of in vitro apoptosis among human cancer cell lines. J Exp Nanosci 8:32–45. doi:10.1080/17458080.2011.557841

MubarakAli D, Thajuddin N, Jeganathan K, Gunasekaran M (2011) Plant extract mediated synthesis of silver and gold nanoparticles and its antibacterial activity against clinically isolated pathogens. Colloids Surf B Biointerfaces 85:360–365. doi:10.1016/j.colsurfb.2011.03.009

Mubayi A, Chatterji S, Rai PM, Watal G (2012) Evidence based green synthesis of nanoparticles. Adv Mat Lett 3:519–525. doi:10.5185/amlett.2012.icnano.353

Mudzviti T, Maponga CC, Khoza S, Ma Q, Morse GD (2012) The impact of herbal drug use on adverse drug reaction profiles of patients on antiretroviral therapy in Zimbabwe. AIDS Res Treat 2012:1–4. doi:10.1155/2012/434171

Muthukumar T, Sudhakumari SB, Aravinthan A, Sastry TP, Kim J (2016) Green synthesis of gold nanoparticles and their enhanced synergistic antitumor activity using $HepG_2$ and MCF7 cells and its antibacterial effects. Process Biochem 51:384–391. doi:10.1016/j.procbio.2015.12.017

Nair S, Varalakshmi KN (2011) Anticancer, cytotoxic potential of *Moringa oleifera* extracts on HeLa cell line. J Nat Pharm 2:138–142. doi:10.4103/2229-5119.86260

Nepolean P, Anitha J, Emilin RR (2009) Isolation, analysis and identification of phytochemicals of antimicrobial activity of *Moringa oleifera* Lam. Curr Biotica 3:33–39

Nithipongvanitch R, Ittarat W, Velez JM, Zhao R, St Clair DK, Oberley TD (2007) Evidence for p53 as guardian of the cardiomyocyte mitochondrial genome following acute adriamycin treatment. J Histochem Cytochem 55:629–639. doi:10.1369/jhc.6A7146. 2007

Pajares MJ, Ezponda T, Catena R, Calvo A, Pio R, Montuenga LM (2007) Alternate splicing: an emerging topic in molecular and clinical oncology. Lancet Oncol 8:349–357. doi:10.1016/S1470-2045(07)70104-3

Patel VR, Pramod S, Rao KS (2014) Cambial activity, annual rhythm of xylem production in relation to phenology and climatic factors and lignification pattern during xylogenesis in drumstick tree (*Moringa oleifera*). Flora 209:556–566. doi:10.1016/j.flora.2014.08.002

Prasad TNVKV, Elumalai EK (2011) Biofabrication of Ag nanoparticles using *Moringa oleifera* leaf extract and their antimicrobial activity. Asian Pac J Trop Biomed 1(6):439–442. doi:10.1016/S2221-1691(11)60096-8

Radhakrishna Pillai G, Srivastava AS, Hassanein T, Chauhan DP, Carrier E (2004) Induction of apoptosis in human lung cancer cells by curcumin. Cancer Lett 208:163–170. doi:10.1016/j.canlet.2004.01.008

Rai M, Yadav A, Gade A (2009) Silver nanoparticles as a new generation of antimicrobials. Biotechnol Adv 27:76–83. doi:10.1016/j.biotechadv.2008.09.002

Rybicki EP, Chikwamba R, Koch M, Rhodes JI, Groenewald J (2012) Plant-made therapeutics: An emerging platform in South Africa. Biotechnol Adv 30:449–459. doi:10.1016/j.biotechadv.2011.07.014

Sanchez-Machado D, Nunez-Gastelum JA, Reyes-Moreno C, Ramirez-Wong B, Lopez-Cervantes J (2010) Nutritional quality of edible parts of *Moringa oleifera*. Food Anal Methods 3:175–180. doi:10.1007/s12161-009-9106-z

Sangeetha V, Ponnuswami V (2014) Synthesis and characterization of silver nanoparticles using Moringa (*Moringa oleifera* Lam.) pod extract. Trends Biosci 7:675–679

Sardar R, Funston AM, Mulvaney P, Murray RW (2009) Gold nanoparticles: past, present, and future. Langmuir 25:13840–13851. doi:10.1021/la9019475

Schlinkert P, Casals E, Boyles M, Tischler U, Hornig E, Tran N, Zhao J, Himly M, Riediker M, Oostingh GJ, Puntes V, Duschl A (2015) The oxidative potential of differently charged silver and gold nanoparticles on three human lung epithelial cell types. J Nanobiotechnol 13:1–18. doi:10.1186/s12951-014-0062-4

Selim ME, Hendi AA (2012) Gold nanoparticles induce apoptosis in MCF-7 human breast cancer cells. Asian Pac J Cancer Prev 13:1617–1620. doi:10.7314/APJCP.2012.13.4.1617

Shankar SS, Rai A, Ahmad A, Sastry M (2004) Rapid synthesis of Au, Ag, and bimetallic Au core—Ag shell nanoparticles using neem (*Azadirachta indica*) leaf broth. J Colloid Interface Sci 275:496–502. doi:10.1016/j.jcis.2004.03.003

Shankarling G, Lynch KW (2010) Living or dying by RNA processing: caspase expression in NSCLC. J Clin Invest 120:3798–3801. doi:10.1172/JCI45037

Sharma H, Parihar L, Parihar P (2011) Review on cancer and anticancerous properties of some medicinal plants. J Med Plant Res 5:1818–1835

Sheridan J, Collins AM (2013) Adult lung cancer in Southern Africa: epidemiology and aetiology. Afr J Respir Med 8:10–12

Shultz JC, Goehe RW, Wijesinghe DS, Murudkar C, Hawkins AJ, Shay JW, Minna JD, Chalfant CE (2010) Alternative splicing of caspase 9 is modulated by the phosphoinositide 3-kinase/Akt pathway via phosphorylation of SRp30a. Cancer Res 70:9185–9196. doi:10.1158/0008-5472.CAN-10-1545

Shultz JC, Goehe RW, Murudkar CS, Wijesinghe DS, Mayton EK, Massiello A, Hawkins AJ, Mukerjee P, Pinkerman RL, Park MA, Chalfant CE (2011) SRSF1 regulates the alternative splicing of caspase 9 via a novel intronic splicing enhancer affecting the chemotherapeutic sensitivity of non-small cell lung cancer cells. Mol Cancer Res 9:889–900. doi:10.1158/1541-7786.MCR-11-0061

Siddiqi NJ, Abdelhalim MAK, El-Ansary AK, Alhomida AS, Ong WY (2012) Identification of potential biomarkers of gold nanoparticle toxicity in rat brains. J Neuroinflammation 9:1–7. doi:10.1186/1742-2094-9-123

Singh BN, Singh BR, Singh RL, Prakash D, Dhakarey R, Upadhyay G, Singh HB (2009) Oxidative DNA damage protective activity, antioxidant and anti-quorum sensing potentials of *Moringa oleifera*. Food Chem Toxicol 47:1109–1116. doi:10.1016/j.fct.2009.01.034

Song K, Xu P, Meng Y, Geng F, Li J, Li Z, Xing J, Chem J, Kong B (2013) Smart gold nanoparticles enhance killing effect on cancer cells. Int J Oncol 42:597–608. doi:10.3892/ijo.2012.1721

Sreelatha S, Jeyachitra A, Padma PR (2011) Antiproliferation and induction of apoptosis by *Moringa oleifera* leaf extract on human cancer cells. Food Chem Toxicol 49:1270–1275. doi:10.1016/j.fct.2011.03.006

Sriram MI, Kanth SBM, Kalishwaralal K, Gurunathan S (2010) Antitumor activity of silver nanoparticles in Dalton's lymphoma ascites tumor model. Int J Nanomedicine 5:753–762. doi:10.2147/IJN.S11727

Tedesco S, Doyle H, Blasco J, Redmond G, Sheehan D (2010) Oxidative stress and toxicity of gold nanoparticles in *Mytilus edulis*. Aquat Toxicol 100:178–186. doi:10.1016/j.aquatox.2010.03.001

Tiloke C, Phulukdaree A, Chuturgoon A (2013) The antiproliferative effect of *Moringa oleifera* crude aqueous leaf extract on cancerous human alveolar epithelial cells. BMC Complement Altern Med 13:1–8. doi:10.1186/1472-6882-13-226

Tiloke C, Phulukdaree A, Chuturgoon A (2016a) The antiproliferative effect of *Moringa oleifera* crude aqueous leaf extract on human esophageal cancer cells. J Med Food 19:398–403. doi:10.1089/jmf.2015.0113

Tiloke C, Phulukdaree A, Anand K, Gengan RM, Chuturgoon A (2016b) *Moringa oleifera* gold nanoparticles modulate oncogenes, tumor suppressor genes, and *Caspase-9* splice variants in A549 cells. J Cell Biochem 9999:1–13. doi:10.1002/jcb.25528

Tomar A, Garg G (2013) Short review on application of gold nanoparticles. Global J Pharmacol 7:34–38. doi:10.5829/idosi.gjp.2013.7.1.66173

Uboldi C, Bonacchi D, Lorenzi G, Hermanns MI, Pohl C, Baldi G, Unger RE, Kirkpatrick CJ (2009) Gold nanoparticles induce cytotoxicity in the alveolar type-II cell lines A549 and NCIH441. Part Fibre Toxicol 6:1–12. doi:10.1186/1743-8977-6-18

Vasantha K, Ilangoa K, MohanKumara R, Agrawala A, Dubeya GP (2014) Anticancer activity of *Moringa oleifera* mediated silver nanoparticles on human cervical carcinoma cells by apoptosis induction. Colloids Surf B Biointerfaces 117:354–359. doi:10.1016/j.colsurfb.2014.02.052

Velazquez-Velazquez JL, Santos-Flores A, Araujo-Melendez J, Sanchez-Sanchez R, Velasquillo C, Gonzalez C, Martínez-Castanon G, Martinez-Gutierrez F (2014) Anti-biofilm and cytotoxicity activity of impregnated dressings with silver nanoparticles. Mater Sci Eng C 49:604–611. doi:10.1016/j.msec.2014.12.084

Vicente-Duenas C, Romero-Camarero I, Cobaleda C, Sanchez-Garcia I (2013) Function of oncogenes in cancer development: a changing paradigm. EMBO J 32:1502–1513. doi:10.1038/emboj.2013.97

Vinoth B, Manivasagaperumal R, Balamurugan S (2012) Phytochemical analysis and antibacterial activity of *Moringa oleifera* Lam. Int J Res Biol Sci 2:98–102

Visaria RK, Griffin RJ, Williams BW, Ebbini ES, Paciotti GF, Song CW, Bischof JC (2006) Enhancement of tumor thermal therapy using gold nanoparticle-assisted tumor necrosis factor-α delivery. Mol Cancer Ther 5:1014–1020. doi:10.1158/1535-7163.MCT-05-0381

Yacoby I, Benhar I (2007) Targeted antibacterial therapy. Infect Disord Drug Targets 7:1–9. doi:10.2174/187152607782109998

Zhou Y, Kong Y, Kundu S, Cirillo JD, Liang H (2012) Antibacterial activities of gold and silver nanoparticles against *Escherichia coli* and bacillus Calmette-Guérin. BMC J Nanobiotechnol 10:19. doi:10.1186/1477-3155-10-19

Nanomaterial in Diverse Biological Applications

14

Deepika Sharma, Rahul Shrivastava, and Gopal Singh Bisht

Abstract

Nanotechnology has revolutionized the world ranging from simple food packaging articles to medicinal therapeutics. Nano-sized objects have found application in most of the fields in sciences. Nanotechnology has been applied to create molecules possessing dimensions according to the type of application desired. Pharmaceutical industry has shown an immense interest in nanotechnology to improve pharmacologic and therapeutic properties of conventional drugs. Pharmaceutical formulations based on nanotechnology have improved the therapy strategies to treat simple as well as complex diseases. Selective targeting potential of these formulations not only deliver drug at active sites, but they also protect healthy cells from harmful effects of active drug. Nanotechnology has provided us with a lot of advantages, but along with these advantages, certain nanomaterials have been found to be toxic to human health as well as the environment. Nanotechnology-based products must be checked for their safety and toxicity profiles before launching them in market. The study of safety aspects and toxicity profile of nanomaterials is emerging as a new area of research. This chapter focuses on pharmacological applications, theranostic potential, safety aspects, and risk factors of various nanomaterial-based products. Along with this, it highlights the role of nanomaterials in bioremediation and controlling pollution.

Keywords

Nanomaterial • Nanotechnology • Drug delivery • Bioremediation • Imaging and diagnostic agent

D. Sharma • G.S. Bisht (✉)
Department of Pharmacy, Jaypee University of Information Technology, Waknaghat, Solan 173 234, Himachal Pradesh, India
e-mail: bisht.gopal@gmail.com; bishtchem2005@gmail.com

R. Shrivastava
Department of Biotechnology & Bioinformatics, Jaypee University of Information Technology, Waknaghat, Solan 173 234, Himachal Pradesh, India

V.C. Kalia, A.K. Saini (eds.), *Metabolic Engineering for Bioactive Compounds*,
DOI 10.1007/978-981-10-5511-9_14

14.1 Introduction

Materials developed to deliver therapeutic molecules at the desired site must have the potential to enhance efficacy and safety of treatment. Nanostructures like nanoemulsions, dendrimers, liposomes, and micelles can be made from a variety of materials having the ability to bind and deliver therapeutic molecules (Korsmeyer 2016). Nanomaterials have an important role in therapeutic field especially as carriers to aid in delivery of bioactive molecules at target site; some even act as protective agents to modulate the deregulated metabolic pathways in a pathological condition (Bianco and Muller 2016). Different types of nanomaterials having application in biomedical field are nanotubes, nanoparticles, quantum dots, micelles, nanosheets, etc. (Peng and Mu 2016). Nowadays, nanotechnology has revolutionized the biomedical field with smart nanomaterials that act according to the environmental conditions at the target site within the cells. Smart nanomaterials sense the changes in the environment and act by either of the two ways:

1. By protecting drug in hostile conditions
2. By releasing the drug at intended target site (Choi et al. 2014)

Nanotechnology-based products like nanoparticles and nanostructures have the potential to enhance sensitivity in fabrication of sensors; over the past decade, the development of such sensors has increased significantly (Bulbul et al. 2015). Similarly, there has been a considerable increase in the development of nanotechnology-based regenerative medicine. Current trend in regenerative medicines includes incorporation of different nanomaterials into nanostructure grafts and scaffolds. Nanotechnology-based regenerative medicines enhance the cellular and tissue regeneration property by mimicking the native cells and tissues. Nanomaterials used in regenerative medicines have shown better efficacy and biocompatibility in damaged tissues. Various scaffolds and grafts have been generated to enhance cell and tissue regeneration properties (Chaudhury et al. 2014).

A nanomaterial required in biomedical field should have excellent conductivity; conductive nanomaterials with a high surface area to volume ratio have shown effective transfer of electrical signal and induction of cellular response. Conductive nanomaterials find applications as biosensors and drug delivery vehicles (Li et al. 2016). DNA nanotechnology is another emerging field in drug delivery systems. DNA has been used as a versatile building material for nanostructures; it has shown specific interactions between base pairs (Kumar et al. 2016). Similarly, self-assembled polymeric nanostructures provide advantages of biocompatibility, nontoxicity, and a high capacity to load a drug or imaging agent. Different nanostructures like nanoparticles, micelles, nanogels, polymersomes, and polymer-grafted liposomes have been used so far to deliver bioactive molecule for therapeutic and diagnostic purposes (Pippa et al. 2016). Improvement in field of nanomedicine will enhance the ability of nanoparticles to deliver drug, and it will continue to have a significant clinical impact in nanotherapeutics (Perry et al. 2015).

Nanomaterials possess various valuable properties that allow nanotechnology to touch human lives in the form of therapeutics, technology, and cosmetics. Along with providing various advantages and applications, they also suffer with a drawback of getting released into the environment in uncontrolled fashion, thus raising issues concerning toxic potential of such materials toward the environment (Golbamaki et al. 2015). Research studies on release of nanomaterial from a nanocomposite may help in assessment of risk associated with a particular nanomaterial. However, research on understanding the condition of release of nanomaterial into the environment is not sufficient enough, and there is not much data available on it (Froggett et al. 2014). Delivery of nanomaterials suffers from one more limitation due to formation of protein corona when nanomaterials interact with endogenous proteins. Thorough studies on protein-nanomaterial interaction are required to inhibit the formation of protein corona for efficient drug delivery in biomedical field (Peng and Mu 2016).

14.2 Nanomaterials as Therapeutics for Metabolic Disorders, Malignant Growth Disorders, Microbial Pathogenesis, Immune Disorders, etc.

Metabolic disorder: A metabolic disorder occurs due to abnormal chemical reactions which alter the normal metabolic process in the body (MedlinePlus). Common metabolic disorders and role of nanodelivery systems in these metabolic disorders are given below.

- *Glucose metabolism disorder*: Diabetes is a metabolic disorder and one of the major challenges nowadays. 171 million people have been affected by diabetes worldwide, and around 4 million people die each year due to diabetes. Complications due to diabetes can be avoided if the blood glucose level of a patient is controlled. That's why it becomes very necessary to monitor the blood glucose level of a patient. Two types of approaches which can be used to detect blood sugar level are enzyme-dependent and nonenzymatic glucose-sensing approaches. Various types of nanostructured glucose sensors have been developed so far having high sensing ability (Kitture et al. 2015; Zaidi and Shin 2016). The antidiabetic potential of an antidiabetic therapeutic can be assessed with the help of inhibition of alpha enzymes: alpha glucosidase and alpha amylase. Enzyme inhibition activity of one such diabetes therapeutic agent, zinc oxide-red sandalwood conjugate, has been studied in murine pancreatic and small intestinal extracts; it holds a great potential in treatment of diabetes (Kitture et al. 2015). Nanodelivery systems have shown promising results in oral delivery of insulin in animal studies (Abbad et al. 2015).
- *Iron metabolism disorder*: Iron deficiency anemia is a nutritional disorder which adversely affects the gastrointestinal tract (Pereira et al. 2014; Hosny et al. 2015), and the treatment of this disease is usually based on soluble iron supplements. There are various disadvantages associated with commercially available tablets;

the two common disadvantages are blood in stool and constipation. Iron solid lipid nanoparticle has shown promising results as a carrier for iron with improved oral absorption (Hosny et al. 2015). Pereira et al. have developed iron hydroxide adipate tartrate-based nanomaterial for the treatment of iron deficiency anemia in cellular and murine models (Pereira et al. 2014). These studies show that iron metabolism disorders can be efficiently treated with nanomaterials.

- *Lipid metabolism disorder*: Fabry disease and dyslipidemia are lipid-based metabolic disorders; solid lipid nanoparticles and nanoliposomes have the potential to treat these diseases (Ruiz de Garibay et al. 2015; Sahebkar et al. 2015). Solid lipid nanoparticles act as vectors of genes to efficiently transfect Fabry patient cell line (Ruiz de Garibay et al. 2015), whereas nanoliposomes with a negative charge have been found to regulate lipid metabolism in hyperlipidemic mouse model. The nanodelivery system improved serum profile, further suggesting its potential to be used in treatment of patients suffering from dyslipidemias (Sahebkar et al. 2015).

14.2.1 Malignant Growth Disorders

- *Liver tumor*: Biodegradable polymer like poly(epsilon-caprolactone) (PCL) has been used to overcome toxicity issues related to the use of polymers. PCL nanofibers have been used for delivery of drugs and therapeutic genes to hinder the growth and prevent metastasis of tumor cells in liver cancer. Temperature-responsive PCL nanofibers loaded with paclitaxel have been used in the treatment of liver cancer (Che et al. 2015).
- *Colorectal cancer*: Lymph node labeled with carbon nanoparticles has been used in patients affected with colorectal cancer for laparoscopic colorectal cancer surgery. It improved the accuracy of postoperative adjuvant therapy by enhancing lymph node detection rate, thus aiding in lymph node dissection (Zhang et al. 2015).
- *Oral cancer*: Conventional delivery of drug in treatment of oral cancer causes physical as well as toxicological damage. Therefore, a drug delivery is required which provides sustained release of drug under physiological conditions. Nanodiamond-based platform has been used for the sustained release of celecoxib in oral cancer lesions (Yen et al. 2016).

Microbial pathogenesis: Most common microbes which cause disease can be inhibited by silver nanoparticles, as they are more effective antimicrobials than antibiotics in some cases (Rai et al. 2015). Similarly another microbe which persists in various areas is biofilm-producing bacteria; it is commonly found in hospitals. Biofilm-forming bacteria are responsible for many diseases, and current antibiotic therapy to eliminate biofilm-protected bacteria is not much efficient. DNase I-coated ciprofloxacin-loaded PLGA (poly lactic-co-glycolic acid) nanoparticle is one such novel therapeutic agent that enhanced antibiotic delivery and controlled the infections occurring due to biofilm-producing bacteria *Pseudomonas aeruginosa* (Baelo et al. 2015). Another common disease which causes deaths in population of developing countries is diarrhea. It is caused by microbes *Vibrio cholerae* and enterotoxic *Escherichia coli*. Antibacterial nanoparticles of silver and zinc prepared by green synthesis approach have been used to treat diarrhea. Administration of silver nanoparticles to mice colonized with abovementioned bacteria reduced the colonization rates of pathogens suggesting antibacterial potential of silver nanoparticles in diarrhea (Salem et al. 2015).

Immune disorders: Nanoparticle-based therapies have shown promising results in preclinical models of immune-mediated diseases (Getts et al. 2015; Serra and Santamaria 2015). Nanomedicine has emerged as a new field in treatment of rheumatoid based on functionalization of solid lipid nanoparticles with antibodies that specifically target macrophages in RA (rheumatoid arthritis). This approach has been used in formulation of solid lipid nanoparticles made up of methotrexate and superparamagnetic iron oxide. This conjugate has theranostic potential in treatment of RA, and it should be further exploited to treat autoimmune disease of joints (Albuquerque et al. 2015). The aim of therapy against autoimmune disorders is to block the pathogen causing immune disease without affecting the other immune mechanisms of the body. Nanodelivery systems can be used for the selective delivery of drugs to inflamed synovium thereby reducing the dose required for treatment (Albuquerque et al. 2015; Gouveia et al. 2015).

14.3 Drug Delivery and Pharmacological Application

14.3.1 Drug delivery

Nanomedicines based on targeted drug delivery systems can be used to improve the delivery of drug, as they increase the concentration of drug at target site, protect the drug from premature degradation, and provide controlled release of drug (Pooja et al. 2015). Drug release from such delivery systems can be tuned accordingly based on microenvironment stimuli in target tissues (Garciafigueroa et al. 2015). Polymeric nanostructures have the potential to be used in treatment of cancers as they deliver the drug at target site; polymeric micelles have been used for targeting drug at tumor site. Polymeric micelle drug complex internalizes via receptor-mediated endocytosis (Kapishon et al. 2016).

Pharmacological applications: Biomaterials, having the ability to maintain their mechanical integrity while delivering the bioactive agents, are required in different cellular applications (Gand et al. 2014). Nanomaterials and nanoproducts have various applications ranging from industrial application in textiles to biomedical applications in tissue engineering, biomedicine, and bioremediation (Eduok et al. 2013). Nanoparticle-based formulations have been developed with the aim to increase efficacy and safety of drugs. Therapy with such formulations provides an option to improve the safety profiles of drugs (Cooper et al. 2014). Nanomaterials find different applications in cancer therapy, targeted drug delivery, tissue engineering, controlled drug delivery, and improving bioavailability. Different nanomaterials that have been used in various drug delivery and pharmacological applications are explained below, and structures of some nanocarriers are given in Fig. 14.1.

Polymersomes: With the advancement in recombinant DNA technology and molecular biology, mass production of pharmaceutical proteins has become easy. Nanostructures which have been used in the delivery of proteins are polymersomes (Pachioni-Vasconcelos Jde et al. 2016).

Nanogel: Nanomaterial-based drug delivery system can be used to enhance oral absorption of drugs. An example is poly-alpha,beta-polyasparthydrazide-based nanogel for delivering anticancer drug, paclitaxel. This nanogel gave sustained release of drug (Guo et al. 2015).

Another example of nanogel is oleic acid-modified nanocarrier in combination with hydroxypropyl cellulose to deliver 1,1-bis(3′-indolyl)-1-(p-chlorophenyl)

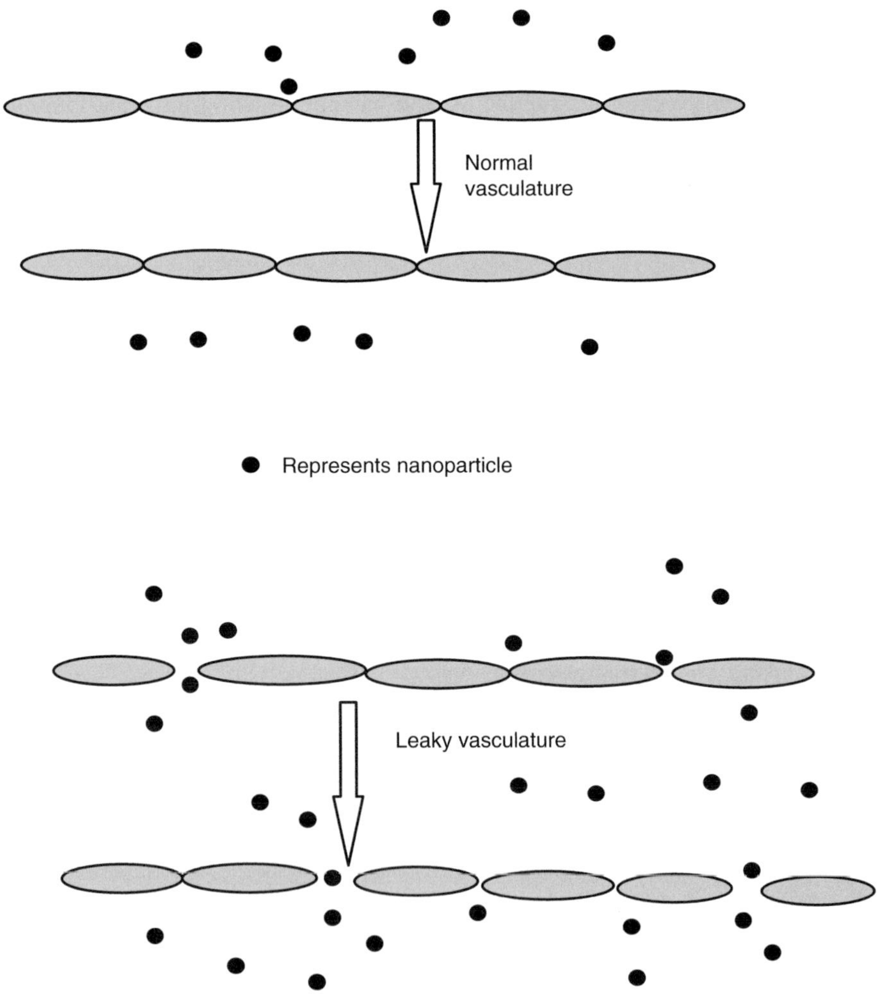

Fig. 14.1 Structures of different nanocarriers (Cho et al. 2008)

methane (DIM-D) in skin cancer treatment using an *in vivo* UVB-induced skin cancer model. DIM-D-N pretreatment was found to decrease damage to skin lipids and proteins caused by UVB rays (Boakye et al. 2015).

DNA origami nanocarrier: DNA nanotechnology is another emerging field in drug delivery systems. DNA has been used as a versatile building material for nanostructures; it has shown specific interactions between base pairs (Kumar et al. 2016). Overexpression of efflux pumps is one of the causes of drug resistance in cancers. Rodlike DNA origami drug carrier loaded with daunorubicin has shown good results in circumventing efflux pump-mediated drug resistance in leukemia cell line model (Halley et al. 2016).

Microemulsion system based on nanostructure: Amphotericin B is widely used in treatment of ocular infections, but its application is limited due to low patient compliance and toxicity of amphotericin B. Keeping disadvantages associated with amphotericin B in mind, a microemulsion system based on nanostructure containing amphotericin B has been designed to deliver this drug in the eye in the form of eye drops for treatment of diseased eye (da Silveira et al. 2015).

Solid lipid nanoparticles: A controlled drug delivery system has been prepared by loading rifampicin in solid lipid nanoparticles conjugated to wheat germ agglutinin for efficient treatment of tuberculosis (Pooja et al. 2015).

Liposomes: Hyaluronic acid is widely used as an anticancer drug as hyaluronic acid receptors are overexpressed on various tumor cells. Various anticancer therapies based on liposomes have been used to target hyaluronic acid toward cancer cells (Dosio et al. 2016).

Nanocapsules: Biodegradable nanoparticles are gaining importance in tumor treatment as they deliver drug in specific tissues. A formulation based on nanoparticles containing methotrexate-loaded lipid core nanocapsules has shown promising results in preclinical evaluation of this drug in glioblastoma model (Figueiro et al. 2015).

Polymeric nanoparticles: Drugs which have been delivered via polymeric nanoparticles and polymeric nanostructures are:

- *Anastrozole*: Anastrozole is an anticancer drug which is used in treatment of breast cancer. Based on biodegradable nanoparticles, different formulations have been developed to deliver anastrozole in rat model. Polymers which can be utilized to prepare these nanoparticles are poly(lactide-co-glycolide), polycaprolactone, and poly(lactic acid) (Shavi et al. 2015).
- *Vaccine*: An adjuvant approved by FDA (Food and Drug Administration) for human use which can be used to deliver vaccines is poly lactic-co-glycolic acid (PLGA). PLGA nanoparticles have been used to deliver a vaccine RI/L12 in mice, an immunodominant antigen of *Brucella* for the treatment of *Brucella* infection in mice as an option of vaccine delivery (Singh et al. 2015).
- *Indole-3-carbinol*: Indole-3-carbinol has been used to treat neurodegenerative diseases, but treatment is limited due to poor ability of this agent to cross blood-brain barrier. Therefore, an oil-in-water emulsion containing indole-3-carbinol-loaded nanoparticles has been used as a strategy to cross blood-brain barrier in damaged neuronal cells (Jeong et al. 2015).

- *Docetaxel and berbamine*: Nanoparticles can also be used to deliver two drugs simultaneously. One such nanoparticle-based delivery system has been developed to deliver docetaxel as well as berbamine (Wu et al. 2015a).
- *Simvastatin*: Simvastatin is used in treatment of breast adenocarcinoma. To control the delivery of simvastatin in breast adenocarcinoma, nanocarrier system consisting of a core made up of cholic acid and star-shaped polymer made up of PLGA has been used to sustain and control the delivery of simvastatin in chemotherapy of BALB/c nude mice (Wu et al. 2015b).

Pharmacological applications of nanomaterials have been explained under:

Cancer therapy: Nanomaterials required in cancer therapy must have very high stability as stability is the major problem which hinders the practical application of nanomaterial in actual practice. Gold nanoclusters acted as nontoxic and ultrastable nanomaterial for anticancer drug delivery in lung cancer cell lines (Zhang et al. 2015).

Targeted drug delivery: Nanomaterial required for targeted drug delivery should have nontoxic and biodegradable nature; nanotechnology offers an advantage of targeted drug delivery. Carbon nanotubes have been utilized as a potential nanovector to deliver drug at the site of brain cancer (Kateb et al. 2007).

Tissue engineering: Nanomaterial required in tissue engineering should have the ability to enhance cell growth. One such example is functionalized polyaniline copolymer-based nanofiber that enhances cell growth as well as stops microbial growth (Gizdavic-Nikolaidis et al. 2010).

Dry powder delivery: Mixture of phospholipids and proteins acts as pulmonary surfactant; such surfactants can be used as absorption enhancers in the delivery of dry powders that are required to be administered via pulmonary route. Nanoparticles made up of these pulmonary surfactants have shown promising results as absorption enhancers in delivery of insulin powder via pulmonary route (Zhang et al. 2009).

Bioavailability enhancement: Drugs which show poor bioavailability upon oral administration can be administered transdermally; nanotransferosomes are a possible solution to solve such problems of bioavailability. One such drug which is effective in treatment of invasive breast cancer is raloxifene hydrochloride, but poor oral bioavailability limits the treatment. Therefore, nanotransferosomes have been used to solve the problem, by delivering the drug transdermally in *ex vivo* studies as an alternative to oral treatment (Mahmood et al. 2014).

Controlled drug delivery: Controlled drug delivery is required in various pharmaceutical applications. Nanomaterials required for such applications must have the capacity to release drug in a controlled manner. Methacrylic acid copolymer S-based nanofibers are one such nanomaterial which has the capability of holding water-soluble drugs and releasing them in a controlled manner (Hamori et al. 2015).

Miscellaneous applications: Other than the above applications, nanomaterials do have miscellaneous pharmacological applications. These applications are given in Table 14.1.

Table 14.1 Miscellaneous applications of nanomaterials

S. No	Nanomaterial system	Application	References
1	Lipid nanocapsule	Dermal application	Abdel-Mottaleb et al. (2011)
2	Titanium dioxide nanoparticles	Application as antibacterial agent	Ahmad et al. (2015)
3	Chitosan nanoparticles	Delivery of drug to eye	Al-Kinani et al. (2015)
4	Nickel-doped biphasic nanoplatelet	Biomedical application	Baradaran et al. (2015)
5	Gold nanostars	As Raman activity enhancer in intracellular applications	Cao et al. (2015)
6	Lecithin/chitosan nanoparticles	Prolonged ocular application	Chhonker et al. (2015)
7	Curcumin-loaded hydrogel	Antimalarial therapy	Dandekar et al. (2010)
8	Nanotemplated polyelectrolyte films	Biomolecule delivery	Gand et al. (2014)
9	Polyaniline-based nanofibers	Tissue engineering	Gizdavic-Nikolaidis et al. (2010)
10	Silver nanoparticles	Retinal therapy	Kalishwaralal et al. (2010)
11	Terbium acetylacetone composite nanoparticles	Determination of salicylic acid in human plasma	Karim et al. (2008)
12	Miconazole-conjugated silver nanoparticles	As an antifungal agent	Kumar and Poornachandra (2015)
13	Silver nanoparticles	Retinal therapy	Kalishwaralal et al. (2010)
14	Iron oxide nanoparticles	Cancer imaging	Lee et al. (2007)

14.3.2 Nanomaterial and Tissue Interaction

Nanomaterials which mimic the topography of native tissues give improved biocompatible responses. Such nanomaterials give better results in tissue engineering as they show better tissue integration in medical implants (Kriparamanan et al. 2006). Biomimetic properties of a nanomaterial can be engineered according to the specific type of application required (Chow et al. 2008). Nanoscale surface modification of such materials influences cell behavior and has been used to enhance the bioactivity of various implants (Bruinink et al. 2014). Implants made up of nanomaterials are required to have a long-term efficiency (Dorkhan et al. 2014). Biomaterial which is used in tissue engineering must have the potential to grow functional tissues *in vitro*, and it should allow integration of biomaterial into existing tissues *in vivo* (Hass et al. 2012).

Polycaprolactone nanofiber implants have been utilized as a scaffold in tissue engineering, as they minimize host responses. Topography of a material influences the cellular behavior; therefore, controlling this parameter is considered as an important parameter in tissue engineering (Cao et al. 2010). For example, in the case of orthopedic implants, topographical modification has been done to guide tissue integration; such modifications have shown to inhibit bone resorption and regulate integrin-mediated cell adhesion (Biggs et al. 2009).

14.3.3 Interaction of Nanomaterial with Different Types of Tissues and Cells

Hepatocytes: A suitable cell source and methodology are required to assemble the cells in vivo in order to maintain cell viability and function in liver tissue engineering. In vivo study of a hydrogel fiber-based scaffold asserted the integration of scaffold with host vasculature, shown by the presence of human albumin in mouse serum (Du et al. 2014).

Vascular cells: In vascular tissue engineering, the scaffolds made up of nanomaterials must have the potential to construct vascular grafts that are nonthrombogenic. Keeping this in mind, biodegradable nanofibrous scaffolds with nanofibers were developed. These scaffolds in combination with bone marrow mesenchymal cells were found to be antithrombogenic allowing infiltration of vascular cells and thereby matrix remodeling (Hashi et al. 2007).

Human mesenchymal stem cells: Nanostructured titanium implants can be utilized in dental and orthopedic implants as they accelerate bone apposition and bonding strength of bones in vivo as well as in vitro. These nanostructures have shown controlled differentiation of cells thereby promoting integration of implants in tissues (Lavenus et al. 2012).

Osteoblasts: A large number of bone disorders are growing day by day, hence becoming a challenge for our society. Various orthopedic implants are used to promote bone repair. Therefore, it becomes necessary to understand the cellular responses to a nanomaterial's topography by carrying out in vitro studies in order to ensure the efficient and effective performance of nanomaterial-based orthopedic devices (Cassidy et al. 2014). Metal-coated nanosprings are efficient bionanomaterial which provide such advantages and enhance the behaviors of normal human osteoblasts to improve the orthointegration of orthopedic implants (Hass et al. 2012).

Human nasoseptal chondrocytes (NC): Bacterial nanocellulose can be used for reconstruction of auricular cartilage as it is a biocompatible material which promotes tissue integration. Scaffolds based on this material are nonpyrogenic as well as noncytotoxic. Such scaffolds have good mechanical strength which provides a porous structure to support cell growth (Martinez Avila et al. 2015).

Bone marrow-derived stromal cells: Bone augmentation is an important aspect of tissue engineering; materials showing this advantage can be used for bone regeneration. Scaffolds containing fibroblast growth factor along with multiwalled carbon nanotubes have the potential as a substituent material for bone regeneration (Hirata et al. 2013).

Soft connective tissue: Autografts and allografts carry the risk of disease transmission; therefore, tissue-engineered implants can serve as potential alternatives. Such nanostructured implants show improved cellular infiltration as well as in vivo integration (James et al. 2011).

Mucosal lining: Electrospun composite/Eudragit nanofibers which deliver two different classes of drugs like aceclofenac and pantoprazole compensate the gastrointestinal toxicity of NSAIDs. These nanofibers have been found to preserve the mucosal architecture of gastric tissue of rats treated with drug-loaded nanofibers (Karthikeyan et al. 2012).

Cardiac tissue: Myocardial regeneration based on stem cell methods can be done with a material which mimics the extracellular residence matrix of cardiac cells in vivo. One such artificial matrix has been prepared by combining autologous cardiosphere-derived cells with nanotopographed hydrogels; from this, cardiac patches were developed, which were implanted at the site of infarct facilitating stem cell-based cardiac tissue engineering (Kim et al. 2012b).

Cartilage tissue: Cartilage repair requires delivery of isolated chondrocytes in a matrix to promote repair of cartilage. In vitro and in vivo cartilage regeneration of partial thickness cartilage defects can be enhanced by preparing a scaffold/hydrogel composite providing advantages of both porous scaffold and hydrogel to deliver chondrocytes to cartilage defects (Kim et al. 2012a).

Carbon nanotubes: Carbon nanotubes have shown promising results and biocompatibility in vivo, as they facilitate cell maturation and osteointegration (Singh et al. 2010). Carbon nanotubes being biocompatible with blood have also been used in cardiac surgeries (Veetil and Ye 2009).

14.3.4 Imaging and Diagnostic Agents

Imaging techniques are used in medical field on routine basis; their use is growing day by day as the number of contrast agents being developed is rising. Nanoparticles have been used as imaging and diagnostic agents; dual-function imaging agents are being developed providing function of imaging agent as well as diagnostic agent on the same nanoparticle (Branca et al. 2014). Nanodelivery systems can be used to deliver magnetic resonance imaging contrast agents; an example based on this type of delivery system is single-walled carbon nanotube functionalized with hyaluronic acid (Hou et al. 2015).

Current trend which has attracted immense interest in cancer therapy is the use of theranostic agents. Such agents have imaging as well as photothermal properties. An example of such theranostic agent is Prussian blue nanocubes which serve both mentioned purposes, for imaging-guided cancer therapy in mice (Cheng et al. 2014). One such carrier is chitosan nanobubble which along with imaging properties was used for codelivery of prednisolone and gadnium III complex. Such theranostic agents have a great potential in enhancing cancer therapy (Cavalli et al. 2015). Similarly materials like magnetite or maghemite have been used to induce hyperthermia and in the delivery of drugs at specific sites. Core shell nanostructures based on iron oxide cores have the potential to be used as theranostic agents (Andrade et al. 2015).

Nanoparticles made up of ferric oxide cores coated with bovine serum albumin can be used as imaging agent. In vivo studies of these magnetic nanoparticles show their potential as magnetic resonance imaging agent in visualization of intracranial glioma (Abakumov et al. 2015). A hybrid nanostructure based on functional quantum dots has been used as an imaging agent in C6 glioma cells in in vitro studies

(Fanizza et al. 2016). Cyanine dye nanoconjugates have been found to enhance the internalization as well as hyperthermia-mediated cytotoxicity in cancer cells as compared to free dye (Fernandez-Fernandez et al. 2014). With the advances in cancer therapy, new classes of materials like nanodiamonds are being used to enhance the imaging and treatment of cancers. Nanodiamonds have uniquely facet surfaces that act as versatile platform carriers. Along with magnetic resonance imaging, nanodiamonds can be used to deliver drug systematically as well as locally (Ho 2015).

14.3.5 Nanomaterial in Environmental Science Bioremediation and Controlling Pollution

Bioremediation is a process in which biotic agents degrade contaminants completely and help in removal of such contaminants from the environment (Biswas et al. 2015). Over the past 50 years, the production of nanomaterials has increased at a great pace which further increases the chances of their uncontrolled release into the environment (Musee 2011). There are a wide range of products which are based on engineered nanoparticles, and the use of such products is often associated with an increased concern for environmental safety. Nanomaterial-based products have been shown to have potential toxicological and adverse effects; there are only a few evidences on environmental fate of engineered nanoparticles in wastewater (Eduok et al. 2013). Studies have shown that nanomaterials enter wastewater treatment plants (WWTPs). Influences of different nanoparticles on the enzyme activities and structures of microbes involved in wastewater treatment facilities have been studied. Long-term exposure to these influences affects the microbial communities and enzyme activities, resulting in lower efficiency of biological nitrogen removal (Zheng et al. 2015).

In the case of carbon nanomaterials also, very little is known about the fate of carbon nanomaterials once they are released into the environment. Fullerene (C60) forms a material that resists biodegradation. The rate of biodegradation of these particles is a slow process, and these particles may not be degraded by microbes to carbon dioxide (Chae et al. 2014). Fate of carbon-based nanomaterials in the body depends on enzymatic degradation by the immune system (Bhattacharya et al. 2016). Studies on the fate of such nanomaterials in soil show that fullerenes released into the environment are likely to persist in soil for extended periods (Avanasi et al. 2014). Release of nanoparticles is expected to increase in wastewater treatment plants as well as in the environment (Liu et al. 2011).

14.3.6 Types of Nanomaterial Used in Bioremediation

Silica nanoparticles: Immobilization is the basic method to remove micropollutants biologically by improving enzyme activity and stability. One such immobilizing nanobiocatalyst which has been used for removing micropollutants is fumed silica nanoparticles immobilized with laccase (Hommes et al. 2012).

Biodegradable nanoparticles: Nanoparticles from lignin are apparently nontoxic for microalgae like *Chlamydomonas reinhardtii* and yeast. These NPs must be used in place of nonbiodegradable NPs which contain potentially toxic nanomaterials (Frangville et al. 2012).

Vault nanoparticles: Enzymatic biodegradation is a bioremediation process which can be used for in situ cleanup of contaminated environment. But its application is restricted as stability of extracellular enzymes is limited. An approach which can be used to increase the enzymatic stability is by packing enzymes into vault nanoparticles. It not only extends their stability, but it serves as a foundation for removing various contaminants from drinking water and groundwater (Wang et al. 2015a).

Copper nanoparticles: Chen et al. have investigated the influence of copper nanoparticles on wastewater biological nutrient removal and nitrous oxide generation. Their study has shown that the total nitrogen removal was enhanced and nitrous oxide generation was reduced at any copper nanoparticles levels investigated, whereas concentration of ammonia and phosphorus was not affected. Absorption of nanoparticles on activated sludge and the release of copper ion from nanoparticles were the main reasons for total nitrogen removal improvement and nitrous oxide reduction (Chen et al. 2012a).

Alumina nanoparticles: Alumina nanoparticles act as contaminants of the environment. Study on interaction of alumina nanoparticle with activated sludge shows that, even after getting adsorbed on sludge, they do not have any adverse effects on activated sludge (Chen et al. 2012b).

Zinc oxide nanoparticles: Zinc oxide nanoparticles synthesized via green route were doped with tetraethoxy silane to inhibit biofilm formation (Krupa and Vimala 2016).

14.3.7 Bioremediation of Different Contaminants in Controlling Pollution

Arsenic: Nanotechnology in combination with biotechnology can effectively remediate arsenic from groundwater. Removal of arsenic can be done with combination of nanomaterial- and arsenic-reducing bacteria. Palladium-akaganeite nanoparticles along with metal-reducing bacteria remove arsenic from water (Sun et al. 2011).

Halogens: Nanopalladium is a biologically derived nano-sized substance achieved by precipitation of palladium on a bacterium. This bio-palladium can be utilized as a catalyst in dehalogenation of contaminants containing halogen from the environment (Hennebel et al. 2011).

Manganese: There is an increasing demand for developing cost-effective green synthesis approaches for synthesis of nanoparticles. Keeping this in mind, a heavy metal-resistant strain of *Bacillus cereus* species (MTCC10650) has been used for production of manganese oxide nanoparticles as well as remediation of manganese from the environment (Sinha et al. 2011).

Uranium: Removal of uranium from effluent can be done by reduction of soluble uranium (IV) to insoluble uranium (VI). Iron nanoparticles immobilized on active carbon help in the removal of uranium from effluents. Use of lactate in such systems supports the growth of uranium-reducing microorganisms in the effluent sample (Baiget et al. 2013).

Dioxin: The dioxin isomer 2,3,7,8-tetrachlorodibenzo-p-dioxin is difficult to degrade due to its highly recalcitrant nature and low bioavailability. Various efforts have been made to degrade it using different single techniques, but none of the techniques provided the desired results. Recent technique which degraded this contaminant is an integrated system of palladized iron nanoparticles and degrading microorganism (Bokare et al. 2012).

Cadmium: *Phanerochaete chrysosporium* is a bioremediation agent; however, its application for bioremediation is limited to its long degradation time and low resistance to pollutants. Its activity was enhanced by attaching nitrogen-doped titanium dioxide nanoparticles to aid the degradation process for pollutants. This system can be utilized to remove cadmium from wastewater (Chen et al. 2013).

Phenol: A nanomaterial should be eco-friendly as well as reusable. Ferromagnetic nanoparticles (MNPs) are eco-friendly as well as reusable nanomaterial. They have been employed for bioremediation of phenol from aqueous solutions (Zhang et al. 2008).

Dyes: Dyes are used to color materials like cotton and rayon; the wastewater which is produced from reactive dyeing process contains up to 50% dye and various electrolytes. Dyes can be removed from wastewater with the help of nanoparticles. Biodegradable hollow zein nanoparticles show promising adsorption potential for removal of dyes from water (Xu et al. 2013).

Heavy metals: Lead, manganese, and chromium act as contaminants for water; removal of such metals from water requires an efficient biosorbent. A biosorbent system which can be used for removal of such contaminants from aqueous system is a combination of gellan gel beads and magnetic nanoparticles (Wang et al. 2009).

Chromium: Palladium nanoparticles synthesized by green chemistry approach help in bioremediation of chromium (Tuo et al. 2013). Some biogenic nanocatalysts from acid mine drainage have also been used along with metal-reducing bacteria for reduction and immobilization of chromium (IV) for the remediation of chromium (IV)-contaminated water (Seo and Roh 2015). Microbe that has been used for Cr(IV) reduction is *Pseudomonas aeruginosa* (Pang et al. 2011).

Petroleum products: Petroleum products act as contaminants in the atmosphere; therefore, it becomes essential to remove such components from the environment. Toluene is a petroleum product which must be removed from the environment; technique which can be used to remove toluene is the oxidation of toluene using calcium peroxide nanoparticles (Qian et al. 2013).

Silver: Silver-based nanomaterials are used as an antimicrobial agent in various products; thc amount of nanomaterial released into water can be determined by assessing the leaching rate of silver in water. Nanosilver can be removed from water by method of biosorption, thereby preventing environmental risks (Pasricha et al. 2012). Another method to control the potential environmental risks associated with silver nanoparticles is the production of silver nanoparticles using microbes, as no toxic chemicals are used in such processes (Poulose et al. 2014).

Copper: Copper nanoparticles can be synthesized extracellularly using dead fungal biomass. Extracellular synthesis and uptake of copper nanoparticles by such biomass containing *Trichoderma koningiopsis* are a green synthesis approach and can also be utilized for bioremediation of copper from wastewater (Salvadori et al. 2014).

Polycyclic aromatic hydrocarbons: They belong to the category of organic pollutants which are very harmful for the environment as they have low water solubility and strong sorption property. Therefore, it becomes essential to remove such contaminants; nanoparticle-supported bilayers made up of silicon dioxide nanoparticles and 1,2-dimyristoyl-sn-glycero-3-phosphocholine as lipid can be used as an effective method for the remediation of polycyclic aromatic hydrocarbons (Wang et al. 2015b).

Trichloroethylene: It causes environmental pollution; an efficient technique which can be used to remove this contaminant from the environment is based on poly(methyl methacrylate) (PMMA)-coated nanoscale zerovalent iron which effectively reduces TCE (Trichloroethylene). In this approach, PMMA not only enhances sorption in chlorination reaction, but it also inhibits corrosive inhibition (Wang et al. 2010).

Saline industrial water: Saline industrial water contains dissolved salts like chlorides and sulfates and is diluted up to 10–20 times to remove such salts. Nanofilters provide an option to remove such salts at a much lower dilution, thereby saving time and energy (Tuin et al. 2006).

Water purification: Bioremediation of contaminated waters can also be done using bioreactive nanofibers made up of polyvinyl alcohol-based cores containing *Escherichia coli* and silica-based sturdy porous shells. It is a low-cost design with high efficiency and flexibility to remove various contaminants (Tong et al. 2014). Similarly, eco-friendly nanogold bioconjugate has shown promising results in purification of contaminated water. Pathogens and pesticides can also be removed from wastewater using this bioconjugate (Das et al. 2009).

14.3.8 Safety Aspects and Risk Factors of Nanomaterial via Nanotoxicology

Nanostructured materials having commercial application must be safe, and it is necessary to ensure their safety before launching them in market. Materials must be characterized into highly risky and least risky depending on their toxicity toward human health and the environment (Godwin et al. 2015). Nano-objects can be released into workplace during manufacture of nanomaterials. Data on experimental studies on exposure of nanoparticles to different workplaces is scarce. New approaches should be designed to conduct exposure studies (Brouwer 2010). The European Commission has funded a project on NanoSafety Cluster for the management of toxicological data of nanomaterials. This cluster is based on an eNanoMapper database that gives information on toxicity of a nanomaterial and related nanomodels (Jeliazkova et al. 2015). Nanomaterials also affect aquatic life, but the data on toxicity of nanomaterials to aquatic life is scarce; therefore, a study on toxicity of synthetic nanoparticles made up of titanium dioxide, carbon nanotubes, copper oxide, and zinc oxide has been carried out which shows that algae and daphnids are the most sensitive organism groups that get affected due to release of nanoparticles into aquatic environment (Kahru and Dubourguier 2010). Since there is limited data on potential toxicities, more investigations are required to be done to verify the reliability of a particular nanomaterial (Kim et al. 2014).

Nanomaterial-based products are also widely used in food industry as they provide protection against microbes responsible for deterioration of food. Most of the nanomaterials used in food industry contain organic compounds; therefore, their toxic potential must be assessed in animals before utilizing them for food articles (Das et al. 2011). Similarly, magnetic nanoparticles have gained a lot of importance in drug delivery, tissue engineering, and cancer therapy. Along with these advantages, bioengineered magnetic nanoparticles do have toxic effects, and their toxicity has become a matter of concern. Current methods of safety assessment of magnetic nanomaterial are of limited use now; therefore, new approaches are required to assess their toxicity potential (Cywinska and Grudzinski 2012). The protocols for safety assessment of engineered nanomaterials should allow screening of multiple toxicants at cellular as well as biomolecular level. Nowadays the toxicological approaches that are used to assess the safety of engineered nanomaterials are alternative test strategies (ATSs). An ATS is an alternative to methods which require animal testing for assessing safety of nanomaterials, hence reducing the number of animals being required in nanotoxicity studies (Nel 2013). Examples of different nanodelivery systems are given in (Table 14.2).

Table 14.2 Examples of different nanodelivery systems

S. No	Nanodelivery system	Use	References
1	Chitosan-modified cationic amino acid nanoparticles	Oral delivery of insulin in diabetic rats	Abbad et al. (2015)
2	Nanofibrous scaffolds	Diabetic wound healing	Gholipour-Kanani et al. (2016)
3	Chonglou saponin VII plus silica nanocomplex	Inhibition of proliferation rate in drug-resistant human ovarian cancer cell line	Yang et al. (2015)
4	Nanodiamonds	In oral cancer treatment	Yen et al. (2016)
5	Quantum dot-labeled polymersomes	In lymph node mapping	Bakalova et al. (2015)
6	Iron oxide nanoparticles	As contrast agent and drug delivery carrier	Chen et al. (2015)
7	Silver nanoparticles	In inhibition of dermatophyte fungi	Ouf et al. (2015)
8	Gold nanoparticles	As contrast agents in X-ray imaging	Cole et al. (2015)
9	Synthetic amorphous silica	In food and cosmetic industry	Fruijtier-Polloth (2012)
10	Zerovalent iron nanoparticles	Bioremediation	Kharangte-Lad et al. (2016)

Conclusion: This chapter covers various biological and nonbiological applications of nanomaterials, their interaction with tissues, risks associated with such materials, and their role in environmental sciences. Nanotechnology-based materials have various cellular applications (Gand et al. 2014); they can be used to treat common diseases like diarrhea (Salem et al. 2015) to complex diseases like cancer (Che et al. 2015). Nanomaterials have various applications in cancer (Zhang et al. 2015), targeted systems (Kateb et al. 2007), tissue engineering (Gizdavic-Nikolaidis et al. 2010), drug delivery (Hamori et al. 2015), and as theranostic agent (Cheng et al. 2014). Nanomaterials are biocompatible in nature (Kriparamanan et al. 2006); therapies with such formulations improve the safety profiles (Cooper et al. 2014) and bioavailability of conventional drugs (Mahmood et al. 2014). These systems provide sustained release of drugs (Yen et al. 2016); materials like polymeric nanostructures can be used to deliver drug at target site of disease thereby protecting normal cells from harmful effects of conventional drugs (Kapishon et al. 2016). Nanomaterial not only helps in controlling diseases, but they can also help in the removal of pollutants from the environment (Wang et al. 2015a). Their safety aspects and risk factors can be studied according to protocols available (Nel 2013), but there is limited data on toxic effects of nanomaterials; therefore, more studies are required to be carried out in order to find out their toxic potential (Kim et al. 2014). Hence, we conclude that nanotechnology-based materials have wide applications due to their versatile characteristics that might turn into big wonders someday.

Opinion: A lot of research has been done in the area of nanotechnology; it is being applied in every aspect of life. This chapter focused on various applications of nanomaterials and toxicity issues related to them. Nanomaterials have immense biological as well as industrial applications. Nanodelivery systems have been developed to reduce drawbacks of traditional drug delivery systems. Nanomaterials allow better control over the drawbacks associated with a drug as they sense changes in biological environment and release the drug content over the diseased area thereby sparing healthy human cells. Nanomaterials have beneficial as well as toxic effects. Biodegradable polymers provide an option for developing drug delivery system with reduced toxic potential. Protocols which are currently available to study toxicity of nanomaterials are not sufficient. Hence, more efforts are required to design *in vivo* testing protocols to evaluate toxic potential of nanodelivery systems. The field of nanotechnology has a lot of potential; it can serve humanity with loads of biological applications.

References

Abakumov MA, Nukolova NV, Sokolsky-Papkov M, Shein SA, Sandalova TO, Vishwasrao HM, Grinenko NF, Gubsky IL, Abakumov AM, Kabanov AV, Chekhonin VP (2015) VEGF-targeted magnetic nanoparticles for MRI visualization of brain tumor. Nanomedicine 11:825–833. doi:10.1016/j.nano.2014.12.011

Abbad S, Zhang Z, Waddad AY, Munyendo WL, Lv H, Zhou J (2015) Chitosan-modified cationic amino acid nanoparticles as a novel oral delivery system for insulin. J Biomed Nanotechnol 11:486–499. doi:10.1166/jbn.2015.1924

Abdel-Mottaleb MM, Neumann D, Lamprecht A (2011) Lipid nanocapsules for dermal application: a comparative study of lipid-based versus polymer-based nanocarriers. Eur J Pharm Biopharm 79:36–42. doi:10.1016/j.ejpb.2011.04.009

Ahmad R, Mohsin M, Ahmad T, Sardar M (2015) Alpha amylase assisted synthesis of TiO(2) nanoparticles: structural characterization and application as antibacterial agents. J Hazard Mater 283:171–177. doi:10.1016/j.jhazmat.2014.08.073

Albuquerque J, Moura CC, Sarmento B, Reis S (2015) Solid lipid nanoparticles: a potential multifunctional approach towards rheumatoid arthritis theranostics. Molecules 20:11103–11118. doi:10.3390/molecules200611103

Al-Kinani AA, Naughton DP, Calabrese G, Vangala A, Smith JR, Pierscionek BK, Alany RG (2015) Analysis of 2-oxothiazolidine-4-carboxylic acid by hydrophilic interaction liquid chromatography: application for ocular delivery using chitosan nanoparticles. Anal Bioanal Chem 407:2645–2650. doi:10.1007/s00216-015-8494-8

Andrade AL, Fabris JD, Domingues RZ, Pereira MC (2015) Current status of magnetite-based Core Shell structures for diagnosis and therapy in oncology short running title: biomedical applications of magnetite shell structures. Curr Pharm Des 21:5417–5433. doi:10.2174/13816 12821666150917093543

Avanasi R, Jackson WA, Sherwin B, Mudge JF, Anderson TA (2014) C60 fullerene soil sorption, biodegradation, and plant uptake. Environ Sci Technol 48:2792–2797. doi:10.1021/es405306w

Baelo A, Levato R, Julian E, Crespo A, Astola J, Gavalda J, Engel E, Mateos-Timoneda MA, Torrents E (2015) Disassembling bacterial extracellular matrix with DNase-coated nanoparticles to enhance antibiotic delivery in biofilm infections. J Control Release 209:150–158. doi:10.1016/j.jconrel.2015.04.028

Baiget M, Constanti M, Lopez MT, Medina F (2013) Uranium removal from a contaminated effluent using a combined microbial and nanoparticle system. New Biotechnol 30:788–792. doi:10.1016/j.nbt.2013.05.003

Bakalova R, Zhelev Z, Nikolova B, Murayama S, Lazarova D, Tsoneva I, Aoki I (2015) Lymph node mapping using quantum dot-labeled polymersomes. Gen Physiol Biophys 34:393–398. doi:10.4149/gpb_2015007

Baradaran S, Moghaddam E, Nasiri-Tabrizi B, Basirun WJ, Mehrali M, Sookhakian M, Hamdi M, Alias Y (2015) Characterization of nickel-doped biphasic calcium phosphate/graphene nanoplatelet composites for biomedical application. Mater Sci Eng C Mater Biol Appl 49:656–668. doi:10.1016/j.msec.2015.01.050

Bhattacharya K, Mukherjee SP, Gallud A, Burkert SC, Bistarelli S, Bellucci S, Bottini M, Star A, Fadeel B (2016) Biological interactions of carbon-based nanomaterials: from coronation to degradation. Nanomedicine 12:333–351. doi:10.1016/j.nano.2015.11.011

Bianco A, Muller S (2016) Nanomaterials, autophagy, and lupus disease. Chem Med Chem 11:166–174. doi:10.1002/cmdc.201500233

Biggs MJ, Richards RG, Gadegaard N, Wilkinson CD, Oreffo RO, Dalby MJ (2009) The use of nanoscale topography to modulate the dynamics of adhesion formation in primary osteoblasts and ERK/MAPK signalling in STRO-1+ enriched skeletal stem cells. Biomaterials 30:5094–5103. doi:10.1016/j.biomaterials.2009.05.049

Biswas B, Sarkar B, Rusmin R, Naidu R (2015) Bioremediation of PAHs and VOCs: advances in clay mineral-microbial interaction. Environ Int 85:168–181. doi:10.1016/j.envint.2015.09.017

Boakye CH, Shah PP, Doddapaneni R, Patel AR, Safe S, Singh M (2015) Enhanced percutaneous delivery of 1,1-bis(3′-indolyl)-1-(p-chlorophenyl) methane for skin cancer chemoprevention. J Biomed nanotech 11:1269–1128. doi:10.1166/jbn.2015.2064

Bokare V, Murugesan K, Kim JH, Kim EJ, Chang YS (2012) Integrated hybrid treatment for the remediation of 2,3,7,8-tetrachlorodibenzo-p-dioxin. Sci Total Environ 435-436:563–566. doi:10.1016/j.scitotenv.2012.07.079

Branca M, Pelletier F, Cottin B, Ciuculescu D, Lin CC, Serra R, Mattei JG, Casanove MJ, Tan R, Respaud M, Amiens C (2014) Design of FeBi nanoparticles for imaging applications. Faraday Discuss 175:97–111. doi:10.1039/c4fd00105b

Brouwer D (2010) Exposure to manufactured nanoparticles in different workplaces. Toxicology 269:120–127. doi:10.1016/j.tox.2009.11.017

Bruinink A, Bitar M, Pleskova M, Wick P, Krug HF, Maniura-Weber K (2014) Addition of nanoscaled bioinspired surface features: a revolution for bone related implants and scaffolds? J Biomed Mater Res A 102:275–294. doi:10.1002/jbm.a.34691

Bulbul G, Hayat A, Andreescu S (2015) Portable nanoparticle-based sensors for food safety assessment. Sensors (Basel) 15:30736–30758. doi:10.3390/s151229826

Cao H, McHugh K, Chew SY, Anderson JM (2010) The topographical effect of electrospun nanofibrous scaffolds on the *in vivo* and *in vitro* foreign body reaction. J Biomed Mater Res A 93:1151–1159. doi:10.1002/jbm.a.32609

Cao X, Shi C, Lu W, Zhao H, Wang M, Tong W, Dong J, Han X, Qian W (2015) Synthesis of au nanostars and their application as surface enhanced Raman scattering-activity tags inside living cells. J Nanosci Nanotechnol 15:4829–4836. doi:10.1166/jnn.2015.9828

Cassidy JW, Roberts JN, Smith CA, Robertson M, White K, Biggs MJ, Oreffo RO, Dalby MJ (2014) Osteogenic lineage restriction by osteoprogenitors cultured on nanometric grooved surfaces: the role of focal adhesion maturation. Acta Biomater 10:651–660. doi:10.1016/j.actbio.2013.11.008

Cavalli R, Argenziano M, Vigna E, Giustetto P, Torres E, Aime S, Terreno E (2015) Preparation and *in vitro* characterization of chitosan nanobubbles as theranostic agents. Colloids Surf B Biointerfaces 129:39–46. doi:10.1016/j.colsurfb.2015.03.023

Chae SR, Hunt DE, Ikuma K, Yang S, Cho J, Gunsch CK, Liu J, Wiesner MR (2014) Aging of fullerene C(6)(0) nanoparticle suspensions in the presence of microbes. Water Res 65:282–289. doi:10.1016/j.watres.2014.07.038

Chaudhury K, Kumar V, Kandasamy J, RoyChoudhury S (2014) Regenerative nanomedicine: current perspectives and future directions. Int J Nanomed 9:4153–4167. doi:10.2147/IJN.S45332

Che HL, Lee HJ, Uto K, Ebara M, Kim WJ, Aoyagi T, Park IK (2015) Simultaneous drug and gene delivery from the biodegradable poly(epsilon-caprolactone) nanofibers for the treatment of liver cancer. J Nanosci Nanotechnol 15:7971–7975. doi:10.1166/jnn.2015.11233

Chen G, Guan S, Zeng G, Li X, Chen A, Shang C, Zhou Y, Li H, He J (2013) Cadmium removal and 2,4-dichlorophenol degradation by immobilized Phanerochaete chrysosporium loaded with nitrogen-doped TiO2 nanoparticles. Appl Microbiol Biotechnol 97:3149–3157. doi:10.1007/s00253-012-4121-1

Chen GJ, Hsu C, Ke JH, Wang LF (2015) Imaging and chemotherapeutic comparisons of iron oxide nanoparticles chemically and physically coated with poly(ethylene glycol)-b-poly(epsilon-caprolactone)-g-poly(acrylic acid). J Biomed Nanotechnol 11:951–963. doi:10.1166/jbn.2015.2012

Chen Y, Su Y, Zheng X, Chen H, Yang H (2012a) Alumina nanoparticles-induced effects on wastewater nitrogen and phosphorus removal after short-term and long-term exposure. Water Res 46:4379–4386. doi:10.1016/j.watres.2012.05.042

Chen Y, Wang D, Zhu X, Zheng X, Feng L (2012b) Long-term effects of copper nanoparticles on wastewater biological nutrient removal and N2O generation in the activated sludge process. Environ Sci Technol 46:12452–12458. doi:10.1021/es302646q

Cheng L, Gong H, Zhu W, Liu J, Wang X, Liu G, Liu Z (2014) PEGylated Prussian blue nanocubes as a theranostic agent for simultaneous cancer imaging and photothermal therapy. Biomaterials 35:9844–9852. doi:10.1016/j.biomaterials.2014.09.004

Chhonker YS, Prasad YD, Chandasana H, Vishvkarma A, Mitra K, Shukla PK, Bhatta RS (2015) Amphotericin-B entrapped lecithin/chitosan nanoparticles for prolonged ocular application. Int J Biol Macromol 72:1451–1458. doi:10.1016/j.ijbiomac.2014.10.014

Cho K, Wang X, Nie S, Chen ZG, Shin DM (2008) Therapeutic nanoparticles for drug delivery in cancer. Clin Cancer Res 14:1310–1316. doi:10.1158/1078-0432.CCR-07-1441

Choi S, Tripathi A, Singh D (2014) Smart nanomaterials for biomedics. J Biomed Nanotechnol 10:3162–3188. doi:10.1166/jbn.2014.1933

Chow EK, Pierstorff E, Cheng G, Ho D (2008) Copolymeric nanofilm platform for controlled and localized therapeutic delivery. ACS Nano 2:33–40. doi:10.1021/nn7000917

Cole LE, Ross RD, Tilley JM, Vargo-Gogola T, Roeder RK (2015) Gold nanoparticles as contrast agents in x-ray imaging and computed tomography. Nanomedicine (Lond) 10:321–341. doi:10.2217/nnm.14.171

Cooper DL, Conder CM, Harirforoosh S (2014) Nanoparticles in drug delivery: mechanism of action, formulation and clinical application towards reduction in drug-associated nephrotoxicity. Expert Opin Drug Deliv 11:1661–1680. doi:10.1517/17425247.2014.938046
Cywinska MA, Grudzinski IP (2012) Modern toxicology of magnetic nanomaterials. Ann Nat. Occup Hyg 63:247–256. doi:10.3390/nano4020505
da Silveira WL, Damasceno BP, Ferreira LF, Ribeiro IL, Silva KS, Silva AL, Giannini MJ, da Silva-Junior AA, de Oliveira AG (2015) Socrates Tabosa do Egito E. Development and characterization of a microemulsion system containing Amphotericin B with potential ocular applications. Curr Neuropharmacol 8:8. doi: 10.2174/1570159x14666151109110734
Dandekar PP, Jain R, Patil S, Dhumal R, Tiwari D, Sharma S, Vanage G, Patravale V (2010) Curcumin-loaded hydrogel nanoparticles: application in anti-malarial therapy and toxicological evaluation. J Pharm Sci 99:4992–5010. doi:10.1002/jps.22191
Das M, Ansari KM, Tripathi A, Dwivedi PD (2011) Need for safety of nanoparticles used in food industry. J Biomed Nanotechnol 7:13–14. doi:10.1166/jbn.2011.1176
Das SK, Das AR, Guha AK (2009) Gold nanoparticles: microbial synthesis and application in water hygiene management. Langmuir 25:8192–8199. doi:10.1021/la900585p
Dorkhan M, Yucel-Lindberg T, Hall J, Svensater G, Davies JR (2014) Adherence of human oral keratinocytes and gingival fibroblasts to nano-structured titanium surfaces. BMC Oral Health 14:75. doi:10.1186/1472-6831-14-75
Dosio F, Arpicco S, Stella B, Fattal E (2016) Hyaluronic acid for anticancer drug and nucleic acid delivery. Adv Drug Deliv Rev 97:204–236. doi:10.1016/j.addr.2015.11.011
Du C, Narayanan K, Leong MF, Wan AC (2014) Induced pluripotent stem cell-derived hepatocytes and endothelial cells in multi-component hydrogel fibers for liver tissue engineering. Biomaterials 35:6006–6014. doi:10.1016/j.biomaterials.2014.04.011
Eduok S, Martin B, Villa R, Nocker A, Jefferson B, Coulon F (2013) Evaluation of engineered nanoparticle toxic effect on wastewater microorganisms: current status and challenges. Ecotoxicol Environ Saf 95:1–9. doi:10.1016/j.ecoenv.2013.05.022
Fanizza E, Iacobazzi RM, Laquintana V, Valente G, Caliandro G, Striccoli M, Agostiano A, Cutrignelli A, Lopedota A, Curri ML, Franco M, Depalo N, Denora N (2016) Highly selective luminescent nanostructures for mitochondrial imaging and targeting. Nanoscale 8:3350–3361. doi:10.1039/c5nr08139d
Fernandez-Fernandez A, Manchanda R, Carvajal DA, Lei T, Srinivasan S, McGoron AJ (2014) Covalent IR820-PEG-diamine nanoconjugates for theranostic applications in cancer. Int J Nanomed 9:4631–4648. doi:10.2147/IJN.S69550
Figueiro F, de Oliveira CP, Rockenbach L, Mendes FB, Bergamin LS, Jandrey EH, Edelweiss MI, Guterres SS, Pohlmann AR, Battastini AM (2015) Pharmacological improvement and preclinical evaluation of methotrexate-loaded lipid-core nanocapsules in a glioblastoma model. J Biomed Nanotechnol 11:1808–1818. doi:10.1166/jbn.2015.2125
Frangville C, Rutkevicius M, Richter AP, Velev OD, Stoyanov SD, Paunov VN (2012) Fabrication of environmentally biodegradable lignin nanoparticles. Chem phys chem 13:4235–4243. doi:10.1002/cphc.201200537
Froggett SJ, Clancy SF, Boverhof DR, Canady RA (2014) A review and perspective of existing research on the release of nanomaterials from solid nanocomposites. Part Fibre Toxicol 11:17. doi:10.1186/1743-8977-11-17
Fruijtier-Polloth C (2012) The toxicological mode of action and the safety of synthetic amorphous silica-a nanostructured material. Toxicology 294:61–79. doi:10.1016/j.tox.2012.02.001
Gand A, Hindie M, Chacon D, Van Tassel PR, Pauthe E (2014) Nanotemplated polyelectrolyte films as porous biomolecular delivery systems. Application to the growth factor BMP-2. Biomatter 4:e28823. doi:10.4161/biom.28823
Garciafigueroa Y, Trucco M, Giannoukakis N (2015) A brief glimpse over the horizon for type 1 diabetes nanotherapeutics. Clin Immunol 160:36–45. doi:10.1016/j.clim.2015.03.016
Getts DR, Shea LD, Miller SD, King NJ (2015) Harnessing nanoparticles for immune modulation. Trends Immunol 36:419–427. doi:10.1016/j.it.2015.05.007

Gholipour-Kanani A, Bahrami SH, Rabbani S (2016) Effect of novel blend nanofibrous scaffolds on diabetic wounds healing. IET Nanobiotechnol 10:1–7. doi:10.1049/iet-nbt.2014.0066

Gizdavic-Nikolaidis M, Ray S, Bennett JR, Easteal AJ, Cooney RP (2010) Electrospun functionalized polyaniline copolymer-based nanofibers with potential application in tissue engineering. Macromol Biosci 10:1424–1431. doi:10.1002/mabi.201000237

Godwin H, Nameth C, Avery D, Bergeson LL, Bernard D, Beryt E, Boyes W, Brown S, Clippinger AJ, Cohen Y, Doa M, Hendren CO, Holden P, Houck K, Kane AB, Klaessig F, Kodas T, Landsiedel R, Lynch I, Malloy T, Miller MB, Muller J, Oberdorster G, Petersen EJ, Pleus RC, Sayre P, Stone V, Sullivan KM, Tentschert J, Wallis P, Nel AE (2015) Nanomaterial categorization for assessing risk potential to facilitate regulatory decision-making. ACS Nano 9:3409–3417. doi:10.1021/acsnano.5b00941

Golbamaki N, Rasulev B, Cassano A, Marchese Robinson RL, Benfenati E, Leszczynski J, Cronin MT (2015) Genotoxicity of metal oxide nanomaterials: review of recent data and discussion of possible mechanisms. Nanoscale 7:2154–2198. doi:10.1039/c4nr06670g

Gouveia VM, Lima SA, Nunes C, Reis S (2015) Non-biologic Nanodelivery therapies for rheumatoid arthritis. J Biomed Nanotechnol 11:1701–1721. doi:10.1166/jbn.2015.2159

Guo J, Ma M, Chang D, Zhang Q, Zhang C, Yue Y, Liu J, Wang S, Jiang T (2015) Poly-alpha,beta-polyasparthydrazide-based nanogels for potential oral delivery of paclitaxel: *In vitro and in vivo* properties. J Biomed Nanotechnol 11:2231–2242. doi:10.1166/jbn.2015.2118

Halley PD, Lucas CR, McWilliams EM, Webber MJ, Patton RA, Kural C, Lucas DM, Byrd JC, Castro CE (2016) Daunorubicin-loaded DNA origami nanostructures circumvent drug-resistance mechanisms in a leukemia model. Small 12:308–320. doi:10.1002/smll.201502118

Hamori M, Shimizu Y, Yoshida K, Fukushima K, Sugioka N, Nishimura A, Naruhashi K, Shibata N (2015) Preparation of methacrylic acid copolymer S nano-fibers using a solvent-based electrospinning method and their application in pharmaceutical formulations. Cheml Pharma Bull 63:81–87. doi:10.1248/cpb.c14-00563

Hashi CK, Zhu Y, Yang GY, Young WL, Hsiao BS, Wang K, Chu B, Li S (2007) Antithrombogenic property of bone marrow mesenchymal stem cells in nanofibrous vascular grafts. Proc Natl Acad Sci U S A 104:11915–11920. doi:10.1073/pnas.0704581104

Hass JL, Garrison EM, Wicher SA, Knapp B, Bridges N, McLlroy D, Arrizabalaga G (2012) Synthetic osteogenic extracellular matrix formed by coated silicon dioxide nanosprings. J Nanobiotechnology 10:6. doi:10.1186/1477-3155-10-6

Hennebel T, Van Nevel S, Verschuere S, De Corte S, De Gusseme B, Cuvelier C, Fitts JP, van der Lelie D, Boon N, Verstraete W (2011) Palladium nanoparticles produced by fermentatively cultivated bacteria as catalyst for diatrizoate removal with biogenic hydrogen. Appl Microbiol Biotechnol 91:1435–1445. doi:10.1007/s00253-011-3329-9

Hirata E, Menard-Moyon C, Venturelli E, Takita H, Watari F, Bianco A, Yokoyama A (2013) Carbon nanotubes functionalized with fibroblast growth factor accelerate proliferation of bone marrow-derived stromal cells and bone formation. Nanotechnology 24:435101. doi:10.1088/0957-4484/24/43/435101

Ho D (2015) Nanodiamond-based chemotherapy and imaging. Cancer Treat Res 166:85–102. doi:10.1007/978-3-319-16555-4_4

Hommes G, Gasser CA, Howald CB, Goers R, Schlosser D, Shahgaldian P, Corvini PF (2012) Production of a robust nanobiocatalyst for municipal wastewater treatment. Bioresour Technol 115:8–15. doi:10.1016/j.biortech.2011.11.129

Hosny KM, Banjar ZM, Hariri AH, Hassan AH (2015) Solid lipid nanoparticles loaded with iron to overcome barriers for treatment of iron deficiency anemia. Drug Des Devel Ther 9:313–320. doi:10.2147/DDDT.S77702

Hou L, Zhang H, Wang Y, Wang L, Yang X, Zhang Z (2015) Hyaluronic acid-functionalized single-walled carbon nanotubes as tumor-targeting MRI contrast agent. Int J Nanomed 10:4507–4520. doi:10.2147/IJN.S78563

James R, Toti US, Laurencin CT, Kumbar SG (2011) Electrospun nanofibrous scaffolds for engineering soft connective tissues. Methods Mol Biol 726:243–258. doi:10.1007/978-1-61779-052-2_16

Jeliazkova N, Chomenidis C, Doganis P, Fadeel B, Grafstrom R, Hardy B, Hastings J, Hegi M, Jeliazkov V, Kochev N, Kohonen P, Munteanu CR, Sarimveis H, Smeets B, Sopasakis P, Tsiliki G, Vorgrimmler D, Willighagen E (2015) The eNanoMapper database for nanomaterial safety information. Beilstein J Nanotechnol 6:1609–1634. doi:10.3762/bjnano.6.165

Jeong JH, Kim JJ, Bak DH, KS Y, Lee JH, Lee NS, Jeong YG, Kim DK, Han SY (2015) Protective effects of Indole-3-Carbinol-loaded poly(lactic-co-glycolic acid) nanoparticles against glutamate-induced neurotoxicity. J Nanosci Nanotechnol 15:7922–7928. doi:10.1166/jnn.2015.11219

Kahru A, Dubourguier HC (2010) From ecotoxicology to nanoecotoxicology. Toxicology 269:105–119. doi:10.1016/j.tox.2009.08.016

Kalishwaralal K, Barathmanikanth S, Pandian SR, Deepak V, Gurunathan S (2010) Silver nano—a trove for retinal therapies. J Control Release 145:76–90. doi:10.1016/j.jconrel.2010.03.022

Kapishon V, Allison S, Whitney RA, Cunningham MF, Szewczuk MR, Neufeld RJ (2016) Oseltamivir-conjugated polymeric micelles prepared by RAFT living radical polymerization as a new active tumor targeting drug delivery platform. Biomater Sci 4:511–521. doi:10.1039/c5bm00519a

Karim MM, Alam SM, Lee SH (2008) Application of a lanthanide composite nanoparticle-sensitized luminescence method for the determination of salicylic acid in pharmaceutical formulations and human plasma. Lumin J Biol Chem Lumin 23:417–423. doi:10.1002/bio.1072

Karthikeyan K, Guhathakarta S, Rajaram R, Korrapati PS (2012) Electrospun zein/eudragit nanofibers based dual drug delivery system for the simultaneous delivery of aceclofenac and pantoprazole. Int J Pharm 438:117–122. doi:10.1016/j.ijpharm.2012.07.075

Kateb B, Van Handel M, Zhang L, Bronikowski MJ, Manohara H, Badie B (2007) Internalization of MWCNTs by microglia: possible application in immunotherapy of brain tumors. Neuro Image 37(Suppl 1):S9–17. doi:10.1016/j.neuroimage.2007.03.078

Kharangate-Lad A, Pereira F, Fernandes J, Bhosle S (2016) Studies on the effects of zerovalent iron nanoparticles on bacteria from the mangrove ecosystem. Environ Sci Pollut Res Int 23:927–931. doi:10.1007/s11356-015-5588-y

Kim M, Hong B, Lee J, Kim SE, Kang SS, Kim YH, Tae G (2012a) Composite system of PLCL scaffold and heparin-based hydrogel for regeneration of partial-thickness cartilage defects. Biomacromolecules 13:2287–2298. doi:10.1021/bm3005353

Kim DH, Kshitiz, Smith RR, Kim P, Ahn EH, Kim HN, Marban E, Suh KY, Levchenko A (2012b) Nanopatterned cardiac cell patches promote stem cell niche formation and myocardial regeneration. Integr Biol (Camb) 4:1019–1033. doi:10.1039/c2ib20067h

Kim YR, Park SH, Lee JK, Jeong J, Kim JH, Meang EH, Yoon TH, Lim ST, JM O, An SS, Kim MK (2014) Organization of research team for nano-associated safety assessment in effort to study nanotoxicology of zinc oxide and silica nanoparticles. Int J Nanomed 9(Suppl 2):3–10. doi:10.2147/IJN.S57915

Kitture R, Chordiya K, Gaware S, Ghosh S, More PA, Kulkarni P, Chopade BA, Kale SN (2015) ZnO nanoparticles-red sandalwood conjugate: a promising anti-diabetic agent. J Nanosci Nanotechnol 15:4046–4051. doi:10.1166/jnn.2015.10323

Korsmeyer R (2016) Critical questions in development of targeted nanoparticle therapeutics. Regen Biomater 3:143–147. doi:10.1093/rb/rbw011

Kriparamanan R, Aswath P, Zhou A, Tang L, Nguyen KT (2006) Nanotopography: cellular responses to nanostructured materials. J Nanosci Nanotechnol 6:1905–1919. doi:10.1166/jnn.2006.330

Krupa AN, Vimala R (2016) Evaluation of tetraethoxysilane (TEOS) sol-gel coatings, modified with green synthesized zinc oxide nanoparticles for combating microfouling. Mater Sci Eng C Mater Biol Appl 61:728–735. doi:10.1016/j.msec.2016.01.013

Kumar V, Palazzolo S, Bayda S, Corona G, Toffoli G, Rizzolio F (2016) DNA nanotechnology for cancer therapy. Theranostics 6:710–725. doi:10.7150/thno.14203

Kumar CG, Poornachandra Y (2015) Biodirected synthesis of Miconazole-conjugated bacterial silver nanoparticles and their application as antifungal agents and drug delivery vehicles. Colloids Surf B Biointerfaces 125:110–119. doi:10.1016/j.colsurfb.2014.11.025

Lavenus S, Trichet V, Le Chevalier S, Hoornaert A, Louarn G, Layrolle P (2012) Cell differentiation and osseointegration influenced by nanoscale anodized titanium surfaces. Nanomedicine (Lond) 7:967–980. doi:10.2217/nnm.11.181

Lee H, MK Y, Park S, Moon S, Min JJ, Jeong YY, Kang HW, Jon S (2007) Thermally cross-linked superparamagnetic iron oxide nanoparticles: synthesis and application as a dual imaging probe for cancer *in vivo*. J Am Chem Soc 129:12739–12745. doi:10.1021/ja072210i

Li X, Zhao T, Sun L, Aifantis KE, Fan Y, Feng Q, Cui F, Watari F (2016) The applications of conductive nanomaterials in the biomedical field. J Biomed Mater Res A 104:322–339. doi:10.1002/jbm.a.35537

Liu G, Wang D, Wang J, Mendoza C (2011) Effect of ZnO particles on activated sludge: role of particle dissolution. Sci Total Environ 409:2852–2857. doi:10.1016/j.scitotenv.2011.03.022

Mahmood S, Taher M, Mandal UK (2014) Experimental design and optimization of raloxifene hydrochloride loaded nanotransfersomes for transdermal application. Int J Nanomed 9:4331–4346. doi:10.2147/IJN.S65408

Martinez Avila H, Feldmann EM, Pleumeekers MM, Nimeskern L, Kuo W, de Jong WC, Schwarz S, Muller R, Hendriks J, Rotter N, van Osch GJ, Stok KS, Gatenholm P (2015) Novel bilayer bacterial nanocellulose scaffold supports neocartilage formation *in vitro* and *in vivo*. Biomaterials 44:122–133. doi:10.1016/j.biomaterials.2014.12.025

MedlinePlus Metabolic disorders. https://www.nlm.nih.gov/. Retrieved 27 July 2015

Musee N (2011) Nanowastes and the environment: potential new waste management paradigm. Environ Int 37:112–128. doi:10.1016/j.envint.2010.08.005

Nel AE (2013) Implementation of alternative test strategies for the safety assessment of engineered nanomaterials. J Intern Med 274(6):561–577. doi:10.1111/joim.12109

Ouf SA, El-Adly AA, Mohamed AA (2015) Inhibitory effect of silver nanoparticles mediated by atmospheric pressure air cold plasma jet against dermatophyte fungi. J Med Microbiol 64:1151–1161. doi:10.1099/jmm.0.000133

Pachioni-Vasconcelos Jde A, Lopes AM, Apolinario AC, Valenzuela-Oses JK, Costa JS, Nascimento Lde O, Pessoa A, Barbosa LR, Rangel-Yagui Cde O (2016) Nanostructures for protein drug delivery. Biomater Sci 4:205–218. doi:10.1039/c5bm00360a

Pang Y, Zeng GM, Tang L, Zhang Y, Liu YY, Lei XX, MS W, Li Z, Liu C (2011) Cr(VI) reduction by Pseudomonas aeruginosa immobilized in a polyvinyl alcohol/sodium alginate matrix containing multi-walled carbon nanotubes. Bioresour Technol 102:10733–10736. doi:10.1016/j.biortech.2011.08.078

Pasricha A, Jangra SL, Singh N, Dilbaghi N, Sood KN, Arora K, Pasricha R (2012) Comparative study of leaching of silver nanoparticles from fabric and effective effluent treatment. J Environ Sci (China) 24:852–859. doi:10.1016/S1001-0742(11)60849-8

Peng Q, Mu H (2016) The potential of protein-nanomaterial interaction for advanced drug delivery. J Control Release 225:121–132. doi:10.1016/j.jconrel.2016.01.041

Pereira DI, Bruggraber SF, Faria N, Poots LK, Tagmount MA, Aslam MF, Frazer DM, Vulpe CD, Anderson GJ, Powell JJ (2014) Nanoparticulate iron(III) oxo-hydroxide delivers safe iron that is well absorbed and utilised in humans. Nanomedicine 10:1877–1886. doi:10.1016/j.nano.2014.06.012

Perry JL, Kai MP, Reuter KG, Bowerman C, Christopher Luft J, DeSimone JM (2015) Calibration-quality cancer nanotherapeutics. Cancer Treat Res 166:275–291. doi:10.1007/978-3-319-16555-4_12

Pippa N, Pispas S, Demetzos C (2016) Polymer self-assembled nanostructures as innovative drug Nanocarrier platforms. Curr Pharm Des 22:2788–2795. doi:10.2174/1381612822666160217141232

Pooja D, Tunki L, Kulhari H, Reddy BB, Sistla R (2015) Characterization, biorecognitive activity and stability of WGA grafted lipid nanostructures for the controlled delivery of rifampicin. Chem Phys Lipids 193:11–17. doi:10.1016/j.chemphyslip.2015.09.008

Poulose S, Panda T, Nair PP, Theodore T (2014) Biosynthesis of silver nanoparticles. J Nanosci Nanotechnol 14:2038–2049. doi:10.1166/jnn.2014.9019

Qian Y, Zhou X, Zhang Y, Zhang W, Chen J (2013) Performance and properties of nanoscale calcium peroxide for toluene removal. Chemosphere 9:717–723. doi:10.1016/j.chemosphere.2013.01.049

Rai M, Ingle AP, Gade A, Duran N (2015) Synthesis of silver nanoparticles by Phoma gardeniae and *in vitro* evaluation of their efficacy against human disease-causing bacteria and fungi. IET Nanobiotechnol 9:71–75. doi:10.1049/iet-nbt.2014.0013

Ruiz de Garibay AP, Solinis MA, del Pozo-Rodriguez A, Apaolaza PS, Shen JS, Rodriguez-Gascon A (2015) Solid lipid nanoparticles as non-viral vectors for Gene transfection in a cell model of Fabry disease. J Biomed Nanotechnol 11:500–511. doi:10.1166/jbn.2015.1968

Sahebkar A, Badiee A, Hatamipour M, Ghayour-Mobarhan M, Jaafari MR (2015) Apolipoprotein B-100-targeted negatively charged nanoliposomes for the treatment of dyslipidemia. Colloids Surf B Biointerfaces 129:71–78. doi:10.1016/j.colsurfb.2015.03.012

Salem W, Leitner DR, Zingl FG, Schratter G, Prassl R, Goessler W, Reidl J, Schild S (2015) Antibacterial activity of silver and zinc nanoparticles against vibrio cholerae and enterotoxic Escherichia coli. Int J Med Microbiol 305:85–95. doi:10.1016/j.ijmm.2014.11.005

Salvadori MR, Ando RA, Oller Do Nascimento CA, Correa B (2014) Bioremediation from wastewater and extracellular synthesis of copper nanoparticles by the fungus Trichoderma koningiopsis. J Environ Sci Health A Tox Hazard Subst Environ Eng 49:1286–1295. doi:10.1080/10934529.2014.910067

Seo H, Roh Y (2015) Biotransformation and its application: biogenic Nano-catalyst and metal-reducing-bacteria for remediation of Cr(VI)-contaminated water. J Nanosci Nanotechnol 15:5649–5652. doi:10.1166/jnn.2015.10446

Serra P, Santamaria P (2015) Nanoparticle-based autoimmune disease therapy. Clin Immunol 160:3–13. doi:10.1016/j.clim.2015.02.003

Shavi GV, Nayak UY, Maliyakkal N, Deshpande PB, Raghavendra R, Kumar AR, Reddya MS, Udupa N, Shrawan B (2015) Nanomedicine of anastrozole for breast cancer: physicochemical evaluation, *in vitro* cytotoxicity on BT-549 and MCF-7 cell lines and preclinical study on rat model. Life Sci 141:143–155. doi:10.1016/j.lfs.2015.09.021

Singh MK, Gracio J, LeDuc P, Goncalves PP, Marques PA, Goncalves G, Marques F, Silva VS, Capela e Silva F, Reis J, Potes J, Sousa A (2010) Integrated biomimetic carbon nanotube composites for *in vivo* systems. Nanoscale 2:2855–2863. doi:10.1039/c0nr00237b

Singh D, Somani VK, Aggarwal S, Bhatnagar R (2015) PLGA (85:15) nanoparticle based delivery of rL7/L12 ribosomal protein in mice protects against Brucella abortus 544 infection: a promising alternate to traditional adjuvants. Mol Immunol 68:272–279. doi:10.1016/j.molimm.2015.09.011

Sinha A, Singh VN, Mehta BR, Khare SK (2011) Synthesis and characterization of monodispersed orthorhombic manganese oxide nanoparticles produced by bacillus sp. cells simultaneous to its bioremediation. J Hazard Mater 192:620–627. doi:10.1016/j.jhazmat.2011.05.103

Sun EY, Kim Y, Park B, Roh Y (2011) Environmental application of nanomaterials and metal-reducing bacteria to remediate arsenic-contaminated groundwater. J Nanosci Nanotechnol 11:1589–1592. doi:10.1166/jnn.2011.3411

Tong HW, Mutlu BR, Wackett LP, Aksan A (2014) Manufacturing of bioreactive nanofibers for bioremediation. Biotechnol Bioeng 111:1483–1493. doi:10.1002/bit.25208

Tuin BJ, Geerts R, Westerink JB, van Ginkel CG (2006) Pretreatment and biotreatment of saline industrial wastewaters. Water Sci Technol 53:17–25. doi:10.2166/wst.2006.072

Tuo Y, Liu G, Zhou J, Wang A, Wang J, Jin R, Lv H (2013) Microbial formation of palladium nanoparticles by Geobacter sulfurreducens for chromate reduction. Bioresour Technol 133:606–611. doi:10.1016/j.biortech.2013.02.016

Veetil JV, Ye K (2009) Tailored carbon nanotubes for tissue engineering applications. Biotechnol Prog 25:709–721. doi:10.1002/btpr.165

Wang M, Abad D, Kickhoefer VA, Rome LH, Mahendra S (2015a) Vault nanoparticles packaged with enzymes as an efficient pollutant biodegradation technology. ACS Nano 9:10931–10940. doi:10.1021/acsnano.5b04073

Wang H, Kim B, Wunder SL (2015b) Nanoparticle-supported lipid bilayers as an in situ remediation strategy for hydrophobic organic contaminants in soils. Environ Sci Technol 49:529–536. doi:10.1021/es504832n

Wang X, Zhao C, Zhao P, Dou P, Ding Y, Xu P (2009) Gellan gel beads containing magnetic nanoparticles: an effective biosorbent for the removal of heavy metals from aqueous system. Bioresour Technol 100:2301–2304. doi:10.1016/j.biortech.2008.10.042

Wang W, Zhou M, Jin Z, Li T (2010) Reactivity characteristics of poly(methyl methacrylate) coated nanoscale iron particles for trichloroethylene remediation. J Hazard Mater 173:724–730. doi:10.1016/j.jhazmat.2009.08.145

Wu J, Bu X, Dou L, Fang L, Shen Q (2015a) Co-delivery of Docetaxel and Berbamine by Chitosan/Sulfobutylether-β-Cyclodextrin Nanoparticles for Enhancing Bioavailability and Anticancer Activities. J Biomed Nanotechnol 11(10):1847–57

Wu Y, Wang Z, Liu G, Zeng X, Wang X, Gao Y, Jiang L, Shi X, Tao W, Huang L, Mei L (2015b) Novel Simvastatin-Loaded Nanoparticles based on Cholic Acid-Core Star-Shaped PLGA for Breast Cancer Treatment. J Biomed Nanotechnol 11(7):1247–60

Xu H, Zhang Y, Jiang Q, Reddy N, Yang Y (2013) Biodegradable hollow zein nanoparticles for removal of reactive dyes from wastewater. J Environ Manag 125:33–40. doi:10.1016/j.jenvman.2013.03.050

Yang R, Qi J, Zhang J, Wang F, Fan L (2015) Effects of Paris polyphylla saponin VII plus silica nano composite on ovarian cancer drug resistance in vitro. Zhonghua Yi Xue Za Zhi 95(23):1859–61.

Yen A, Zhang K, Daneshgaran G, Kim HJ, Ho D (2016) Chemopreventive Nanodiamond Platform for Oral Cancer Treatment. J Calif Dent Assoc 44(2):121–7.

Zaidi SA, Shin JH (2016) Recent developments in nanostructure based electrochemical glucose sensors. Talanta 149:30–42. doi:10.1016/j.talanta.2015.11.033

Zhang X, FG W, Liu P, Wang HY, Gu N, Chen Z (2015) Synthesis of ultrastable and multifunctional gold nanoclusters with enhanced fluorescence and potential anticancer drug delivery application. J Colloid Interf Sci 455:6–15. doi:10.1016/j.jcis.2015.05.029

Zhang Y, Zhu J, Tang Y, Chen X, Yang Y (2009) The preparation and application of pulmonary surfactant nanoparticles as absorption enhancers in insulin dry powder delivery. Drug Dev Ind Pharm 35:1059–1065. doi:10.1080/03639040902769628

Zhang J, Zhuang J, Gao L, Zhang Y, Gu N, Feng J, Yang D, Zhu J, Yan X (2008) Decomposing phenol by the hidden talent of ferromagnetic nanoparticles. Chemosphere 73:1524–1528. doi:10.1016/j.chemosphere.2008.05.050

Zheng X, Huang H, Su Y, Wei Y, Chen Y (2015) Long-term effects of engineered nanoparticles on enzyme activity and functional bacteria in wastewater treatment plants. Water Sci Technol 72:99–105. doi:10.2166/wst.2015.194

Nanomaterial-Enabled Immunotherapeutic Applications

15

Reena V. Saini and Reena Kumari

Abstract

Nanomaterials are widely used for various biomedical applications to the diseased site for enhanced therapy and reduced toxicity. The ability of nanomaterials to activate or modulate the immune system can be used for immunotherapy, which is an encouraging option for curing various diseases like cancer, allergy, autoimmune diseases, etc. Nanoformulations have been shown to augment efficacy of vaccines, therapeutics, and adoptive cell therapies via targeted delivery of these immunostimulatory materials. Various nanocarriers like polymer-drug conjugates, liposomes, emulsions, and solid lipid nanoparticles have been used to transport immune effectors for active and passive nanoimmunotherapy. However, inappropriate immune responses generated by nanomaterial exposure could lead to autoimmunity and destruction to the host system. So, the need of the hour is to comprehend the impact of nanomaterials on the cells that participate in the innate and adaptive immune responses.

Keywords

Immunotherapy • Nanotechnology • Biopolymers • Liposomes • Cancer • Autoimmune disorders • HIV • Allergy • Mucosal immunity

R.V. Saini (✉) • R. Kumari
Animal Biotechnology Laboratory, Faculty of Applied Sciences and Biotechnology, Shoolini University, Solan, Himachal Pradesh, India
e-mail: reenavohra10@gmail.com

V.C. Kalia, A.K. Saini (eds.), *Metabolic Engineering for Bioactive Compounds*,
DOI 10.1007/978-981-10-5511-9_15

15.1 Introduction

15.1.1 The Immune System and Nanoimmunotherapeutics

Immune system is a collection of organs, immune cells, and molecules that protect us from infections and some other diseases. Innate immune system, the first respondents, consists of neutrophils and macrophages; these respond immediately to invading pathogen/foreign particles by expressing receptors that recognize conserved molecular signature of bacteria, viruses, and fungi, to quickly phagocytose these microbes and secrete reactive oxygen species or cytokines that provide an immediate response to invading pathogen/foreign particles. The adaptive immune system include $CD4^+$ helper T cells that secrete cytokines to direct the functions of innate cells, killer cells, and B cells and $CD8^+$ killer T cells that recognize and destroy infected or transformed cells. B cells produce antibodies that neutralize the ability of microbes to invade host cells and/or promote their phagocytosis (Charles et al. 2001). Antigen-presenting cells (APCs) particularly dendritic cells present in all peripheral tissues activate immature T cells (and in some cases B cells) when come in contact with antigen (Mellman and Steinman 2001; Steinman and Banchereau 2007). Dendritic cells (DCs) and other immune cells express receptors that specifically recognize antigens to trigger their activation; the most studied among these receptors are the Toll-like receptors (Kawai and Akira 2011).

Immunotherapy is based on modulation of the person's immune system either by stimulating host's immune system or by giving immune system molecules like recombinant immune proteins to work harder against diseases. Immunotherapies have been pursued for a broad range of diseases, but arguably the greatest effort has been invested in the development of therapeutics that prime the immune system to attack cancer (Mellman et al. 2011; Restifo et al. 2012; Couzin-Frankel 2013). The major challenge for therapies based on traditional single drug remedies is the complexity of signaling pathways that regulate functioning of the immune system. Recently, nanomaterials are being studied immensely as delivery agents and vaccine adjuvants (Reddy et al. 2006; Hubbell et al. 2009; Kasturi et al. 2011; Moon et al. 2011), diagnostic tools (Cho et al. 2011; Noh et al. 2011), and ex vivo systems for treatment of cancer and infectious diseases by therapeutic immune cells (Hellstrom et al. 2001; Steenblock and Fahmy 2008; Kalos et al. 2011)..Nanoparticles (NPs) enable detection and treatment of diverse conditions like cancer, infectious disease, and autoimmune disorders with major effects on global health.

The advancement of nanoengineered delivery vehicles provides new tools that could improve the efficacy of current immunotherapies having the following benefits:

- Nanovehicles protect encapsulated biomolecules and bioactive materials from enzymatic destruction and early clearance from the circulatory system.
- Deliver bioactive materials and molecules with high accuracy and efficiency.
- Labeling nanocarriers with ligands lead to better specificity for cellular targets.
- Appropriate size and surface properties of nanocarriers with low circulation in reticuloendothelial system result in prolonged half-life and change the availability of the biomolecules (Li et al. 2012, 2013a, B39).

15.2 Various Nanoparticles and their Effects

15.2.1 Polymeric Nanoparticles

Polymeric nanoparticles are synthesized by biodegradable polymers, dendrimers, and micelles which act as nanoscale drug delivery vehicles. The nanoparticles of the biodegradable polymers like polylactides or poly lactic-co-glycolic acid (PLGA) are used in vaccine development by encapsulating the antigen in these polymers (Mundargi et al. 2008). These polymers offer various advantages as they enhance the half-life of the bioactive molecules by protecting them from hostile environment and enzymatic attack; they provide sustained release and targeted delivery by specific ligand tagging, and finally they also possess adjuvant effects.

15.2.2 Nanoliposomes

Nanoliposomes are enclosed structures of aqueous compartments that are surrounded by one or more amphiphilic lipid bilayers (US Food and Drug Administration 2002). From immunological prospective, two types of nanoliposomes have been reported: (1) antigen-encapsulated liposomes designed to enhance immune response and (2) polymer-coated liposomes to prevent immune recognition. Currently, numerous nanoliposomal formulations have been accepted for treatment of cancer, infections, and meningitis (Zolnik and Sadrieh 2009), and others are being examined as therapeutic vaccines against HIV, hepatitis A, malaria, prostate cancer, and colorectal cancer (Peek et al. 2008). Additionally, several manipulations can be done on the surface charge of nanoliposomes to increase antigen delivery, for example, (1) incorporation of the cationic lipids that are capable of generating more potent immune reaction than anionic or neutral liposomes (Nakanishi et al. 1999) and (2) tagging of antibodies to liposomes for targeted antigen delivery to dendritic cells (Mellman and Steinman 2001).

15.2.3 Nanoemulsions

Nanoemulsions are heterogeneous dispersions of two immiscible liquids (oil in water or water in oil) with droplet size in the nanometer scale. Nanoemulsions are ideal drug delivery systems as they are site specific, dissolve large quantity of hydrophobics, and protect the drug from degradation with long-term stability (Lovelyn and Attama 2011). It has been studied earlier that the nano-sized emulsions are more competent than larger-sized emulsions to invade the nasal mucosa and to deliver antigens to the APCs (Makidon et al. 2008).

15.2.4 Solid Lipid Nanoparticles (SLNs)

SLNs are known as colloidal drug vehicles (Mehnert and Mader 2001; Muller and Keck 2004; Castelli et al. 2005) but unlike nanoemulsions the SLNs utilized solid lipids having high melting point like glycerides or waxes (Manjunath and Venkateswarlu 2005). SLN-encapsulated antisense oligodeoxyribonucleotide G3139 demonstrated higher immunoenhancing potential and antitumor property than unencapsulated G3139. The small particle size of SLNs enhanced their uptake by macrophages and dendritic cells residing in tumor microenvironment (Pan et al. 2009). SLN-based targeted gene therapy for lung cancer has revealed that protein expression of p53 was better in H1299 lung cancer cells when transfected with SLN-carrying p53 as compared to Lipofectin Transfection Reagent (Choi et al. 2008). In a recent report, researchers successfully used SLNs with a stable core/shell nanocompartment encapsulating quantum dots and paclitaxel complexed with anionic Bcl-2 siRNA molecules for anticancer theranostics (Bae et al. 2013).

15.3 Immunotherapy for Various Diseases

15.3.1 Cancer Nanoimmunotherapy

Cancer is one of the most concerning diseases which is affecting populations across the globe. Presently available treatments like chemotherapy or radiotherapy have side effects as they are toxic to normal cells; therefore, new therapeutic approaches need to be explored. Scientific interventions have enhanced the knowledge about the tumor biology and immunology that headed toward discovery of specific and nontoxic immunotherapies. The immunotherapies currently being used to treat cancer are monoclonal antibodies, adjuvants, vaccines, and adoptive cellular therapies. Targeted immunotherapies are being explored to overcome the immune inhibition induced by tumor microenvironment. Among immune cells, the cytotoxic T lymphocytes (CTL) and NK cells can specifically recognize and target cancerous cells. However, efficacy of a cancer vaccine depends on collaboration among T helper (Th) cells and CTL for a specific tumor antigen (Fong and Engleman 2000; Palucka and Banchereau 2012). Combination of ovalbumin antigen-loaded PLGA nanoparticles and monophosphoryl lipid A as adjuvants resulted in promising $CD4^+$ and $CD8^+$ T-cell responses (Hamdy et al. 2007). Recently, chitosan nanoparticles modified with folate and coated with human IP-10 gene were found to stimulate $CD8^+$ CTL leading to tumor growth inhibition and enhanced life span of nude mice (Duan et al. 2016). The protective and therapeutic cancer vaccines also indulge dendritic cell (DC) activity to stimulate a specific immune response toward tumor. For example, biodegradable poly-(γ-glutamic acid) nanoparticles carrying tumor-associated antigens were efficiently engulfed by immature DCs, leading to DC maturation (Vacchelli et al. 2012).

Passive immunotherapy depends on the indulgence of antibodies and CTLs leading to enhanced therapeutic effects with very few adverse effects. Currently, monoclonal antibodies (MABs) are leading immunotherapeutic specialists that are used in clinic for the treatment of solid tumors and lymphomas (Krishnamachari et al. 2011). For instance, trastuzumab is being utilized for the treatment of $HER2^+$ breast cancer and adenocarcinoma, while alemtuzumab is being employed for treating chronic lymphocytic leukemia (Li et al. 2013a, b). The high cost of production and other constrains, for example, insufficient pharmacokinetics and tissue approachability, are the main hindrances while utilizing antibodies in a clinical trial. Nevertheless, nanoparticle-based passive nanoimmunotherapy improves the retention and stability of the MABs within the tumor while bringing down tissue disruption and other pathophysiological effects (Fuchs et al. 2012). Additionally, hollow mesoporous silica (HMS) capsules are also promising vehicles for targeted delivery of the MABs. Lim et al. synthesized cell membrane permeable HMS capsules which efficiently conveyed the antibodies into the HeLa cells (Lim et al. 2012). The in vivo localized administration of CTLA4 antibodies captured in functionalized mesoporous silica displayed highly significant anticancerous effect as compared to free antibodies injected systemically (Lei et al. 2010).

Nanoformulations with herbal materials are promising as they improve biopharmaceutic properties of these plant products, especially in cancer diagnostic and healing. The combination of curcumin-polyethylene glycol conjugate and Trp2 (tumor-associated antigen) peptide vaccine revealed an interactive antitumor impact when administered intravenously in a melanoma-induced mouse. Furthermore, it was observed that the combination therapy significantly stimulated CTL response and IFN-γ production in the immune organs (Lu et al. 2015). Roy et al. studied combined chemo-immunotherapeutic effect of PLGA-conjugated paclitaxel and lipopolysaccharide (TLNP) nanoformulation, and they found enhanced anticancer effect of TLNP as compared to paclitaxel in a cell-splenocyte coculture assay system. The in vivo experiments showed high amount of paclitaxel in the tumor microenvironment of the mice injected with TLNP as compared to mice treated with paclitaxel alone (Roy et al. 2010). Several paclitaxel nanoformulations have been affirmed by the Food and Drug Administration (FDA) for cancer treatment, for example, paclitaxel albumin-stabilized nanoparticle formulation in combination with gemcitabine for metastatic adenocarcinoma of the pancreas, paclitaxel protein-bound particles combined with carboplatin for local and metastatic non-small cell lung cancer, and paclitaxel protein-bound particles for breast cancer treatment (http://www.cancer.gov/about-cancer/treatment/drugs/paclitaxel).

15.3.2 Autoimmune Diseases

Immune system ailments, for example, rheumatoid arthritis and type 1 diabetes, are results of inadequate activation of T cells, B cells, or both, such that harm to one or more organ systems happens (Davidson and Diamond 2001). Immune homeostasis and immune system provocation are frequently maintained by regulatory T cells (Tregs). Nanotechnology-based methodologies are intended to enhance immune tolerance via different components that include local or systemic inhibition of antigen-presenting cells, development of antigen-specific Treg cells, and removal of antigen-specific T cells (Serra and Santamaria 2015). Folic acid (FA)-conjugated dendrimers loaded with methotrexate have been utilized for arthritis treatment by aiming stimulated macrophages (expressing folate receptor beta) in a rat model of arthritis induced by type II collagen (Thomas et al. 2011). Likewise, azabisphosphonate-capped dendrimers (Hayder et al. 2011) were employed in rheumatoid arthritis murine models to provoke an anti-inflammatory response in mice monocytes. Recently, Clemente-Casares et al. synthesized nanoparticles carrying self-peptides associated with MHC class II proteins that were capable of activating Tregs which in turn suppress autoimmune reactions (Clemente-Casares et al. 2016).

15.3.3 Immunotherapy for HIV/AIDS

Lately, there has been expanding enthusiasm for the treatment of HIV/AIDS by using immunotherapy so that the immune system responsibilities could be reestablished (Gandhi and Walker 2002; Cohen 2007; Rinaldo 2009). The different immunotherapeutic strategies utilized for HIV/AIDS treatment are dependent on delivering various antigens and cytokines like as IL-2, IL-7, and IL-15 (Gandhi and Walker 2002; Pett 2009).

Tragically, despite many clinical trials for HIV/AIDS that demonstrated increased immune reactions, most of these reports have been steadily unsuccessful to give clinical advancements to the AIDS sufferers (Bourinbaiar et al. 2006). The in vitro studies done in this context showed that nanoparticles of the PLGA were capable of delivering antigens to DCs derived from mice bone marrow, suggesting their promising role in therapy (Elamanchili et al. 2004). More recently, surfactant-free anionic poly (D, L-lactide) (PLA) nanoparticles coated with HIV p24 protein were found to be proficiently engulfed by mouse DCs, which leads to increased mucosal and cellular immune reactions in mice (Aline et al. 2009). DermaVir Patch is a nanoformulation which consists of HIV antigen-coding plasmid that helps in the transportation of the nanoparticle to Langerhans cells (LC). Furthermore epidermal Langerhans cells entrap these nanoparticles and mature as exceptionally immunogenic antigen-presenting cells capable of activating T cells in the lymph nodes. Because of the safety and tolerability depicted by DermaVir Patch in Phase I clinical trials, this nano-immunoformulation for HIV/AIDS is the first one to reach Phase II trials (Lori et al. 2007).

15.3.4 Allergen Immunotherapy

The main reasons for the allergic diseases are inappropriate stimulation of $CD4^+$ T helper cells (T_H1 and T_H2) and T regulatory (T_{Reg}) cells which perform a central role in immunological responses stimulated toward allergens (Romagnani 2004). Garaczi et al. developed DermAll, a unique immunotherapy against IgE-mediated allergies exhibiting safe and immunogenic responses in rhinitis model for allergic reactions. DermAll-OVA is a nanoformulation that overexpress ovalbumin (OVA) as an allergen. They showed that DermAll ASIT suppressed T_H2-mediated allergic reactions; furthermore, the acute allergy responses generated in mice were weakened by using DermAll nanoformulation (Garaczi et al. 2013).

15.3.5 Mucosal Immune Diseases

Cells, molecules, and lymphoid organs of the immune system provide protection against mucosal pathogens which are responsible for mucosal immune diseases. The best defense against these pathogens are antibiotics and mucosal vaccines which induce systemic and mucosal immunity. But the major obstacle is that these antibiotics and vaccines are less effective in combating emerging infectious diseases (Ginkel et al. 2000). Over the past few years, nanotechnology is growing fastly in pharmaceutical industries and has a huge impact on mucosal immunology (McNeil 2011). Current research on mucosal immunology aimed to synthesize biocompatible and biodegradable nanoparticles that transfer anti-inflammatory molecules and other substances to diseased mucosal tissues, like the lung airways (Roy and Vij 2010), the gastrointestinal tract (Laroui et al. 2011), and the eye (du Toit et al. 2011). Laroui et al. showed that alginate/chitosan polymeric hydrogel containing polylactic acid nanoparticles with anti-inflammatory tripeptide (Lys-Pro-Val) has been used for the treatment of mucosal inflammation in mice (Laroui et al. 2011).

15.4 Immunotoxicological Impact of Nanomaterial Exposure

Nanomaterials are extensively used in biomedical applications that lead to the enhanced probability of human exposures. To evaluate overall health and safety of nanoformulations, it is necessary to understand the mechanism of interaction between nanomaterials and constituents of the immune system. The use of nanoformulations as drug delivery system has many advantages as they increase the efficacy of drug by reaching the molecular targets penetrating physiologic obstructions which can be attributed to their small size characteristics (Christopher 2010; Misra et al. 2010; Zhang et al. 2010; Grimm and Scheinberg 2011; Liu et al. 2011). In spite of these benefits, there are some off-target effects like unsuitable activation of immune system or the development of hypersensitivity, generation of allergic responses, autoimmune diseases, and some types of malignancies (Wang et al. 2013). For instance, nondegradable nanoparticles can amass in organs and cells as they cannot be cleared properly from the body that leads to toxicity. However, the degraded constituents of the biocompatible materials may cause lethal outcomes (Fischer and Chan 2007). Metal/metal oxide nanoparticles and other nanomaterials such as fullerenes, quantum dots, and fibrous particles have also shown adverse effects on genetic material leading to aberrations in gene expression (Singh et al. 2009). Therefore, for the safer usage of nanoparticles, it is necessary to understand the interaction mechanism between nanoparticles and immune system, so that hostile reactions could be prevented.

Conclusion

The development of nanomaterials has provided new ways to enhance specific immune responses for prevention or treatment of diseases. Numerous nanomaterials are designed for delivery of conventional therapeutics to initiate or enhance the immune system against various diseases. Some nanoformulations are as of now in clinical trials, though several other formulations are in different phases of preclinical improvement. Further research about the mechanism of interaction between nanoparticles and immune cells is necessary to make better and safer nanoformulations as sometimes it may stimulate molecular responses that can have toxic effects and lead to greater susceptibility to infectious diseases.

Acknowledgment The authors acknowledge Shoolini University, Himachal Pradesh, for providing necessary funds and all the required conveniences.

References

Aline F, Brand D, Pierre J, Roingeard P, Severine M, Verrier B, Dimier-Poisson I (2009) Dendritic cells loaded with HIV-1 p24 proteins adsorbed on surfactant free anionic PLA nanoparticles induce enhanced cellular immune responses against HIV-1 after vaccination. Vaccine 27:5284–5291. doi:10.1016/j.vaccine.2009.05.028

Bae KH, Lee JY, Lee SH, Park TG, Nam YSS (2013) Optically traceable solid lipid nanoparticles loaded with siRNA and paclitaxel for synergistic chemotherapy with in situ imaging. Adv Healthc Mater 2:576–584. doi:10.1002/adhm.201200338

Bourinbaiar AS, Root-Bernstein RS, Abulafia-Lapid R, Rytik PG, Kanev AN, Jirathitikal V, Orlovsky VG (2006) Therapeutic AIDS vaccines. Curr Pharm Des 12:2017–2030. doi:10.2174/138161206777442119

Castelli F, Puglia C, Sarpietro MG, Rizza L, Bonina F (2005) Characterization of indomethacin-loaded lipid nanoparticles by differential scanning calorimetry. Int J Pharm 304:231–238. doi:10.1016/j.ijpharm.2005.08.011

Charles AJJ, Travers P, Walport M, Shlomchik MJ (2001) Principles of innate and adaptive immunity immunobiology, 5th edn. Garland Science, New York

Cho NH, Cheong TC, Min JH, Wu JH, Lee SJ, Kim D, Yang JS, Kim S, Kim YK, Seong SY (2011) A multifunctional core-shell nanoparticle for dendritic cell-based cancer immunotherapy. Nat Nanotechnol 6:675–682. doi:10.1038/nnano.2011.149

Choi SH, Jin SE, Lee MK, Lim SJ, Park SJ, Kim BG, Ahn WS, Kim CK (2008) Novel cationic solid lipid nanoparticles enhanced p53 gene transfer to lung cancer cells. Eur J Pharm Biopharm 68:545–554. doi:10.1016/j.ejpb.2007.07.011

Christopher C (2010) The immune effects of naturally occurring and synthetic nanoparticles. J Autoimmun 34:J234–J246. doi:10.1016/j.jaut.2009.11.009

Clemente-Casares X, Blanco J, Ambalavanan P, Yamanouchi J, Singha S, Fandos C, Tsai S, Wang J, Garabatos N, Izquierdo C, Agrawal S, Keough MB, Yong VW, James E, Moore A, Yang Y, Stratmann T, Serra P, Santamaria P (2016) Expanding antigen-specific regulatory networks to treat autoimmunity. Nature 530(7591):434–440. doi: 10.1038/nature16962

Cohen J (2007) Building an HIV proof immune system. Science 317:612–614. doi:10.1126/science.317.5838.612

Couzin-Frankel J (2013) Breakthrough of the year 2013. Cancer Immunother Sci 342:1432–1433. doi:10.1126/science.342.6165.1432

Davidson A, Diamond B (2001) Autoimmune diseases. N Engl J Med 345:340–350. doi:10.1056/NEJM200108023450506

du Toit LC, Pillay V, Choonara YE, Govender T, Carmichael T (2011) Ocular drug delivery-a look towards nanobioadhesives. Expert Opin Drug Deliv 8:71–94. doi:10.1517/17425247.2011.542142

Duan S, Song M, He J, Zhou N, Zhou S, Zhao J, Fang Y, Yi P, Huang X, Luo G, Lai C, Yu X, Zhang Z, Xie Y, Zhao Y, Lu X (2016) Folate-modified chitosan nanoparticles coated interferon-inducible protein-10 Gene enhance cytotoxic T lymphocytes' responses to hepatocellular carcinoma. J Biomed Nanotechnol 12:700–709. doi:10.1166/jbn.2016.2216

Elamanchili P, Diwan M, Cao M, Samuel J (2004) Characterization of poly (D, L-lactic-co-glycolic acid) based nanoparticulate system for enhanced delivery of antigens to dendritic cells. Vaccine 22:2406–2412. doi:10.1016/j.vaccine.2003.12.032

Fischer HC, Chan WC (2007) Nanotoxicity: the growing need for *in vivo* study. Curr Opin Biotechnol 18:565–571. doi:10.1016/j.copbio.2007.11.008

Fong L, Engleman EG (2000) Dendritic cells in cancer immunotherapy. Annu Rev Immunol 18:245–273. doi:10.1146/annurev.immunol.18.1.245

Fuchs S, Klier J, May A, Winter G, Coester C, Gehlen H (2012) Towards an inhalative *in vivo* application of immunomodulating gelatin nanoparticles in horse-related preformulation studies. J Microencapsul 29:615–625. doi:10.3109/02652048.2012.668962

Gandhi RT, Walker BD (2002) Immunologic control of HIV1. Annu Rev Med 53:149–172. doi:10.1146/annurev.med.53.082901.104011

Garaczi E, Szabo K, Francziszti L, Csiszovszki Z, Lorincz O, Toke ER, Molnar L, Bitai T, Janossy T, Bata-Csorgo Z, Kemeny L, Lisziewicz J (2013) DermAll nanomedicine for allergen-specific immunotherapy. Nanomedicine 9:1245–1254. doi:10.1016/j.nano.2013.05.011

Ginkel FWV, Nguyen HH, McGhee JR (2000) Attenuated viral vaccines for mucosal immunity to combat emerging infectious diseases. Emerg Infect Dis 6:123–132. doi:10.3201/eid0602.000204

Grimm J, Scheinberg DA (2011) Will nanotechnology influence targeted cancer therapy? Semin Radiat Oncol 21:80–87. doi:10.1016/j.semradonc.2010.10.003

Hamdy S, Elamanchili P, Alshamsan A, Molavi O, Satou T, Samuel J (2007) Enhanced antigen-specific primary $CD4^+$ and $CD8^+$ responses by codelivery of ovalbumin and toll like receptor ligand monophosphoryl lipid a in poly (D, L-lactic- co-glycolic acid) nanoparticles. J Biomed Mater Res A81:652–662. doi:10.1002/jbm.a.31019

Hayder M, Poupot M, Baron M, Nigon D, Turrin CO, Caminade AM, Majoral JP, Eisenberg RA, Fournie JJ, Cantagrel A, Poupot R, Davignon JL (2011). A phosphorus-based dendrimer targets inflammation and osteoclastogenesis in experimental arthritis. Sci Transl Med 3:81ra35. doi: 10.1126/scitranslmed.3002212

Hellstrom I, Ledbetter JA, Scholler N, Yang Y, Ye Z, Goodman G, Pullman J, Hayden-Ledbetter M, Hellstrom KE (2001) CD3-mediated activation of tumor-reactive lymphocytes from patients with advanced cancer. Proc Natl Acad Sci U S A 98:6783–6788. doi:10.1073/pnas.021557498

Hubbell JA, Thomas SN, Swartz MA (2009) Materials engineering for immunomodulation. Nature 462:449–460. doi:10.1038/nature08604

Kalos M, Levine BL, Porter DL, Katz S, Grupp SA, Bagg A, June CH (2011) T cells with chimeric antigen receptors have potent antitumor effects and can establish memory in patients with advanced leukemia. Sci Transl Med 3:95ra73. doi:10.1126/scitranslmed.3002842

Kasturi SP, Skountzou I, Albrecht RA, Koutsonanos D, Hua T, Nakaya HI, Ravindran R, Stewart S, Alam M, Kwissa M, Villinger F, Murthy N, Steel J, Jacob J, Hogan RJ, Garcia-Sastre A, Compans R, Pulendran B (2011) Programming the magnitude and persistence of antibody responses with innate immunity. Nature 470:543–547. doi: 10.1038 /nature 09737

Kawai T, Akira S (2011) Toll-like receptors and their crosstalk with other innate receptors in infection and immunity. Immunity 34:637–650. doi:10.1016/j.immuni.2011.05.006

Krishnamachari Y, Geary SM, Lemke CD, Salem AK (2011) Nanoparticle delivery systems in cancer vaccines. Pharm Res 28:215–236. doi: 10.1007/s11095-010-0241-4

Laroui H, Wilson DS, Dalmasso G, Salaita K, Murthy N, Sitaraman VS, Merlin D (2011) Nanomedicine in GI. Am J Physiol Gastrointest Liver Physiol 300:G371–G383. doi:10.1152/ajpgi.00466.2010

Lei C, Liu P, Chen B, Mao V, Engelmann H, Shin Y, Jaffar J, Hellstrom I, Liu J, Hellstrom KE (2010) Local release of highly loaded antibodies from functionalized nanoporous support for cancer immunotherapy. J Am Chem Soc 132:6906–6907. doi:10.1021/ja102414t

Li W, Feng SS, Guo Y (2013a) Polymeric nanoparticulates for cancer immunotherapy. Nanomedicine (Lond) 8:679–682. doi:10.2217/nnm.13.28

Li W, Zhang L, Zhang G, Wei H, Zhao M, Guo S, Gao J, Kou G, Li B, Dai J, Wang H, Guo Y (2013b) The finely regulating well-defined functional polymeric nanocarriers for antitumor immunotherapy. Mini Rev Med Chem 13:643–652. doi:10.2174/1389557511313050003

Li W, Zhao H, Qian W, Li H, Zhang L, Ye Z, Zhang G, Xia M, Li J, Gao J, Li B, Kou G, Dai J, Wang H, Guo Y (2012) Chemotherapy for gastric cancer by finely tailoring anti- Her2 anchored dual targeting immunomicelles. Biomaterials 33:5349–5362. doi:10.1016/j.biomaterials.2012.04.016

Lim JS, Lee K, Choi JN, Hwang YK, Yun MY, Kim HJ, Won YS, Kim SJ, Kwon H, Huh S (2012) Intracellular protein delivery by hollow mesoporous silica capsules with a large surface hole. Nanotechnology 23:085101. doi:10.1088/0957-4484/23/8/085101

Liu Z, Robinson JT, Tabakman SM, Yang K, Dai H (2011) Carbon materials for drug delivery & cancer therapy. Mater Today 14:316–323. doi:10.1016/S1369-7021(11)70161-4
Lori F, Calarota SA, Lisziewicz J (2007) Nano chemistry based immunotherapy for HIV-1. Curr Med Chem 14:1911–1919. doi:10.2174/092986707781368513
Lovelyn C, Attama AA (2011) Current state of nanoemulsions in drug delivery. J Biomater Nanobiotechnol 2:626–639. doi:10.4236/jbnb.2011.225075
Lu Y, Miao L, Wang Y, Xu Z, Zhao Y, Shen Y, Xiang G, Huang L (2015) Curcumin micelles remodel tumor microenvironment and enhance vaccine activity in an advanced melanoma model. Mol Ther 24:364–374. doi:10.1038/mt.2015.165
Makidon PE, Bielinska AU, Nigavekar SS, Janczak KW, Knowlton J, Scott AJ, Mank N, Cao Z, Rathinavelu S, Beer MR, Wilkinson JE, Blanco LP, Landers JJ, Baker JR Jr (2008) Pre-clinical evaluation of a novel nanoemulsion-based hepatitis B mucosal vaccine. PLoS One 3:e2954. doi:10.1371/journal.pone.0002954
Manjunath K, Venkateswarlu V (2005) Pharmacokinetics, tissue distribution and bioavailability of clozapine solid lipid nanoparticles after intravenous and intraduodenal administration. J Control Release 107:215–228. doi:10.1016/j.jconrel.2005.06.006
McNeil SE (2011) Unique benefits of nanotechnology to drug delivery and diagnostics. Methods Mol Biol 697:3–8. doi:10.1007/978-1-60327-198-1-1
Mehnert W, Mader K (2001) Solid lipid nanoparticles: production, characterization and applications. Adv Drug Deliv Rev 47:165–196. doi:10.1016/S0169-409X (01)00105-3
Mellman I, Coukos G, Dranoff G (2011) Cancer immunotherapy, comes of age. Nature 480:480–489. doi:10.1038/nature10673
Mellman I, Steinman RM (2001) Dendritic cells: specialized and regulated antigen processing machines. Cell 106:255–258. doi:10.1016/S0092-8674(01)00449-4
Misra R, Acharya S, Sahoo SK (2010) Cancer nanotechnology: application of nanotechnology in cancer therapy. Drug Discov Today 15:842–850. doi:10.1016/j.drudis.2010.08.006
Moon JJ, Suh H, Bershteyn A, Stephan MT, Liu H, Huang B, Sohail M, Luo S, Um SH, Khant H, Goodwin JT, Ramos J, Chiu W, Irvine DJ (2011) Interbilayer-crosslinked multilamellar vesicles as synthetic vaccines for potent humoral and cellular immune responses. Nat Mater 10:243–251. doi:10.1038/nmat2960
Muller RH, Keck CM (2004) Challenges and solutions for the delivery of biotech drugs - a review of drug nanocrystal technology and lipid nanoparticles. J Biotechnol 113:151–170. doi:10.1016/j.jbiotec.2004.06.007
Mundargi RC, Babu VR, Rangaswamy V, Patel P, Aminabhavi TM (2008) Nano/micro technologies for delivering macromolecular therapeutics using poly (D, L-lactide-co-glycolide) and its derivatives. J Control Release 125:193–209. doi:10.1016/j.jconrel.2007.09.013
Nakanishi T, Kunisawa J, Hayashi A, Tsutsumi Y, Kubo K, Nakagawa S, Nakanishi M, Tanaka K, Mayumi T (1999) Positively charged liposome functions as an efficient immune adjuvant in inducing cell-mediated immune response to soluble proteins. J Control Release 61:233–240. doi:10.1016/S0168-3659(99)00097-8
Noh YW, Jang YS, Ahn KJ, Lim YT, Chung BH (2011) Simultaneous *in vivo* tracking of dendritic cells and priming of an antigen-specific immune response. Biomaterials 32:6254–6263. doi:10.1016/j.biomaterials
Palucka K, Banchereau J (2012) Cancer immunotherapy via dendritic cells. Nat Rev Cancer 12:65–277. doi:10.1038/nrc3258
Pan X, Chen L, Liu S, Yang X, Gao JX, Lee RJ (2009) Antitumor activity of G3139 lipid nanoparticles (LNPs). Mol Pharm 6:211–220. doi:10.1021/mp800146j
Peek LJ, Middaugh CR, Berkland C (2008) Nanotechnology in vaccine delivery. Adv Drug Deliv Rev 60:915–928. doi:10.1016/j.addr.2007.05.017
Pett SL (2009) Immunotherapies in HIV1 infection. Curr Opin HIV AIDS 4:188–193. doi:10.1097/COH.0b013e328329d090
Reddy ST, Rehor A, Schmoekel HG, Hubbell JA, Swartz MA (2006) *In vivo* targeting of dendritic cells in lymph nodes with poly (propylene sulfide) nanoparticles. J Control Release 112:26–34. doi:10.1016/j.jconrel.2006.01.006

Restifo NP, Dudley ME, Rosenberg SA (2012) Adoptive immunotherapy for cancer: harnessing the T cell response. Nat Rev Immunol 12:269–281. doi:10.1038/nri3191

Rinaldo CR (2009) Dendritic cell based human immunodeficiency virus vaccine. J Intern Med. 265:138–158. doi:10.1111/j.1365-2796.2008.02047.x

Romagnani S (2004) Immunologic influences on allergy and the TH1/TH2 balance. J Allergy Clin Immunol 113:395–400. doi:10.1016/j.jaci.2003.11.025

Roy A, Singh MS, Upadhyay P, Bhaskar S (2010) Combined chemo-immunotherapy as a prospective strategy to combat cancer: a nanoparticle based approach. Mol Pharm 7:1778–1788. doi:10.1021/mp100153r

Roy I, Vij N (2010) Nanodelivery in airway diseases: challenges and therapeutic applications. Nanomedicine 6:237–244. doi:10.1016/j.nano.2009.07.001

Serra P, Santamaria P (2015) Nanoparticle-based autoimmune disease therapy. Clin Immunol 160:3–13. doi:10.1016/j.clim.2015.02.003

Singh N, Manshian B, Jenkins GJ, Griffiths SM, Williams PM, Maffeis TG, Wright CJ, Doak SH (2009) NanoGenotoxicology: the DNA damaging potential of engineered nanomaterials. Biomaterials 30:3891–3914. doi:10.1016/j.biomaterials.2009.04.009

Steenblock E, Fahmy T (2008) A comprehensive platform for *ex vivo* T-cell expansion based on biodegradable polymeric artificial antigen-presenting cells. Mol Ther 16:765–772. doi:10.1038/mt.2008

Steinman RM, Banchereau J (2007) Taking dendritic cells into medicine. Nature 449:419–426. doi:10.1038/nature06175

Thomas TP, Goonewardena SN, Majoros IJ, Kotlyar A, Cao Z, Leroueil PR, Baker JR Jr (2011) Folate-targeted nanoparticles show efficacy in the treatment of inflammatory arthritis. Arthritis Rheum 63:2671–2680. doi:10.1002/art.30459

U.S. Food and Drug Administration (2002) Guidance for industry: liposome drug products. USFDA, Rockville

Vacchelli E, Martins I, Eggermont A, Fridman WH, Galon J, Sautes-Fridman C, Tartour E, Zitvogel L, Kroemer G, Galluzzi L (2012) Trial watch: peptide vaccines in cancer therapy. Oncoimmunology 1:1557–1576. doi:10.4161/onci.22428

Wang X, Reece PS, Brown MJ (2013) Immunotoxicological impact of engineered nanomaterial exposure: mechanisms of immune cell modulation. Toxicol Mech Methods 23:168–177. doi:10.3109/15376516.2012.757686

Zhang Y, Bai Y, Yan B (2010) Functionalized carbon nanotubes for potential medicinal applications. Drug Discov Today 15:428–435. doi:10.1016/j.drudis.2010.04.005

Zolnik BS, Sadrieh N (2009) Regulatory perspective on the importance of ADME assessment of nanoscale material containing drugs. Adv Drug Deliv Rev 61:422–427. doi:10.1016/j.addr.2009.03.006

Biosynthesis of Nanoparticles and Their Application in Pharmaceutical Industry

16

Bisma Malik, Tanveer Bilal Pirzadah, Manoj Kumar, and Reiaz Ul Rehman

Abstract

Nanotechnology emerges from the different fields of science, viz. physical, chemical, biological and engineering sciences, where novel approaches are being unfolded to investigate and apply single atoms and molecules for various applications in different fields of scientific world. In this technology, nanoparticle, a minute object, functions as an entire unit in terms of its transport and characteristics. The nanosystem that involves science and engineering technology is one of the most emergent and time-demanding areas of research in nanotechnology. Currently, due to the advancement in science and technology, researchers made it possible to synthesize nanoparticles of size 100 nm, and this attempt opens and widens the scope of nanoparticle research due to its multiple applications in different fields of science and technology. The progress and the development of technology related to mankind are directly linked with the advancements and achievements of material science and processing technology. Currently the research and the progress in nanotechnology and the validation based on several specified size effects in nanomaterials describe that most of the new findings and designs of the future will be based on qualities of nanomaterials. In this field of technology, nanoparticle has massive scope for pharmaceutical industries which include health-care products and much more such as burn dressings, antimicrobial applications, medical devices and scaffolds. Several approaches have been employed for the synthesis of nanomaterials which includes chemical reduction, photochemical reactions, electrochemical techniques and green chemistry route.

B. Malik • T.B. Pirzadah • R.U. Rehman (✉)
Department of Bioresources, University of Kashmir, Srinagar 190 006, Jammu and Kashmir, India
e-mail: reiazrehman@yahoo.co.in; rreiazbiores@gmail.com

M. Kumar
Department of Microbiology, Amity University, Noida, Uttar Pradesh, India

V.C. Kalia, A.K. Saini (eds.), *Metabolic Engineering for Bioactive Compounds*,
DOI 10.1007/978-981-10-5511-9_16

Among different types of nanoparticles, the metallic nanoparticles such as silver, gold, zinc, iron and metal oxide have shown considerable improvement in the field of biomedical utilization, not only because of their high surface area to volume ratio but also because they display several medicinal properties. It is in this perspective that the current review will focus on the biosynthesis of nanoparticles and their application in pharmaceutical industry and also try to overview the most recent developments in this field.

Keywords
Nanotechnology • Biomedicines • Nanoparticles

16.1 Introduction

The term "nano" is originated from the Greek word "nanos" which means tiny, and it is used as the prefix for one billionth parts (10^{-9}). The definition of nanoparticles according to American Society for Testing and Materials International (2006) as those nanomaterials which have more than two dimensions and are in the range of 1–100 nm in size (Alanazi et al. 2010). These particles have unique and modified physical and chemical properties because of their particular electronic structure, large reactive, open surface area and quantum size effects. A nanoparticle nowadays finds its application in many research areas such as biomedicine, electronics, cosmetics, textiles, etc. (Di Guglielmo et al. 2010; Botha et al. 2015). It has been found that due to expeditious industrialization and urbanization, an extensive amount of dangerous and unsafe chemicals, gases or substances are liberated into the environment, thus degrading its quality, and now it is the need of the hour to unravel the secrets of the nature and its byproducts which results in the development of the nanoparticle synthesis process. It is noteworthy to mention that the application of nanotechnology is highly convenient to biological samples due to their intensive properties. These biological samples undergo extremely controlled assembly for making them convenient for the synthesis of metal nanoparticle which is found to be trustworthy and eco-friendly (Harekrishna et al. 2009; Govindarajan et al. 2016). In addition to this, the process of synthesis of semiconductor nanoparticles is becoming an upcoming area of research due to its exclusive applications in various fields which leads in the developmental growth of novel. Besides, nanotechnology as a field of research opens new insights in research field specifically in the field of material science. This is because the nanoparticle exhibits highly new or modified properties which include its shape, size, distribution and external features of the particles. That is why the nanoparticles or nanomaterials find its application in almost every field and emerge as one of the active research areas in modern material science (Kaviya et al. 2011; Shiekh et al. 2016). Nanotechnology as a field of research is progressing day by day creating an impact in almost every sphere of human lives and generating a growing excitement in the biological science field particularly biotechnology and biomedical science (Prashanth et al. 2011; Eatemadi et al. 2016). As nanoparticles show improved properties on specific attributes such as distribution, size and shape, nanocrystalline particles have found immense applications in the field of high-sensitivity bimolecular detection and diagnostics, therapeutics and antimicrobials (Sridhara et al. 2012; Yohan and Chithrani 2014) and catalysis and microelectronics (Veera et al. 2012; Wong et al. 2014). Besides these extraordinary applications of nanoparticles, there is a need of the hour to find out the clean routes (biological route) for the synthesis of nanoparticles which is commercially viable and eco-friendly in nature (Lekshmi et al. 2012; Govindarajan et al. 2016). To address this concern, a number of approaches are available, for example, solution reduction, photochemical and chemical reactions, thermal decomposition of nanoparticle compounds (Akl et al. 2012; Padmavathy and Vijayaraghavan 2016), electrochemical, radiation-assisted and microwave-assisted process and currently by using biological samples (green chemistry) (Ravindra et al. 2012; González-Sánchez et al. 2015) for the synthesis of nanoparticles. Due to certain limitations and ill-effects of chemical synthesis of nanoparticles, the utilization of environmentally benign material like plant extract (leave, flower, bark, seed, peels, etc.), fungi, bacteria and enzyme for the synthesis of nanoparticles provides various advantages like eco-friendliness and compatibility for biomedical and other pharmaceutical applications as these synthesis procedures do not utilize any toxic chemicals (Gokulakrishnan et al. 2012). Various synthesis procedures (green chemistry route) have been already started for the synthesis of metal nanoparticles such as the use of fungi like *Fusarium oxysporum* (Nelson et al. 2005), *Penicillium* sp. (Hemanth et al. 2010) and bacteria such as *Bacillus subtilis*, etc. (Elumalai et al. 2010). In addition to this, nanoparticle synthesis by employing plant sources or extracts is the highly accepted approach of green, environment-friendly production of nanoparticles as it exhibited a unique advantage, that is, the plants are widely distributed and readily available and much secure to deal with, and also it acts as a factory of various important metabolites (Ankamwar et al. 2005). Hence, synthesis of nanoparticles through green chemistry route holds a great promise not only in terms of multiple applications in diverse fields of science but also for the healthy environment. As far as nanoparticle applications are concerned, it has been long identified that these are having inhibitory effects on microbes present in industrial and medical processes (Nasrollahi et al. 2011). Nanoparticles nowadays find its application in every important field of science which includes the high sensitive areas like genomics, immune modulation, biosensors, clinical-chemistry, control of microorganism and diagnostic procedures (detection and targeted drug delivery) (Diva et al. 2012). Furthermore, these biosynthesized nanoparticles were found to be extremely lethal against various multidrug-resistant human pathogens. Besides this, the work is in progress to answer the questions that limits the biosynthesis and application of nanoparticles to diverse systems in one or other way. There are different types of nanoparticles, but the metallic nanoparticles such as gold, silver zinc, iron and metal oxide nanoparticles proved to be of great importance in biomedical application not only because of their large surface area to volume ratio (Bhattacharya and Mukherjee 2008) but also because of their unique properties that show multifunctional aspect of these particles in diverse fields (Fig. 16.1) (Hussain and Ferguson 2006).

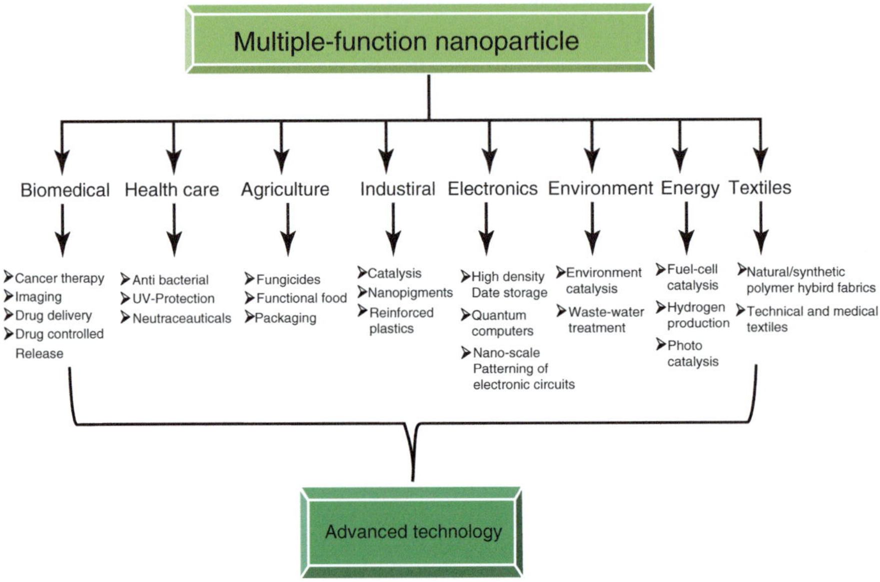

Fig. 16.1 Advanced functions of metal nanoparticle in multiple fields

16.2 Why Green Synthesis

Green synthesis provides various advantages over physical and chemical method of nanoparticle synthesis as it is cost-effective and eco-friendly and has easy modification procedures for high-scale production of nanoparticles as this method of synthesis does not use any hazardous chemicals, high pressure, temperature and high amount of energy (Fig. 16.2) (Ravindra et al. 2012).

16.2.1 The Green Route for Biosynthesis of Nanoparticles

Several different procedures for the synthesis of nano- and micro-length nanomaterials which contribute to the advancement of relatively novel and highly unexplored area of research depends upon the procedures of biosynthesis of nanomaterials (Salam et al. 2012). It is very important to mention that the biosynthesis of nanoparticles should be cost-effective and eco-friendly so that the synthesis and the applications do not remain restricted to one or other area of research. The best way of synthesizing the nanoparticle is through green chemistry route that makes use of eco-friendly, non-toxic and safe reagent. There are different methods of extracting a metal from its source that depend upon the type of protocol followed. The term "phytomining" refers to the uses of hyper-accumulating plants to extract a metal from the biomass to return an economic profit (Lamb et al. 2001). Hyper-accumulating species have endogenous defence mechanism to regulate the

Fig. 16.2 Advantages of green synthesis

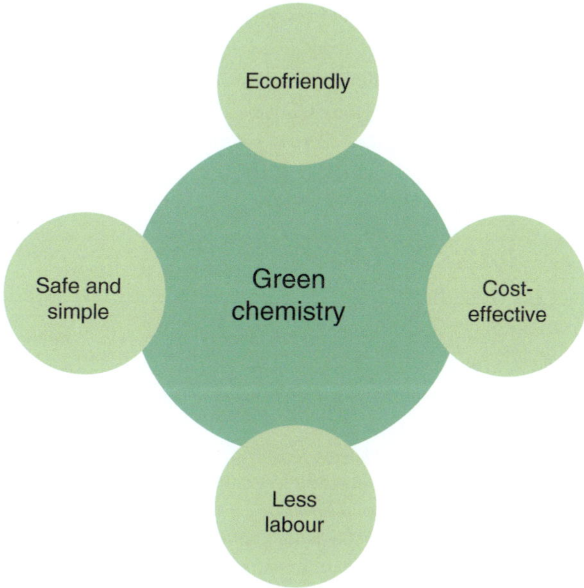

concentration of metals in the soil. Phytoremediation concept in plants may be in association with the mechanism of biosynthesis of nanoparticles in plants (Haverkamp et al. 2007).

16.2.2 Biological Method

Extract from biological organism may act both as reducing and capping agent in the synthesis process of nanoparticles. In the biological method of synthesis, the reduction of metal ions takes place by the combination of biomolecules found in these extracts such as protein/enzymes, polysaccharide, vitamins and amino acids that are environmentally benign but yet are chemically complex and diverse. Several studies have reported the successful synthesis of nanomaterials by using bio-organic compounds (Sahayaraj and Rajesh 2011). Perhaps, the biological methods are termed as safer, cost-effective, eco-friendly and sustainable procedure for the synthesis of nanoparticles, yet much more is still unanswered (Diva et al. 2012). Currently, biological method for the synthesis of nanoparticles is progressing day by day and becomes the thrust area of research due to its sustainable and eco-friendly nature. Though the nanoparticle biosynthesis by plants such as *Alfalfa*, *Emblica officinalis*, *Lemongrass*, *Aloe vera*, *Tamarindus indica* and *Cinnamomum camphora* has been described, the benefits of these plants as biological materials for the nanoparticle synthesis are yet to be unexplored (Yogeswari et al. 2012). Biosynthesis of nanoparticles with their shape, size, natural resources and plant part used is given in Table 16.1 A generalized scheme of nanoparticle synthesis by using biological materials is given in Fig. 16.3.

Table 16.1 Biosynthesis of nanoparticles with their size, shape and plant parts used

S.No.	Material	Natural resource	Part used	Component	Size (nm)	Shape
1.	AgNps	*Ocimum sanctum*	Leaves	$AgNO_3$	20	Spherical
2.	AgNps $AgNO_3$	Onion (*Allium cepa*)	Onion (bulb)	$AgNO_3$	33.6	Spherical
3.	AgNps	*Costus speciosus*	Callus culture	$AgNO_3$	420	Spherical
4.	AgNps	*Oryza sativa*, *Vigna radiata*, *Brassica campestris* L.	Seeds	$AgNO_3$	25	Spherical
5.	AgNps	*Tridax procumbens*	Leaves	$AgNO_3$	55	–
6.	AgNps	*Euphorbia hirta*	Leaves	$AgNO_3$	50	Spherical
7.	AgNps	*Ficus benghalensis*	Leaves	$AgNO_3$	44	Spherical
8.	AgNps	*Svensonia hyderobadensis*, *Boswellia ovalifoliolata*, *Shorea tumbuggaia*	Leaves, bark	$AgNO_3$	430	–
9.	AgNps	*Withania somnifera*	Leaves	$AgNO_3$	40	Spherical
10.	AgNps	*Ocimum sanctum*	Leaves	$AgNO_3$	50	–
11.	AgNps	*Ocimum sanctum*	Leaves	$AgNO_3$	30	Spherical
12.	AgNps	*Euphorbia prostrata*	Leaves	$AgNO_3$	80	Rod Shape
13.	AgNps	*Trianthema decandra*	Roots	$AgNO_3$	50	Hexagonal
14.	AgNps	*Mulberry*	Leaves	$AgNO_3$	40	Spherical
15.	AgNps	*Cassia auriculata*	Leaves	$AgNO_3$	–	Spherical
16.	AgNps	*Solanum xanthocarpum*	Berry	$AgNO_3$	406	Spherical
17.	AgNps	*Vitex negundo*	Leaves	$AgNO_3$	30	Spherical
18.	AgNps	*Elaeagnus latifolia*	Leaves	$AgNO_3$	50	Spherical
19.	AgNps	*Azadirachta indica*, *Triphala*	Leaves, powder	$AgNO_3$	59	Spherical
20.	AgNps	*Cleome viscosa*	Leaves	$AgNO_3$	50	Spherical
21.	AgNps	*Saururus chinensis*	Leaves	$AgNO_3$	54	Spherical
22.	AgNps	*Ocimum basilicum*	Leaves	$AgNO_3$	89	Spherical
23.	AgNps	*Memecylon umbellatum*	Leaves	$AgNO_3$	20	Spherical
	Au-Nps			$HAuCl_4$		
24.	AgNps	*Punica granatum*	Peels	$AgNO_3$	25	Spherical
	Au–Nps			$HAuCl_4$		
25.	CdO–Nps	*Achillea wilhelmsii*	Flowers	$CdCl_2$	10, 35	Spherical
26.	ZnO–Nps	*Calotropis procera*	Milky latex	$\{(CH_3CO)_2Zn_2.H_2O\}$	40	Sphercial
27.	AgNps	*Punica granatum*	Seeds	$AgNO_3$	400	Spherical
28.	AgNps	*Punica granatum*, *Rosa damascena*	Peels, Petals	$AgNO_3$	21	Spherical
29.	AgNps	*Punica granatum*	Seeds	$AgNO_3$	20	Spherical

Table 16.1 (continued)

S.No.	Material	Natural resource	Part used	Component	Size (nm)	Shape
30.	AgNps	*Loquat*	Leaves	$AgNO_3$	18	Spherical
31.	Cr_2O_3–Nps	*Arachis hypogaea*	Leaves	K_2CrO4	80	Hexagonal
32.	Au-Nps	*Putranjiva roxburghii*	Leaves	$HAuCl_4$	38	Spherical
33.	Au-Nps	*Caesalpinia pulcherrima*	Flowers	$HAuCl_4$	50	Spherical
34.	AgNps	*Nerium oleander*	Leaves	$AgNO_3$	67	-
35.	Au-Nps	*Tagetes erecta*	Flowers	$HAuCl_4$	10	Spherical
36.	AuNps	Onion (*Allium cepa*)	Bulb	$HAuCl_4$	54	Spherical

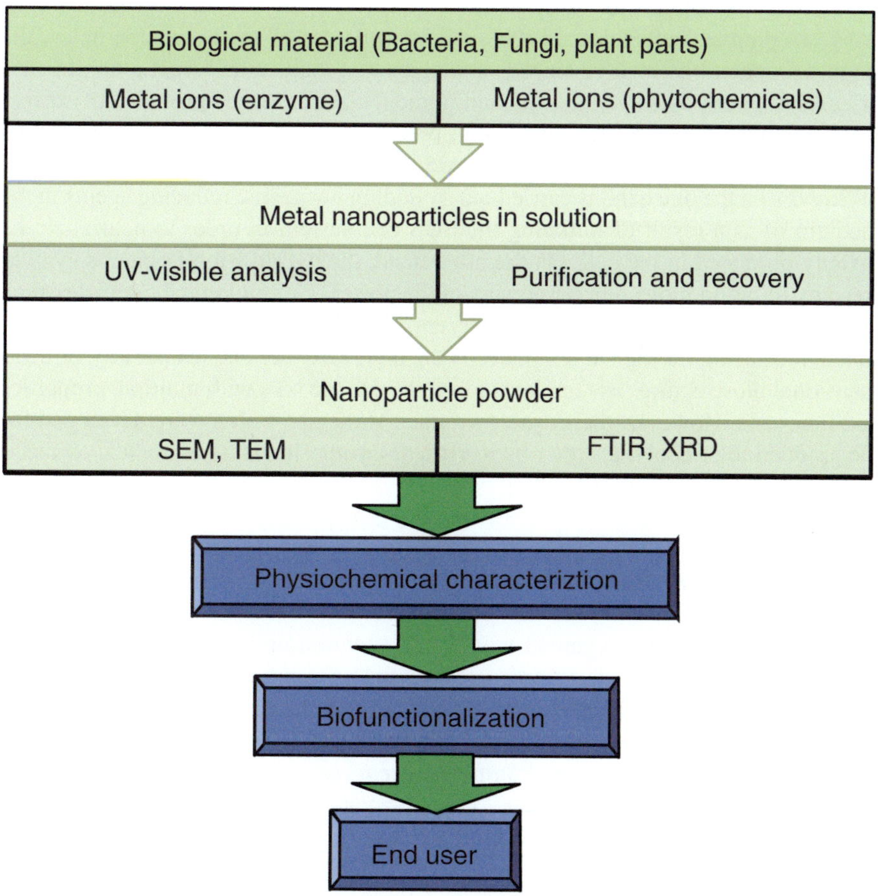

Fig. 16.3 Flow chart of biological synthesis of nanoparticles

16.3 Biosynthesis of Nanoparticles (Overview and Applications)

Green synthesis approach of nanoparticles is more advantageous over the chemical approach because of simple, cost-effective and eco-friendly nature. Biosynthesis of nanoparticles can be categorized into organic, inorganic and hybrid. Among the biosynthetic approach of nanoparticles, most of biomolecules (macromolecules and cellular substructures) have the ability to form nanoparticles which are based on environmental factors and sample conditions. The green synthesis of nanoparticles has already been reported using several living species ranging from yeast, fungi, plants, algae to bacteria (eubacteria, cyanobacteria), recently reviewed by Duran et al. (2007), (2011), Mohanpuria et al. (2008), Rodriguez-Carmona and Villaverde (2010), Narayanan and Sakthivel (2010), Lloyd et al. (2011), Narayanan and Sakthivel (2011), Duran and Marcato (2012), and Gurrappa and Binder (2008). It has also been described that the green synthesis of nanoparticles can be done utilizing various macromolecules such as carbohydrates, lipids, DNA, proteins and mixtures of complex molecules. In the synthesis process of green nanoparticles, the sample preparations involve various strategies that include cell fractioning in case of cell lysis such as solvents, salts and homogenization in the presence of surfactants, chaotropic agents and differential precipitation. Besides, inorganic BioNPs have been categorized as metallic and oxide NPs, and the process for the synthesis of BioNPs on the one hand is carried out by adding unspecific reducing agents in the medium or as a result of initiating the SOS cell system in order to neutralize the toxicity produced in the cell. On the other hand, the hybrid BioNPs can be synthesized by utilizing molecular precursors in the presence of biological templates such as DNA and proteins (Contescu and Putyera 2009; Silpa 2016). The most important function of using biological templates is the high diversity and availability of tridimensional biostructures as templates to synthesize the NPs with multiple properties and functions. However, there were two main strategies which were developed for the production of BioNPs, either by in vitro biosynthesis (using biological extracts) or in vivo biosynthesis (using living cells).

16.3.1 Silver Nanoparticles

Silver nanoparticles have gained worldwide attention and opens a wider area of research interest due to their immense potential as antimicrobial agents, catalytic agents and surface-enhanced Raman scattering (Gokulakrishnan et al. 2012). Apart from physiochemical methods, AgNPs can be produced by a green approach, i.e. using of supernatant of culture of *Staphylococcus aureus* for the reduction of aqueous Ag ions (Nanda and Saravanan 2009). Biosynthesis of AgNPs involves the use of microbial or plant extract that acts as a reducing agent and reduces Ag^{2+} to Ag^{0}, and the spectra is measured by UV-vis spectrophotometer at a resolution of 1 nm. Shahverdi et al. (2009) reported the use of fungus *Alternaria alternata* to produce AgNPs. There are many hypotheses that describe the antimicrobial activity of Ag

NPs, for example, Ag NPs undergo rapid breakdown leading to the release of Ag ions in the medium which inhibits important bacterial enzymes by attacking the essential –SH groups. In addition to this, Ag ions inhibit replication of bacterial DNA, damage bacterial cytoplasm membranes, reduce intracellular adenosine triphosphate (ATP) level and ultimately cause apoptosis (Parveen et al. 2012). Silver has a very long history in the medical sector as it was used to treat many ailments such as venereal infections, epilepsy, leg ulcers and acnes. Furthermore, foil made up of silver was found to be more effective in wound healing and also decrease the onset of post-operative infections, while wart removal and ulcer debridement were done by using silver and lunar caustic, i.e. pencil containing silver nitrate mitigated with potassium nitrate (Klasen 2000). However, the most common and well-documented application of AgNPs is its use in the dressings of wounds (Leaper 2006). Sibbald et al. (2007) reported that AgNP's dressing has a potential effect on wounds as it protects the wound site from bacterial infections. Besides, medicine sector AgNP finds immense potential in cosmetic sector because of its antimicrobial as well as soothing effect. The reason is that the AgNPs have high surface to volume ratio which in turn increases their contact with microbes, thereby increasing the dissolution of silver ions which ultimately improves its antimicrobial activity. Dhrutika et al. (2013) reported that the ability of the AgNPs to release Ag^{3+} is a key to their antimicrobial activity (Ahmed et al. 2016). Some antimicrobial activities of agent are extremely toxic and irritant, and the need of the hour is to investigate the novel ways and formulations leading to the discovery of safe and cost-effective antimicrobial agents or effective biocidal materials (Dhrutika et al. 2013). Chemical synthesis of AgNPs involves the commonly used borohydrides and other stabilizing chemicals that act as capping agents that bind to the nanoparticle surface which increases water solubility and stability in order to prevent aggregation (e.g. water-soluble polymers, oligosaccharides and polysaccharides, glycolipids and sodium dodecyl sulphate) (Sondi and Sondi 2004). Chemosynthetic approach of AgNPs possessing some drawbacks such as extremely expensive use of certain toxic substances thus poses various ill effects to environment.

16.3.2 Gold Nanoparticles

Nanoparticles are extensively and particularly exploited in microorganism because of their biocompatibility; gold nanoparticles are considered to be biologically inactive but can be modified to exhibit photochemical and chemical properties (Prasson and Chittaranjan 2012). Depending upon the physical properties (size, shape, etc.), gold nanoparticles are categorized into various subtypes like gold nano-cages, nanospheres, nano-needles and nanorods with characteristics of near-infrared irradiation absorption that can damage cancer cells and bacteria through the process of photothermal heating (Jing et al. 2014; Hasna et al. 2012). Among these subtypes, nanospheres are the earliest studied nanoparticles that are synthesized under controlled reduction of aqueous $HAuCl_4$ solution by utilizing various reducing agents like citric acid, which possesses the ability to generate monodisperse gold nanospheres (Turkevich et al. 1951; Frens 1973). However, the size of these nanospheres can be regulated by using different concentrations of citric acid/gold ratio. Usually little amount of citric acid results in the formation of nanospheres with larger size; besides, another important drawback of this method is the minimum yield and the limited use of water as the solvent. Faraday (1857) proposed a two-phase system that has the capability of generating thermally and air-stable gold nanospheres with reduced disparity and controlled size (10 nm) as reported in 1993 (Giersig and Mulvaney 1993). Later this technique was also improved by using phase-transfer reagents, viz. tetraoctyl ammonium bromide (Brust et al. 1994), but the physical properties (size) still rely on ratio of reducing agent and gold concentration. Later, researchers also investigated other protocols for the synthesis of gold nanospheres by using several reductant or ligands (Leff et al. 1996; Weare et al. 2000; Hiramatsu and Osterloh 2004). Several reports reveal the use of dendrimers as templates or stabilizers for gold nanosphere synthesis (Esumi et al. 1998; Garcia et al. 1999; Manna et al. 2001; Kim et al. 2004; Scott et al. 2005; Shi et al. 2006, 2008). Another gold nanoparticle subtype is the nanorods, which is synthesized by using numerous strategies, such as template method that is based on electrochemical deposition of gold salt within the pores of nanoporous polycarbonate or alumina template membranes (Martin 1994; Reetz and Helbig 1994; Yu et al. 1997; van der Zande et al. 1997; Chang et al. 1999). However, the seed-mediated synthesis of gold nanorods is well-established protocol that has gained wide importance as compared to other protocols (Jana et al. 2001b; Busbee et al. 2003). Jana et al. (2001a, 2002) reported that gold nanorods can be synthesized in a quantitative yield when silver nitrate is added to it. In addition to above-mentioned methods, numerous other strategies have also been examined especially for the fabrication of gold nanorods that includes bio-reduction (Canizal et al. 2001), growth on mica surface (Mieszawska and Zamborini 2005) and photochemical synthesis (Kim et al. 2002). Besides, gold nanoparticles in association with photosensitizers can be used for photodynamic antimicrobial chemotherapy and NIR photothermal radiation (example includes the killing of methicillin-resistant *Staphylococcus aureus* (MRSA)) (Tikariha et al. 2012). Gold nanoparticles have been successfully synthesized using various fungi and bacteria. Gold nanoparticles with antibodies and antibiotics also have been used for selective photothermal killing of bacteria and protozoa.

16.3.3 Zinc Oxide Nanoparticles

Green synthesis of spherical-shaped zinc nanoparticles was first carried out by using milky latex of *Calotropis procera* that acts as reducing as well as stabilizing agent. Ravindra et al. (2012) carried out biosynthesis of zinc oxide nanoparticles by using *Aloe vera* extract, and later these nanoparticles were characterized by means of various standard techniques. One of the most used techniques was the precipitation method followed by controlling, freezing and then drying processes, and the resulting material was thermally treated at different temperatures, and the effect of temperature on the various properties of materials such as the texture, the morphology and the structure can be investigated by using the techniques of SEM (scanning electron microscope), TEM (transmission electron microscopy) and thermal analysis (Haritha et al. 2011). Zinc nanoparticles find great application in the pharmaceutical sector as they emerge as novel antimicrobial agents because of high surface area to volume ratio. In the recent year, the interest in nanoparticles increases due to the growing microbial resistances against metal ions, antibiotic development of resistant strains, etc. (Haritha et al. 2011). Currently, zinc oxide nanoparticles are being used in the field of porous and nanometric materials prepared by nonconventional processes. Besides, these are also used in solar cells, ceramics, catalysts and cosmetic and gas sensors (Dhrutika et al. 2013).

16.3.4 Copper Oxide Nanoparticles

Green synthesis of copper oxide nanoparticles using antimicrobial activity is an important area of research in bio nanotechnology (Akhtar et al. 2016). This is an emerging cost-effective, eco-friendly science of well-defined, shape, size and controlled monodispersity. One of the recent studies carried out by the researchers proposed biosynthesis of extraneous production of copper oxide nanoparticles. Soheyla et al. (2012) reported that copper nanoparticles are synthesized and stabilized by using *Penicillium aurantiogriseum*, *Penicillium citrinum* and *Penicillium waksmanii* isolated from the soil. Copper nanoparticles can easily oxidize to form copper oxide. Oxidation of copper nanoparticle can be inhibited by coating or encapsulating the copper nanoparticles with carbon (organic) and silica (inorganic) material (Abdul et al. 2009). The various biological-reducing agents used in the synthesis of copper nanoparticles are bacteria, actinomycetes, fungi, yeast and plants. Green synthesis of nanoparticles by utilizing various microbes to form inorganic materials, either internally or externally, should have the characteristics that are same to biological and chemical materials (Amrut et al. 2010). Although employing different biological agents for the synthesis of nanoparticles has not been yet formulated, some studies suggested that several biomolecules are responsible for the synthesis of copper nanoparticles. But in the intracellular and extracellular biosynthesis of nanoparticles by fungi, numerous capping and reducing agents are employed that are possibly involved and also affected by the shape and size of these nanoparticles which needs to be fully examined (Shloma et al. 2010).

16.3.5 Titanium Dioxide Nanoparticles

Green synthesis of nanoparticles currently has immense potential in various sectors because of environment-friendly nature and cost-effectiveness (Kapilashrami et al. 2014). Synthesis of nanoparticles depends upon numerous parameters such as size, shape, morphology, nature and composition of the material used. TiO_2 finds great application in water and air purification system because of their immense oxidation capability, huge photostability and non-toxic nature. Besides, these nanoparticles possess other unique features like optical and chemical stability and catalytic activity and therefore possess immense industrial applications like photocatalysts and in pigments (Parthasarathi 2009). However, green synthesis of nanoparticles depends on various factors such as choice of the solvent and reduction and stabilization of agent. Currently, the biosynthesis of nanoparticles was achieved by microorganisms like bacteria and actinomycetes. In addition to this, leaf extract of *Nyctanthes* was used to synthesize TiO_2 nanoparticles because of its antinociceptive, antioxidant, anti-inflammatory, antidiabetic, antifungal and antimicrobial activity. However, the biosynthesis of titanium dioxide nanoparticles using *Nyctanthes* has not been fully explored yet (Sundrarajan and Gowri 2011).

16.4 Nanoparticles: The Real "Metal Bullets" in Clinical Medicines

Nowadays, nanoparticles possess immense potential especially in the medical sector because of its strong biocidal effectiveness. Besides, nanoparticles possess various attributes and multiple functions which make them unique and important for developing diagnostics tools and other antimicrobial materials (Ravindra et al. 2012). Moreover, they play a pivotal role in drug-delivery system and are also used as hyperthermic agents because they act as mediators of drug release (Khlebtsov and Dykman 2010). The various functions of metal nanoparticles particularly to clinical medicines are shown in Fig. 16.4. Due to certain unique properties of

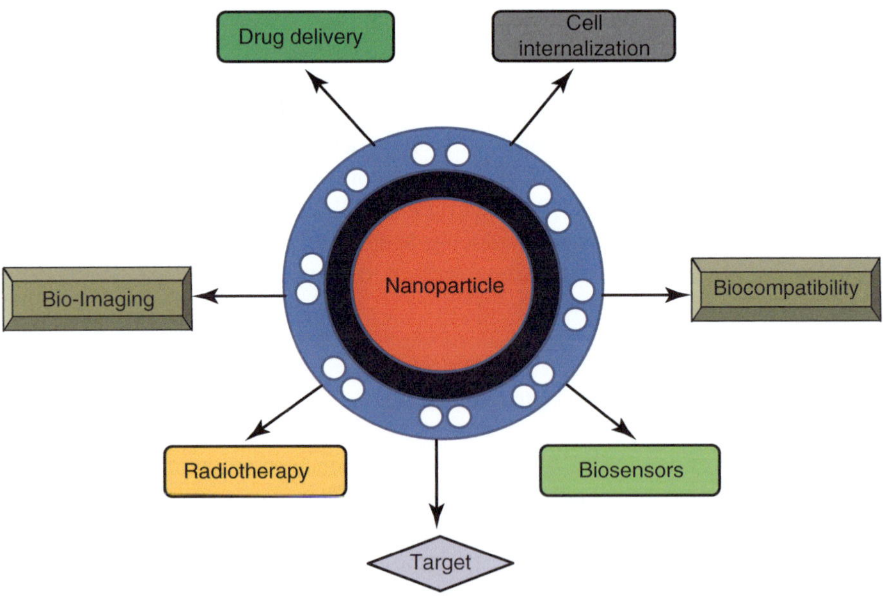

Fig. 16.4 Multiple functions of nanoparticle in clinical medicines

nanoparticles such as small size (1–100 nm), large surface to volume ratio, greater affinity towards target candidate and other special features, these act as a potential probe of biological markers (Sahayaraj and Rajesh 2011; Jain and Aggarwal 2012). Presently, the main focus of the scientists is to control the structural dynamics (e.g. size and shape) of these nanoparticles because it has a great influence on its properties (catalytic, magnetic as well as optical). Due to the nanosize of these particles, they possess immense potential in tumour diagnosis and treatment (Cai and Chen 2007). The most well-studied nanoparticles include quantum dots (Cai et al. 2006, 2007), carbon nanotubes (Liu et al. 2007), paramagnetic nanoparticles (Thorek et al. 2006), liposomes (Park et al. 2004), gold nanoparticles (Huang et al. 2007) and many others (Ferrari 2005; Grodzinski et al. 2006). Besides, these nanoparticles especially metal oxides possess a significant role in biomedical sensors, fabrication, fuel cells and other medical gadgets (Chen 1998). Among inorganic materials, oxides like ZnO, MgO, CaO and TiO_2 are of particular interest because their stability is maintained even in harsh conditions and are generally regarded as safe materials to organisms (Stoimenov et al. 2002; Fu et al. 2005). Nanoparticles, especially silver and zinc, act as potential antitopical agents to cure infectious diseases due to antimicrobial properties (Saha et al 2011). These metal oxides possess a strong inhibitory and bactericidal effects against fungi, virus and bacteria; that is why they have been used since ancient times to cure infection and burns (Saha et al. 2011). In medical sector due to continuous use of spurious and over-the-counter drugs, the resistance of microorganisms against antibiotics has been increased (Nithya and Ragunathan 2009). Therefore, developing novel, cost-effective and safe drug formulations is the need of the hour to cure such type of infectious diseases (Hasna et al. 2012). Saha et al. (2011) evaluated various antimicrobial activities of silver nanoparticles against different microorganisms. This property is considered as the most exploited nature of nanoparticles in the medical field (Geethalakshmi and Sarada 2010). Oza et al. (2012) describes the use of these nanoparticles in the formulation of dental resins, in ion-exchange fibres and in coating of medical equipment. In addition to this, metal nanoparticles replace the use of conventional fluorescent dyes as they exhibit photoelectric effect which counteracts photobleaching effect.

16.4.1 Summary and Outlook

Nanotechnology in the current scenario is a revolutionary field, and its advancement leads to the utilization of particles of the nanoscale dimensions. This is the beginning in the field of nanosystems, the next step in the coming decade being its association with the green chemistry route. This field comprises of several strategies that involve extensive strain selection, cultivation modes, recombinant DNA technology, metabolic engineering, protein designing and re-engineering and predictive modelling which allows us to create nanobioreactors resulting in the creation of a new nanobiotechnological field exhibiting huge impact in several areas of research. The application of nanomaterials has opened novel therapeutic arena, and the studies related to nanoparticles such as silver and gold reveal the huge benefits and multiple functions in various fields of biological and chemical science. The nanoparticle system presently approved in many cases increased the therapeutic index of drugs by decreasing the drug toxicity or elevating the efficacy by targeting ligands such as antibodies, pesticides and aptamers which may further improve the therapeutic nanoparticle system in the next generations. Multifunctionality of nanoparticles becomes a more complex system as they are capable of targeting, imaging and therapy which may be the subject of the future research. As the functionality of nanoparticles becomes more complex, there is a trend to redesign the nanoparticles with optimally physiochemical and biological properties to achieve the desired function. Indeed, this was the biggest hindrance for the nanoparticles to enter the clinical practices, the reason being that various targeted liposome was explored before 20 years, but not yet anyone of them has been approved for use. By introducing safer nanoparticles integrated with novel emerging techniques, we expect that more number of multifunctional nanomaterials enter into the clinical practices in the near future. As synthesis of nanoparticles is considered, the green chemistry approach is of great importance due to eco-friendly and cost-effective approach with a wide range of applications such as in nanomedicines, catalysis medicines, nano-optoelectronics and many more. This new and emerging field of research in the current scenario with day by day developments holds a good promise for the bright future in this field. This green chemistry route involves the synthesis of nanoparticles that have several advantages including the ease of the scaled-up processes, economic viability, etc. It was concluded that the plant-based synthesis of nanoparticles like silver and gold possesses potential clinical benefits, and the nanoparticle so obtained in nano-dimensions would be equally effective in medicinal applications as that of antibiotics and several other drugs. The application of these plant-mediated nanoparticles in drug delivery system might be the thrust area of research in nanomedicines. This green chemistry route for the synthesis of nanoparticles includes bacteria, yeast, fungi, several plant biomass or plant extracts which have tremendous application in biology with continued research and developmental efforts; nanotechnology is expected to have a tremendous impact on medicines for decades to come. The ongoing research was focused on the safety measures of nanomaterials, which are a critical point to be addressed for further development in the advancement of technology. There are many challenges ahead that need to be addressed and solved in order to make nanoparticle-based products commercially viable. Now, the need of the hour is to develop the cost-effective procedures for producing reproducible, biocompatible and stable nanoparticles from bioresources.

References

Abdul R, Amri I, Desi J, Stella M, Hanggara S (2009) Synthesis of copper oxidenanoparticles by using *Phormidium cyanobacterium*. Int J Chem 9:355–360

Ahmed S, Ahmad M, Swami BL, Ikram S (2016) A review on plants extract mediated synthesis of silver nanoparticles for antimicrobial applications: A green expertise. J Adv Res 7:17–28. doi:10.1016/j.jare.2015.02.007

Akhtar MJ, Kumar S, Alhadlaq HA, Alrokayan SA, Abu-Salah KM, Ahamed M (2016) Dose-dependent genotoxicity of copper oxide nanoparticles stimulated by reactive oxygen species in human lung epithelial cells. Toxicol Ind Health 32:809–821. doi:10.1177/0748233713511512

Akl M, Awwad M, Nida S (2012) Green synthesis of silver nanoparticles by mulberry Leaves extract. J Nanosci Nanotechnol 2:125–128. doi:10.5923/j.nn.20120204.06

Alanazi FK, Radwan AA, Alsarra IA (2010) Biopharmaceutical applications of nanogold. Saudi Pharm J 18:179–193. doi:10.1016/j.jsps.2010.07.002

Amrut SL, Satish JS, Ramchandara BP, Raghumani SN (2010) Synthesis and opticalcharacterization of copper oxide nanoparticles. Adv Appl Sci Res 1:36–40

Ankamwar B, Damle C, Ahmad A, Sastry M (2005) Biosynthesis of gold and silver nanoparticles using Emblica Officinalis Fruit extract and their Phase Transfer and Transmetallation in an Organic Solution. J Nanosci Nanotechnol 5:1665–1671. doi:10.1166/jnn.2005.184

Bhattacharya R, Mukherjee P (2008) Biological properties of naked metal nanoparticles. Adv Drug Deliv Rev 60:1289–1306. doi:10.1016/j.addr.2008.03.013

Botha TL, Tanyn EJ, Wepener V (2015) Comparative aquatic toxicity of gold nanoparticles and ionic gold using a species sensitivity distribution approach. J Nanomater 501:986902. doi:10.1155/2015/986902

Brust M, Walker M, Bethell D, Schiffrin DJ, Whyman R (1994) Synthesis of thiol-derivatized gold nanoparticles in a two-phase liquid-liquid system. J Chem Soc Chem Commun 7:801–802. doi:10.1039/C39940000801

Busbee BD, Obare SO, Murphy CJ (2003) An improved synthesis of high aspect- ratio gold nanorods. Adv Mater 15:414–416. doi:10.1002/adma.200390095

Cai W, Chen X (2007) Nanoplatforms for targeted molecular imaging in living subjects. Small 3:1840–1854. doi:10.1002/smll.20070035

Cai W, Hsu AR, Li ZB, Chen X (2007) Are quantum dots ready for in vivo imaging in human subjects? Nanoscale Res Lett 2:265–281. doi:10.1007/s11671-007-9061-9

Cai W, Shin DW, Chen K, Gheysens O, Cao Q, Wang SX, Gambhir SS, Chen X (2006) Peptide-labeled near-infrared quantum dots for imaging tumor vasculature in living subjects. Nano Lett 6:669–676. doi:10.1021/nl052405t

Canizal G, Ascencio JA, Gardea-Torresday J, Yacamán MJ (2001) Multiple twinned gold nanorods grown by bio-reduction techniques. J Nanopart Res 3:475–481. doi:10.1023/A:1012578821566

Chang SS, Shih CW, Chen CD, Lai WC, Chris Wang CR (1999) The shape transition of gold nanorods. Langmuir 15:701–709. doi:10.1021/la980929l

Chen YB, Bagnall DM, Koh HJ, Park KT, Hiraga K, Zhu Z, Yao T (1998) Plasma assisted molecular beam epitaxy of ZnO on c-plane sapphire: Growth and characterization. J Appl Phys 34:3912–3918. doi:10.1063/1.368595

Contescu CI, Putyera K (2009) Dekker encyclopedia of nanoscience and nanotechnology, vol I-VI. ISBN 13:978-0-8493-9639-7. Taylor & Francis, New York

Dhrutika P, Miral P, Krishnamurthy R (2013) Silver nanoparticles biosynthesis and its antimicrobial activity. CIBtech J Bio-protocol 2:50–57

Di Guglielmo C, Lopez DR, De Lapuente J, Mallafre JM, Suarez MB (2010) Embryotoxicity of cobalt ferrite and gold nanoparticles: a first in vitro approach. Reprod Toxicol 30:271–276. doi:10.1016/j.reprotox.2010.05.001

Diva B, Lingappa K, Dayanand A (2012) Antibacterial activity of nano gold particles synthesized by *Bacillus* sps. J Ecobiotechnol 4(1):43–45

Duran N, Marcato PD (2012) Biotechnological routes to metallic nanoparticles production: Mechanistics aspects, antimicrobial activity, toxicity and industrial applications. In: Rai M, Cioffi N (eds) Nano-antimicrobials: progress and prospects, vol 3. Springer, Berlin, pp 337–374

Duran N, Marcato PD, De Souza GIH, Alves OL, Esposito E (2007) Antibacterial effect of silver nanoparticles produced by fungal process on textile fabrics and their effluent treatment. J Biomed Nanotechnol 3:203–208. doi.org/10.1166/jbn.2007.022

Duran N, Marcato PD, Duran M, Yadav A, Gade A, Rai M (2011) Mechanistic aspects in the biogenic synthesis of extracellular metal nanoparticles by peptides, bacteria, fungi and plants. Appl Microbiol Biotechnol 90:1609–1624. doi:10.1007/s00253-011-3249-8

Eatemadi A, Daraee H, Zarghami N, Melat Yar H, Akbarzadeh A (2016) Nanofiber: synthesis and biomedical applications. Artif Cells Nanomed Biotechnol 44:111–121. doi:10.3109/21691401.2014

Elumalai EK, Prasad TNVKV, Hemachandran J, Vivivan Therasa S, Thirumalai T, David E (2010) Extracellular synthesis of silver nanoparticles using leaves of Euphorbia hirta and their antibacterial activities. J Pharm Sci 2:549–554

Esumi K, Suzuki A, Aihara N, Usui K, Torigoe K (1998) Preparation of gold colloids with UV irradiation using dendrimers as stabilizer. Langmuir 14:3157–3159. doi:10.1021/la980162x

Ferrari M (2005) Cancer nanotechnology: opportunities and challenges. Nat Rev Cancer 5:161–171. doi:10.1038/nrc1566

Frens G (1973) Controlled nucleation for the regulation of the particle size in monodisperse gold suspensions. Nature 241:20–22

Fu L, Liu Z, Liu Y, Han B, Hu P, Cao L, Zhu D (2005) Beaded Cobalt oxide nanoparticles along carbon nanotubes: towards more highly integrated electronic devices. Adv Mater 17:217–221. doi:10.1002/adma.200400833

Garcia ME, Baker LA, Crooks RM (1999) Preparation and characterization of dendrimer-gold colloid nanocomposites. Anal Chem 71:256–258. doi:10.1021/ac980588g

Geethalakshmi R, Sarada DLV (2010) Synthesis of plant-mediated silver nanoparticles using *Trianthema decandra* extract and evaluation of their antimicrobial activities. Int J Eng Sci Technol 2:970–975

Giersig M, Mulvaney P (1993) Preparation of ordered colloid monolayers by electrophoretic deposition. Langmuir 9:3408–3413. doi:10.1021/la00036a014

Gokulakrishnan R, Ravikumar S, Raj JA (2012) In vitro antibacterial potential of metal oxide nanoparticles against antibiotic resistant bacteria pathogens. Asian Pacific J Tropical Dis 2:411–413. doi:10.1016/S2222-1808(12)60089-9

González-Sánchez MI, Perni S, Tommasi G, Morris NG, Hawkins K, López-Cabarcos E, Prokopovich P (2015) Silver nanoparticle based antibacterial methacrylate hydrogels potential for bone graft applications. Mater Sci Eng A 50:332–340

Govindarajan M, Rajeswary M, Veerakumar K, Muthukumaran U, Hoti SL, Mehlhorn H, Barnard DR, Benelli G (2016) Novel synthesis of silver nanoparticles using Bauhinia variegata: a recent eco-friendly approach for mosquito control. Parasitol Res 115:723–733

Grodzinski P, Silver M, Molnar LK (2006) Nanotechnology for cancer diagnostics: promises and challenges. Expert Rev Mol Diagn 6:307–318. doi:10.1586/14737159.6.3.307

Gurrappa I, Binder L (2008) Electrodeposition of nanostructured coatings and their characterization—a review. Sci Technol Adv Mater 9(4):043001

Harekrishna B, Dipak KB, Gobinda SP, Priyanka S, Sankar PD (2009) Green synthesis of silver nanoparticles using latex of *Jatropha curcas*. Colloids Surf A Physicochem Eng Asp A39:134–139. doi:10.1016/j.colsurfa.2009.02.008

Haritha M, MeenaV SCC, Srinivasa RB (2011) Synthesis and characterization of Zinc oxide nanoparticles and its antimicrobial activity against *Bacillus subtilis* and *Escherichia coli*. Rasavan J Chem 4:217–222

Hasna ASRP, Kamaraj M, Jagadeeswaran P, Sangeetha G, Rajeshwari S (2012) Plants: Green route for nanoparticles Synthesis. Int Res J Biol Sci 1:85–90

Haverkamp RG, Marshall AT, Agterveld DV (2007) Pick your carats: Nanoparticles of gold- silver – copper alloy produced *In-vivo*. J Nanopart Res 9:697–700. doi:10.1007/s11051-006-9198-y

Hemanth NKS, Kumar G, Karthik L, Bhaskara RKV (2010) Extracellular biosynthesis of silver nanoparticles using the filamentous fungus Penicillium sp. Arch Appl Sci Res 2:161–167

Hiramatsu H, Osterloh FE (2004) A simple large-scale synthesis of nearly monodisperse gold and silver nanoparticles with adjustable sizes and with exchangeable surfactants. Chem Mater 16:2509–2511. doi:10.1021/cm049532v

Huang X, Jain PK, El-Sayed IH, El-Sayed MA (2007) Gold nanoparticles: interesting optical properties and recent applications in cancer diagnostics and therapy. Nanomedicine (Lond) 2:681–693. doi:10.2217/17435889.2.5.681

Hussain S, Ferguson C (2006) Best evidence topic report. Silver sulfadiazine cream in burns. Emerg Med J 23:929–932. doi:10.1136/emj.2006.043059

Jain P, Aggarwal V (2012) Synthesis, characterization and antimicrobial effects of silver nanoparticles from microorganism. Int J Nano Mater Sci 1:108–120

Jana NR, Gearheart L, Murphy CJ (2001a) Seed-mediated growth approach for shape-controlled synthesis of spheroidal and rod-like gold nanoparticles using a surfactant template. Adv Mater 13:1389–1393. doi:10.1002/1521-4095(200109)13:18<1389

Jana NR, Gearheart L, Murphy CJ (2001b) Wet chemical synthesis of high aspect ratio cylindrical gold nanorods. J Phys Chem B 105:4065–4067. doi:10.1021/jp0107964

Jana NR, Gearheart L, Obare SO, Murphy CJ (2002) Anisotropic chemical reactivity of gold spheroids and nanorods. Langmuir 18:922–927. doi:10.1021/la0114530

Jing L, Liang X, Deng Z, Feng S, Li X, Huang M, Li C, Dai Z (2014) Prussian blue coated gold nanoparticles for simultaneous photoacoustic/CT bimodal imaging and photothermal ablation of cancer. Biomaterials 35(22):5814–5821

Kapilashrami M, Zhang Y, Liu YS, Hagfeldt A, Guo J (2014) Probing the optical property and electronic structure of TiO2 nanomaterials for renewable energy applications. Chem Rev 114:9662–6707

Kaviya S, Santhanalakshmi J, Viswanathan B (2011) Green synthesis of silver nanoparticles using Polyalthia longifolia leaf extract along with D-Sorbitol. J Nanotechnol 1–5

Khlebtsov NG, Dykman LA (2010) Optical properties and biomedical applications of plasmonic nanoparticles. J Quantitat Spectroscop Radiat Transf 111:1–35. doi:10.1016/j.jqsrt.2009.07.012

Kim YG, Oh SK, Crooks RM (2004) Preparation and characterization of 1–2 nm dendrimer-encapsulated gold nanoparticles having very narrow size distributions. Chem Mater 16:167–172. doi:10.1021/cm034932o

Kim F, Song JH, Yang P (2002) Photochemical synthesis of gold nanorods. J Am Chem Soc 124:14316–14317. doi:10.1021/ja028110o

Klasen HJ (2000) Historical review of the use of silver in the treatment of burns II. Renewed interest for silver. Burns 26:131–138. doi:10.1016/S0305-4179(99)00116-3

Lamb AE, Anderson CWN, Haverkamp RG (2001) The extraction of gold from plants and its application to phytomining. Chem N Z 65:31–33

Leaper DJ (2006) Silver dressing: the role in wound management. Int Wound J 3:282–294. doi:10.1111/j.1742-481X.2006.00265

Leff DV, Brandt L, Heath JR (1996) Synthesis and characterization of hydrophobic, organically soluble gold nanocrystals functionalized with primary amines. Langmuir 12:4723–4730. doi:10.1021/la960445u

Lekshmi NCJP, Sumi SB, Viveka S, Jeeva S, Brindha JR (2012) Antibacterial activity of nanoparticles from *Allium* sp. J Microbiol and Biotechnol Res 2:115–119

Liu Z, Cai W, He L, Nakayama N, Chen K, Sun X, Chen X, Dai H (2007) In vivo biodistribution and highly efficient tumour targeting of carbon nanotubes in mice. Nat Nanotechnol 2:47–52. doi:10.1038/nnano.2006.170

Lloyd JR, Byrne JM, Coker VS (2011) Biotechnological synthesis of functional nanomaterials. Curr Opin Biotechnol 22:509–515. doi:10.1016/j.copbio.2011.06.008

Manna A, Imae T, Aoi K, Okada M, Yogo T (2001) Synthesis of dendrimer-passivated noble metal nanoparticles in a polar medium: comparison of size between silver and gold particles. Chem Mater 13:1674–1681. doi:10.1021/cm000416b

Martin CR (1994) Nanomaterials: a membrane-based synthetic approach. Science 266:1961–1996. doi:10.1126/science.266.5193.1961

Mieszawska AJ, Zamborini FP (2005) Gold nanorods grown directly on surfaces from microscale patterns of gold seeds. Chem Mater 17:3415–3420. doi:10.1021/cm050072v

Mohanpuria P, Rana NK, Yadav SK (2008) Biosynthesis of nanoparticles: Technological concepts and future applications. J Nano Res 3:507–517. doi: 1007/s11051-007-9275-x

Nanda A, Saravanan M (2009) Biosynthesis of silver nanoparticles from *Staphylococcus aureus* and its antimicrobial activity against MRSA and MRSE. Nanomed Nanotechnol Biol Med 5:452–456. doi:10.1016/j.nano.2009.01.012

Narayanan KB, Sakthivel N (2010) Biological synthesis of metal nanoparticles by microbes. Adv Colloid Interface Sci 156:1–13. doi:10.1016/j.cis.2010.02.001

Narayanan KB, Sakthivel N (2011) Green synthesis of biogenic metal nanoparticles by terrestrial and aquatic phototrophic and heterotrophic eukaryotes and biocompatible agents. Adv Colloid Interface Sci 169:59–79. doi:10.1016/j.cis.2011.08.004

Nasrollahi A, Paurshamsian KH, Mansourkiaee P (2011) Antifungal activity of silver nanoparticles on some of fungi. Int J Nano Dim 1:233–239. doi:10.7508/ijnd.2010.03.007

Nelson D, Priscyla DM, Oswaldo LA, Gabriel IHDS, Elisa E (2005) Mechanical aspects of biosynthesis of silver nanoparticles by several Fusarium oxysporum strains. J Nanobiotechnology 3:8. doi:10.1186/1477-3155-3-8

Nithya R, Ragunathan R (2009) Synthesis of silver nanoparticles using Pleurotus Sajorcaju and its antimicrobial activity. Dig J Nanomater Biostruct 4:623–629

Oza V, Pandey S, Shah V, Sharon M (2012) Extracellular fabrication of silver nanoparticles using *Pseudomonas aeruginosa* and its antimicrobial assay. *Pelagia* Res Lib. Adv Appl Sci Res 3:1778–1783

Padmavathy N, Vijayaraghavan R (2016) Enhanced bioactivity of ZnO nanoparticles—an antimicrobial study. Sci Technol Adv Mater 9:035004. doi:10.1088/1468-6996/9/3/035004

Park JW, Benz CC, Martin FJ (2004) Future directions of liposome and immunoliposome-based cancer therapeutics. Semin Oncol 31:196–205. doi:10.1053/j.seminoncol.2004.08.009

Parthasarathi M et al (2009) Synthesis and characterization of titanium dioxide nanoparticles and their applications to textiles for microbe resistance. J Textile Apparel Technol Manage 6:1–8

Parveen A, Ashish SR, Srinath R (2012) Biosynthesis and characterization of silver nanoparticles from *Cassia auriculata* leaf extract and in-vitro evaluation of antimicrobial activity. Int J Appl Biol Pharma Technol 3:222–228

Prashanth S, Menaka I, Muthezhilan R, Navin KS (2011) Synthesis of plant mediated silver nanoparticles using medical plant extract and evaluation of its antimicrobial activities. Int J Eng Sci Technol 3:6235–6250. doi.org/10.1155/2011/573429

Prasson PS, Chittaranjan B (2012) Green synthesis of gold nanoparticles and silver nanoparticles from leaves and bark of *Ficus carica* for nanotechnology application. IJSRP 2:1–4

Ravindra BM, Seema LN, Neelambika TM, Gangadhar SM, Nataraja K, Vijaya KS (2012) Silver nanoparticles synthesized by in-vitro derived plants and *Callus* culture of *Clitoria ternatea*; evaluation of antimicrobial activity. Res Biotechnol 3:26–38

Reetz MT, Helbig W (1994) Size-selective synthesis of nanostructured transition metal clusters. J Am Chem Soc 116:7401–7402. doi:10.1021/ja00095a051

Rodriguez-Carmona E, Villaverde A (2010) Nanostructured bacterial materials for innovative medicines. Trends Microbiol 18:423–430. doi:10.1016/j.tim.2010.06.007

Saha S, Chattopadhyay D, Acharya K (2011) Preparation of silver nanoparticles by bioproduction using *Nigrospora oryzae* culture filtrate and its antimicrobial activity. Digest J Nano Mater Biostruct 6:1519–1528

Sahayaraj K, Rajesh S (2011) Bionanoparticles: synthesis and antimicrobial applications, science against microbial pathogens: communicating current research and technological advances. In: Méndez-Vilas A (ed). FORMATEX, pp 228–44

Salam HA, Rajiv P, Kamaraj M, Jagadeeswaran P, Gunalan S, Sivaraj R (2012) Plant: Green route for nanoparticles synthesis. Inter. Res J Biol Sci 1:85–90

Scott RWJ, Wilson OM, Crooks RM (2005) Synthesis, characterization and applications of dendrimer-encapsulated nanoparticles. J Phys Chem B 109:692–704. doi:10.1021/jp0469665

Shahverdi M, Kesharwani J, Ingle A (2009) Fungus-mediated synthesis of silver nanoparticles and their activity against pathogenic fungi in combination with fluconazole. Nanomed Nanotechnol Biol 5:382–386. doi:10.1016/j.nano.2009.06.005

Shi X, Ganser TR, Sun K, Balogh LP, Baker JR (2006) Characterization of crystalline dendrimer-stabilized gold nanoparticles. Nanotechnology 17:1072–1078. doi:10.1088/0957-4484/17/4/038

Shi X, Lee I, Baker JR (2008) Acetylation of dendrimer-entrapped gold and silver nanoparticles. J Mater Chem 18:586–593. doi:10.1039/B714133E

Shiekh FA, Farooq O, Mian SH, Bautista RL, Arja SB, Andrabi KI (2016) The pitfalls of growing nanomaterials. Nanomedicine (Lond) 11:1635–1638. doi:10.2217/nnm-2016-0043

Shloma M, Grouchko M, Kamyshny A (2010) Copper nanoparticles for printed electronics: Routes towards achieving oxidation stability. Materials 3:4626–4638. doi:10.3390/ma3094626

Sibbald RG, Contreras-Ruiz J, Coutts P, Fierheller M, Rothman A, Woo K (2007) Bacteriology, inflammation, and healing: a study of nanocrystalline silver dressings in chronic venous leg ulcers. Adv Skin Wound Care 20:549–458. doi:10.1097/01.ASW.0000294757.05049.85

Silpa S (2016) Nanotechnology-present revolutionary biotechnology. Int J Pharma Res Health Sci 4:1261–1267

Soheyla H, Hamed B, Eshrat GF, Farzaneh N (2012) Green synthesis of copper oxide nanoparticles using *Penicillium aurantiogriseum, Penicillium citrinum* and *Penicillium waksmami*. Dig J Nanomater Biostruct 7:7999–1005

Sondi I, Sondi B (2004) Silver nanoparticles as antimicrobial agent: A case study on *E coli* as a model for gram negative bacteria. J Colloid Interface Sci 275:175–182. doi:10.1016/j.jcis.2004.02.012

Sridhara V, Ali B, Shaziya K, Satapathy LN, Khandelwal P (2012) Biosynthesis and antibacterial activity of silver nanoparticles. Res. J Biotechnol 8:11–17. doi: 10.1016/j.nano.2011.05.007

Stoimenov PK, KRL MGL, Klabunde KJ (2002) Metal oxide nanoparticles as bactericidal agents. Langmuir 18:6679–6686. doi:10.1021/la0202374

Sundrarajan M, Gowri S (2011) Green synthesis of titanium dioxide nanoparticles nyctanthes arbor-tristis leave extracts. Chalcogenide Lett 8(8):447–451. doi:10.1016/S1995-7645(14)60171-1

Thorek DL, Chen AK, Czupryna J, ANDREW T (2006) Super-paramagnetic iron oxide nanoparticle probes for molecular imaging. Ann Biomed Eng 34:23–38. doi:10.1007/s10439-005-9002-7

Tikariha S, Singh S, Banerjee S, Vidyarthi AS (2012) Biosynthesis of gold nanoparticles, Scope and application: A review. Int J Pharm Sci Res 3:1603–1615

Turkevich J, Stevenson PC, Hillier J (1951) The study of the nucleation and growth processes in the synthesis of colloidal gold. Discuss Faraday Soc 11:55–75. doi:10.1039/DF9511100055

van der Zande BMI, Boehmer MR, Fokkink LGJ, Schönenberger C (1997) Aqueous gold sols and rod-shaped particles. J Phys Chem B 101:852–854. doi:10.1021/jp963348i

Weare WW, Reed SM, Warner MG, Hutchison JE (2000) Improved synthesis of small (dCORE ≈1.5 nm) phosphine-stabilized gold nanoparticles. J Am Chem Soc 122:12890–13901. doi:10.1021/ja002673n

Wong EH, Lam SJ, Nam E, Qiao GG (2014) Biocompatible single-chain polymeric nanoparticles via organo-catalyzed ring-opening polymerization. ACS Macro Lett 3:524–528

Yogeswari R, Sikha B, Akshya KO, Nayak PC (2012) Green synthesis of silver nanoparticles using *Ocimum sanctum* (Tulashi) and study of their antibacterial and antifungal activities. J Microbiol Antimicrobial 4:103–109. doi:10.5897/JMA11.060

Yohan D, Chithrani BD (2014) Applications of nanoparticles in nanomedicine. J Biomed Nanotechnol 10:2371–2392. doi:10.1166/jbn.2014.2015

Yu YY, Chang SS, Lee CL, Chris Wang CR (1997) Gold nanorods: electrochemical synthesis and optical properties. J Phys Chem B 101:6661–6664. doi:10.1021/jp971656q

Factories for Antibody Generation

17

Gopal Singh Bisht, Kinam Gupta, and Rahul Shrivastava

Abstract

Antibodies are gathering increasing importance as diagnostic tools and as therapeutic agents. The major limiting factor is their low quality and productivity. Progress in exploiting molecular biology tools has provided platforms for developing recombinant antibodies. The major focus is on (1) performing complicated engineering by simpler techniques, (2) building up binders and fusing to diverse effectors and tags, (3) manufacturing the constructs economically using microbes, and (4) retaining stable material clonality for higher reproducibility. Among the most regularly used expression systems used for generating artificial antibody fragments are microalgae, bacteria, and mammalian cells. Each expression system has its own merits and demerits; however, mammalian cell expression systems have quite a few advantages: for recombinant protein production is their ability to induce proper folding of the protein, posttranslational changes, and generating the product—factors necessary for proper biological activity.

Keywords

Antibody • Antigen • Affinity maturation • Enzymes • Inhibitors • Expression systems

---

G.S. Bisht • K. Gupta • R. Shrivastava (✉)
Department of Biotechnology and Bioinformatics, Jaypee University of Information Technology, Waknaghat, Solan 173 234, Himachal Pradesh, India
e-mail: juit.rahul@gmail.com; rahulmicro@gmail.com

V.C. Kalia, A.K. Saini (eds.), *Metabolic Engineering for Bioactive Compounds*,
DOI 10.1007/978-981-10-5511-9_17

17.1 Introduction

Antibodies are gaining importance in diagnostics and therapy. However, their production is facing difficulties in terms of quantity and quality. In order to resolve these issues, there has been a progression in formats and production platforms. The rising attention toward recombinant antibodies is exclusively because of:

- Performing complicated engineering by simpler techniques
- Building up binders and fusing to diverse effectors and tags
- Manufacturing the constructs economically using microbes
- Retaining stable material clonality for higher reproducibility

Presently there are many expression systems being used for generating artificial antibody fragments. Among them, microalgae are environment-friendly protein factories fueled by photosynthesis and work without external supply of CO_2. Microalgae have many advantages of classical expression systems such as rapid growth rates, easy handling, and most importantly; it enables eukaryotic posttranslational modifications. But bacterial expression system also remains popular for recombinant protein expression and represents the standard for producing antibodies on a small scale. The main advantages are simplicity, the accessibility of well-tested reagents, and tremendous experience (protocols, common expertise). It is also seen that there is an increased application of mammalian cells for producing proteins. The main advantage of using mammalian cell expression systems for recombinant protein production is their ability to induce proper folding of the protein, posttranslational changes, and generating the product—factors necessary for proper biological activity (de Marco 2015).

17.2 Mechanism of B Cell Development

B cells are manufactured and matured to form a specialized defense cell in the bone marrow. Antibody remains attached to B cells unless their production is triggered by a pathogen. B cell matures into plasma cell whenever it encounters any nonself substance, resulting in production of antibodies. Development of B cell occurs through a highly regulated process in which progenitor B cell matures through an organized sequence arrangements of Ig genes. Modifications in the gene lead to highly diverse products. At the end of gene rearrangements, immunocompetent B cell has variable regions in the functional heavy and light chains, which provide it specificity toward a particular epitope.

Rearrangement in constant region gene segment occurs after the B cell is stimulated by an antigen. This changes its biological effector functions without altering its specificity. Hence, mature B cell chromosomal DNA is different from germline DNA due to rearrangements.

17.2.1 B Cell Maturation

B cell generation first occurs in the yolk sac, liver, and bone marrow of the fetus in the embryonic stage, and then the generation of mature B cell continues in the bone marrow cells. Development of B cell begins with the transformation of lymphoid stem cells into the progenitor B cell (pro-B cell). Microenvironment provided by the bone marrow stromal cells is necessary for the development of pro-B cells into precursor B cells and the secretion of diverse cytokines, remarkably IL-7, which supports the developmental process.

Cell adhesion molecules like VCAM-1 and VLA-4 and its ligand are required for direct contact between stromal cells and pro-B cells. A receptor c-Kit on the pro-B cell surface gets activated by interaction with the stromal cell surface (SCF) molecule. Activation of c-Kit leads to division and differentiation of pro- into pre-B cells and initiates expression of a receptor of IL-7. IL-7 downregulates the adhesion molecules on pre-B cells to enable detachment of the proliferating cells from those of the stromal cells.

The pre-B cell receptor is formed when μ heavy chain and light chain along with Ig-α/Ig-β heterodimer appear on pre-B cell. This complex then proceeds to maturation. A speculation has been made about the recognition of an unidentified ligand on the stromal cell membrane by the pre-B cell receptor which leads to allelic exclusion as the signal transmitted by the receptor-ligand binding prevents V_H to D_H J_H rearrangement of other allele of the heavy chain. Hereafter, multiple divisions of B cell produce 32 to 64 progenies, each of which rearranges different light chain gene segments leading to an increment in the overall antibody diversity.

Changing pattern of surface markers characterizes the development of progenitor to mature B cell. At pro-B stage, the cells express CD45R, the signal-transducing molecules Ig-α/Ig-β, CD19 (which is a component of B cell co-receptor), CD43, and CD24 (heat stable antigen). Pro-B cell also expresses c-Kit, a receptor for a growth-promoting ligand. As the cell develops to pre-B stage, many of the markers they express are similar to that present during pro-B stage; however c-Kit and CD43 are no longer expressed. Instead they begin to express CD25 and α chain of IL-2 receptor. The salient feature of pre-B cell stage is the expression of their receptor (pre-BCR). When the expression of CD25 and pre-BCR is ceased and surface immunoglobulin appears, then cells are classified as immature B cell.

It is estimated that in mouse, only 10% (i.e., 5×10^6 B cells/day) of the total (i.e.,˜5×10^7 B cells/day) B cells, which are produced in bone marrow, actually enter into the circulating B cell pool, and the remaining 90% die within the bone marrow. This loss includes immature B cells which express auto-antibodies and, hence, attributable to negative selection and clonal deletion. However, it was studied that negative selection does not always result in immediate deletion of immature B cells; rather its maturation is stopped until the B cell amends the light chain gene.

17.2.2 Proliferation and Activation of B Cell

Activation and development of B cells require antigen and take place in the periphery lymphoid organs once they are exported from bone marrow. Naive B cells usually have short half-life and perish within a couple of weeks by apoptosis if they are not triggered by the inducing antigens. B cell gets activated via different routes, which depend on the characteristics of the antigen; one depends upon TH cells, the other independent of it. Direct contact with TH cells is necessary for B cell response to thymus-dependent (TD) antigens. B cells can even be triggered by antigens in the absence of TH cells and are termed as thymus-independent (TI) antigens.

TI antigens are split into two types, which activate B cells in two systems. Example of type 1 thymus-independent (TI-1) antigens is a few bacterial cell wall components, lipopolysaccharide (LPS), whereas TI-2 is polymeric protein—bacterial flagellin or cell wall polysaccharides. TI-1 antigens are generally polyclonal B cell activators, and they also can stimulate B cells irrespective of their antigenic specificity. TI-1 antigens at high concentration stimulate proliferation and antibody secretion. B cell specific for epitopes of the antigen get triggered on exposure to low antigen concentrations. TI-2 antigens stimulate B cells by thoroughly cross-linking the mIg receptor, and they differ from TI-1 in the following aspects:

- They are not B cell mitogens; therefore, they are not polyclonal activators.
- They stimulate B cells, whereas TI-2 antigens stimulate only mature B cells.
- B cell responds to TI-2 antigens without TH cells; however cytokines produced from TH cells are necessary for effective B cell proliferation as well as for switching to isotypes apart from IgM.

17.2.3 Antigen Recognition and B Cell Activation

B cell receptor (BCR) is produced by association of membrane Ig with disulfide-linked heterodimer Ig-α/Ig-β, and when BCR is associated with an antigen, it activates intracellular signaling pathways. B cells can proliferate on activation with membrane proteins from activated T_H cells but differentiate only in the presence of cytokines. T-independent humoral response is simple and involves IgG and IgM, while the response of T_H cell-dependent humoral ones is highly specialized and requires immunoglobulins of various classes and subclasses.

Antigens entering the body from different locations are siphoned to specific secondary lymphoid organ, e.g., antigens from blood are drawn into spleen and; antigens from respiratory, gastrointestinal and urogenital tract are drawn into Mucosa Associated Lymphoid Tissue (MALT). B cells develop the capability to identify antigens in bone marrow, but interaction of mature B cells with forcign antigens occurs in peripheral lymphoid tissues.

B cell activation is initiated by binding of B cell surface receptors—IgD and IgM with specific antigen. The B lymphocyte antigen receptor helps in delivery of biochemical signals to the B cells by antigen-induced clustering of receptors and the binding of protein antigen with the receptor and its internalization into endosomal vesicles for its processing and presentation to T_H cells with the help of MHC II molecules. BCR delivers activating signals to the cells when multivalent antigens cross-link two or more receptor molecules. Two membrane proteins, Ig-α and Ig-β, are associated together by disulfide linkage, and when these two molecules are covalently linked to the membrane Ig, they form the B lymphocyte antigen receptor complex which leads to transduction of the signals generated due to clustering of surface receptors. Cascade of signaling events eventually leads to the activation of transcription factors which induce the expression of those genes necessary for activation of B cells. Complement proteins provide another set of signals required for B cell activation. On the complement receptor 2 (CR2) of B cells, a split-up component of complement binds and acts as an important indicator for B cell activation. Complement protein C3d is produced by the proteolysis of C3 and binds to CR2. Then, C3d binds to the antigen-antibody complex or microbe. For antigen recognition by surface Ig, the complex of antigen-C3d binds to the CR2 present on B cell surface. Initiation of signaling pathways occurs by antigen binding, and their expansion happens due to binding of C3d to B cell CR2.

B cells are prepared for proliferation and differentiation by cellular events which are stimulated by binding of antigen to their receptor complex. The events are:

- Resting B cells enter into cell cycle, leading to increase in cell size, ribosomes, and cytoplasmic RNA.
- Half-life of the B cell is enhanced due to stimulation of various anti-apoptotic genes.
- Increase in the expression of MHC class II molecules and their associated stimulators.
- Enhanced expression of receptors for cytokines.

B cell activation also depends on the nature of antigen as its nature alters the signal generated by BCR for B cell activation. For example, T-independent antigens, like glycolipids and polysaccharides, express multiple identical epitopes on each molecule, and hence, they can easily cross-link to surface receptors and initiate response without being recognized by T_H cells. Quite the opposite, only one copy of epitope/molecule is expressed in case of naturally occurring globular proteins, and thus, they cannot bind and cross-link antigen receptors and do not deliver activating signals. They can induce B cell activation only when they are recognized by T_H cells after binding to previously produced antibodies or complement components.

17.2.4 Response to T-Dependent Antigens

Recognition of antigen by helper T cells is necessary for antibody response against antigen. Initially, antigen is presented by B cells to T cells which activate T_H cells, resulting in expression of various molecules, which are required for B cell activation. These steps lead to the clonal expansion of B cell, affinity maturation, switching of isotype, and their differentiation into memory cells.

T-dependent antibody response to antigen occurs in the lymphoid organs. The initial response consists of proliferation of B cell, secretion of antibody, and isotype switching whereas late phase results in affinity maturation and production of memory B cell. Early phase occurs in the T cell area and primary follicles, and late phase takes place in the germinal center of the lymphoid follicles. Naive CD4+ T cells present in T cell area recognize antigen within 1 or 2 days of its administration and get activated. B cell present in the primary follicle moves to T cell area after their encounter with antigen. The initiation of the reaction between the antigen-activated T and B cells happens at the interface of follicles and T cell and takes place 3–7 days after the exposure of antigen. Humoral response against antigen is generated by collaboration of activated B cells and T cells.

Processing of antigen is carried out in endosomal vesicles after the B cells bind to antigen at the surface Ig receptors. After processing, the peptide fragment of the antigen is presented to the T cells with the help of class II-MHC molecules. Antibody response remains specific to conformational determinants of the native protein even though B cell can present multiple different peptides to different T cells.

The expression of surface co-stimulators is enhanced by antigen binding to membrane Ig. CD28 of T_H cell binds to the B7 molecule expressed on the B cell and results in stimulation of T_H cell proliferation. After the activation, T cell expresses CD40L on its surface that binds to CD40 on the exterior of B cell. This initiates an enzyme cascade that leads to increased expression of B7 on B cell, leading to increased activation of T cell. B cell expresses cytokine receptors after recognizing the antigen. Cytokines are secreted by activated T_H cell to stimulate proliferation of B cell. Major functions in antibody responses are served by cytokines:

- By proliferating and differentiating B cell, they provide amplification mechanism.
- By promoting isotype-class switching, they determine type of antibodies produced.

High concentration of cytokines is present around the B cell if they are in close contact to the activated T cells. IL-2, IL-4, and IL-5 induce B cell proliferation. Ig gene transcription is turned on by all the stimuli that a B cell receives. Antibodies are secreted by B cells when they differentiate into effector cells. Secreted antibodies have almost similar specificity as the Ig receptor. Cytokines can also affect RNA processing to enhance the Ig production. Cells secreting antibody are mainly present in extrafollicular positions of lymphoid tissue such as medulla of the lymph node and red pulp of the spleen. At 2–3 weeks after antigen exposure, these cells also migrate to bone marrow, and it becomes the primary site for antibody production. Cells that secrete antibody do not circulate well. Quite a few B cells transform into plasma cells. Plasma cells are distinct from B cells and their main function is to produce antibody abundantly. Initially the heavy chain μ (IgM) isotype antibody is being secreted predominantly. Heavy chain isotype switching occurs in activated B cells after stimulation from CD40 engagement and cytokines and leads to production of different classes of antibodies—ε (IgE), γ (IgG), and α (IgA). IgD is not found in plasma as the secretory δ heavy chain is synthesized very rarely. Apart from CD40 signaling, cytokines also regulate the heavy chain isotype switching. For instance, IL-4 activates switch to IgE. Helper T cells generate different cytokines in response to microbes that regulate heavy chain class switching. TGF-β and T cell-derived IL-5 stimulate generation of IgA, which results in local immunity. Cytokines are also antagonistic like IFN-γ, which inhibits IL-4-mediated B cell switching to IgE. Switch recombination is the process of isotype switching.

Germinal center is formed when some activated B cells move into the follicle and multiply. Each completely formed germinal center has cells derived from clones of antigen-specific B cells. Germinal center B cells are stimulated by complement receptors (CR1–3), Fc receptors, and CD40L which are located in lymphoid follicles only. Centrocytes are the smaller cells which are descendants of proliferating B cells. These cells gather in the germinal center, which has fewer follicular dendritic cells. These nondividing B cells move to lighter zone.

Affinity maturation process results in enhanced affinity of antibodies toward a specific antigen with the progression of T-dependent humoral response. This occurs as a consequence of mutations in the Ig genes. This follows selective survival of B cells, which produce antibodies having highest affinity. Affinity maturation is observed only in the immunogenicity produced by T-dependent antigen as T_H cells and interactions between CD40 and CD40L are necessary for maturation. High rates of point mutations are shown by the proliferating germinal center B cells in their heavy and light chain genes. Antibodies with high affinities for the antigen are generated through somatic mutations. And those B cells which attach with high affinity to antigens displayed by follicular dendritic cells are selected to survive.

Antigen signals are required to rescue small B cells from apoptosis as in these cells Ig genes have encountered point mutations. Follicular dendritic cells express receptors which can bind and show antigens which interact with antibodies. Antigen binding to Ig prevents B cell apoptosis by generating signals. The end result is the selection of population of B cells which can produce antibodies having significantly enhanced affinities. After exiting the germinal center, cells mature into antibody secretors with high affinity. There after they move to the bone marrow and continuously secrete antibodies for a long period. A few antigen-activated B cells transform into memory cells which have the ability to survive and generate rapid antibody responses subsequently.

17.3 Bacteria as Antibody Factory

In recent decades, monoclonal antibodies (mAbs) have gotten to be a standout among the most helpful and essential tools with a huge number of therapeutic and diagnostic applications. Out of the 151 unique engineered therapeutics that were affirmed by the Food and Drug Administration, 33% are monoclonal antibodies, with numerous more in innovative work conductor. Most of the mAbs that were affirmed for curative applications are produced in various cultures, for example, Chinese hamster ovary (CHO) cells. These are utilized as a part of research and in generating remedial proteins and mouse cell lines (created from tissues such as neural sections, liver, tumorigenic cells, fibroblasts, and furthermore the blood.). The hybridomas and the two antigen-binding fragments (Fabs) are developed in *E. coli* periplasmically.

Antibodies produced by recombinant technology work as ideal agents for diagnostics and therapeutic purposes. Moreover, the best example for this is *E. coli* that relies on the periplasmic space for antibody synthesis. A small amount of this material is exuded into the medium, due to the outer bacterial membrane. Recently, *Bacillus megaterium* was observed to exude rec-proteins—antibody fragments into the outer medium (Reichert 2012).

However, *E. coli* remains a system of choice which is utilized as a part of both industry and the educated community on an expansive scale and wide generation of recombinant proteins. A noteworthy test happens while confronting the utilization of *E. coli* as an antibody expression system in the creation of mAbs, having proper disulfide bonds. The arrangement of disulfide bonds in *E. coli* can be catalyzed either in the cytoplasm of genetically engineered strains or naturally oxidative periplasmic compartment. Numerous fragments from monoclonal antibody, for instance, Fab that is single-chain Fv (scFv), Fc, and a scFv–Fv combination, were expressed in *E. coli*. Others have expressed functional IgGs in *E. coli* which is constrained by the absence of and the small volume of adenosine triphosphate (ATP)-dependent chaperones and also by the requirement for substantial scale optimization to proficiently discharge both the IgG chains over the firmly secure cytoplasmic membrane. These impediments were addressed, and several groups have endeavored to create solvent IgGs in the cytoplasm of *E. coli*, yet none of them were fruitful. Best case scenario, these endeavors brought about the misfolded of IgG chains present inside the inclusion bodies.

So these present techniques for creating immunoglobulin G (IgG) in designed cells fundamentally need refolding and also their secretion. It is being depicted as a powerful expression system for synthesis of IgG antibodies in the *E. coli*. Heavy and light chains, which are artificially synthesized in nature, lack recognized export signals. IgGs with clinically pertinent antigen and effectors restricting exercises are enthusiastically created in the *E. coli* by transferring antigen-specific variable domains, which assist the binding to Fc receptors. The IgGs in the cytoplasm also called as "cyclonals" effectively bypass the rate-limiting steps of membrane translocations (chromosome abnormality caused by rearrangement of parts between nonhomologous chromosomes) and glycosylation (Robinson et al. 2015).

Nowadays, medicines have to deal with the multidrug-resistant bacteria. Well the past experiences of the pre-antibiotic with serum therapy have suffered with high rate in the clinics basically due to the lack of efficacy.

17.4 Plant Can Work as Antibody Factory

Disease-fighting antibodies could be turned into a substantial scale agriculture crop. Molecular researchers are required to deliver a lot of specifically designed antibody molecules, which can be utilized to treat distinctive diseases. The plant-made antibodies are popularly called as "plantibodies" (a plantibody is characterized as a counteracting agent delivered from the plant that is hereditarily designed with the creature DNA) that firmly and specifically bind to one or a couple sorts of particles. Regularly, plants do not synthesize or produce antibodies. To get plants to synthesize antibodies, numerous researchers have used couple ventures to transport mouse genes coding for an antibody, into tobacco plant cells. The remote genes get embedded into the cells genome. The two genes present in the cells are derived from an altered mouse immune system; cell line created by the Scripps was created to deliver uncommon catalytic antibodies, which bind to particular targets known as carboxylic esters but also dissect them.

The researchers utilize the genetically modified cells to develop whole plants, for example, tobacco, some of which produce one section or the other part of the two antibodies. At that point crossing over the two plants occurred, after that each makes part of the antibody. Scientists then located a specific proportion of the offspring plants to produce complete antibodies. The outcomes demonstrated that the reactivity is preserved in plantibodies.

This is another innovation equipped for creating antibodies from a plant, yet it is appropriate to some quite vital issue like ecological contamination and conceivably tumor treatment. There is a plausibility of bringing into the plant a practically boundless measure of disease resistance for securing the supplements included inside it. These plant-made antibodies cost less than $100, compared to $2 million-plus cost for the proportional heap of antibodies prepared utilizing the technology of monoclonal antibody. Growing plants with ability to produce antibodies is an upscaling on industrial level and can cater to many researchers in the field. This could be quite similar to plant containing dioxin-binding antibodies, which can act as biofilters for cleaning soils polluted with dioxins, etc. Transferring genes for producing antibodies into plant cells can lead to attacking cancer in humans.

For that researchers ought to beat the difficulties before plantibodies get standardized. As indicated by Hiatt Scripps immunologist, "Nobody has built up the capacity to purify antibodies in huge amounts." So to make the process simpler, the specialists are attempting to change nontoxic or less dangerous plant cells, for example, soybean or alfalfa, into antibody producers. Another issue confronted amid antibody production is in the likelihood of plantibodies evoking autoimmune and different life-threatening reactions in individuals. Specialists likewise found that methodology for producing protein, for example, antibodies, is slower than microorganism-based procedures; however, they concurred that plants may prove to be more appropriate for the expansive scale applications.

17.4.1 Moving from Antisera to Antibodies

ZMapp's (genes encoding for the antibodies) origin started with a 100-year-old treatment for infections with no prevention and cure: antiserum. For the most part, antiserum was derived from the serum of humans or animals, which had survived the infection. The serum is the clean and clear portion of blood. It has no RBCs or clotting proteins but contains numerous proteins, such as antibodies. Antiserum treatment includes using the survivor's serum and infusing it to the individuals who are recently exposed to the same disease. The serum antibodies rapidly activate the immune system of the person who is recently infected.

Prior to the revelation of antimicrobials such as penicillin, passive immunization in anticipation of infection in which antisera were utilized to treat infections, including diverse diseases like pneumonia, diphtheria, cholera. Presently, antisera that can fight toxins are utilized to treat patients from tetanus infections, rabies, and snakebites.

Horses, infected rats, and rabbits have been valuable for creating antisera. However, these all were expensive and incompetent to utilize. Modern way of understanding the structure and genetics of antibodies helped to refine the antiserum idea. Rather than utilizing a wide range of polyclonal antibodies or pAb, specialists pick only the effective ones, from which they reproduce the genetic code for one particular antibody—monoclonal antibody or mAb.

The genes for those monoclonal antibodies are inserted into cultures of animal cells or bacteria, which then synthesize large amounts of mAbs. Since no serum or animal products are included in this procedure, it wipes out the risk of working with infectious agents in animals which can have fatal impact.

The family of tobacco is usually utilized as a part of these novel procedures. Specialists modified tobacco-like plants which were developed to produce antibodies against Ebola to create ZMapp. *N. benthamiana* is grown indoors like in processing plant like ranches, so it does not put commercial tobacco or other different crops at danger.

N. benthamiana is a well-established plant utilized for the procedure of agroinfiltration. The desired antibody gene is added to *Agrobacterium tumefaciens* which is an engineered bacterium. Now it has been hitched up to saturate a plant with useful DNA. Plants are immersed entirely into a solution of *A. tumefaciens* and the leaves ingest the bacteria. Then, the plants start producing the desired protein in mass.

The tobacco plants' leaf is the fundamental place for storing proteins. Consequently, purification processes must start promptly after harvesting. Extraction of the desired protein is as yet costly and tedious. Tobacco and its relatives possess poisonous alkaloids which must be expelled before administering the protein as a drug.

The production of large amount of the ZMapp monoclonal antibody cocktails is still distant. The procedure should be scaled up to fulfill the demand of a huge number of dosages. ZMapp is yet to be tested on human beings to figure out whether it is safe and effective (Both et al. 2013).

Like rabies, many life-threatening diseases kill many people around the world. The murine monoclonal antibody (mAb) has been recognized for its future forthcoming use in rabies postexposure prophylaxis (PEP). The objective of this methodology was to set up a plant-based system for generating a chimeric mouse-human mAb, to describe the anti-recombinant agent, and to analyze its interaction with rabies virus glycoprotein at a molecular level. Chimeric antibody was recently and successfully expressed in *Nicotiana benthamiana plant closely related to tobacco* (Both et al. 2012). Glycosylation was inspected for its usefulness by antigen

ELISA. Epitope characterization was achieved by utilizing pseudotype virus by expressing mutagenized rabies glycoprotein. Purified mAb exhibited strong viral neutralization at the concentration of 500 IU/mg. The contribution of the antigenic site I of the glycoprotein and two amino acids residues (K226 furthermore the G229) inside site I was identified with respect to mAb neutralization. Pseudotype viruses which expressed glycoprotein from lyssaviruses, which cannot be neutralized by antibody, were taken as controls. The outcomes here provided the molecular basis for generating mob for rabies PEP (Goudsmit et al. 2006). They also establish the basis for developing cheap or low prize plant-based antibody product to benefit medial class or poor families in developing countries.

The discovery in the field of plant biotechnology and producing cell factories can synthesize functional human antibody for new revolution (Giddings et al. 2000).

17.5 Diatom that Acts as Plasma Cells

Microalgae are of great significance since they are the major contributors of O_2 and help in C fixation. Its biotechnical applications include its tremendous potential and utilization in food and cosmetic industry especially as source of vitamins, omega-3 fatty acids, antioxidants, and pigments. Particularly in the most recent past, they are viewed as sources of fuel mainly biodiesel.

Algae are being utilized as model system for expressing recombinant proteins. Despite the fact that there are various expression systems, for example, yeast, bacteria, and mammalian cells, a considerable number of these experience the ill effects of deficiencies, for example, low yield, expensive, complexity of purification, and contamination with pathogenic organisms. Microalgae provide various advantages of established expression systems as they have fast development rate, are easy to handle, and allow posttranslational modifications. Moreover, microalgae are minimal effort, environment-friendly protein processing plants fueled by photosynthesis and work without external supply of CO_2.

17.5.1 *Phaeodactylum tricornutum*

Until now medically significant proteins like antibodies, hormones, and immunizations were being created productively in *Chlamydomonas reinhardtii* cells. And lately, it was uncovered that different species such as *Phaeodactylum tricornutum* can heterologously express proteins with high productivity including those which need posttranslational changes and the assembly of numerous subunits can be synthesized. It was seen that *P. tricornutum* was able to synthesize much larger amounts of a human IgG against the hepatitis B, which accounted for 9% of its soluble protein content. Moreover, efficient production of the bio-plastic poly-3-hydroxybutyrate (PHB) with the introduction of the bacterial PHB pathway indicates that algae can not only produce protein but other bio-products as well.

Effective protein expression is a key concern, but before acquiring the final product, tedious and extensive processes including cell harvesting and its lysis and purification are also important. Consequently, the ideal expression system produces rec-proteins with high efficiency. It then secretes the proteins, thereby eliminating numerous costly purification steps. In these studies, nitrate-inducible promoter system provided the advantage to tightly control the antibody production and best induction periods for highest efficiency, with the ability to identify the best functionality.

Human IgG antibody was expressed against the surface protein (CL4mAb) of hepatitis B virus, at the C-end of both antibody chains without the ER-retention signal (DDEL). This alteration remarkably helped in secretion of the completely assembled antibody and their accumulation in large amounts in the media.

Expression of rec-proteins was initially induced to monitor antibody secretion for 2 days. Thereafter, the cells were harvested, and the remaining cells were removed by filtering the supernatant. Proteins were subsequently concentrated and precipitated. Gel electrophoresis analysis and staining revealed that both heavy and light chains of antibody are present in high quantity and were having high purity as checked by silver staining. It showed very high antibody concentration in the medium compared to cells expressing the antibody with an ER-retention signal. Gel electrophoresis was carried out with the samples prepared in nonreduced conditions to check the assembly of both the chains. With the Western blot, it was revealed that both chains cannot be detected independently. It gave high M.Wt. of about 170 kDa, i.e., a completely assembled IgG complex.

Subsequently, to check if the secretion of antibody is occurring accidently in some cell lines or it is occurring characteristically, a set of transfectants were analyzed. Cell lines were cultivated without induction and incubated in a medium having nitrate for 2 days, after which the proteins were concentrated and precipitated for Western blot analysis. All the cell line expression and secretion of the antibody varied with cell lines. Western blot analysis with an antibody against α-tubulin proved the absence of intracellular proteins and detected antibody was the result of secretion. Consequently, four cell lines with high secretion efficiency were selected for checking functionality and quantity of the secreted antibody (Hempel and Maier 2012).

17.5.2 Quality and Quantity of Secreted Antibodies

P. tricornutum secreting a fully assembled antibody and not the antibody chains separately was demonstrated through Western blot analyses. However, the most important question of course would be to analyze i) whether the secreted antibodies are well designed to identify the target antigen and ii) would the algal expression system be able to secrete antibodies in amounts that are required to meet commercial demand.

Thus, the secreted antibodies were checked for their quality and quantity for 5 days through ELISA. It verified that the antibodies secreted are functional and can act on the target. The binding efficiency matched with the quantity of antibodies secreted during 48 h of expression (T1, T2) to 450–850 ng Ab/mL. Thereafter, antibody and its binding efficiency decreased due to low nitrate levels. But, antibody expression was enhanced to remarkable levels of 1550–2550 ng/ml by replacing the medium and providing fresh source of nitrate (Hempel and Maier 2012).

17.5.3 Conclusion

Fascinatingly, this alga appears to be quite similar to mammalian cells, which ensures that heavy chains are associated with light chains before they leave the cell and, hence, act like a human plasma cell. This process is mediated by a stable and strong association with BiP; thereafter, the light chains bind to the mammalian cells. Complex molecules like IgG can be produced in the alga system and secreted with high efficiency. This results in easy downstream purification steps which often are problematic and expensive. It is true that the algal system cannot compete with mammalian systems presently, as the engineering of mammalian systems has been advanced to such a remarkable level. Nevertheless, in the future, in order to broaden the application spectra, optimization of expression of other antibodies and their secretion may be done, and cells may be manipulated such that human-specific modifications—glycosylation patterns—are achieved.

17.6 Fungi as Antibody Factory

The scope of diversity of fungi and their related interspecies having a genetic variability offers the field for searching useful and important strains that are producing antibiotics, enzymes, or metabolites of scientific as well as practical importance. For such practices fungi have been developed as chemical and cell factories for secreting several classes of enzymes which further break the polymeric constituents that are dead matter into soluble forms which are very helpful for absorption and utilization as sources of energy and carbon. An additive or supplements to the success in absorptive mode of nutrition result in the production of chemicals that are antagonistic which enable the fungi to survive. They protect them from microflora and microfauna which comprise different predators, parasites, and so on. The diversity of fungi that is approximately 1.5 million offers a large resource for searching a useful strain required for producing antibody for biotransformation of a particular substrate or a metabolite.

Nowadays, filamentous fungi are used by many researchers for producing many commercial enzymes (antibodies) and certain organic compounds. These types of fungi have several advantages for protein production because in the case of fungi the high level of secretion is a common attribute of their decomposer lifestyle. Filamentous fungi become a vehicle of choice in industrial level production of recombinant protein in eukaryotes. The complexities of understanding the physiologies of such fungi have rapid development and produce efficient cell factories as compared to bacteria.

The antibody components and its fusion proteins produced by fungi are widely used by researchers for further studies and for developing the antibodies from them. The production of single-chain antibody fragment (scFv and V_{HH}) and their applications are seen to be very helpful. The "magic bullet" approach which is the fusion protein strategy is used in industrial applications.

17.7 Antibody Fragments in Human Medicine

17.7.1 The Smaller Fragments Are Productive

The application smaller antibody fragments is seen to be better for diagnosis and therapy in medicine. The use of antibodies as the cancer-targeting reagent has been well established (Hazra et al. 1995; Von Mehren and Weiner 1996; Hudson 1999). The smaller-sized antibody fragments have quicker approval rate in blood circulation and can easily perforate tissues and solid tumors when compared to whole antibodies (Yokota et al. 1992) which recently also were shown for V_{HH}s (Cortez-Retamozo et al. 2002). Nowadays, there are also encouraging trials with antibody fragments as therapeutic and diagnostic agents (Hudson 1999; Maynard et al. 2002). Another use of these antibody fragments is to spoon up viral infections with intrabodies. These antibodies are synthesized intracellularly and targeted specifically to inactivate specific proteins within the cell (Marasco 1995).

17.7.2 "Magic Bullets" in Medicine

Bifunctional molecules limit the dose of drug as it is only active wherever it is required and, thus, have minimum side effects or less immunogenic response. Fusion proteins are known as ideal immune agents used in the treatment for cancer diagnosis (Spooner et al. 1994). Bifunctional antibodies carrying cytotoxic molecules specific to cancerous cells continuously eliminate these cells but do no harm to the normal cells (Boleti et al. 1995).

17.7.3 Applications of V_{HH}s

17.7.3.1 As Drug Carriers

In terms of small size, stability, and feasible production, the V_{HH}s are relevant molecules in diagnosis and therapy in human medicine (Muyldermans 2001). V_{HH}s could specifically target tumor cells with the generation of bispecific constructs (Conrath et al. 2001a). These comprise the major interest for cancer therapy.

17.7.3.2 As Delivery Carriers in the Brain

Treatment of disease associated with the brain is very painful and difficult as antibodies being water-soluble compounds are not able to reach it due to the blood-brain barrier. It has been shown that V_{HH} can selectively transmigrate across the BBB in an in vitro BBB human model and partly in vivo in mice. This characteristic can be used for delivering macromolecules across the human BBB and for treating brain-related diseases by developing antibody carriers (Muruganandam et al. 2002).

17.7.3.3 As Enzyme Inhibitors

Hypervariable regions in V_{HH}s extend deep into the active sites of enzymes and bind to novel epitopes which are otherwise not recognized by conventional antibodies (Desmyter et al. 1996; Vu et al. 1997; Lauwereys et al. 1998; Transue et al. 1998; Harmsen et al. 2000). V_{HH}s were thus considered as better enzyme inhibitors (Lauwereys et al. 1998; Conrath et al. 2001b; Desmyter et al. 2002).

The host that responds to fungal infection is basically the result of a complex interaction that happens between the pathogen and the host's innate immunity and also the adaptive immune system. For the successful fungal infections, cell-mediated immunity is considered to be critical. The hosts that are used for the production of proteins, produced genetically, are basically high protein-secreting mutants that are selected for different purposes, for example, production of cellulose degrading enzymes. Basically they show that the protein secretion occurs through the hyphal tip. There are now evidences to support the statement that the secretion of the protein also takes place in the subapical region in filamentous fungi.

Many attempts have been made to increase the correct folding and the yield of heterologous proteins in the fungi by co-expression of cellular chaperons. The physiological changes that were seen in the fungal strains undergoing stress through protein overexpression under strong gene promoter reflect the challenges to the host organism (Nevalainen and Peterson 2014).

17.8 Mammalian Cell as Antibody Factories

Today, proteins are required to be produced in an appropriate quantity and quality. It is seen that mammalian cells have wide applications in the production of proteins. The main advantage of using mammalian cell expression systems for recombinant protein production is their ability to allow good folding of the protein, posttranslational modifications, and assembly of the product, which are essential for full biological activity. However, it differs from species to species.

Among the many mammalian cell lines used for expressing proteins, Chinese hamster ovary (CHO) cells, HEK 293 (Human embryonic kidney), and mouse myeloma cells, such as NS0 and Sp2/0 cells, are predominant. At present, CHO pro3 and CHO-K1 have given rise to the DUKX-X11 and DG44, which are generally used cell lines in bioprocessing. These two cell lines lack dihydrofolate reductase activity. Most of the recombinant proteins are made in CHO cells, such as CHO-K1, DG44, and DUXB11 lineages (Jones et al. 2003). PER.C6® cells are new types, which are derived from human embryonic retina cells—immortalized by transfecting the E1 genes from DNA of adenovirus 5. Thus, they can proliferate continuously in suspension using serum-free conditions. Other cell lines being used are COS and Vero, which are green African monkey kidney and HeLa (human cervical cancer) cell lines.

Polyethylenimine (PEI) or calcium phosphate can be used to transfect the cell lines. Virus-induced transduction by techniques like the BacMam system, which uses recombinant baculoviruses, can also be used to accomplish protein expression in mammalian cells. Cell lines such as NS0 are tough to transfect, and so electroporation method is generally being used provided the cell line production is stable.

17.8.1 Production of Monoclonal Antibody

Therapeutic proteins such as monoclonal antibody (mAbs) are primarily produced in mammalian cells. African green monkey kidney (COS) cells are used to produce small quantities of mAbs as they lose this ability over time. Hence, murine lymphoid cells (NS0, SP2/0) (which are non-immunoglobulin-secreting types) and CHO cells are preferred over COS cells. Murine NS0 cells are cholesterol auxotrophs; however, cholesterol-independent NS0 cells are also available. NS0 cells do not have glutamine synthetase (GS) enzyme activity, which makes them suitable as a selectable marker for expression of recombinant antibody. It has been shown that non-GS NS0 cell lines have higher productivity of antibodies. A study states that the use of heterologous promoters, enhancers, and amplifiable genetic markers also contributes to an increment in the yield of antibody and antibody fragments.

17.8.2 Cell Line Development

Development of stable cell line is initiated with the construction of expression vector and transfection with plasmids containing genes including those coding for antibody, light and heavy chain. Then the cells are checked for productivity and stability. The major limiting factor is to screen and select such clones in a short period of time.

Mammalian expression vectors are composed of two cassettes: (1) for antibody genes and selectable marker gene(s) and (2) for the genes enabling plasmid multiplication. To obtain high expression, strong promoter like the cytomegalovirus (CMV) and elongation factor alpha (EF1α) are employed. To enhance the transcribed mRNA export to cytoplasm, a sequence in the 5′ untranslated region and a 3′ polyadenylation signal sequences are necessary. Polyadenylation signal sequences commonly used are SV40 and the bovine growth hormone. Translation and secretion are also required in addition to transcription, for antibody production. In order to enhance translation initiation, usually a consensus Kozak sequence is generated by introducing GCC GCC(A/G)CC before the initiation codon, while antibody secretion is assured by placing a signal ahead of the mature antibody peptide. For stable insertion of vector DNA into mammalian cells, a range of transfection methods have been developed but most commonly used are electroporation and lipofection. Selection of transfected cells depends on different selectable markers that can be divided into two groups: metabolic selectable markers (glutamine synthase (GS) and dihydrofolate reductase (DHFR)) and antibiotic selectable markers (puromycin acetyltransferase, Bleomycin resistance gene, hygromycin phosphotransferase, zeocin resistance gene, histidinol dehydrogenase, aminoglycoside phosphotransferase, and blasticidin deaminase).

Recently, "double selection" approach was being explored, in which simultaneous transfection of two plasmids into CHO cells was being carried out for stable antibody cell line production. In this approach both plasmids have their own genes for heavy and light chains, as well as one selectable marker. Two different selective reagents are used to select transfected cells. It was noticed that the productivity of such clones was much higher than the clones transfected with either plasmid alone.

The primary screening is usually an ELISA assay. Additional assays are performed to measure productivity and select the top 12–24 clones, which are checked in a fed-batch cell culture models, while high producers are cultured for scale-up processes. Accreditable features like glycosylation profile, variations in charge, levels of aggregation, heterogeneity of protein sequence, and other culture features including growth, productivity, and stability, which attributes to product quality, are analyzed to select four to six clones.

17.8.3 Engineering Cell Lines

A lot of research work has been carried out to genetically amend or upgrade the product quality and to enhance the robustness of the host cell. Glycosylation control is given crucial attention because glycan structures considerably affect clearance rate and activity of the antibodies. Antibodies produced in CHO cells have low or no bisecting N-acetylglucosamine and high levels of fucosylation. N-Acetylglucosaminyl transferase III can enhance antibody-dependent cytotoxicity by increasing the contribution of bisecting GlcNAc on antibodies. Apart from it, ADCC activity can be enhanced by reduction of fucose on antibody using RNAi and gene deletion technologies. Generally, precise control of glycosylation is difficult in mammalian cells, since it is dependent upon many factors. Other factors of host cell manipulation include methods to reduce apoptosis, lactate accumulation, and engineer cell growth. Culture viability can be enhanced by overexpression of anti-apoptotic gene(s) or knocking down the expression of apoptotic gene(s) which would ultimately consequence into improved productivity. Lactate is one of the by-product of cell culture process, and at times, there is accumulation of this lactate because of the very low efficiency of CHO and continuously cultured cell lines to oxidize glucose into CO_2 and water and thus acidifying the medium resulting in high osmolarity, ultimately leading to low viability. Efforts to reduce lactate accumulation have been done; however, the effectiveness of this approach is dependent on the clone.

There has also been a focus on employing inducible expression systems, which decouples cell growth from the product. Significant increase in cell specific productivity was reported in the growth arrested cultures of NS0 by inducing the expression of p21. Randomly integrating the gene leads to variation in expression levels and has resulted in time-consuming and tedious screening strategy of efficient clones. To overcome the positional dependency and integration of genes, systems like the Cre-lox from P1 virus and Flp/FRT from *S. cerevisiae* are used. Transfection of DHFR-deficient CHO cells with genes encoding loxP-GFP fusion and DHFR, FACS was employed (Kito et al. 2002).

Today, systems like *Pichia pastoris* and *E. coli* have also been found to be promising hosts for mAb secretion. Efficacy of mAbs can be improved by controlling the structure of IgGs and glycan composition. GlycoFi (of Merck & Co) have produced mAbs with highly specific glycoforms using glyco-engineered *P. pastoris* cell lines (Li et al. 2006). They reported a tenfold enhancement in their binding affinity and ADCC activity with the glycan-engineered proteins compared to Rituximab. High-fidelity generation of glycosylated mAbs can be assured by employing *P. pastoris* coupled with high cell density cultures, and upscaling fermentation facility.

Apart from mammalian cell expression system, *E. coli* has also been employed for producing antibody fragments—Fabs—which are used in cases where Fc-mediated effector function is either not needed or prove deleterious (Li et al. 2006). In one of the study, it was demonstrated that complete IgGs can be assembled in the *E. coli* periplasm if there is an effective secretion of heavy and light chains in appropriate ratio. The screening of antibody libraries was then performed by flow cytometry.

In *Aspergillus niger*, mAbs or antibody fragments have also been produced (Ward et al. 2004). With the advancement in technology, utilization of cell-free protein synthesis for producing engineered proteins is also emerging. These strategies have also been used to produce proteins, antibody fragments, and vaccine fusions with *E. coli* cell extracts (Goerke and Swartz 2008).

17.8.4 Limitation of Mammalian Expression System

Proteins obtained through gene expression in mammalian cells are functional due to appropriate glycosylation, but this system is highly expensive and demands complicated technology. Moreover, the possibilities of contamination with animal viruses have been a huge barrier for its use in commercial production. It was initially believed that N-glycolylneuraminic acid (NGNA) produced by mouse-derived cell lines, including NSO, induces a potential immunogenicity concern in humans, but these concerns were alleviated when later it was seen that CHO cells also express low levels of NGNA and it is incorporated into protein from the available diet by humans. Alpha-gal linkages mainly produced by murine cells, including NSO, also trigger human immune response. Thus, industrial use of NSO cells to produce therapeutic antibodics has been limited due to these immunogenicity aspects.

17.8.4.1 Improvement of Mammalian Expression System

There are certain guidelines involved in the improvement of mammalian expression system which includes:

- Designing the vector properly
- Use of strong promoter
- Use of appropriate signal peptide
- Optimization of the selected introns and product gene codon
- Using transcription control regions

Approaches employed to generate cell lines for producing antibodies include gene amplification which is induced by markers like DHFR57 or GS. Genes are expressed at a higher level when vectors possess promoters like cytomegalovirus promoter. Processing of mRNA and its secretion can be enhanced by optimization of codon and signal sequence for the target cell type and GC:AT ratio. In order to improve final production, attachment regions and chromatin-opening elements have also been included into the process of optimization of vector, in addition to gene-targeting technology.

Acknowledgments The authors wish to thank the administration of Jaypee University of Information Technology Waknaghat, Solan (H.P.), for providing necessary infrastructure and facilities. Technical input provided by M. Pradhan and K. Sharma is also acknowledged. Financial assistance provided by DST-SERB project (Sanction Order No. SR/FT/LS-117/2012) is also gratefully acknowledged.

References

Boleti E, Deonarain M, Spooner R, Smith A, Epenetos A, George A (1995) Construction, expression and characterisation of a single chain anti-tumour antibody (scFv)-EL-2 fusion protein. Ann Oncol 6:945–947. doi:10.1155/2015/460264

Both L, Banyard AC, van Dolleweerd C, Horton DL, Ma JK, Fooks AR (2012) Passive immunity in the prevention of rabies. Lancet Infect Dis 12:397–407. doi:10.1016/S1473-3099(11)70340-1

Conrath KE, Lauwereys M, Wyns L, Muyldermans S (2001a) Camel single-domain antibodies as modular building units in bispecific and bivalent antibody constructs. J Biol Chem 276:7346–7350. doi:10.1074/jbc.M007734200

Conrath KE, Lauwereys M, Galleni M, Matagne A, Frère J-M, Kinne J, Wyns L, Muyldermans S (2001b) β-actamase inhibitors derived from single-domain antibody fragments elicited in the camelidae. Antimicrob Agents Chemother 45:2807–2812. doi:10.1128/AAC.45.10.2807-2812.2001

Cortez-Retamozo V, Lauwereys M, Gobert M, Conrath K, Muyldermans S, De Baetselier P, Revets H (2002) Efficient tumor targeting by single-domain antibody fragments of camels. Int J Cancer 98:456–462. doi:10.1002/ijc.10212

de Marco A (2015) Recombinant antibody production evolves into multiple options aimed at yielding reagents suitable for application-specific needs. Microb Cell Factories 14:125. doi:10.1186/s12934-015-0320-7

Desmyter A, Spinelli S, Payan F, Lauwereys M, Wyns L, Muyldermans S, Cambillau C (2002) Three camelid VHH domains in complex with porcine pancreatic α-amylase inhibition and versatility of binding topology. J Biol Chem 277:23645–23650. doi:10.1074/jbc.m202327200

Desmyter A, Transuei TR, Ghahroudil MA, Thil M-HD, Poortmans F, Hamersz R, Muyldermans S, Wyns L (1996) Crystal structure of a camel single—domain VH antibody fragment in complex. Nat Struct Biol 3(9):803–811. doi:10.1038/nsb0996-803

Giddings G, Allison G, Brooks D, Carter A (2000) Transgenic plants as factories for biopharmaceuticals. Nature Biotechnol 18:1151–1155. doi:10.1038/81132

Goerke AR, Swartz JR (2008) Development of cell-free protein synthesis platforms for disulfide bonded proteins. Biotechnol Bioeng 99:351–367. doi:10.1002/bit.21567

Harmsen MM, Ruuls RC, Nijman IJ, Niewold TA, Frenken LG, de Geus B (2000) Llama heavy-chain V regions consist of at least four distinct subfamilies revealing novel sequence features. Mol Immunol 37:579–590. doi:10.1016/s0161-5890(00)00081-x

Hazra D, Britton K, Lahiri V, Gupta A, Khanna P, Saran S (1995) Immunotechnological trends in radioimmuno targeting: From 'magic bullet' to 'smart bomb'. Nucl Med Commun 16:66–75

Hempel F, Maier UG (2012) An engineered diatom acting like a plasma cell secreting human IgG antibodies with high efficiency. Microb Cell Factories 11:126. doi:10.1186/1475-2859-11-126

Hudson PJ (1999) Recombinant antibody constructs in cancer therapy. Curr Opin Immunol 11:548–557. doi:10.1016/s0952-7915(99)00013-8

Jones D, Kroos N, Anema R, Van Montfort B, Vooys A, Svd K, Evd H, Smits S, Schouten J, Brouwer K (2003) High-level expression of recombinant IgG in the human cell line PER. C6. Biotechnol Prog 19:163–168. doi:10.1021/bp025574h

Kito M, Itami S, Fukano Y, Yamana K, Shibui T (2002) Construction of engineered CHO strains for high-level production of recombinant proteins. Appl Microbiol Biotechnol 60:442–448. doi:10.1007/s00253-002-1134-1

Lauwereys M, Ghahroudi MA, Desmyter A, Kinne J, Hölzer W, De Genst E, Wyns L, Muyldermans S (1998) Potent enzyme inhibitors derived from dromedary heavy-chain antibodies. EMBO J 17:3512–3520. doi:10.1093/emboj/17.13.3512

Li H, Sethuraman N, Stadheim TA, Zha D, Prinz B, Ballew N, Bobrowicz P, Choi B-K, Cook WJ, Cukan M (2006) Optimization of humanized IgGs in glycoengineered Pichia Pastoris. Nature biotechnol 24:210–215. doi:10.1038/nbt1178

Marasco WA (1995) Intracellular antibodies (intrabodies) as research reagents and therapeutic molecules for gene therapy. Immunotechnology 1:1–19

Maynard JA, Maassen CB, Leppla SH, Brasky K, Patterson JL, Iverson BL, Georgiou G (2002) Protection against anthrax toxin by recombinant antibody fragments correlates with antigen affinity. Nature biotechnol 20:597–601. doi:10.1038/nbt0602-597

Muruganandam A, Tanha J, Narang S, Stanimirovic D (2002) Selection of phage-displayed llama single-domain antibodies that transmigrate across human blood–brain barrier endothelium. FASEB J 16:240–242. doi:10.1096/fj.01-0343fje

Muyldermans S (2001) Single domain camel antibodies: current status. Rev Mol Biotechnol 74:277–302. doi:10.1016/s1389-0352(01)00021-6

Nevalainen H, Peterson R (2014) Making recombinant proteins in filamentous fungi-are we expecting too much? Front Microbiol 5:75. doi:10.3389/fmicb.2014.00075

Reichert JM (2012) Marketed therapeutic antibodies compendium. mAbs 4:413–415. doi:10.4161/mabs.19931

Robinson M-P, Ke N, Lobstein J, Peterson C, Szkodny A, Mansell TJ, Tuckey C, Riggs PD, Colussi PA, Noren CJ (2015) Efficient expression of full-length antibodies in the cytoplasm of engineered bacteria. Nat Commun 6:8072. doi:10.1038/ncomms9072

Spooner RA, Murray S, Rowlinson-Busza G, Deonarain MP, Chu A, Epenetos AA (1994) Genetically engineered antibodies for diagnostic pathology. Hum Pathol 25:606–614

Transue TR, De Genst E, Ghahroudi MA, Wyns L, Muyldermans S (1998) Camel single-domain antibody inhibits enzyme by mimicking carbohydrate substrate. Proteins Struct Funct Genet 32:515–522

von Mehren M, Weiner LM (1996) Monoclonal antibody-based therapy. Curr Opin Oncol 8:493–498

Vu KB, Ghahroudi MA, Wyns L, Muyldermans S (1997) Comparison of llama V H sequences from conventional and heavy chain antibodies. Mol Immunol 34:1121–1131. doi:10.1016/s0161-5890(97)00146-6

Ward M, Lin C, Victoria DC, Fox BP, Fox JA, Wong DL, Meerman HJ, Pucci JP, Fong RB, Heng MH (2004) Characterization of humanized antibodies secreted by aspergillus niger. Appl Environ Microbiol 70:2567–2576. doi:10.1128/aem.70.5.2567-2576.2004

Yokota T, Milenic DE, Whitlow M, Schlom J (1992) Rapid tumor penetration of a single-chain Fv and comparison with other immunoglobulin forms. Cancer Res 52:3402–3408

Remedial Role of Nanocomposite as Photocatalysts, Adsorbents, and Disinfectants in Aqueous System and Their Biomedical Applications

18

Arush Sharma, Swadeep Sood, and Deepak Pathania

Abstract

With increase in population, the development of inventive materials and technologies that provide drinkable water are major research areas with emphasis mainly on public health and environmental quality. Recently, the progresses in nanoscale and engineering science recommend that many of the issues involving water contamination could be easily resolved or improved using nanomaterials. A nanoparticle epitomizes the crucial material for the fabrication of nanomaterials. Their basic properties like small size and high surface area afford improved method for the sequestration of different pollutants from water. Nanoparticles were amalgamated into various biomaterials. The primary advantage of these materials includes mosaic-like structure and porosity which contains hydrophilic and hydrophobic phases and facilitates in the elimination of polar and nonpolar substances from aquatic system. Thus, nanocomposite consisting of inorganic-organic components signifies a fast-rising area of research and shows improved properties as compared to bare nanoparticles. The amalgamation of inorganic nanofiller into the organic matrix affords remarkable thermal, mechanical, rheological, electrical, photocatalytic, adsorptional, optical, and antibacterial properties. This chapter explains a brief review on the different modes of synthesis and diverse applications of nanocomposites such as photocatalysis, adsorption, and disinfectants for water detoxification and their character in biomedical science.

Keywords

Nanocomposite • Nanoparticles • Photocatalysts • Adsorbents • Disinfectants Detoxification

A. Sharma • S. Sood
School of Chemistry, Shoolini University, Solan 173 212, Himachal Pradesh, India

D. Pathania (✉)
Department of Environmental Sciences, Central University of Jammu, Jammu, Jammu and Kashmir 181143, India
e-mail: dpathania74@gmail.com

V.C. Kalia, A.K. Saini (eds.), *Metabolic Engineering for Bioactive Compounds*,
DOI 10.1007/978-981-10-5511-9_18

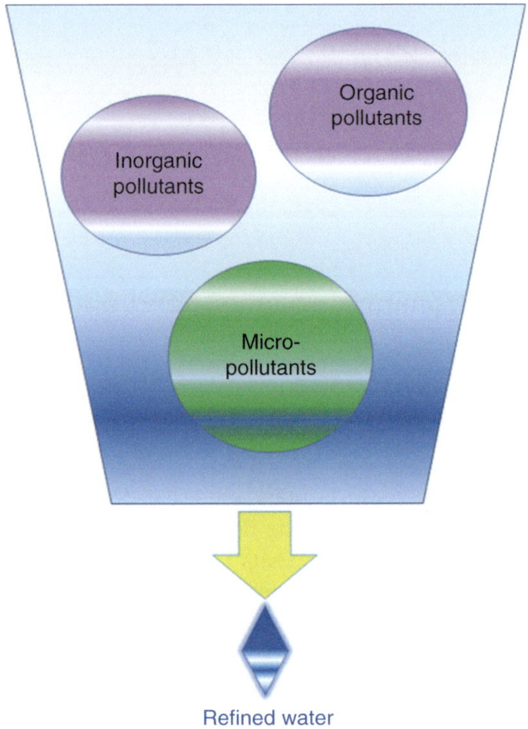

Fig. 18.1 Principal pollutants in water system

18.1 Introduction

The incredible advancement in science has absolutely elevated the standards of the human life and health recently but also resulted in the congregation of certain problems. Water contamination is one of the leading and extreme disturbing problems that need attention. In developing countries, the nonavailability of economical water treatment methods additionally intensifies the condition (Cardoso et al. 2011; Gupta et al. 2012). Consequently, environmental adulteration has been emerged as a most exigent problem and reached a stage where it should be seriously examined. It is pervasively connected with entire biosphere and poses a severe peril to human being as well as aquatic system (Pathania and Sharma 2012; Gupta et al. 2014). However, among the various types of pollutions, various researchers have been paying attention to water pollution (Fig. 18.1). Due to utmost eminence of water quality, the necessity of its improvement and preservation is growing continuously. The point and nonpoint sources are unceasingly contaminating our valuable resources. The contamination diminishes its utility for intended use by altering its physicochemical and biological parameters. The water quality of resources is worsening day by day due to the continuous addition of detrimental pollutants (Siddiqi and Pathania 2003; Pathania et al. 2016; Sharma et al. 2016). The chief sources of water pollution include industrialization, civilization, agricultural activities, and various global changes. Since industrial development proves to be beneficial in several ways but caused a nuisance to us in the form of pollution, which is gaining widespread civic attention (Sharma et al. 2014a, b; Awual et al. 2015). Inorganic and organic pollutants usually exist in the contaminated surface and groundwater system. These are regarded as serious threats to the environmental safety and public health issue, even when present in very low concentration (Hamdy et al. 2012; Kumar et al. 2014). Thus, the effluents emitted from various industrial activities containing tenacious pollutants are root for water pollution. It reduces the light penetration that hampers photosynthetic activity in the aquatic system.

If the level of these pollutants in water exceeds the permissible limit, then water is considered unsafe, which is aesthetically objectionable for drinking, irrigation, industrial applications, and other purposes (Chowdhury et al. 2011; Sharma et al. 2014a, b). The contamination of aquatic environment by pollutants like polycyclic aromatic hydrocarbons, phenolic compounds, agricultural chemicals (organic pesticides and organic herbicides), synthetic dyes, and heavy metals is a serious problem, due to their negative toxicological effects (Kant et al. 2013; Pathania et al. 2014). The presence of hazardous heavy metals initiates a variety of ailments in nervous system, secretory system, skin, liver, and bones (Nanseu-Njiki et al. 2010; Chen et al. 2011; Pathania et al. 2011). Although enormous initiatives have already been taken to tackle aforesaid problem, highly rigorous research is still required in this direction. Therefore, the solution of this pollution problem usually focuses on the decrease in contaminants at the source level and treatment of it prior to use (Gupta et al. 2013; Pathania et al. 2015). With the rising awareness, such hazardous effluents often need to be treated before being discharged into water sources.

So, to tackle the aforementioned intimidation, various materials have been invented by researchers. Among these, nanomaterials are of immense interest due to their utility in different fields of chemical, biological, and environmental sciences (Ahmad et al. 2010; Chen et al. 2011). Nanocomposites signify materials with enhanced performance (Katwal et al. 2015). Modifications on the surface of inorganic nanoparticles result in interface between inorganic and organic parts (Shahat et al. 2015). These nanocomposite materials act as good sensors, membrane electrodes, adsorbents, ion exchangers, and photocatalysts (Peng et al. 2006; Taniguchi et al. 2008; Jeon and Baek 2010; Sanghavi et al. 2013). Therefore, this article focuses on methods of synthesis and surface amendment of nanoparticles leading to improved organic-inorganic nanocomposite. The application of nanocomposites for water remediation and their role as photocatalyst, adsorbent, and disinfectants in aqueous system are also discussed.

18.2 Methods of Synthesizing of Nanoparticles

18.2.1 Sol-Gel Technique

Various nanoparticles have been synthesized by this technique. This process involves transformation of liquid "sol" phase to a solid "gel" phase (Zhang et al.

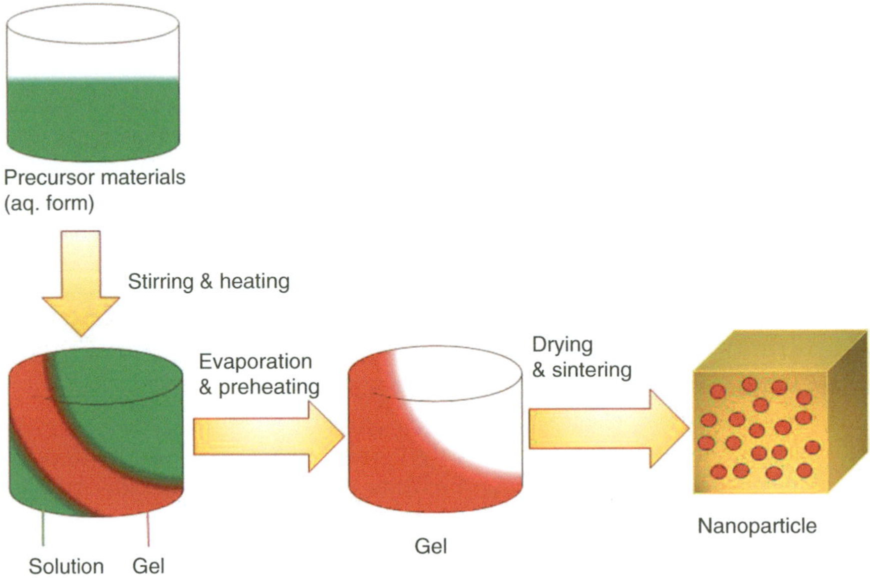

Fig. 18.2 Synthesis of nanoparticles by sol-gel technique

2004). Sol-gel method was preferred when a large production of nanomaterials is required (Qin, 2007). Nanoparticles can be processed inside the polymer and result in interpenetrating networks which exist between inorganic and organic parts. The network enhances the compatibility between constituents and interaction of the phases. Spherically shaped zinc oxide nanoparticles were reported using sol-gel method in the presence of TEA (triethanolamine) as a surfactant (Vafaee and Ghamsari, 2007). The consequence of various factors such as AcOH to water ratios, calcination temperature, time, etc. on the photocatalytic activity was evaluated (Behnajady et al. 2011). This process has been magnificently used for the preparation of nanocomposites in diverse polymer matrices (Brinker and Scherer, 2013). In this method, the polymerization of organic functional groups with nanomaterial using free radical or cationic polymerization occurs. A sol-gel method was used to produce the PS/silica nanocomposites (Hsiue et al. 2000a, b). The condensation of a precursor, such as tetraethyl oxysilane (TEOS), tetrabutyl titanate, or aluminum isopropoxide, has been carried out with functional organic polymer like polyvinyl acetate, polymethyl methacrylate, polyether imide, polyvinyl alcohol, polyamides, polyimide, etc. (Suzuki et al. 1990; Du et al. 2007; Silveira et al. 1995; Nunes et al. 1999; Hsiue et al. 2000a, b; Crivello et al. 2001; Sengupta et al. 2005). Du et al. (2007) reported ZnO/polyvinyl pyrrolidone nanocomposite thin film using sol-gel as a sensor. Hu and Marand (1999) described in situ synthesis of TiO_2 nanodomains within poly(amide-imide) by a sol-gel technique. The composite films showed an outstanding optical transparency. Nylon-6/SiO_2 nanocomposites were investigated by using a sol-gel method (Garcia et al. 2004). Figure 18.2 indicated the basic mechanism involved in sol-gel technique.

18.2.2 Hydrothermal Method

This method is carried out at high vapor pressure of 0.3–4 MPa and temperature (130–250 °C) (Wu et al. 2008). In this process the nanoparticles formed possess better crystallinity as compared to obtained in other methods. The particle size is found to increase with time, temperature, concentration of stabilizer, and with decrease in pH (Williams et al. 2007). The semiconductor nanoparticles such as copper sulfide, CdTe, ZnO, SnO_2, CdS, and CdS/ZnO have been prepared by hydrothermal technique method (Yan et al. 2002; Yang et al. 2006; Aneesh and Jayaraj 2010), etc.

18.2.3 Microemulsion Method

It is a thermodynamically stable system which includes a surfactant molecule along with immiscible components (water and oil). The surfactant helps in lowering the tension between water and oil and thus forms a transparent solution. The water-in-oil microemulsion is a method where the dispersion of aqueous phase nanosized droplets is covered by a phase of surfactant molecules. These droplets show a dynamic exchange of the contents and thus facilitate the reaction. The controlled crystallites were synthesized using this micellar exchange. The semiconductor nanoparticles – CdS, PbS, CuS, Cu_2S, and CdSe – are prepared by microemulsion method (Eastoe and Warne, 1996; Malik et al. 2012).

18.2.4 Chemical Precipitation

Chemical precipitation is used to prepare various semiconductor nanoparticles. ZnO nanoparticles were prepared by simple coprecipitation in the presence of thioglycerol and methanol, which acted as a capping agent and solvent, respectively (Kripal et al. 2011). The crystal structure determined by X-ray diffractometry is found to be hexagonal in nature (Sankara Reddy et al. 2013). SnO_2 tetragonal rutile nanoparticles and titanium dioxide-doped nanoparticles were prepared by coprecipitation (Chauhan et al. 2010; Rao et al. 2011; Kumar et al. 2013; Naje et al. 2013).

18.2.5 In Situ Polymerization

Nylon-6/potassium titanate and nylon-6/silica, transparent polymer nanocomposites containing ZnS nanoparticles and nanocomposites were prepared by this approach (Ou et al. 1998; Guan et al. 2009; Cheng et al. 2008, 2009). Colloidal suspension of the magnetic nanocomposites of $NiO.5ZnO.5Fe_2O_4$ prepared by wet chemical method was followed by in situ polymerization (Jiang, 2007). In situ approach for the synthesis of Ag nanoparticles was presented in Fig. 18.3.

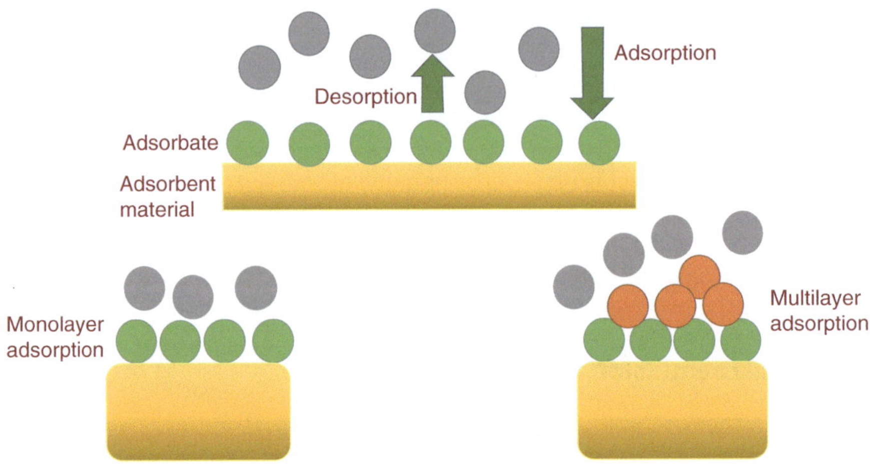

Fig. 18.3 Pictorial demonstration for adsorption process

18.2.6 Other Methods

The other methods such as sol-gel autocombustion, solid-state metathesis, chemical aerosol, spray pyrolysis, and electrochemical for the synthesis of semiconductor nanoparticles have also been reported. Sol-gel autocombustion preparation of samarium-doped TiO_2 nanoparticles and their photocatalytic activity was attempted by Xiao et al. (2007). Didenko and Suslick (2005) studied the chemical aerosol flow technique for preparation of CdS, CdSe, and CdTe nanoparticles. Treece et al. (1993) explored the solid-state metathesis method for synthesis of III–V semiconductors.

Hwang et al. (2007) investigated the GaP nanoparticles from reduction method. Other methods included for synthesis of group III–V (GaN, GaAs, InP, GaP, etc.), group IV [Si (silica coated), Ge], and group II–VI (CdS, ZnO, CdSe, CdTe, etc.) nanoparticles include combustion of silane, phase solution reaction ultrasonic reduction, gas-aerosol, thermolysis, and arrested precipitation were also been used for (Kornowski et al. 1993; Jegier et al. 1998).

18.2.7 Solvent Casting

A widely used method for producing cellulose-based nanocomposites is the solvent casting, which was initially reported in 2008 (Fu et al. 2008). In water-soluble polymers, the strong interactions between the nanocellulose and their matrices are because of the presence of polar constituents. Solid nanocomposite film was produced by solvent evaporation technique (casting). CNC enhanced the mechanical strength of the different matrices (Williams et al. 2008a, b).

18.2.8 Nanocomposites

Combining nanotechnology with biology results in enhancing the possibilities of developing new materials in the nanorange (1–100 nm) that find potential applications in diverse fields. These are the multiphase structures, and precursor materials may be organic, inorganic, or both. Generally, nanoparticles are modified with polymers to enhance their properties.

18.2.8.1 Inorganic Nanocomposite

The surface of nanoparticles can be altered using chemical method, and it frequently magnifies the dispersion stability of nanoparticles. The surface of metal oxide nanoparticles was usually reformed by reactions with various metal alkoxides, epoxides, aryl isocyanates, and propylene oxide. Guo et al. (2008) reformed the silica nanoparticles with a silane coupling agent such as 3-(trimethoxysilyl) propyl methacrylate (MPS) and perceived that grafting ratio of MPS onto nanosilica surface was augmented with MPS amount.

18.2.8.2 Carbon Nanocomposites

Carbon nanostructures such as carbon nanotubes, nanorods, fullerene, and graphene have been found to be excellent candidate materials that possess the projecting mechanical, thermal, electrical, and structural physiognomies which have been drawn copious attention. They lead to the foundation of nanocomposites with amended catalytical activities. Hence, graphene is a numerous atomic-layered graphite, captivating a novel stream of two-dimensional carbon nanocomposite. Their synthesis not only entails the good-quality production of graphene nanosheets (GNSs), but effective integration of desirable matrices was also required. The reduced graphene nanosheets tend to aggregate because of their robust van der Waals interactions. Therefore, electrostatic stabilization and chemical functionalization have been efficiently employed in order to inhibit the aggregation of exfoliated graphene nanosheets (Wang et al. 2008a, b).

18.2.8.3 Quantum Dots

Quantum dots (QDs) are usually semiconductor nanoparticles, which hold nanosize of less than 10 nm diameter (Mansur 2010). So, their optical and electronic properties were different from larger particles. They were first discovered in 1983 as small spheres in a colloidal solution. Quantum dots display remarkable tunable properties that are found in between of bulk semiconductor material and discrete molecules. Their optoelectronic properties usually change with their size and shape. Thus, recently various researchers have presented a large number of investigations on QDs (Gaponik and Rogach 2010). Various approaches have been exploited for the synthesis of QDs. While, colloidal preparation technique is widely explored for fabricating the QDs. Selenium compound in organic solvents, inoculated into reaction vessel at room temperature, caused the supersaturation of CdSe solution mixture. If temperature dribs to 290 °C, then nucleation of new crystals stops and existing crystals propagate. After some time, lesser quantity of zinc sulfide was inserted into the reaction vessel for coating the QDs and averts their reaction with environment. Mädler et al. (2002) exploited the synthesis of ZnO nanocrystallites (1.5 nm) using spray combustion method. These nanocrystallites showed a quantum size effect, and QDs agreeably obeyed the linearity between particle size and optical bandgap.

18.2.8.4 Magnetic Nanocomposite

Recently, magnetic separation technique has been employed for various environmental dilemmas. Magnetic nano- and microparticles have been selectively detached from the aqueous phase by an external magnetism. Nanoparticles and magnetic nanocomposite must be characterized as "stimuli-responsive substance" because these composite materials showed the response to applied magnetic field. The magnetic nanoparticles have been found copious of application in various sectors such as medicine, biotechnology, and environmental detoxification. The mainstream of biological materials possesses diamagnetism which permits the proficient departure of magnetic and magnetically reformed materials. Magnetic nanocomposites have been targeted and kept there through applied magnetism (Safarik and Safarikova, 2009).

The magnetic carbon nanocomposite has exposed to be very auspicious technique for the departure of solid-aqueous phase and received a great concern due to their probable adsorption applications for the amputation of disastrous pollutants (Zhang et al. 2011). Carbon matrix eludes the accumulation of metal oxide nanoparticles, and magnetic separability method has been found to be a substitute over conventional techniques such as filtration or centrifugation, and it also avoids the loss of nanocomposite (Zhu and Diao, 2011). Magnetic separation has been considered as quick and effective technique. Functionalized magnetic nanocomposites have distinct advantages over conventional materials due to their selective adsorptivity, strong magnetic responsiveness, favorable water dispersibility, and benign biocompatibility. Magnetic nanocomposites have been utilized in several fields such as magnetic resonance imaging, bioscience, analytical chemistry, and pollutant deduction (Guo et al. 2009; Arsalani et al. 2010; Zhang et al. 2010; Erathodiyil and Ying, 2011). Carbon-based magnetic nanocomposite (Fe_3O_4–C) has a distinct structure with functionalized carbon on outer side, and magnetic material forms an inner core. The polysaccharide surface layer contains abundant functional sites such as carboxylic, formyl, and hydroxyl groups. This dense outer layer protects the inner Fe_3O_4 microspheres from dissolved acidic solutions (Wang et al. 2010a, b). The magnetic nanocomposites have been employed as cellular labeling, reclamation of biological fluids, selective drug delivery, magnetic resonance imaging, hyperthermia, tissue reparation, and magnetofection (Yuan et al. 2008). The nanocomposite efficiently diminishes the wastage of drug, frequency of drug administration, and side effects including continued drug delivery to anticipated organ.

18.2.9 The Role of Nanocomposite as Photocatalyst

Since the revolutionary work in the 1970s, the semiconductor-based nanocomposites have revealed high proficiency for the catalytic degradation of pollutants from aqueous sector (Chen et al. 2008; Mai et al. 2008; Lu et al. 2009). Photocatalysis using heterogeneous semiconductor appeared as the most agreeable mean and is found to be economical and nontoxic, with higher surface area, broad absorption spectra with greater absorption, and capability for multi-electron transfer.

Photocatalysis is an operative and competent advanced oxidation processes (AOPs) which can be employed to detoxify the wastewaters emitted from various industrial processes with negligible waste. Photocatalysis typically ensued with the illumination of solar light. The process of semiconductor photodegradation fundamentally involves the ensuing steps. The solar radiation corresponding to certain wavelength is allowed to incident onto semiconductor. If the energy of incident radiation is equivalent to band energy of photocatalyst, electrons might be excited from the valence to the conduction band of photocatalyst and holes left in the valence shell. These electrons and holes endure successive oxidation and reduction reactions with active compounds, which were adsorbed onto the semiconductor surface to produce the required products (Kumar et al. 2015).

Various heterogeneous semiconductors including TiO_2, V_2O_5, ZnO, CeO_2, WO_3, ZrO_2, Fe_2O_3, and sulfides have been used as photocatalyst (Junwu et al. 2006; Tayade et al. 2009). Among these, titanium dioxide (TiO_2) has been emerged as most persuasive because of its photostability, ease of availability, biological inertness, low energy consumption, low operation temperature, high photocatalytic activity, higher chemical stability, water insolubility, and avoidability of the development of unwanted side products (Li et al. 2006a, b, c; Zhang et al. 2008; Skocaj et al. 2011). The surface properties like crystal structure, particle size, surface area, porosity, hydroxyl density, and band energy gap play a vital role for the photocatalytic activity of TiO_2. The size of photocatalyst is rightly connected to the efficacy of a catalyst. Lesser size of photocatalyst intensifies the specific surface area, and hence it augments the number of active sites, leading to higher catalytical activity (Birnie et al. 2006). Figure 18.4 demonstrates the mechanism for photocatalytic oxidation of organic pollutants under UV light. Photocatalytic activities depend on a number of variables.

An analysis of influential parameters revealed that light intensity distribution, dissolved oxygen, photocatalyst loading, temperature, and hydrogen peroxide and air flow rates have positive influence on the removal efficiency (Saquib and Muneer 2002; Kasprzyk-Hordern et al. 2003; Sahoo et al. 2005; Gupta et al. 2006; Bahnemann et al. 2007). On the other hand, factors such as initial concentration of reactants, light wavelength, irradiation time, feed flow rate, and pH displayed negative effects (Galindo et al. 2001; Georgiou et al. 2003; Naldoni et al. 2011). More clarification on their effect onto performance of photocatalyst has been presented in the succeeding sections.

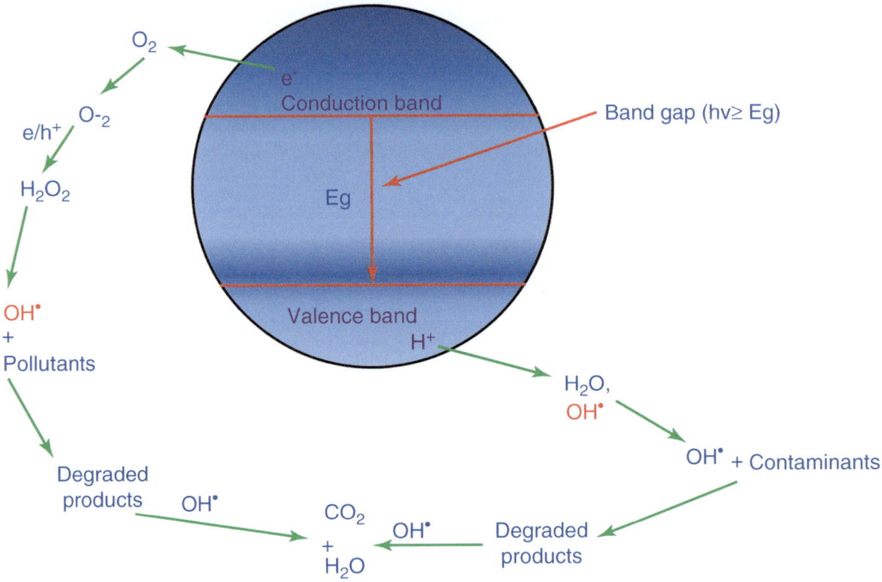

Fig. 18.4 Schematic diagram for photocatalysis

18.2.9.1 UV Radiation

The intensity of light decides the magnitude of radiation absorbed by semiconductor catalyst. The rate of photocatalysis and electron-hole creation is mainly dependent on light intensity (Cassano and Alfano 2000). The intensity of UV radiation is mainly altered by their distance from UV source. Principally, catalytically degradation competency boosted with the intensity of UV radiation, so that greater intensity of light heightened the excitation of photocatalyst to produce electron-hole pairs. The holes can decompose the impurities adsorbed on the surface of photocatalyst and oxidize the molecules of water into free radicals. Furthermore, hydroxyl radicals eradicate the pollutants from aqueous phase. Photocatalytic efficiency mainly depends on the radiation absorbed by surface of photocatalyst (Gaya and Abdullah 2008). UV radiation governs the quantity of photon absorbed by a catalyst. As power of UV lamp increases, catalyst absorbs more photons from solution mixture, and more electron-hole pairs are produced onto catalyst surface. So, concentration of hydroxyl radical increases and, consequently, degradation rate improved.

18.2.9.2 Influence of Dissolved Oxygen

Dissolved oxygen (DO) delivers an imperative fact in the degradation of obstinate pollutants. It has been observed that existence of oxygen might hamper the photodegradation rate, but it depends on the mechanism involved in the degradation process. Oxygen does not affect the adsorption onto TiO_2 surface, as reduction reaction ensued on different site from where oxidation occurs (Gaya and Abdullah 2008). The DO intensifies the separation of photo-generated electrons, holes, and OH radical. Pure oxygen has greater impact than air for the augmentation of refinement (Nam et al. 2002).

18.2.9.3 Influence of Flow Rate of the Feed

Feed flow rate has reverse effect with the retention time and rightly disturbs the degradation rate. Mostly, enhanced flow rate declines the rate of the reaction, though, in photocatalytic reactors, other parameters as cited above direct the degradation rate, and in particular circumstances, the flow rate has no influence, while for the degradation of indole, enhancement in flow rate from 0.5 to 4.03 mL/s drops the degradation of indole from 93% to 58% (Nam et al. 2002). It also affects the accumulation of catalyst particles under nonsupported media. It has been observed that with increase in flow rate, progress of degradation reaction was facilitated due to diffusion between pollutants and TiO_2 catalyst. Nevertheless, the effect of reduction in residence time appears to be predominant than the enhancement of diffusion process, causing decline in proficiency.

18.2.9.4 Influence of Pollutant Concentration

Photocatalytic process mainly ensued on the surface of photocatalyst. It is assumed that augmented concentration of pollutants influences the quantity of light which penetrates into the contaminant solution. Therefore, at higher initial concentration, penetration of light was condensed and fewer photons could reach the surface of the catalyst. The amount of contaminants adsorbed on the surface of photocatalyst is crucial, as it only contributes to the photocatalytic process (Guettai and Amar 2005). It has been investigated that increase in the initial concentration of MeO from 5 to 75 mg/L diminished the photocatalytic removal of MeO from 100 to 70%. Similar outcomes have been stated for the subtraction of RO16 from aqueous system (Li et al. 2006a, b, c).

18.2.9.5 Consequences of Irradiation Time

Photodegradation reaction linearly fitted using pseudo-first-order kinetics and degradation rate drops with time. In addition, competition for degradation occurred between reactants and the intermediate products. Slow degradation rate after certain time is primarily accredited to the hindered reaction of aliphatic compounds with $^{\bullet}OH$ radicals and lesser lifetime of photocatalyst. The active sites are usually deactivated by deposition of strong side products (Herrmann 1999).

18.2.9.6 Influence of Ozonation

A synergetic amalgamation of $UV/TiO_2/O_3$ is another form of photocatalytic oxidation. In this route, O_3 is utilized for the decontamination, and it possesses the higher scavenging effect. Another affirmative impact of ozone must be attributed to decline in the de-excitation rate of electrons and eradication of deposits. Moreover, mutual process delivers the sterilizing conditions. In the occurrence of ozone, process of obliteration became hastened. Therefore, synergistic influence has been observed upon organic pollutants. Degradation rate in the occurrence of ozone utilizing visible radiation has gained lesser impact (Colombo et al. 2012).

18.2.9.7 Effect of Hydrogen Peroxide

Hydrogen peroxide (H_2O_2) absorbs the UV radiation and O–O bond is broken. It leads to the generation of $^{\bullet}OH$ radical and atomic oxygen. Addition of H_2O_2 intensifies the formation of hydroxyl radical in the solution mixture, which impedes the electron-hole recombination process. But higher H_2O_2 amount declines the degradation rate, because hydroxyl radical possesses scavenging effect (Tayade et al. 2009).

18.2.9.8 Effect of pH

pH of the water system regulates the surface charge of photocatalyst (Bahnemann et al. 2007). The pH of aqueous phase affects the surface properties of photocatalyst and ionization (pK_a) of organic pollutants. The protonation and deprotonation of titania has been ensued under acidic or alkaline conditions. The titania surface gets negatively charged in alkaline medium (pH > 6.9) and positively charged under acidic medium (pH < 6.9). Titanium dioxide (TiO_2) possesses enhanced oxidizing properties at lower pH, while at low pH, surplus H^+ ions are reducing the degradation rate. Electrostatic interactive mechanism generated between solvent molecules, semiconductor, and radicals during the process of photocatalysis is vigorously reliant onto pH of aqueous solution.

18.2.9.9 Influence of Air Flow Rate

It has a positive influence upon the obliteration of pollutants. Therefore, the mixing between photocatalyst and contaminants is boosted, and higher mass transfer rates were attained. But, in a fluidized bed reactor, when the flow rate is amplified to 1.5 L/min, then degradation efficiency is deteriorated. The large amount of bubbles obstructs the absorbance of UV light by photocatalyst (Nam et al. 2002). While air flow rate is connected to the quantity of DO in solution mixture, it helps to elude the recombination of electron-hole pairs (Ling et al. 2004).

18.2.9.10 Effect of Temperature

The photocatalytic process is meticulously associated with the temperature. With increase in temperature, the generation of free radicals increases. It also caused the effective competition between electron-hole and oxidation rate, thus increasing at interface. But, it declines the adsorption capacity allied with organics and dissolved oxygen (Wang and Shui 2007). The best temperature range for the photodegradation of obstinate pollutants was found to be 20–80 °C.

18.2.10 Photocatalytic Activity of Nanocomposite

Nanocomposites based on hybrid organic-inorganic have received great attention due to their integrity and deliver the reciprocated feature of organic counterpart with inorganic component. The dual advantages of inorganic matter such as rigidity, thermal stability, and organic material such as flexibility, dielectric, and ductility have been encountered during organic-inorganic nanocomposite. However, in polymer nanocomposites, the lesser size of fillers proliferates the interfacial area in comparison to the conventional composites. This interfacial area additionally produces the substantial volume, and their properties are usually different from the bulk polymer (Schadler et al. 2007; Schaefer and Justice 2007). The surfaces of metal oxide nanoparticles are generally hydrophilic, while polymeric materials possess hydrophobicity. Thus, surface of inorganic nanoparticles needs alterations through coupling agents. So, compatibility between polymeric and inorganic phase has been indorsed. Polymeric materials are the excellent precursor materials for metal oxide nanoparticles and semiconductor photocatalyst. These hybrid nanomaterials retain various characteristics including better strength, stiffness, robustness, corrosion resistance, low density, and thermal protection (Carbonaro et al. 2013).

Recently, graphene-based nanocomposite which is a single two-dimensional carbon sheet having same structure as graphite has been initiated due to astonishing electrical, thermal, and mechanical properties. Several graphene-inorganic nanocomposites have been efficiently synthesized and demonstrate the excellent combinations of properties that are not found in individual components (Paek et al. 2008; Williams et al. 2008a, b; Wang et al. 2010a, b).

Recently, huge numbers of graphene metal oxide nanocomposite including ZnO, WO_3, CuO, Cu_2O, Mn_3O_4, Mn_2O_3, SnO_2, Bi_2WO_6, $ZnWO_4$, Bi_2MoO_6, $BiVO_4$, $BaCrO_4$, $CoFeO_4$, etc. have been studied for photodegradation of numerous noxious pollutants (An et al. 2012; Chandra et al. 2012; Gao et al. 2012; Min et al. 2012a, b; Zhou et al. 2012; Yang and Xu 2013; Yao et al. 2013; Yusoff et al. 2013). The graphene composites with metal nanoparticles exhibited higher photocatalytic activity. Wang et al. (2013) verified that graphene-SnO_2 nanocomposite enhanced the photocatalytic activity of the SnO_2 for the removal of pendimethalin, which was ascribed to the electron transfer from the excited pendimethalin onto SnO_2. Xu et al. (2011) explored the ZnO/graphene nanocomposite for their efficient photocatalytic activity.

18.2.11 Nanocomposite as an Adsorbent

Adsorption techniques have been extensively utilized for the amputation of various noxious wastes from water system. Often, various biomaterials have been established as effective and less expensive adsorbents. However, these low-cost bio-adsorbents often possess low adsorption capacity. So, there is a need to develop inexpensive, easily available, durable, and highly operative adsorbents. The schematic design for adsorption method has been provided in Fig. 18.4.

Currently, activated carbons (AC) possess greater surface area and admirable adsorption capacity, which has engrossed copious attention in the field of environmental reclamation (Ania et al. 2005; Lin et al. 2013). Owing to the efficiency, low cost, and environmental protection, AC has demonstrated as effective adsorbent for the elimination of various pollutants from wastewater. But isolation of AC from aqueous phase poses a great problematic concern, when it was exhausted in wastewater system. An outmoded separation technique, such as filtration process, caused the blockage of filters, and it leads to secondary pollution. Hence, AC-magnetic nanocomposite has engrossed the widespread civic courtesy from globally circulated water, which might be detached using magnetic separation techniques (Zhang et al. 2007; Park et al. 2013). Different types of carbon nanocomposites are used in different fields such as adsorption, batteries, etc. Metal oxide nanoparticles have plenty of applications in various fields such as environmental, catalysis, batteries, sensors, magnetic storage media, and magnetic resonance imaging. Different morphologies of metal oxides such as nano-flowers, nanorods, nano-wires, nanosheets, and nanotubes have been investigated. The individual properties of metal oxide have been altered by encapsulating them into carbon or organic materials such as activated carbon, carbon nanotube, etc.

In order to uphold the higher adsorption capacity of AC-magnetic composite, magnetic medium must be entrenched into the surface of activated carbon. Ferromagnetic substance holds the higher saturation magnetization (Ms), and these adsorbents easily separated from water system using external magnet. If the ferromagnetism were hosted into the AC, then composite material indorses higher magnetic response as compared to the other magnetic sources.

Various techniques have been explored to insert the magnetic mediums into activated carbon such as chemical coprecipitation, solution impregnating, thermal treatment, heterogeneous nucleation, and hydrothermal technique (Yang et al. 2008; Nguyen et al. 2011; Faulconer et al. 2012; Konicki et al. 2013). Ferrimagnetic oxides possess chemical and physical stability, appropriate saturation magnetization (Ms), and very small coercive force (Hc) (Mohan et al. 2008; Deng et al. 2013). Numerous magnetic nanocomposites have been probed as adsorbents for the amputation of pollutants. Shao et al. (2012) synthesized the $MnFe_2O_4$/activated carbon using coprecipitation scheme for exclusion of tetracycline from water system. Nguyen et al. (2011) explored the Fe_2MO_4 (M:Fe,Mn)-AC nanocomposite using impregnating method for the catalytic activity of methyl orange. The magnetic Fe_2O_3-AC composites have been investigated for the amputation of methylene blue (Kadirova et al. 2014). The magnetic nanocomposites ($NiFe_2O_4$-AC) were synthesized by hydrothermal scheme for the removal of pollutants such as MB, rhodamine B, and malachite green from aqueous system (Feng et al. 2012).

Qu et al. (2008) explored the carbon nanotubes filled with Fe_2O_3 particles as magnetic adsorbent for the deduction of organic pollutants. Gong et al. (2009) probed the magnetic nanocomposite as an adsorbent for the obliteration of cationic dyes. Magnetic Fe_3O_4-activated carbon has been utilized for the amputation of methylene blue and magnetic alginate beads encapsulated with activated carbon employed for the elimination of persistent dyes from water system (Rocher et al. 2008; Yang et al. 2008).

18.2.11.1 The Role of Graphene-Metal Oxide Nanoparticles as Adsorbent

Graphene is a two-dimensional material, which displays admirable mechanical, thermal, optical, and electrical properties. So, there is undeniably impelled interest for synthesizing the innovative graphene-based nanocomposite for numerous technological applications.

Graphene demonstrates to be a huge-specific surface area nearly equal to 2600 $m^2 g^{-1}$, which makes it as smart adsorbent. Thus, the graphene and its products were used as adsorbents for the elimination of hazardous toxins from aqueous phase (Min et al. 2012a, b; Pu et al. 2013). Adsorption efficiency of graphene has been improved by encapsulating them into nanomaterials. Nanomaterials possess a more surface to volume ratio, and, henceforth, the composite of graphene attained increased the adsorption capacity. Fe_3O_4-graphene nanocomposites have been employed for water sanitization. The composite of graphene with magnetic nanomaterials such as Fe_3O_4 has been broadly exploited for amputation of pollutants (Wu et al. 2011; Yao et al. 2012). Fe_3O_4 is most commonly used nanomaterials for the purification of water due to its nontoxicity. After adsorption, Fe_3O_4-graphene nanocomposites were easily separated using the magnet. However, poor stability and ease of further oxidation to Fe_2O_3 α-Fe_2O_3 and λ-Fe_2O_3 restricts its application (Chandra et al. 2010). These complications might be overwhelmed by adding the Fe_3O_4 nanoparticles into the graphene matrix. The materialization of nanocomposite not only augments the lifetime of Fe_3O_4 nanoparticles but also permits the repossession after the completion of adsorption. The robustness of nanocomposite as an adsorbent is uttered by the binding of Fe_3O_4 with graphene matrix, which certifies the more lifetime of the adsorbent during water remediation.

Chandra et al. (2010) employed a magnetite-graphene oxide nanocomposite for the exclusion of arsenic ions from aqueous system. Adsorption proficiency depends on the pH of the solution. The electrostatic attraction has been deliberated between positively charged nanocomposite and negatively charged arsenic-arsenous acid (H_3AsO_4). At lower pH the graphene oxide nanocomposite attained positively charged, thus, attract more As (V) anions. It results in hastened adsorption of As (V) anions. At high pH the positive charge on nanocomposite gets reduced, but anionic charge increased due to As (III) anions, thus increasing the rate of adsorption. Liu et al. (2011a, b) studied the consequence of pH on the adsorption of Co (II) using the magnetite-graphene hybrid. Adsorption increased abruptly with pH ranging from 6.0 to 8.5.

He et al. (2010) prepared Fe_3O_4-graphene composite using tetraethyl orthosilicate and triethoxysilane. Additionally, active amino groups entwined into Fe_3O_4 surface. Thus, Fe_3O_4-graphene nanocomposites displayed admirable adsorption capacities for methylene blue and neutral red, and adsorption uptake was found to be 190.14 and 140.79 mg g^{-1}, respectively. Wang et al. (2011) stated the 96% adsorption of organic dye onto magnetic nanocomposite within short contact time. The rapid adsorption of fuchsine dye using graphene was ascribed due to the van der Waals interactions between dye molecule and carbon atoms. Additionally π-π stacking interaction occurred between aromatic dye and delocalized π-π system of graphene sheet. The magnetite-graphene oxide nanocomposites have been prepared by solvothermal technique (Sun et al. 2011). It was verified that 91% of rhodamine B dye and 94% of malachite green dyes have been removed from aqueous system. The parameters which affect the performance of composites include filling of Fe_3O_4

particles and pH of solution mixture. If Fe_3O_4 particles are loaded in the composite beyond certain amount, drip in the removal efficiency has been observed. It may be due to the lesser specific surface area of prepared nanocomposite.

18.2.12 Magnetic-Graphene Composite for the Detection of Peril Pollutants in Aqueous System

Graphene-nanocomposite-based adsorbents were not only exploited for the decontamination, but it has been employed for the extraction of minor toxins from aqueous system. Graphene-magnetic nanocomposite delivers a prevalent material for the detoxification process. However, a number of reports are available for the extraction of minor pollutants using magnetic graphene composites. Fe_3O_4-graphene composite was employed for the extraction of various sulfonamides from the aqueous system. The reaction conditions such as pH, amount of graphene, and extraction time were tested and optimized. The optimum conditions were found to be pH 3.0, graphene amount 0.3 mg, and extraction time of 4 min (Kroto et al. 1985). The graphene-magnetic composite has been investigated for the pre-concentration of pesticides such as carbamate from water system (Iijima 1991). These magnetic composites presented virtuous adsorption and robust magnetism. The magnetic composite must be recycled up to 12th time without substantial loss of adsorption efficiency.

18.2.13 Hydrogel Nanocomposite

Hydrogels are the networks of cross-linked hydrophilic polymer and can be classified as anionic, cationic, and ampholytic on the basis of charge present in structure. Hydrogels are defined as homopolymer and copolymer and interpenetrating according to monomer unit. Hydrogels are termed as copolymer when two or more monomeric units are present in the structure. The interpenetrating hydrogels involve two independent cross-linked networks in which one polymeric network swells in the presence of another polymer material. They are polymeric materials which do not dissolve in aqueous medium at physiological temperature and pH but swell significantly in aqueous medium. Hydrogels are capable of swallowing the large volumes of water or biological fluids. They hold numerous applications such as sustained and prolonged action, then conventional drug delivery, less side effects, improved drug utilization, and drug targeted to definite sites such as colon.

The majority of hydrogels respond to environmental stimulants such as pH of solution, ionic strength, solvent composition, and light and electrical field (Ma et al. 2007; Xiang and Chen 2007; Xu et al. 2008). Their unique features have created widespread applications such as tissue engineering, artificial muscles, wound dressing, enzyme biosensor, contact lens, drug delivery, separation devices, sensors, preparation of nanoparticle, pollutants adsorption, filters, and catalysis (Haraguchi & Matsuda, 2005; Haraguchi and Matsuda 2005; Li et al. 2006a, b, c; Haraguchi,

2007; Kokabi et al. 2007; Murthy et al. 2008; Liu et al. 2008). The individual properties of hydrogel can be enhanced by forming a nanocomposite. Additionally, incorporation of metal nanoparticles offers a cost-effective approach to improve the properties of hydrogels.

Recently, nanocomposite of polymer-clay hydrogel has gained pronounced courtesy due to their low cost and higher sequestration capacity for various organic pollutants (Wang et al. 2008a, b). These hydrogels possess special structure and various ionic or nonionic functional groups that permit the diffusion of solutes into the inner network, which absorb the dye molecules from water system. However, Jang et al. (2007) and Cheng et al. (2010) investigated that metal-binding biopolymers deliver good adsorbent because of their chemical stability and improved adsorption efficiency. The iron-chitosan composites and cross-linked Fe-chitosan have been investigated for the removal of toxic inorganic pollutants Cr (VI), As (III), and As (V) from wastewater (Klepka et al. 2008). Several natural materials such as cellulose, starch, chitosan, and guar gum were altered for the controlled drug delivery and medicinal applications (Kaith et al. 2012; Kaith and Ranjta 2010). Hydrogels have been used to cure multidrug-resistant infections and specific protein delivery. Chitosan-grafted hydrogels have been successfully used for oral insulin delivery (Mukhopadhyay et al. 2014).

18.2.14 Strategies Adopted for Refining Adsorption Capacity of Nanocomposites

Graphene sheet possesses low density and higher surface area as compared to other adsorbents. Therefore, graphene metal oxide nanoparticles must be synthesized in exact weight ratio, as it plays a vital role for the removal process. In photocatalytic techniques, semiconductor particle regulates the photodegradation activity of composite. Nanoparticles possess lesser surface area to weight ratio than graphene sheet. Therefore, excessive loading of metal oxide nanoparticles in composites must decrease the removal competency.

Adsorption performance of composites for organic contaminants is extremely reliant on pH of solution. At higher pH, concentration of hydroxyl ions increases, which hampers the removal of inorganic pollutants. Acidic pH increases the H^+ ions concentration in aqueous solution and competes with pollutants for the active sites on the surface of adsorbent. The inorganic pollutants including chromium and arsenic produce diverse metastable species under different pH. The weak physical adhesion and bonding between nanoparticles and graphene sheets lead to the leaching of metal oxide nanoparticles into purified water. Additionally, it diminishes the lifetime of adsorbent and caused the cross-adulteration of water. Therefore, the bonding strength between nanoparticles and graphene plays a crucial role. The synthetic procedure, reaction optimization, and reactants result in the stable inactive bonding in metal oxide-graphene nanocomposites. Hence, exterior of metal oxide-graphene nanocomposites must be improved with some functional groups that offer the strong bonding between metal oxide nanoparticles and graphene.

Since adsorption is a superficial process, surface properties of adsorbent play a vital role for obliteration of pollutants. It was observed that functional groups such as –COOH and $^{-}$OH onto graphene enhance the adsorption capacity. Moreover, the surface of graphene sheet must be improved with other functional groups such as $^{-}NH_2$, which eagerly combine with metal ions in water system. A noble adsorbent not only swiftly eliminates the pollutants but also validates the complete desorption of contaminants during regeneration procedure. The regeneration of an adsorbent elects its widespread economic value. Desorption of contaminants was usually done by regulating the pH of aqueous solution. The removal of metal ions using H^+ ions by acidic action efficiently regenerates the absorbent. For organic contaminants, regeneration was achieved by solvent washing in order to dissolve the contaminants. While adsorbent molecules are found to be stable at higher temperature, then thermal treatment offers a striking option for the regeneration of adsorbent.

Advantages of Graphene-Supported Metal Oxide Nanocomposite as an Adsorbent

1. Graphene increases the specific surface area of nanocomposite.
2. Graphene provides support to the metal oxide nanoparticles and avoids the leaching of metal oxide during water remediation.
3. Graphene enhances the mechanical strength of adsorbent.
4. It enables the decontamination of organic pollutants which has aromatic structure, as it caused π-π interaction with aromatic ring of organic pollutants.

Advantages of Graphene-Supported Metal Oxide Nanocomposite as a Photocatalyst

1. It diminishes the electron-hole recombination by acting as electron accepter.
2. It facilitates the interaction between organic pollutants and metal oxide by π-π stacking.
3. Graphene keeps the metal nanoparticles in dispersed phase, hence preventing the agglomeration of nanocomposite.
4. Graphene expands the absorption range of metal oxide from UV to visible spectrum.

18.2.15 Biomedical Application of Nanocomposite

Cancer is proven to be a foremost cause of mortality, and it is a multifactorial disorder which occurred by multifaceted combination of genetic and environmental aspects. Various developments have been initiated for the better consideration of cancer at various levels such as molecular, genetic, or cellular levels, thus providing novel tactics for rehabilitation. However, these advancements need to be successfully translated into functional types of diagnostics. But efficiency of anticancer drugs has been restricted due to their incapability to grasp the targeted site in appropriate concentrations (Peer et al. 2007).

Nanoscale technology delivers an immense revolt for medicinal and healthcare treatment. However, nanotechnology proposed a healthy tool to detect and treat the cancer such as multifunctional, new imaging agents and targeted devices that are proficient in bypassing biological barriers to transport therapeutic agents directly to the cells and/or tissues which are engaged in cancerous growth and metastasis (Heath and Davis 2008). Thus, therapies based on nanotechnology for treating cancer with negligible adverse effects and higher specificity play an important role that deliver such system for molecular therapy, which are proficient for circulating in the bloodstream, recognize the desirable target organ, and act as signal for effective drug delivery.

Inorganic nanoparticles, specifically noble metals, have widespread biomedical applications such as biosensing contrast agents for tissue engineering, magnetic resonance imaging, and tumor destruction. Nanoparticles are generally indorsed with numerous sensing agents like DNA, RNA, antibodies, and peptides, to targeted cells and biocompatible polymers (Zhang et al. 2002). The biocompatibilities of metal nanoparticles are usually heightened through conjugation or integration of ligands on their surface such as permeation enhancers, targeting agents, and therapeutic agents.

The superparamagnetic nanoparticles were modified using folic acid and PEG to increase the intracellular uptake and capability for targeted cells (Gillich et al. 2012). The preparation of biocompatible TiO_2 nanoparticles comprising NH_2 or SH groups has been carried out (Cheyne et al. 2011). These nanoparticles are functionalized with various organic materials for biological applications including medical imaging and radio-immunotherapy, where nanoparticles are crucial for quick renal clearance. Kohler et al. (2005) described the formation of methotrexate (MTX) restrained on nanoparticle for the sustained delivery of MTX in brain and breast tumor cells. The methotrexate was covalently bonded to amine groups of metal oxide nanoparticles by amide bonds to confirm the constancy of drug conjugate at intravenous situation. It has been examined that polyethyleneimine-modified magnetic nanoparticles are successfully employed for the distribution of drugs and genes to brain tumor cell.

18.2.16 Nanocomposites as Antimicrobial Agents

Various nanocomposites have been probed as antimicrobial agents. The graphene composites with heavy metals and their oxides including Ag, Fe, Fe_3O_4, and TiO_2 have been investigated as antimicrobial agent for the decontamination of water. The Fe_3O_4 graphene composite has been synthesized and exhibits disinfection properties against *E. coli* (Gollavelli et al. 2013). The antibacterial properties of magnetic graphene composite were ascribed to the penetrating of cell membrane and high surface roughness. It has been reported that TiO_2-graphine oxide thin films toughened at 400 °C could photo-inactivate the *Escherichia coli* (nearly 25%) with higher efficiency, as compared to TiO_2 (Akhavan et al. 2009). Thus, graphene oxide obliged as an e^- acceptor for the elimination of e^- from e^- hole pair is created due to exposure material under solar illumination. Concurrently, quantum efficiency of photocatalytic degradation was improved by increasing the lifetime of hole and declines in the e^--hole recombination process. The graphene-TiO_2 thin films showed ominously enhanced antibacterial activity as compared to unsupported TiO_2, Ag–SiO_2, Ag nanorods, and Ag–TiO_2–Ag-a–TiO_2 films, respectively. Nanorods have been employed for the photocatalytic deduction of microorganisms. Bactericidal activity has been found in TiO_2 nanorod with graphene oxide under solar radiation for E. coli colony (Liu et al. 2011a, b). TiO_2 nanorod-graphene composite has greater bactericidal performance than TiO_2-graphene nanocomposite. TiO_2-nanorod graphene composite deactivated the E. coli (90%) during 27 min. TiO_2-graphene composite with 4.2 wt% graphene displayed photoactivity under solar radiation, which was credited to the formation of Ti–C bond between TiO_2 and graphene composite (Cao et al. 2013). The TiO_2-graphene composite was tried photocatalytically for *E. coli*. It became inactive in the absence of solar illumination. But, under solar illumination, the composite exhibited outstanding antibacterial activity on 12 hours of exposure, and bacterial cell viability was abridged up to 9.5%.

The bactericidal performance of ZnO-graphene nanocomposite has been tested for *E. coli* (Kavitha et al. 2012). Further, other semiconductor nanoparticles amalgamated with graphene have been explored as antimicrobial agents such as CdS, Ag_3PO_4, Bi_2MoO_6, etc. (Gao et al. 2013; Zhang et al. 2013). The antibacterial performance of CdS-graphene oxide nanocomposites obstructs the progress of Gram-positive and Gram-negative bacteria. These nanocomposites destroyed almost all *E. coli* within 25 min exposure under visible light. CdS nanoparticles without graphene oxide destroy only 55% of bacteria during same time interval. The CdS-GO exhibited greater antibacterial action against Gram-positive than Gram-negative bacteria. The degradation of protein and ribonucleic acid in the efflux of viruses has been investigated using WO_3-graphene composites (Akhavan et al. 2012). It was recorded that WO_3 itself does not reveal any substantial effect on viruses, even in the presence of solar illumination, but composite of WO_3 with graphene showed efficient photodegradation efficiency for viral proteins. The photocatalytic activity exhibited by WO_3-graphene composite was mainly due to the establishment of W–O–C and W–C bonds between WO_3 and graphene composite, which assisted the flow of electron from the WO_3 to graphene, hence prolonging the lifetime of charge carriers and preventing the e^--hole recombination process. Recently, silver nanoparticles with graphene or graphene oxide have been recognized to act as antimicrobial agents (Tang et al. 2013). Ag itself acts as admirable antibacterial agents, and composites of silver with graphene must afford the enriched antimicrobial activity. Other metal oxides including ZnO, TiO_2, and Fe_3O_4 have also been studied for their antimicrobial action.

Conclusion

In this article, various types of materials have been mentioned for photocatalytic, adsorptional, and disinfectant applications. The amalgamation of graphene oxide with semiconductors, especially photoactive metal oxide, synergistically increases the photocatalytic degradation of various waterborne pollutants. Additionally, an inimitable property of graphene with inorganic materials induces high adsorption, prolonged light absorption range, and amended charge separation properties with higher solidity. Nanocomposite possesses distinctive properties as to offer inventive materials for researchers as well as for industrial applications. These composites are encapsulated with inimitable features of inorganic and organic components in single matrix. Their multifunctional nature permits them to apply in various technological sectors such as coverings from corrosion protection, abrasion resistance, catalysts and adsorbents for peril pollutants, biomaterials for ophthalmic agents, and osteo-reconstructive surgery with optoelectronic and magnetic features for telecommunications process. For developing these nanocomposites, inorganic nanoparticles have a robust propensity to aggregates. Hence, to increase their compatibility and dispersion stability with organic matrix, their surfaces must be modified. Surface alteration expands the interfacial interaction between inorganic fillers and organic matrix. Hence, it results in enhanced characteristics such as higher mechanical toughness (lesser loadings of inorganic materials); gas-barrier, electronic, optical, and flame-retardation characteristics; etc.

18.3 Future Perspectives

Wastewater treatment techniques based on nanocomposite deliver a decent way for the treatment of the contaminated water, but still there are few queries which are yet to be answered. However, it means that recyclability and disposal of these nanocomposites prove to be a problematic aspect. Therefore, approaches for reusability and clearance of used nanocomposites will produce simultaneously along with their applications for detoxification. The post-treatment retrieval of nanocomposites from aqueous phase is a critical issue because these nanocomposites possibly pollute the treated water. However, the composites of magnetic nanoparticles with biomaterial provide the alternate clarification. The magnetic nanocomposite must be simply separated from aqueous system using an external magnet, but mostly Fe_3O_4 and Fe_2O_3 have been probed as magnetic materials, and they do not display significant photocatalytic activity. Hereafter, there is a necessity of planning the heterogeneous metal oxide photocatalyst which can show excellent magnetic and photocatalytic activity. Progress of techniques for the creation of organic-inorganic composites with metal nanoparticles without dropping their surface area is also a foremost challenge. However, graphene-based composites have exhibited massive opportunities as photocatalyst, but still there are some problems which need to be resolved. The nature of bonding mechanism for metal nanoparticles and graphene is not copiously understood. So, there is an inevitability of comprehensive examination on the optimization of graphene amount in nanocomposite in order to accomplish the supreme performance from nanocomposite. The synthesis of adsorbents using nanoparticle-graphene composites should be compared with other biomaterial such as activated carbons in terms of expense, availability, efficiency, ease of synthesis, and environmental hazards.

References

Ahmad T, Rafatullah M, Ghazali A, Sulaiman O, Hashim R, Ahmad A (2010) Removal of pesticides from water and wastewater by different adsorbents: a review. J Environ Sci Health C 28:231–271. doi:10.1080/10590501.2010.525782

Akhavan O, Ghaderi E (2009) Photocatalytic reduction of graphene oxide nanosheets on TiO_2 thin film for photoinactivation of bacteria in solar light irradiation. J Phys Chem C 113:20214–20220. doi:10.1021/jp906325q

Akhavan O, Abdolahad M, Abdi Y, Mohajerzadeh S (2009) Synthesis of titania/carbon nanotube heterojunction arrays for photoinactivation of E. coli in visible light irradiation. Carbon 47:3280–3287. doi:10.1016/j.carbon.2009.07.046

Akhavan O, Choobtashani M, Ghaderi E (2012) Protein degradation and RNA efflux of viruses photocatalyzed by graphene-tungsten oxide composite under visible light irradiation. J Phys Chem C 116:9653–9659. doi:10.1021/jp301707m

An X, Jimmy CY, Wang Y, Hu Y, Yu X, Zhang G (2012) WO_3nanorods/graphene nanocomposites for high-efficiency visible-light-driven photocatalysis and NO_2 gas sensing. J Mater Chem 22:8525–8531. doi:10.1039/C2JM16709C

Aneesh PM, Jayaraj MK (2010) Red luminescence from hydrothermally synthesized Eu-doped ZnO nanoparticles under visible excitation. Bull Mater Sci 33:227–231. doi:10.1007/s12034-010-0035-7

Ania CO, Parra JB, Menendez JA, Pis JJ (2005) Effect of microwave and conventional regeneration on the microporous and mesoporous network and on the adsorptive capacity of activated carbons. Micropor Mesopor Mater 85:7–15. doi:10.1016/j.micromeso.2005.06.013

Arsalani N, Fattahi H, Nazarpoor M (2010) Synthesis and characterization of PVP-functionalized superparamagnetic Fe_3O_4 nanoparticles as an MRI contrast agent. Express Polym Lett 4:329–338. doi:10.3144/expresspolymlett.2010.42

Awual MR, Hasan MM, Shahat A, Naushad M, Shiwaku H, Yaita T (2015) Investigation of ligand immobilized nano-composite adsorbent for efficient cerium (III) detection and recovery. Chem Eng J 265:210–218. doi:10.1016/j.cej.2014.12.052

Bahnemann W, Muneer M, Haque MM (2007) Titanium dioxide-mediated photocatalysed degradation of few selected organic pollutants in aqueous suspensions. Catal Today 124:133–148. doi:10.1016/j.cattod.2007.03.031

Behnajady MA, Eskandarloo H, Modirshahla N, Shokri M (2011) Sol-gel low-temperature synthesis of stable Anatase-type TiO_2 nanoparticles under different conditions and its photocatalytic activity. Photochem Photobiol 87:1002–1008. doi:10.1111/j.1751-1097.2011.00954.x

Birnie M, Riffat S, Gillott M (2006) Photocatalytic reactors: design for effective air purification. Int J Low Carbon Technol 1:47–58. doi:10.1093/ijlct/1.1.47

Brinker CJ, Scherer GW (2013) Sol-gel science: the physics and chemistry of sol-gel processing. Academic press

Cao B, Cao S, Dong P, Gao J, Wang J (2013) High antibacterial activity of ultrafine TiO_2/graphene sheets nanocomposites under visible light irradiation. Mater Lett 93:349–352. doi:10.1016/j.matlet.2012.11.136

Carbonaro S, Sugihara MN, Strathmann TJ (2013) Continuous-flow photocatalytic treatment of pharmaceutical micropollutants: activity, inhibition, and deactivation of TiO_2photocatalysts in wastewater effluent. Appl Catal Environ 129:1–2. doi:10.1016/j.apcatb.2012.09.014

Cardoso NF, Lima EC, Pinto IS, Amavisca CV, Royer B, Pinto RB, Alencar WS, Pereira SF (2011) Application of cupuassu shell as biosorbent for the removal of textile dyes from aqueous solution. J Environ Manage 92:1237–1247. doi:10.1016/j.jenvman.2010.12.010

Cassano AE, Alfano OM (2000) Reaction engineering of suspended solid heterogeneous photocatalytic reactors. Catal Today 58:167–197. doi:10.1016/S0920-5861(00)00251-0

Chandra V, Park J, Chun Y, Lee JW, Hwang IC, Kim KS (2010) Water-dispersible magnetite-reduced graphene oxide composites for arsenic removal. ACS Nano 4:3979–3986. doi:10.1021/nn1008897

Chandra S, Das P, Bag S, Bhar R, Pramanik P (2012) Mn_2O_3 decorated graphene nanosheet: an advanced material for the photocatalytic degradation of organic dyes. Mater Sci Eng B 177:855–861. doi:10.1016/j.mseb.2012.04.006

Chang C, Duan B, Cai J, Zhang L (2010) Superabsorbent hydrogels based on cellulose for smart swelling and controllable delivery. European Polym J 46:92–100. doi:org/10.1016/j.eurpolymj.2009.04.033

Chauhan R, Kumar A, Chaudhary RP (2010) Synthesis and characterization of copper doped ZnO nanoparticles. J Chem Pharm Res 2:178–183

Chen CC, Fan HJ, Jan JL (2008) Degradation pathways and efficiencies of acid blue 1 by photocatalytic reaction with ZnO nanopowder. J Phys Chem C112:11962–11972. doi:10.1021/jp801027r

Chen B, Yuan M, Liu H (2011) Removal of polycyclic aromatic hydrocarbons from aqueous solution using plant residue materials as a biosorbent. J Hazard Mater 188:436–442. doi: 10.1016/j.jhazmat.2011.01.114

Cheng Y, Lü C, Lin Z, Liu Y, Guan C, Lü H, Yang B (2008) Preparation and properties of transparent bulk polymer nanocomposites with high nanophase contents. J Mater Chem 18:4062–4068. doi:10.1039/B803237H

Cheng R, Ou S, Xiang B, Li Y, Liao Q (2009) Equilibrium and molecular mechanism of anionic dyes adsorption onto copper (II) complex of dithiocarbamate-modified starch. Langmuir 26:752–758. doi:10.1021/la903948

Cheyne RW, Smith TA, Trembleau L, Mclaughlin AC (2011) Synthesis and characterization of biologically compatible TiO_2 nanoparticles. Nanoscale Res Lett 6:1–6. doi:10.1186/1556-276X-6-423

Chowdhury ZZ, Ahmed AA, Khan RA, Zain SM (2011) Equilibrium kinetics and isotherm studies of cu (II) adsorption from waste water onto alkali activated oil palm ash. Am J Appl Sci 8:230–237

Colombo A, Cappelletti G, Ardizzone S, Biraghi I, Bianchi CL, Meroni D, Pirola C, Spadavecchia F (2012) Bisphenol a endocrine disruptor complete degradation using TiO_2 photocatalysis with ozone. Environ Chem Let 10:55–60. doi:10.1007/s10311-011-0328-0

Crivello JV, Song KY, Ghoshal R (2001) Synthesis and photoinitiated cationic polymerization of organic-inorganic hybrid resins. Chem Mater 13:1932–1942. doi:10.1021/cm0100671

Deng J, Shao Y, Gao N, Tan C, Zhou S, Hu X (2013) $CoFe_2O_4$ magnetic nanoparticles as a highly active heterogeneous catalyst of oxone for the degradation of diclofenac in water. J Hazard Mater 262:836–844. doi:10.1016/j.jhazmat.2013.09.049

Didenko YT, Suslick KS (2005) Chemical aerosol flow synthesis of semiconductor nanoparticles. J Am Chem Soc 127:12196–12197. doi:10.1021/ja054124t

Du T, Song H, Ilegbusi OJ (2007) Sol-gel derived ZnO/PVP nanocomposite thin film for superoxide radical sensor. Mater Sci Eng C 27:414–420. doi:10.1016/j.msec.2006.05.040

Eastoe J, Warne B (1996) Nanoparticle and polymer synthesis in microemulsions. Curr Opin Colloid Interface Sci 1:800–805. doi:10.1016/S1359-0294(96)80084-7

Eisa WH, Abdel-Moneam YK, Shabaka A, Hosam AE (2012) In situ approach induced growth of highly monodispersed ag nanoparticles within free standing PVA/PVP films. Spectrochimica Acta A Mol Biomol Spectrosc 95:341–346. doi:10.1016/j.saa.2012.03.085

Erathodiyil N, Ying JY (2011) Functionalization of inorganic nanoparticles for bioimaging applications. Acc Chem Res 44:925–935. doi:10.1021/ar2000327

Faulconer EK, von Reitzenstein NV, Mazyck DW (2012) Optimization of magnetic powdered activated carbon for aqueous hg (II) removal and magnetic recovery. J Hazard Mater 199:9–14. doi:10.1016/j.jhazmat.2011.10.023

Feng L, Hu F, Liu C, Chen F, Xu N, Liu S, Chen Z (2012) Photocatalytic properties of activated carbon-NiFe sub (2) O sub (4) magnetic catalyst. Cuihua Xuebao Chin J Catal 33(8)

Fu H, Xu T, Zhu S, Zhu Y (2008) Photocorrosion inhibition and enhancement of photocatalytic activity for ZnO via hybridization with C60. Environ Sci Technol 42:8064–8069. doi:10.1021/es801484x

Galindo C, Jacques P, Kalt A (2001) Photochemical and photocatalytic degradation of an indigoid dye: a case study of acid blue 74 (AB74). J Photochem Photobiol A Chem 141:47–56. doi:10.1016/S1010-6030(01)00435-X

Gao Z, Liu J, Xu F, Wu D, Wu Z, Jiang K (2012) One-pot synthesis of graphene–cuprous oxide composite with enhanced photocatalytic activity. Solid State Sci 14:276–280. doi:10.1016/j.solidstatesciences.2011.11.032

Gao P, Ng K, Sun DD (2013) Sulfonated graphene oxide-ZnO-ag photocatalyst for fast photodegradation and disinfection under visible light. J Hazard Mater 262:826–835. doi:10.1016/j.jhazmat.2013.09.055

Gaponik N, Rogach AL (2010) Thiol-capped CdTe nanocrystals: progress and perspectives of the related research fields. Phys Chem Chem Phys 12:8685–8693. doi:10.1039/C000916D

Garcia M, van Vliet G, ten Cate MG, Chávez F, Norder B, Kooi B, van Zyl WE, Verweij H, Blank DH (2004) Large-scale extrusion processing and characterization of hybrid nylon-6/SiO_2 nanocomposites. Polymer Adv Tech 15:164–172. doi:10.1002/pat.458

Gaya UI, Abdullah AH (2008) Heterogeneous photocatalytic degradation of organic contaminants over titanium dioxide: a review of fundamentals, progress and problems. J Photochem Photobiol C Photochem Rev 9:1–2. doi:10.1016/j.jphotochemrev.2007.12.003

Georgiou D, Aivazidis A, Hatiras J, Gimouhopoulos K (2003) Treatment of cotton textile wastewater using lime and ferrous sulfate. Water Res 37:2248–2250. doi:10.1016/S0043-1354(02)00481-5

Gillich T, Acikgöz C, Isa L, Schlüter AD, Spencer ND, Textor M (2012) PEG-stabilized core-shell nanoparticles: impact of linear versus dendritic polymer shell architecture on colloidal properties and the reversibility of temperature-induced aggregation. ACS Nano 7:316–329. doi:10.1021/nn304045q

Gollavelli G, Chang CC, Ling YC (2013) Facile synthesis of smart magnetic graphene for safe drinking water: heavy metal removal and disinfection control. ACS Sustain Chem Eng 1:462–472. doi:10.1021/sc300112z

Gong JL, Wang B, Zeng GM, Yang CP, Niu CG, Niu QY, Zhou WJ, Liang Y (2009) Removal of cationic dyes from aqueous solution using magnetic multi-wall carbon nanotube nanocomposite as adsorbent. J Hazard Mater 164:1517-1522. doi: 10.1016/j.Jhazmat.2008.09.072

Guan C, Lü C, Cheng Y, Song S, Yang B (2009) A facile one-pot route to transparent polymer nanocomposites with high ZnS nanophase contents via in situ bulk polymerization. J Mater Chem 19:617–621. doi:10.1039/B818876A

Guettai N, Amar HA (2005) Photocatalytic oxidation of methyl orange in presence of titanium dioxide in aqueous suspension. Part I: Parametric study. Desalination 185:427–437. doi:10.1016/j.desal.2005.04.048

Guo Y, Wang M, Zhang H, Liu G, Zhang L, Qu X (2008) The surface modification of nanosilica, preparation of nanosilica/acrylic core-shell composite latex, and its application in toughening PVC matrix. J Appl Polym Sci 107:2671–2680. doi:10.1002/app.27310

Guo QZ, Mei B, Zhou SX, Shi ZG, Feng YQ, Wu JY, Yan GP, Li L (2009) Synthesis, characterization and application of magnetic-zirconia nanocomposites. J Non Cryst Solids 355:922–925. doi:10.1016/j.jnoncrysol.2009.04.030

Gupta AK, Pal A, Sahoo C (2006) Photocatalytic degradation of a mixture of crystal violet (basic violet 3) and methyl red dye in aqueous suspensions using ag^+ doped TiO_2. Dyes Pigments 69:224–232. doi:10.1016/j.dyepig.2005.04.001

Gupta VK, Pathania D, Agarwal S, Singh P (2012) Adsorptional photocatalytic degradation of methylene blue onto pectin-CuS nanocomposite under solar light. J Hazard Mater 243:179–186. doi:10.1016/j.jhazmat.2012.10.018

Gupta VK, Pathania D, Sharma S, Agarwal S, Singh P (2013) Remediation and recovery of methyl orange from aqueous solution onto acrylic acid grafted Ficus carica fiber: isotherms, kinetics and thermodynamics. J Mol Liq 177:325–334. doi:org/10.1016/j.molliq.2012.10.007

Gupta VK, Pathania D, Agarwal S, Sharma S (2014) Amputation of congo red dye from waste water using microwave induced grafted Luffa cylindrica cellulosic fiber. Carbohydr Polym 111:556–566. doi:10.1016/j.carbpol.2014.04.032

Hamdy MS, Nickels P, Abd-Elmaksood IH, Zhou H, El-Mossalamy ES, Alyoubi AO, Lynch S, Nathan A, Thornton G (2012) Parameters controlling the photocatalytic performance of ZnO/Hombikat TiO_2 composites. J Photochem Photobiol A Chem 228:1–7. doi:10.1016/j.carbpol.2013.02.045

Haraguchi K (2007) Nanocomposite gels: new advanced functional soft materials. Macromol Symposia 256:120–130. doi:10.1002/masy.200751014

Haraguchi K, Li HJ (2005) Control of the coil-to-globule transition and ultrahigh mechanical properties of PNIPA in nanocomposite hydrogels. Angew Chem Int Ed 44:6500–6504. doi:10.1002/anie.200502004

Haraguchi K, Matsuda K (2005) Spontaneous formation of characteristic layered morphologies in porous nanocomposites prepared from nanocomposite hydrogels. Chem Mater 17:931–934. doi:10.1021/cm048093x

He F, Fan J, Ma D, Zhang L, Leung C, Chan HL (2010) The attachment of Fe_3O_4 nanoparticles to graphene oxide by covalent bonding. Carbon 48:3139–3144. doi:10.1016/j.carbon.2010.04.052

Heath JR, Davis ME (2008) Nanotechnology and cancer. Annu Rev Med 59:251. doi:10.1146/annurev.med.59.061506.185523

Herrmann JM (1999) Heterogeneous photocatalysis: fundamentals and applications to the removal of various types of aqueous pollutants. Catal Today 53:115–129. doi:10.1016/S0920-5861(99)00107-8

Hsiue GH, Kuo WJ, Huang YP, Jeng RJ (2000a) Microstructural and morphological characteristics of PS-SiO_2 nanocomposites. Polym 41:2813–2825. doi:10.1016/S0032-3861(99)00478-4

Hsiue GH, Chen JK, Liu YL (2000b) Synthesis and characterization of nanocomposite of polyimide–silica hybrid from nonaqueous sol-gel process. J Appl Polym Sci 76:1609–1618. doi:10.1002/(SICI)1097-4628(20000613)76:11<1609::AID-APP1>3.0.CO;2-W

Hu Q, Marand E (1999) In situ formation of nanosized TiO_2 domains within poly (amide-imide) by a sol–gel process. Polym 40:4833–4843. doi:10.1016/S0032-3861(98)00264-X

Hwang H, Kim MG, Cho J (2007) Li reaction behaviour of gap nanoparticles prepared by a sodium naphthalenide reduction method. J Phys Chem C 111:1186–1193. doi:10.1021/jp065560k

Iijima S (1991) Helical microtubules of graphitic carbon. Nature 354:56–58

Jang M, Min SH, Park JK, Tlachac EJ (2007) Hydrous ferric oxide incorporated diatomite for remediation of arsenic contaminated groundwater. Environ Sci Technol 41:3322–3328. doi:10.1021/es062359e

Jegier JA, McKernan S, Gladfelter WL (1998) Poly (imidogallane): synthesis of a crystalline 2-D network solid and its pyrolysis to form nanocrystalline gallium nitride in supercritical ammonia. Chem Mater 10:2041–2043. doi:10.1021/cm980257k

Jeon IY, Baek JB (2010) Nanocomposites derived from polymers and inorganic nanoparticles. Materials 3:3654–3674. doi:10.3390/ma3063654

Jiang J (2007) Ultrasonic-assisted synthesis of PMMA/Ni0.5Zn0.5Fe_2O_4 nanocomposite in mixed surfactant system. Eur Polym J 43:1724–1728. doi:10.1016/j.eurpolymj.2007.02.011

Junwu L, Zhixiang Z, Kaihui Z, Yucheng W (2006) Preparation and characterization of Fe^{3+}doped nanometer TiO_2photocatalysts. J Wuhan Univ Technol Mater Sci Ed 21:57–60. doi:10.1007/BF02840880

Kadirova ZC, Hojamberdiev M, Katsumata KI, Isobe T, Matsushita N, Nakajima A, Sharipov K, Okada K (2014) Preparation of iron oxide-impregnated spherical granular activated carbon-carbon composite and its photocatalytic removal of methylene blue in the presence of oxalic acid. J Environ Sci Health A Tox Hazard Subst Environ Eng 49:763–769. doi:10.1080/10934529.2014.882185

Kaith BS, Ranjta S (2010) Synthesis of pH-thermosensitive gum Arabic based hydrogel and study of its salt resistant swelling behaviour for saline water treatment. Desalin Water Treat 24:28–37. doi:10.5004/dwt.2010.1145

Kaith BS, Jindal R, Mittal H, Kumar K (2012) Synthesis, characterisation and swelling behaviour evaluation of hydrogels based on gum ghatti and acrylamide for selective absorption of saline from different petroleum fraction-saline emulsions. J Appl Polym Sci 124:2037–2047. doi:10.1002/app.35238

Kant S, Kalia S, Kumar A (2013) A novel nanocomposite of polyaniline and Fe 0.01 Ni 0.01 Zn 0.98 O: photocatalytic, electrical and antibacterial properties. J Alloys Compd 578:249–256. doi:10.1016/j.jallcom.2013.05.120

Karger-Kocsis J, Kmetty Á, Lendvai L, Drakopoulos SX, Bárány T (2014) Water-assisted production of thermoplastic nanocomposites: a review. Mater 8:72–95. doi:10.3390/ma8010072

Kasprzyk-Hordern B, Ziółek M, Nawrocki J (2003) Catalytic ozonation and methods of enhancing molecular ozone reactions in water treatment. Appl Catal Environ 46:639–669. doi:10.1016/S0926-3373(03)00326-6

Katwal R, Kaur H, Sharma G, Naushad M, Pathania D (2015) Electrochemical synthesized copper oxide nanoparticles for enhanced photocatalytic and antimicrobial activity. J Ind Eng Chem 31:173–184. doi:10.1016/j.jiec.2015.06.021

Kavitha T, Gopalan AI, Lee KP, Park SY (2012) Glucose sensing, photocatalytic and antibacterial properties of graphene-ZnO nanoparticle hybrids. Carbon 50:2994–3000. doi:10.1016/j.carbon.2012.02.082

Klepka MT, Nedelko N, Greneche JM, Lawniczak-Jablonska K, Demchenko IN, Slawska-Waniewska A, Rodrigues CA, Debrassi A, Bordini C (2008) Local atomic structure and magnetic ordering of iron in Fe-chitosan complexes. Biomacromolecules 9:1586–1594. doi:10.1021/bm800044y

Kohler N, Sun C, Wang J, Zhang M (2005) Methotrexate-modified superparamagnetic nanoparticles and their intracellular uptake into human cancer cells. Langmuir 21:8858–8864

Kokabi M, Sirousazar M, Hassan ZM (2007) PVA-clay nanocomposite hydrogels for wound dressing. Eur Polym J 43:773–781. doi:10.1021/la0503451

Konicki W, Sibera D, Mijowska E, Lendzion-Bieluń Z, Narkiewicz U (2013) Equilibrium and kinetic studies on acid dye acid red 88 adsorption by magnetic $ZnFe_2O_4$ spinel ferrite nanoparticles. J Colloid Interface Sci 398:152–160. doi:10.1016/j.jcis.2013.02.021

Kornowski A, Giersig M, Vogel R, Chemseddine A, Weller H (1993) Nanometer-sized colloidal germanium particles: wet-chemical synthesis, laser-induced crystallization and particle growth. Adv Mater 5:634–636. doi:10.1002/adma.19930050907

Kripal R, Gupta AK, Srivastava RK, Mishra SK (2011) Photoconductivity and photoluminescence of ZnO nanoparticles synthesized via co-precipitation method. Spectrochim Acta A Mol Biomol Spectrosc 79:1605–1612. doi:10.1016/j.saa.2011.05.019

Kroto HW, Heath JR, O'Brien SC, Curl RF, Smalley RE (1985) C 60: buckminsterfullerene. Nature 318:162–163

Kumar SS, Venkateswarlu P, Rao VR, Rao GN (2013) Synthesis, characterization and optical properties of zinc oxide nanoparticles. Int Nano Lett 3:1–6. doi:10.1186/2228-5326-3-30

Kumar A, Sharma G, Naushad M, Singh P, Kalia S (2014) Polyacrylamide/Ni0.02Zn0.98O nanocomposite with high solar light photocatalytic activity and efficient adsorption capacity for toxic dye removal. Ind Eng Chem Res 53:15549–15560. doi:10.1021/ie5018173

Kumar A, Sharma G, Naushad M, Thakur S (2015) SPION/β-cyclodextrin core-shell nanostructures for oil spill remediation and organic pollutant removal from waste water. Chem Eng J 280:175–187. doi:10.1016/j.cej.2015.05.126

Li J, Li L, Zheng L, Xian Y, Jin L (2006a) Determination of chemical oxygen demand values by a photocatalytic oxidation method using nano-TiO_2 film on quartz. Talanta 68:765–770. doi:10.1016/j.talanta.2005.06.012

Li H, Liu J, Wang K, Fan L, Yang S (2006b) Synthesis and characterization of novel fluorinated polyimides derived from 4,4′-[2,2,2-trifluoro-1-(3,5-ditrifluoromethylphenyl) ethylidene] diphthalic anhydride and aromatic diamines. Polym 47:1443–1450. doi:10.1016/j.polymer.2005.12.074

Li Y, Li X, Li J, Yin J (2006c) Photocatalytic degradation of methyl orange by TiO_2-coated activated carbon and kinetic study. Water Res 40:1119–1126. doi:10.1016/j.watres.2005.12.042

Lin L, Zhai SR, Xiao ZY, Song Y, An QD, Song XW (2013) Dye adsorption of mesoporous activated carbons produced from NaOH-pretreated rice husks. Bioresour Technol 136:437–443. doi:10.1016/j.biortech.2013.03.048

Ling CM, Mohamed AR, Bhatia S (2004) Performance of photocatalytic reactors using immobilized TiO_2 film for the degradation of phenol and methylene blue dye present in water stream. Chemosphere 57:547–554. doi:10.1016/j.chemosphere.2004.07.011

Liu KH, Liu TY, Chen SY, Liu DM (2008) Drug release behavior of chitosan–montmorillonite nanocomposite hydrogels following electrostimulation. Acta Biomater 4:1038–1045. doi:10.1016/jactbio200801012

Liu J, Liu L, Bai H, Wang Y, Sun DD (2011a) Gram-scale production of graphene oxide-TiO_2 nanorod composites: towards high-activity photocatalytic materials. Appl Catal Environ 106:76–82. doi:10.1016/j.apcatb.2011.05.007

Liu M, Chen C, Hu J, Wu X, Wang X (2011b) Synthesis of magnetite/graphene oxide composite and application for cobalt (II) removal. J Phys Chem C 115:25234–25240. doi:10.1021/jp208575m

Lu C, Wu Y, Mai F, Chung W, Wu C, Lin W, Chen C (2009) Degradation efficiencies and mechanisms of the ZnO-mediated photocatalytic degradation of basic blue 11 under visible light irradiation. J Mol Catal A Chem 310:159–165. doi:10.1016/j.molcata.2009.06.011

Ma J, Xu Y, Fan B, Liang B (2007) Preparation and characterization of sodium carboxymethylcellulose/poly (N-isopropylacrylamide)/clay semi-IPN nanocomposite hydrogels. Eur Polym J 43:2221–2228. doi:10.1016/j.eurpolymj.2007.02.026

Mädler L, Stark WJ, Pratsinis SE (2002) Rapid synthesis of stable ZnO quantum dots. J Appl Phys 92:6537–6540. doi:10.1063/1.1518132

Mai FD, Chen CC, Chen JL, Liu SC (2008) Photodegradation of methyl green using visible irradiation in ZnO suspensions: determination of the reaction pathway and identification of intermediates by a high-performance liquid chromatography–photodiode array-electrospray ionization-mass spectrometry method. J Chromatogr A 1189:355–365. doi:10.1016/j.chroma.2008.01.027

Malik MA, Wani MY, Hashim MA (2012) Microemulsion method: a novel route to synthesize organic and inorganic nanomaterials: 1st nano update. Arab J Chem 5:397–417. doi:10.1016/j.arabjc.2010.09.027

Mansur HS (2010) Quantum dots and nanocomposites. Wiley Interdiscip Rev Nanomed Nanobiotechnol 2:113–129. doi:10.1002/wnan.78

Min Y, Zhang K, Chen L, Chen Y, Zhang Y (2012a) Ionic liquid assisting synthesis of ZnO/graphene heterostructure photocatalysts with tunable photoresponse properties. Diamond Relat Mater 26:32–38. doi:10.1016/j.diamond.2012.04.003

Min Y, Zhang K, Chen L, Chen Y, Zhang Y (2012b) Sonochemical assisted synthesis of a novel TiO_2/graphene composite for solar energy conversion. Synth Met 162:827–833. doi:10.1016/j.synthmet.2012.03.014

Mohan M, Chandra V, Manoharan SS (2008) Nano body-centered cubic $CoFe_2$ alloy precursor for cobalt ferrite via sonoreduction process. J Mater Res 23:1849–1853. doi:10.1557/JMR.2008.0230

Mukhopadhyay P, Sarkar K, Bhattacharya S, Bhattacharyya A, Mishra R, Kundu PP (2014) pH sensitive N-succinyl chitosan grafted polyacrylamide hydrogel for oral insulin delivery. Carbohydr Polym 112:627–637. doi:10.1016/j.carbpol.2014.06.045

Murthy PK, Mohan YM, Varaprasad K, Sreedhar B, Raju KM (2008) First successful design of semi-IPN hydrogel–silver nanocomposites: a facile approach for antibacterial application. J Colloid Interface Sci 318:217–224. doi:10.1016/j.jcis.2007.10.014

Naje AN, Norry AS, Suhail AM (2013) Preparation and characterization of SnO_2 nanoparticles. Int J Innov Res Sci Eng Technol 2(12):7068–7072

Naldoni A, Schiboula A, Bianchi CL, Bremner DH (2011) Mineralisation of surfactants using ultrasound and the advanced Fenton process. Water Air Soil Pollut 215:487–495. doi:10.1007/s11270-010-0493-y

Nam W, Kim J, Han G (2002) Photocatalytic oxidation of methyl orange in a three-phase fluidized bed reactor. Chemosphere 47:1019–1024. doi:10.1016/S0045-6535(01)00327-7

Nanseu-Njiki CP, Dedzo GK, Ngameni E (2010) Study of the removal of paraquat from aqueous solution by biosorption onto Ayous (Triplochitonschleroxylon) sawdust. J Hazard Mater 179:63–71. doi:10.1016/j.jhazmat.2010.02.058

Nguyen TD, Phan NH, Do MH, Ngo KT (2011) Magnetic Fe_2MO_4 (M: Fe, Mn) activated carbons:fabrication, characterization and heterogeneous Fenton oxidation of methyl orange. J Hazard Mater 185:653–661. doi:10.1016/j.jhazmat.2010.09.068

Nunes SP, Peinemann KV, Ohlrogge K, Alpers A, Keller M, Pires AT (1999) Membranes of poly (ether imide) and nanodispersed silica. J Membr Sci 157:219–226. doi:10.1016/S0376-7388(98)00379-2

Ou Y, Yang F, ZZ Y (1998) A new conception on the toughness of nylon 6/silica nanocomposite prepared via in situ polymerization. J Polym Sci B 36:789–795

Paek SM, Yoo E, Honma I (2008) Enhanced cyclic performance and lithium storage capacity of SnO_2/graphene nanoporous electrodes with three-dimensionally delaminated flexible structure. Nano Lett 9:72–75. doi:10.1021/nl802484w

Park J, Hung I, Gan Z, Rojas OJ, Lim KH, Park S (2013) Activated carbon from biochar: influence of its physicochemical properties on the sorption characteristics of phenanthrene. Bioresour Technol 149:383–389. doi:10.1021/nl802484w

Pathak M, Sharma M, Ojha H, Kumari R, Sharma N, Roy B, Jain G (2016) Green synthesis, characterization and antibacterial activity of silver nanoparticles. Green Chem Technol Lett 2:103–109

Pathania D, Sharma S (2012) Effect of surfactants and electrolyte on removal and recovery of basic dye by using Ficus carica cellulosic fibers as biosorbent. Tenside Surfactant Deterg 49:306–314. doi:10.3139/113.110196

Pathania D, Sarita S, Rathore BS (2011) Synthesis, characterization and photocatalytic application of bovine serum albumin capped cadmium sulphide nanoparticles. Chalcogenide Lett 8:396–404

Pathania D, Sharma G, Kumar A, Kothiyal NC (2014) Fabrication of nanocomposite polyaniline zirconium (IV) silicophosphate for photocatalytic and antimicrobial activity. J Alloys Compd 588:668–675. doi:10.1016/j.jallcom.2013.11.133

Pathania D, Sharma G, Thakur R (2015) Pectin@ zirconium (IV) silicophosphate nanocomposite ion exchanger: photo catalysis, heavy metal separation and antibacterial activity. Chem Eng J 267:235–244. doi:10.1016/j.cej.2015.01.004

Pathania D, Sharma A, Siddiqi ZM (2016) Removal of congo red dye from aqueous system using Phoenix dactylifera seeds. J Mol Liq 219:359–367. doi:10.1016/j.molliq.2016.03.020

Peer D, Karp JM, Hong S, Farokhzad OC, Margalit R, Langer R (2007) Nanocarriers as an emerging platform for cancer therapy. Nat Nanotechnol 2:751–760. doi:10.1038/nnano.2007.387

Peng Q, Lai DM, Kang ET, Neoh KG (2006) Preparation of polymer-silicon (100) hybrids via interface-initiated reversible addition-fragmentation chain-transfer (RAFT) polymerization. Macromolecules 39:5577–5582. doi:10.1021/ma0607362

Pu X, Zhang D, Gao Y, Shao X, Ding G, Li S, Zhao S (2013) One-pot microwave-assisted combustion synthesis of graphene oxide-TiO_2 hybrids for photodegradation of methyl orange. J Alloys Compd 551:382–388. doi:10.1016/j.jallcom.2012.11.028

Qin J (2007) Nanoparticles for multifunctional drug delivery systems-licentiate thesis. The Royal Institute of Technology, Stockholm

Qu S, Huang F, Yu S, Chen G, Kong J (2008) Magnetic removal of dyes from aqueous solution using multi-walled carbon nanotubes filled with Fe_2O_3 particles. J Hazard Mater 160:643–647. doi:10.1016/j.jhazmat.2008.03.037

Rao BS, Kumar BR, Reddy VR, Rao TS (2011) Preparation and characterization of CdS nanoparticles by chemical co-precipitation technique. Chalcogenide Lett 8:177–185

Rocher V, Siaugue JM, Cabuil V, Bee A (2008) Removal of organic dyes by magnetic alginate beads. Water Res 42:1290–1298. doi:10.1016/j.jhazmat.2008.03.037

Safarik I, Safarikova M (2009) Magnetic nano-and microparticles in biotechnology. Chem Pap 63:497–505. doi:10.2478/s11696-009-0054-2

Sahoo C, Gupta AK, Pal A (2005) Photocatalytic degradation of methyl red dye in aqueous solutions under UV irradiation using ag^+ doped TiO_2. Desalination 181:91–100. doi:10.1016/j.desal.2005.02.014

Sanghavi BJ, Sitaula S, Griep MH, Karna SP, Ali MF (2013) Swami NS. Real-time electrochemical monitoring of adenosine triphosphate in the picomolar to micromolar range using graphene-modified electrodes. Anal Chem 85:8158–8165. doi:10.1021/ac4011205
Sankara Reddy B, Venkatramana Reddy S, Koteeswara Reddy N, PramodaKumari J (2013) Synthesis, structural, optical properties and antibacterial activity of co doped (ag, co) ZnO nanoparticles. Res. J Mater Sci 1:11–20
Saquib M, Muneer M (2002) Semiconductor mediated photocatalysed degradation of an anthraquinone dye, Remazol brilliant blue R under sunlight and artificial light source. Dyes Pigments 53:237–249. doi:10.1016/S0143-7208(02)00024-4
Schadler LS, Brinson LC, Sawyer WG (2007) Polymer nanocomposites: a small part of the story. JOM 59:53–60. doi:10.1007/s11837-007-0040-5
Schaefer DW, Justice RS (2007) How nano are nanocomposites? Macromolecules 40:8501–8517. doi:10.1021/ma070356w
Sengupta R, Bandyopadhyay A, Sabharwal S, Chaki TK, Bhowmick AK (2005) Polyamide-6, 6/in situ silica hybrid nanocomposites by sol–gel technique: synthesis, characterization and properties. Polym 46:3343–3354
Shahat A, Awual MR, Naushad M (2015) Functional ligand anchored nanomaterial based facial adsorbent for cobalt (II) detection and removal from water samples. Chem Eng J 271:155–163. doi:10.1016/j.cej.2015.02.097
Shao L, Ren Z, Zhang G, Chen L (2012) Facile synthesis, characterization of a $MnFe_2O_4$/activated carbon magnetic composite and its effectiveness in tetracycline removal. Mater Chem Phys 135:16–24. doi:10.1016/j.matchemphys.2012.03.035
Sharma G, Pathania D, Naushad M, Kothiyal NC (2014a) Fabrication, characterization and antimicrobial activity of polyaniline Th (IV) tungstomolybdophosphate nanocomposite material: efficient removal of toxic metal ions from water. Chem Eng J 251:413–421. doi:10.1016/j.cej.2014.04.074
Sharma G, Pathania D, Naushad M (2014b) Preparation, characterization and antimicrobial activity of biopolymer based nanocomposite ion exchanger pectin zirconium (IV) selenotungstophosphate: application for removal of toxic metals. J Indust Eng Chem 20:4482–4490. doi:10.1016/j.jiec.2014.02.020
Sharma A, Sharma G, Kumar A, Siddiqi ZM, Pathania D (2016) Exclusion of organic dye using neoteric activated carbon prepared from Cornulaca monacantha stem: Equilibrium and thermodynamics studies. Mater Sci Forum 875:1–15. doi:10.4028/www.scientific.net/MSF.875.1
Siddiqi ZM, Pathania D (2003) Studies on titanium (IV) tungstosilicate and titanium (IV) tungstophosphate. II. Separation and estimation of heavy metals from aquatic environments. Acta Chromatogr 13:172–184
Silveira KF, Yoshida IV, Nunes SP (1995) Phase separation in PMMA/silica sol-gel systems. Polymer 36:1425–1434. doi:10.1016/0032-3861(95)95921-M
Singh P, Rathore BS, Chauhan P (2013) Cellulose acetate–zirconium (IV) phosphate nanocomposite with enhanced photo-catalytic activity. Carbohydr Polym 95:434–440. doi:10.1016/j.carbpol.2013.02.045
Skocaj M, Filipic M, Petkovic J, Novak S (2011) Titanium dioxide in our everyday life; is it safe? Radiol Oncol 45:227–247. doi:10.2478/v10019-011-0037-0
Sun C, Lee JS, Zhang M (2008) Magnetic nanoparticles in MR imaging and drug delivery. Adv Drug Deliv Rev 60:1252–1265. doi:10.1016/j.addr.2008.03.018
Sun H, Cao L, Lu L (2011) Magnetite/reduced graphene oxide nanocomposites: one step solvothermal synthesis and use as a novel platform for removal of dye pollutants. Nano Res 4:550–562. doi:10.1007/s12274-011-0111-3
Suzuki F, Onozato K, Kurokawa Y (1990) A formation of compatible poly (vinyl alcohol)/alumina gel composite and its properties. J Appl Polym Sci 39:371–381. doi:10.1002/app.1990.070390214
Tang J, Chen Q, Xu L, Zhang S, Feng L, Cheng L, Xu H, Liu Z, Peng R (2013) Graphene oxide-silver nanocomposite as a highly effective antibacterial agent with species-specific mechanisms. ACS Appl Mater Interfaces 5:3867–3874. doi:10.1021/am4005495

Taniguchi Y, Ogawa M, Gang W, Saitoh H, Fujiki K, Yamauchi T, Tsubokawa N (2008) Preparation of hyperfunctional carbon black by grafting of hyperbranched polyester onto the surface. Mater Chem Phys 108:397–402. doi:10.1016/j.matchemphys.2007.10.013

Tayade RJ, Natarajan TS, Bajaj HC (2009) Photocatalytic degradation of methylene blue dye using ultraviolet light emitting diodes. Ind Eng Chem Res 48:10262–10267. doi:10.1021/ie9012437

Treece RE, Macala GS, Rao L, Franke D, Eckert H, Kaner RB (1993) Synthesis of III-V semiconductors by solid-state metathesis. Inorg Chem 32:2745–2752. doi:10.1021/ic00064a027

Vafaee M, Ghamsari MS (2007) Preparation and characterization of ZnO nanoparticles by a novel sol-gel route. Mater Lett 61:3265–3268. doi:10.1016/j.matlet.2006.11.089

Wang ZC, Shui HF (2007) Effect of PO_4^{3-} and PO_4^{3-}-SO_4^{2-} modification of TiO_2 on its photocatalytic properties. J Mol Catal A Chem 263:20–25. doi:10.1016/j.molcata.2006.08.017

Wang X, Zhi L, Tsao N, Tomović Ž, Li J, Müllen K (2008a) Transparent carbon films as electrodes in organic solar cells. Angew Chem Int Ed Engl 120:3032–3034. doi:10.1002/ange.200704909

Wang L, Zhang J, Wang A (2008b) Removal of methylene blue from aqueous solution using chitosan-g-poly (acrylic acid)/montmorillonite super-adsorbent nanocomposite. Colloids Surf A Physicochem Eng Asp 322:47–53. doi:10.1016/j.colsurfa.2008.02.019

Wang D, Kou R, Choi D, Yang Z, Nie Z, Li J, Saraf LV, Hu D, Zhang J, Graff GL, Liu J (2010a) Ternary self-assembly of ordered metal oxide-graphene nanocomposites for electrochemical energy storage. ACS Nano 4:1587–1595. doi:10.1021/nn901819n

Wang H, Sun YB, Chen QW, YF Y, Cheng K (2010b) Synthesis of carbon-encapsulated superparamagnetic colloidal nanoparticles with magnetic-responsive photonic crystal property. Dalton Trans 39:9565–9569. doi:10.1039/C0DT00621A

Wang C, Feng C, Gao Y, Ma X, Wu Q, Wang Z (2011) Preparation of a graphene-based magnetic nanocomposite for the removal of an organic dye from aqueous solution. Chem Eng J 173:92–97. doi:10.1016/j.cej.2011.07.041

Wang Z, Du Y, Zhang F, Zheng Z, Zhang X, Feng Q, Wang C (2013) Photocatalytic degradation of pendimethalin over Cu_2O/SnO_2/graphene and SnO_2/graphene nanocomposite photocatalysts under visible light irradiation. Mater Chem Phys 140:373–381. doi:10.1016/j.matchemphys.2013.03.052

Williams JV, Adams CN, Kotov NA, Savage PE (2007) Hydrothermal synthesis of CdSe nanoparticles. Ind Eng Chem Res 46:4358–4362. doi:10.1021/ie061413x

Williams G, Seger B, Kamat PV (2008a) TiO_2-graphene nanocomposites. UV-assisted photocatalytic reduction of graphene oxide. ACS Nano 2:1487–1491. doi:10.1021/nn800251f

Wu W, He Q, Jiang C (2008) Magnetic iron oxide nanoparticles: synthesis and surface functionalization strategies. Nanoscale Res Lett 3:397. doi:10.1007/s11671-008-9174-9

Wu Q, Zhao G, Feng C, Wang C, Wang Z (2011) Preparation of a graphene-based magnetic nanocomposite for the extraction of carbamate pesticides from environmental water samples. J Chromatogr A 1218:7936–7942. doi:10.1016/j.chroma.2011.09.027

Xiang Y, Chen D (2007) Preparation of a novel pH-responsive silver nanoparticle/poly (HEMA-PEGMA-MAA) composite hydrogel. Eur Polym J 43:4178–4187. doi:10.1016/j.eurpolymj.2007.08.005

Xiao Q, Si Z, Yu Z, Qiu G (2007) Sol–gel auto-combustion synthesis of samarium-doped TiO_2 nanoparticles and their photocatalytic activity under visible light irradiation. Mater Sci Eng B 137:189–194. doi:10.1016/j.mseb.2006.11.011

Xu K, Wang J, Chen Q, Yue Y, Zhang W, Wang P (2008) Spontaneous volume transition of polyampholyte nanocomposite hydrogels based on pure electrostatic interaction. J Colloid Interface Sci 321:272–278. doi:10.1016/j.jcis.2008.02.024

Xu T, Zhang L, Cheng H, Zhu Y (2011) Significantly enhanced photocatalytic performance of ZnO via graphene hybridization and the mechanism study. Appl Catal Environ 101:382–387. doi:10.1016/j.apcatb.2010.10.007

Yan C, Sun L, Fu X (2002) Liao C. PEKING UNIV BEIJING (CHINA), Preparation of CdS/ZnO Core/shell Structured Nanoparticles by Hydrothermal Method

Yang MQ, Xu YJ (2013) Selective photoredox using graphene-based composite photocatalysts. Phys Chem Chem Phys 15:19102–19118. doi:10.1039/C3CP53325E

Yang R, Yan Y, Mu Y, Ji W, Li X, Zou M, Fei Q, Jin Q (2006) A rapid and facile method for hydrothermal synthesis of CdTe nanocrystals under mild conditions. J Nano Sci Nanotechnol 6:215–220. doi:10.1166/jnn.2006.050

Yang N, Zhu S, Zhang D, Xu S (2008) Synthesis and properties of magnetic Fe_3O_4-activated carbon nanocomposite particles for dye removal. Mater Lett 62:645–647. doi:10.1016/j.matlet.2007.06.049

Yao Y, Miao S, Liu S, Ma LP, Sun H, Wang S (2012) Synthesis, characterization, and adsorption properties of magnetic Fe_3O_4@ graphene nanocomposite. Chem Eng J 184:326–332. doi:10.1016/j.cej.2011.12.017

Yao Y, Xu C, Yu S, Zhang D, Wang S (2013) Facile synthesis of Mn_3O_4-reduced graphene oxide hybrids for catalytic decomposition of aqueous organics. Ind Eng Chem Res 52:3637–3645. doi:10.1021/ie303220x

Yuan Q, Venkatasubramanian R, Hein S, Misra RD (2008) A stimulus-responsive magnetic nanoparticle drug carrier: magnetite encapsulated by chitosan-grafted-copolymer. Acta Biomater 4:1024–1037. doi:10.1016/j.actbio.2008.02.002

Yuchun O, Feng Y, Jin C (1997) Interfacial interaction and mechanical properties of nylon 6-potassium titanate composites prepared by in-situ polymerization. J Appl Polym Sci 64:2317–2322

Yusoff N, Huang NM, Muhamad MR, Kumar SV, Lim HN, Harrison I (2013) Hydrothermal synthesis of CuO/functionalized graphene nanocomposites for dye degradation. Mater Lett 93:393–396. doi:10.1016/j.matlet.2012.10.015

Zhang Y, Kohler N, Zhang M (2002) Surface modification of superparamagnetic magnetite nanoparticles and their intracellular uptake. Biomaterials 23:1553–1561. doi:10.1016/S0142-9612(01)00267-8

Zhang K, Zhang X, Chen H, Chen X, Zheng L, Zhang J, Yang B (2004) Hollow titania spheres with movable silica spheres inside. Langmuir 20:11312–11314. doi:10.1021/la047736k

Zhang G, Qu J, Liu H, Cooper AT, Wu R (2007) $CuFe_2O_4$/activated carbon composite: a novel magnetic adsorbent for the removal of acid orange II and catalytic regeneration. Chemosphere 68:1058–1066. doi:10.1016/j.chemosphere.2007.01.081

Zhang S, Chen Z, Li Y, Wang Q, Wan L (2008) Photocatalytic degradation of methylene blue in a sparged tube reactor with TiO_2 fibers prepared by a properly two-step method. Cat Com 9:1178–1183. doi:10.1016/j.catcom.2007.11.009

Zhang S, Niu H, Hu Z, Cai Y, Shi Y (2010) Preparation of carbon coated Fe_3O_4 nanoparticles and their application for solid-phase extraction of polycyclic aromatic hydrocarbons from environmental water samples. J Chromatogr A 1217:4757–4764. doi:10.1016/j.chroma.2010.05.035

Zhang Y, Xu S, Luo Y, Pan S, Ding H, Li G (2011) Synthesis of mesoporous carbon capsules encapsulated with magnetite nanoparticles and their application in wastewater treatment. J Mater Chem 21:3664–3671. doi:10.1039/C0JM03727C

Zhang Y, Zhu Y, Yu J, Yang D, Ng TW, Wong PK, Jimmy CY (2013) Enhanced photocatalytic water disinfection properties of Bi_2MoO_6-RGO nanocomposites under visible light irradiation. Nanoscale 5:6307–6310. doi:10.1039/C3NR01338C

Zhou X, Shi T, Zhou H (2012) Hydrothermal preparation of ZnO-reduced graphene oxide hybrid with high performance in photocatalytic degradation. Appl Surf Sci 258:6204–6211. doi:10.1016/j.apsusc.2012.02.131

Zhu M, Diao G (2011) Magnetically recyclable Pd nanoparticles immobilized on magnetic Fe_3O_4@C nanocomposites: preparation, characterization and their catalytic activity toward Suzuki and Heck coupling reactions. J Phys Chem C 115:24743–24749. doi:10.1021/jp206116e

Part VI

Challenges and Alternative Approaches

Potential Challenges and Alternative Approaches in Metabolic Engineering of Bioactive Compounds in Industrial Setup

19

Adesh K. Saini and Vipin Chandra Kalia

Abstract

One of the greatest boons of biotechnology is to provide ways and means to produce different byproducts by using living cells like *Escherichia coli*, *Bacillus*, *Corynebacterium glutamicum*, yeast and many others. These products include enzymes of industrial applications, secondary metabolites for clinical applications, antibodies, L-amino acids, chemicals or fuels. By using tools of bioinformatics and genetic engineering, researchers are now able to dissect the biosynthetic pathways, which help them not only to understand the biology of metabolite production but also to modulate these pathways for improving quality and quantity of economically important bioactive compounds of clinical and industrial importance. Having said that, it is not easy to tame an organism to produce the desired byproduct. In the present chapter, we are discussing some of the potential challenges and available alternative approaches.

Keywords

Bioactive compounds • Byproducts • Genetic engineering • Biomedicine • Biofactories

A.K. Saini (✉)
Faculty of Applied Sciences and Biotechnology, Shoolini University, Solan 173 229, Himachal Pradesh, India
e-mail: sainiade@gmail.com

V.C. Kalia
Microbial Biotechnology and Genomics, CSIR - Institute of Genomics and Integrative Biology (IGIB), Delhi University Campus, Mall Road, New Delhi 110007, India
e-mail: vckalia@igib.res.in; vc_kalia@yahoo.co.in

V.C. Kalia, A.K. Saini (eds.), *Metabolic Engineering for Bioactive Compounds*, DOI 10.1007/978-981-10-5511-9_19

19.1 Introduction

Natural products have been an integral part of our daily life. A large number of compounds have unique applications in areas like nutrition, medicinal, cosmetics, etc. (Begum et al. 2016; Hernández-Saldaña et al. 2016; Saini and Keum 2017; Thakur et al. 2017). These bioactive molecules have been found to be produced by plants, microbes and marine organisms (Arasu et al. 2015; Balakrishnan et al. 2015; Bandopadhyay et al. 2015; Peña-Yam et al. 2016; Azman et al. 2017; Sanchart et al. 2017). However, it seems these bioactive molecules still hold promise for improving our health and environment (Go et al. 2015; Karumuri et al. 2015; Park et al. 2015; Shiva Krishna et al. 2015; Jeyanthi and Velusamy 2016; Pessione et al. 2017). The scope of these molecules is now extended to treating cancers (Varsha et al. 2016). Metabolic engineering and nanotechnological techniques are being exploited to produce highly efficient bioactive compounds (Bose and Chatterjee 2015; Dobrucka and Długaszewska 2015; Szweda et al. 2015; Wadhwani et al. 2016; Ahiwale et al. 2017).

Recently, yeast cells have been engineered to synthesize opioids which are primary drugs used in managing the pain and mollifying relief. For this Smolke's group has bring the biosynthetic pathways from *Papaver somniferum* (opium poppy) plant to *Saccharomyces cerevisiae* also known as Baker's yeast for meeting the industrial and medical demands which is otherwise met from around one hundred thousand hectares of poppy fields (Galanie et al. 2015). This is a typical case study where bioengineers use the state-of-the-art enzyme-discovery, enzyme-engineering, elucidation of pathways and optimization of strains that led to the production of thebaine from 21 genes and hydrocodone from 23 genes. After this total biosynthesis of opiates was done using engineered *Escherichia coli* (Nakagawa et al. 2016). Similar approach has been taken up to produce an important precursor of antimalarial drug artemisinin (Ro et al. 2006).

Besides yeast, bacterial strains like *Escherichia coli* and *Corynebacterium glutamicum* have also been used as factories for producing L-amino acids of industrial production. The amino acids produced from *E. coli* like L-threonine, L-phenylalanine, L-valine and related compound putrescine are used in food and animal feed, hygiene, health, packaging and agriculture (Becker and Wittmann 2012).

Another area which is likely to gain importance as bioactive molecules in the near future is the microbial polymer (Singh et al. 2015). A large number of microbes undergo metabolic diversion under stress conditions. During normal conditions bacteria metabolize the feed and generate energy; however, under nutritional stress conditions where there is an excess of C and limitation of nutrients such as N, O, P, K, Mg, etc., bacteria divert the tricarboxylic acid (TCA) cycle (Reddy et al. 2003; Kalia et al. 2007; Singh et al. 2009). This pathway terminates with the production of biopolymers such as polyhydroxyalkanoates (PHAs) (Kalia et al. 2003, 2016; Kumar et al. 2009, 2013, 2014, 2015a, b, 2016; Patel et al. 2015a,b, 2016; Singh et al. 2013). Although a large number of bacteria are known to produce PHAs, *Bacillus* spp. have turned out to be among the most favourite because of their abilities to withstand a diversity of stress conditions (Kumar et al. 2013, 2015c). When bacteria sense that normal physiological conditions are available, PHAs are degraded to (R)-3-hydroxyoctanoic acid (3HA) (Ray and Kalia 2017a, b) for re-entering the TCA cycle and generate energy. These simpler molecules like 3HA can be transformed into halo and unsaturated esters. Because of their carboxylic group, these have been found to possess antimicrobial activity. These bioactive molecules also have anti-proliferative effect on (mammalian) cell lines. The diversity of their activities extends also to inhibit quorum sensing system of *Pseudomonas aeruginosa*, an opportunistic pathogen (Radivojevic et al. 2016).

These important discoveries and others support the theme of bioengineering which emphasize on using simple organisms to produce bioactive products that have great industrial value. But engineering these organisms is not easy and has some challenges that need to be understood.

19.2 Pathway Complexity

The microbes that look very simple have very complex and diverse lifestyle governed by metabolic and regulatory controls that need to be understood for efficient metabolic engineering (Me-Eng). System biology (SB), against the reductionist approach which relies on focusing on one gene or one protein, helped in building a comprehensive information related to the gene regulation from DNA to protein modifications as well as in drafting a mathematical or calculative model that helps in the meaningful analysis of huge amount of information coming from genomics, transcriptomics and proteomics. These three interplaying approaches are required to understand the organism at the phenotypic level which is required to modulate the simple organism to an industrial workhouse.

A great challenge before Me-Eng is to acquire control over microorganism of interest by cutting down its complexity or by using targeted SB approach which relies on genome scale forecast and its alteration in the host microbe. Targeted SB approach is a great solution to get around the complexity challenge. It is a rational approach against evolutionary concept, and in this case accurate modification of genome is done to get a precise modified strain. This system metabolic engineering approach helped in the production of L-amino acids like threonine and valine in *E. coli* (Lee et al. 2011). This has greatly increased the production of L-amino acids to around 25%. For this *E. coli* has been modified to get overproduction of enzymatic products and removal of genes which cause feedback inhibition. Using similar approach *C. glutamicum* strain was engineered which could produce a yield of 55% for L-lysine production (Becker et al. 2011). To take these yields to the next level, a sophisticated understanding of regulatory networks that could allow us to alter these in a targeted fashion is required.

To answer the pathway complexity hurdle, scientists also utilize an approach where prior knowledge for the pathways is not a prerequisite. It is a somewhat evolutionary approach called non-targeted systems-level engineering which is widely used because of its less complexity. It is an approach where a researcher looks for a desired phenotype with higher level of yield without worrying about the pathways. The host organism is subjected to random mutagenesis using mutagens or UV treatment which leads to random modifications in DNA and finally leads to modified enzymes (Flibotte et al. 2010). Once we get a better strain in terms of industrial production, then deep sequencing can be done to know the changes or modifications in the genome to better understand the genotype-phenotype coordination. But in this case it is also needed to understand that interpreting the changed networks needs further validation through targeted approach. This can be answered by multiplex automated genome engineering (MAGE) where many selected genes can be altered simultaneously. As an example production of lycopene was increased to fivefold within few days by altering the expression of more than 20 genes in one go (Wang et al. 2009, 2012; Isaacs et al. 2011). Many other examples of MAGE have been reviewed recently (Fehér et al. 2012).

Another way to answer the complexity problem is by reducing the genome size and to have developed host organisms which have minimal genomes to carry out the desired results. This will greatly reduce the complexity and will facilitate both targeted and non-targeted SB approach. The two options which are available for this are as follows: (a) genome reduction by using molecular surgery to the level which is essential to maintain the cell's vital function as in the case of *Mycoplasma genitalium*, which proved that it requires only 387 genes (Glass et al. 2006), and (b) chemical synthesis of reduced genome which is essential to maintain the cell's vital function (Cello et al. 2002; Gibson et al. 2010).

19.3 Xenobiology for Orthogonalization as an Alternative to Improve Workhorses

Every cell is an independent entity and could have different pathways which can convert a precursor to a product. Thus it suggests that different cells could have orthogonal pathways to achieve the same goal. These subsystems can be manipulated to improve the production in an industrial setup. Me-Eng of host by putting orthogonal system has been shown to improve the productivity of cells like T7-based protein production (Temme et al. 2012) or the use of group II introns (Lambowitz and Zimmerly 2011). Synthetic xenobiology is gaining grounds in this direction.

The above mentioned approach of SB or we can say systematic ME has improved the production of microbe based metabolites like L-amino acids, chemicals like butanol or succinic acid, drugs like opioids and artemisinin or fuels. The strategy for improving industrially important byproducts using microbes includes many aspects like project design which takes care of all aspects of regulatory networks, economic benefits and even the legal aspects which are often at the back seat. Selection of host strain and reconstruction of metabolic pathway are the next steps which should avoid expensive and exhaustive computation and inaccurate predictions to correlate phenotype with alterations in genome. Another important aspect is to look at improving the host tolerance towards exogenous product. Availability of cofactors and precursor molecules needs to be optimized through rerouting the fluxes to remove competitive pathways (Lee and Kim, 2015). Once these strains or hosts are engineered, one of the challenges faced by researchers is to identify the desired mutant out of many by using screening.

As ME and SB are improving, we expect that in the coming years plethora of information will be sitting with metabolic engineers to galvanize the production of metabolites.

Acknowledgements AKS is supported by Grant from DST (INT/NZ/P-001/2013) and NMHS-MoEF&CC (NMHS/2015-16/SG09/09) Govt. of India. The author (V.C.K.) wish to thank the Director of CSIR-Institute of Genomics and Integrative Biology, Delhi, India, OLP1126 (CSIR-HRD ES21(1022)/16/EMR-II), for providing the necessary funds, facilities and moral support.

References

Ahiwale SS, Bankar AV, Tagunde S, Kapadnis BP (2017) A bacteriophage mediated gold nanoparticle synthesis and their anti-biofilm activity. Indian J Microbiol 57(2):188–194. doi:10.1007/s12088-017-0640-x

Arasu MV, Al-Dhabi NA, Rejiniemon TS, Lee KD, Huxley VAJ, Kim DH, Duraipandiyan V, Karuppiah P, Choi KC (2015) Identification and characterization of *Lactobacillus brevis* P68 with antifungal, antioxidant and probiotic functional properties. Indian J Microbiol 55:19–28. doi:10.1007/s12088-014-0495-3

Azman CA-S, Othman I, Fang C-M, Chan K-G, Goh B-H, Lee L-H (2017) Antibacterial, anticancer and neuroprotective activities of rare actinobacteria from mangrove forest soils. Indian J Microbiol 57(2):177–187. doi:10.1007/s12088-016-0627-z

Balakrishnan D, Bibiana AS, Vijayakumar A, Santhosh RS, Dhevendaran K, Nithyanand P (2015) Antioxidant activity of bacteria associated with the marine sponge *Tedania anhelans*. Indian J Microbiol 55:13–18. doi:10.1007/s12088-014-0490-8

Bandyopadhyay P, Mishra S, Sarkar B, Swain SK, Pal A, Tripathy PP, Ojha SK (2015) Dietary *Saccharomyces cerevisiae* boosts growth and immunity of IMC *Labeo rohita* (ham.) juveniles. Indian J Microbiol 55:81–87. doi:10.1007/s12088-014-0500-x

Becker J, Wittmann C (2012) Systems and synthetic metabolic engineering for amino acid production - the heartbeat of industrial strain development. Curr Opin Biotechnol 23:718–726. doi:10.1016/j.copbio.2011.12.025

Becker J, Zelder O, Häfner S, Schröder H, Wittmann C (2011) From zero to hero--design-based systems metabolic engineering of *Corynebacterium glutamicum* for L-lysine production. Metab Eng 13:159–168. doi:10.1016/j.ymben.2011.01.003

Begum IF, Mohankumar R, Jeevan M, Ramani K (2016) GC–MS analysis of bioactive molecules derived from *Paracoccus pantotrophus* FMR19 and the antimicrobial activity against bacterial pathogens and MDROs. Indian J Microbiol 56:426–432. doi:10.1007/s12088-016-0609-1

Bose D, Chatterjee S (2015) Antibacterial activity of green synthesized silver nanoparticles using Vasaka (*Justicia adhatoda* L.) leaf extract. Indian J Microbiol 55:163–167. doi:10.1007/s12088-015-0512-1

Cello J, Paul AV, Wimmer E (2002) Chemical synthesis of poliovirus cDNA: generation of infectious virus in the absence of natural template. Science 297:1016–1018. doi:10.1126/science.1072266

Dobrucka R, Długaszewska J (2015) Antimicrobial activities of silver nanoparticles synthesized by using water extract of *Arnicae anthodium*. Indian J Microbiol 55:168–174. doi:10.1007/s12088 015-0516-x

Fehér T, Burland V, Pósfai G (2012) In the fast lane: large-scale bacterial genome engineering. J Biotechnol 160:72–79. doi:10.1016/j.jbiotec.2012.02.012

Flibotte S, Edgley ML, Chaudhry I, Taylor J, Neil SE, Rogula A, Zapf R, Hirst M, Butterfield Y, Jones SJ, Marra MA, Barstead RJ, Moerman DG (2010) Whole-genome profiling of mutagenesis in *Caenorhabditis elegans*. Genetics 185:431–441. doi:10.1534/genetics.110.116616

Galanie S, Thodey K, Trenchard IJ, Filsinger Interrante M, Smolke CD (2015) Complete biosynthesis of opioids in yeast. Science 349:1095–1100. doi:10.1126/science.aac9373

Gibson DG, Glass JI, Lartigue C, Noskov VN, Chuang RY, Algire MA, Benders GA, Montague MG, Ma L, Moodie MM, Merryman C, Vashee S, Krishnakumar R, Assad-Garcia N, Andrews-Pfannkoch C, Denisova EA, Young L, Qi ZQ, Segall-Shapiro TH, Calvey CH, Parmar PP, Hutchison CA 3rd, Smith HO, Venter JC (2010) Creation of a bacterial cell controlled by a chemically synthesized genome. Science 329:52–56. doi:10.1126/science.1190719

Glass JI, Assad-Garcia N, Alperovich N, Yooseph S, Lewis MR, Maruf M, Hutchison CA 3rd, Smith HO, Venter JC. (2006) Essential genes of a minimal bacterium. Proc Natl Acad Sci U S A 103:425-430. doi: 10.1073/pnas.0510013103

Go T-H, Cho K-S, Lee S-M, Lee O-M, Son H-J (2015) Simultaneous production of antifungal and keratinolytic activities by feather-degrading *Bacillus subtilis* S8. Indian J Microbiol 55:66–73. doi:10.1007/s12088-014-0502-8

Hernández-Saldaña OF, Valencia-Posadas M, de la Fuente-Salcido NM, Bideshi DK, Barboza-Corona JE (2016) Bacteriocinogenic bacteria isolated from raw goat milk and goat cheese produced in the Center of México. Indian J Microbiol 56:301–308. doi:10.1007/s12088-016-0587-3

Isaacs FJ, Carr PA, Wang HH, Lajoie MJ, Sterling B, Kraal L, Tolonen AC, Gianoulis TA, Goodman DB, Reppas NB, Emig CJ, Bang D, Hwang SJ, Jewett MC, Jacobson JM, Church GM (2011) Precise manipulation of chromosomes in vivo enables genome-wide codon replacement. Science 333:348–353. doi:10.1126/science.1205822

Jeyanthi V, Velusamy P (2016) Anti-methicillin resistant *Staphylococcus aureus* compound isolation from halophilic *Bacillus amyloliquefaciens* MHB1 and determination of its mode of action using electron microscope and flow cytometry analysis. Indian J Microbiol 56:148–157. doi:10.1007/s12088-016-0566-8

Kalia VC, Chauhan A, Bhattacharyya G, Rashmi (2003) Genomic databases yield novel bioplastic producers. Nat Biotechnol 21:845–846. doi:10.1038/nbt0803-845

Kalia VC, Lal S, Cheema S (2007) Insight into the phylogeny of polyhydroxyalkanoate biosynthesis: horizontal gene transfer. Gene 389:19–26. doi:10.1016/j.gene.2006.09.010

Kalia VC, Prakash J, Koul S (2016) Biorefinery for glycerol rich biodiesel industry waste. Indian J Microbiol 56:113–125. doi:10.1007/s12088-016-0583-7

Karumuri S, Singh PK, Shukla P (2015) *In silico* analog design for terbinafine against *Trichophyton rubrum*: a preliminary study. Indian J Microbiol 55:333–340. doi:10.1007/s12088-015-0524-x

Kumar T, Singh M, Purohit HJ, Kalia VC (2009) Potential of *Bacillus* sp. to produce polyhydroxybutyrate from biowaste. J Appl Microbiol 106:2017–2023. doi:10.1111/j.1365-2672.2009.04160.x

Kumar P, Patel SKS, Lee JK, Kalia VC (2013) Extending the limits of *Bacillus* for novel biotechnological applications. Biotechnol Adv 31:1543–1561. doi:10.1016/j.biotechadv.2013.08.007

Kumar P, Singh M, Mehariya S, Patel SKS, Lee JK, Kalia VC (2014) Ecobiotechnological approach for exploiting the abilities of *Bacillus* to produce co-polymer of polyhydroxyalkanoate. Indian J Microbiol 54:151–157. doi:10.1007/s12088-014-0457-9

Kumar P, Mehariya S, Ray S, Mishra A, Kalia VC (2015a) Biodiesel industry waste: a potential source of bioenergy and biopolymers. Indian J Microbiol 55:1–7. doi:10.1007/s12088-014-0509-1

Kumar P, Mehariya S, Ray S, Mishra A, Kalia VC (2015b) Biotechnology in aid of biodiesel industry effluent (glycerol): biofuels and bioplastics. In: Kalia VC (ed) Microbial factories waste treatment. Springer, New Delhi, pp 105–119. doi:10.1007/978-81-322-2598-0

Kumar P, Ray S, Kalia VC (2016) Production of co-polymers of polyhydroxyalkanoates by regulating the hydrolysis of biowastes. Bioresour Technol 200:413–419. doi:10.1016/j.biortech.2015.10.045

Kumar P, Ray S, Patel SKS, Lee JK, Kalia VC (2015c) Bioconversion of crude glycerol to polyhydroxyalkanoate by *Bacillus thuringiensis* under non-limiting nitrogen conditions. Int J Biol Macromol 78:9–16. doi:10.1016/j.ijbiomac.2015.03.046

Lambowitz AM, Zimmerly S (2011) Group II introns: mobile ribozymes that invade DNA. Cold Spring Harb Perspect Biol 3:a003616

Lee JW, Kim TY, Jang YS, Choi S, Lee SY (2011) Systems metabolic engineering for chemicals and materials. Trends Biotechnol 29:370–378. doi:10.1016/j.tibtech.2011.04.001

Lee SY, Kim HU (2015) Systems strategies for developing industrial microbial strains. Nat Biotechnol 33:1061–1072. doi:10.1038/nbt.3365

Nakagawa A, Matsumura E, Koyanagi T, Katayama T, Kawano N, Yoshimatsu K, Yamamoto K, Kumagai H, Sato F, Minami H (2016) Total biosynthesis of opiates by stepwise fermentation using engineered *Escherichia coli*. Nat Commun 7:10390. doi:10.1038/ncomms10390

Park J-M, Radhakrishnan R, Kang S-M, Lee I-J (2015) IAA producing *Enterobacter* sp. I-3 as a potent bio-herbicide candidate for weed control: a special reference with lettuce growth inhibition. Indian J Microbiol 55:207–211. doi:10.1007/s12088-015-0515-y

Patel SKS, Kumar P, Singh S, Lee JK, Kalia VC (2015a) Integrative approach for hydrogen and polyhydroxybutyrate production. In: Kalia VC (ed) Microbial factories waste treatment. Springer, New Delhi, pp 73–85. doi:10.1007/978-81-322-2598-0_5

Patel SKS, Kumar P, Singh S, Lee JK, Kalia VC (2015b) Integrative approach to produce hydrogen and polyhydroxybutyrate from biowaste using defined bacterial cultures. Bioresour Technol 176:136–141. doi:10.1016/j.biortech.2014.11.029

Patel SKS, Lee JK, Kalia VC (2016) Integrative approach for producing hydrogen and polyhydroxyalkanoate from mixed wastes of biological origin. Indian J Microbiol 56:293–300. doi:10.1007/s12088-016-0595-3

Peña-Yam LP, Esaú Rúiz-Sánchez E, Barboza-Corona JE, Reyes-Ramírez A (2016) Isolation of Mexican *Bacillus* species and their effects in promoting growth of chili pepper (*Capsicum annuum* L. cv Jalapeño). Indian J Microbiol 56:375–378. doi:10.1007/s12088-016-0582-8

Pessione E, Maxxoli R, Riedel K (2017) Bioactive compounds from microbes. Front Microbiol 8:392. doi:10.3389/fmicb.2017.00392

Radivojevic J, Skaro S, Senerovic L, Vasiljevic B, Guzik M, Kenny ST, Maslak V, Nikodinovic-Runic J, O'Connor KE (2016) Polyhydroxyalkanoate-based 3-hydroxyoctanoic acid and its derivatives as a platform of bioactive compounds. Appl Microbiol Biotechnol 100:161–172. doi:10.1007/s00253-015-6984-4

Ray S, Kalia VC (2017a) Biological significance of degradation of polyhydroxyalkanoates. In: Kalia VC, Kumar P (eds) Microbial applications, Bioremediation and bioenergy, vol 1. Springer, Berlin. ISBN: 978-3-319-52665-2

Ray S, Kalia VC (2017b) Microbial cometabolism and polyhydroxyalkanaote co-polymers. Indian J Microbiol 57:39–47. doi:10.1007/s12088-016-0622-4

Reddy CSK, Ghai R, Kalia VC (2003) Polyhydroxyalkanoates: an overview. Bioresour Technol 87:137–146. doi:10.1016/S0960-8524(02)00212-2

Ro DK, Paradise EM, Ouellet M, Fisher KJ, Newman KL, Ndungu JM, Ho KA, Eachus RA, Ham TS, Kirby J, Chang MC, Withers ST, Shiba Y, Sarpong R, Keasling JD (2006) Production of the antimalarial drug precursor artemisinic acid in engineered yeast. Nature 440:940–943. doi:10.1038/nature04640

Saini RK, Keum Y-S (2017) Progress in microbial carotenoids production. Indian J Microbiol 57:129–130. doi:10.1007/s12088-016-0637-x

Sanchart C, Rattanaporn O, Haltrich D, Phukpattaranont P, Maneerat S (2017) *Lactobacillus futsaii* CS3, a new GABA-producing strain isolated from Thai fermented shrimp (*Kung–Som*). Indian J Microbiol 57(2):211–217. doi:10.1007/s12088-016-0632-2

Shiva Krishna P, Sudheer Kumar B, Raju P, Murty MSR, Prabhakar Rao T, Singara Charya MA, Prakasham RS (2015) Fermentative production of pyranone derivate I from marine *Vibrio* sp. SKMARSP9: isolation, characterization and bioactivity evaluation. Indian J Microbiol 55:292–301. doi:10.1007/s12088-015-0521-0

Singh M, Kumar P, Patel SKS, Kalia VC (2013) Production of polyhydroxyalkanoate co-polymer by *Bacillus thuringiensis*. Indian J Microbiol 53:77–83. doi:10.1007/s12088-012-0294-7

Singh M, Kumar P, Ray S, Kalia VC (2015) Challenges and opportunities for customizing polyhydroxyalkanoates. Indian J Microbiol 55:235–249. doi:10.1007/s12088-015-0528-6

Singh M, Patel SKS, Kalia VC (2009) *Bacillus subtilis* as potential producer for polyhydroxyalkanoates. Microb Cell Fact 8:38. doi:10.1186/1475-2859-8-38

Szweda P, Gucwa K, Kurzyk E, Romanowska E, Dzierzanowska-Fangrat K, Jurek AZ, Kuś PM, Milewski M (2015) Essential oils, silver nanoparticles and Propolis as alternative agents against fluconazole resistant *Candida albicans, Candida glabrata* and *Candida krusei* clinical isolates. Indian J Microbiol 55:175–183. doi:10.1007/s12088-014-0508-2

Temme K, Hill R, Segall-Shapiro TH, Moser F, Voigt CA (2012) Modular control of multiple pathways using engineered orthogonal T7 polymerases. Nucleic Acids Res 40:8773–8781. doi:10.1093/nar/gks597

Thakur R, Sharma KC, Gulati A, Sud RK, Gulati A (2017) Stress-tolerant *Viridibacillus arenosi* strain IHB B 7171 from tea rhizosphere as a potential broad-spectrum microbial inoculant. Indian J Microbiol 57(2):195–200. doi:10.1007/s12088-017-0642-8

Varsha KK, Nishant G, Sneha SM, Shilpa G, Devendra L, Priya S, Nampoothiri KM (2016) Antifungal, anticancer and aminopeptidase inhibitory potential of a phenazine compound produced by *Lactococcus* BSN307. Indian J Microbiol 56:411–416. doi:10.1007/s12088-016-097-1

Wadhwani SA, Shedbalkar UU, Singh R, Vashisth P, Pruthi V, Chopade BA (2016) Kinetics of synthesis of gold nanoparticles by *Acinetobacter* sp. SW30 isolated from environment. Indian J Microbiol 56:439–444. doi:10.1007/s12088-016-0598-0

Wang HH, Isaacs FJ, Carr PA, Sun ZZ, Xu G, Forest CR, Church GM (2009) Programming cells by multiplex genome engineering and accelerated evolution. Nature 460:894–898. doi:10.1038/nature08187

Wang HH, Kim H, Cong L, Jeong J, Bang D, Church GM (2012) Genome-scale promoter engineering by coselection MAGE. Nat Methods 9:591–593. doi:10.1038/nmeth.1971

MIX
Papier aus verantwortungsvollen Quellen
Paper from responsible sources
FSC® C105338
FSC
www.fsc.org

If you have any concerns about our products,
you can contact us on
ProductSafety@springernature.com

In case Publisher is established outside the EU,
the EU authorized representative is:
Springer Nature Customer Service Center GmbH
Europaplatz 3, 69115 Heidelberg, Germany

Printed by Libri Plureos GmbH
in Hamburg, Germany